Structure and Bonding in Crystalline Materials

One of the motivating questions in materials research today is: how can elements be combined to produce a solid with specified properties? One part of the answer to this question lies in the fundamental relationship between the composition, structure and bonding in crystalline materials. This book is intended to acquaint the reader with established principles of crystallography and bonding that are needed to understand this relationship.

The book starts with an introduction to periodic trends and then describes the atomic structure of crystalline solids, the experimental interrogation of crystalline structure, the origin of the cohesive forces that stabilize crystalline structures, and how these cohesive forces vary with the elements in the solid. The book finishes by describing a number of models for predicting phase stability and structure.

Containing a large number of worked examples, exercises, and detailed descriptions of numerous crystal structures, this book is primarily intended as an advanced undergraduate or graduate level textbook for students of materials science who are preparing to conduct research. However, it will also be useful to scientists and engineers who work with solid materials.

GREGORY S. ROHRER is a Professor of Materials Science and Engineering at Carnegie Mellon University. Prof. Rohrer was born in Lancaster, PA, in 1962. He received his bachelor's degree in Physics from Franklin and Marshall College in 1984 and his Ph.D. in Materials Science and Engineering from the University of Pennsylvania in 1989. At CMU, Prof. Rohrer is the director of the NSF sponsored Materials Research Science and Engineering Center. His research is directed toward understanding how the properties of surfaces and internal interfaces are influenced by their geometric and crystallographic structure, their stoichiometry, and their defect structure. Prof. Rohrer is an associate Editor for the Journal of the American Ceramic Society and his research earned a National Science Foundation Young Investigator Award in 1994.

Structure and Bonding in
Crystalline Materials

GREGORY S. ROHRER

CAMBRIDGE
UNIVERSITY PRESS

CAMBRIDGE UNIVERSITY PRESS
Cambridge, New York, Melbourne, Madrid, Cape Town, Singapore, São Paulo, Delhi

Cambridge University Press
The Edinburgh Building, Cambridge CB2 8RU, UK

Published in the United States of America by Cambridge University Press, New York

www.cambridge.org
Information on this title: www.cambridge.org/9780521663793

First published 2001

A catalogue record for this publication is available from the British Library

ISBN 978-0-521-66328-1 hardback
ISBN 978-0-521-66379-3 paperback

Transferred to digital printing 2009

Contents

CONTENTS

Preface

This book resulted from lecture notes that I compiled while teaching a course of the same name in the Department of Materials Science and Engineering at Carnegie Mellon University. When I began teaching this class in the early 1990s, there were already excellent textbooks on crystallography, solid state physics, and structural solid state chemistry. However, none of these books by themselves were entirely appropriate for the course I intended to teach to graduate students in materials science and engineering. Therefore, I have attempted to combine the subject matter in a way that would be both appealing and useful for materials scientists and engineers. Included in the book are compilations of data that are a useful resource for students and researchers considering basic structural problems. Much of the material in the book is derived from secondary sources and, to the best of my ability, I have assigned credit to these books in the last section of each chapter, under 'References and Sources for Further Study'. Books by Burger (Contemporary Crystallography), Sands (Introduction to Crystallography), Harrison (Electron Structure and the Properties of Solids), West (Solid State Chemistry and its Applications), Wells (Structural Inorganic Chemistry), Kittel (Introduction to Solid State Physics), and Ashcroft and Mermin (Solid State Physics) were especially useful and it is appropriate that I draw attention to them at the outset.

At Carnegie Mellon University, this course is taught during a 14 week semester consisting of approximately 52 hours of lecture. However, by prioritizing material according to the goals of an individual class, it should be possible to use this book as the basis for abbreviated courses.

This book is the outcome of a National Science Foundation Grant. Specifically, the development of this book was the educational component of a Young Investigator Award (DMR-9458005) that supported my research and educational activities for five years. Assistance also came from the more than 200 students who have been enrolled in my course over the years. The students continually helped me refine the text by pointing out errors and ambiguities. Dr Matt Willard deserves special mention for providing me with extensive detailed comments on an early draft while studying for his Ph.D. qualifying exam. Several other students who worked with me at CMU (Prof. Richard L. Smith, Dr Jennifer B. Lowekamp, and David M. Saylor) contributed figures for this book. My wife, Dr C. Lane Rohrer, was the greatest sustained source of editorial comment. Cathy edited numerous drafts of this book and even contributed several of the sections where my knowledge was inadequate. While the input I received from Cathy and others has made this a better book, I remain responsible for its deficiencies and any errors that might remain. Finally, I thank my father, C.E. Rohrer, who initially inspired my career in science and to whom I dedicate this book.

G.S. Rohrer
Pittsburgh
February, 2001

Chapter 1
Introduction

(A) Introduction

Every active field of scientific investigation has a central, fundamental question that motivates continued research. One way to phrase the motivating question in materials research is: how can elements be combined to produce a solid with specified properties? This is, of course, a complicated question, and it is appropriate to break it up into at least three separate issues. First, when any given elements are combined under some controlled conditions, will they be immiscible, will they dissolve in one another, or will they react to form a compound and, if so, in what atomic ratio? Second, what structure will the product of this combination have and how is it influenced by the processing conditions? While this book deals almost exclusively with the atomic structure of the crystals, it is equally important to be able to specify the defect structure, the microstructure, and the mesoscale structure. Third, given the product phase or phases and the structure (at each length scale), what are the properties of this material? Addressing these fundamental questions in a systematic way requires familiarity with established principles of thermodynamics, kinetics, chemistry, physics, and crystallography. The present book is intended to provide a set of necessary (but not sufficient) skills to conduct materials research. Specifically, the scope of the course encompasses the description of the structure of crystalline matter, the experimental interrogation of crystalline structure, the origin of the cohesive forces that stabilize crystalline structures, and how these cohesive forces vary with the elements in the solid.

In this introductory chapter, the primary goal is to review the elementary ideas that are used to understand the links between chemical bonding, crystal structures, and physical properties. The secondary goal is to motivate the study of more advanced models throughout the remainder of the book by exploring the limitations of the elementary concepts. We begin this chapter by describing the periodic chart and the principles of its arrangement.

Table 1.1. *A comparison of properties predicted by Mendeleev (1871) with those currently accepted for germanium [1]*

Property	Predicted	Current
Color	dark gray	grayish-white
Atomic weight	72	72.59
Density (g cm^{-3})	5.5	5.35
Atomic volume (cm^3 g-atom^{-1})	13	13.5
Specific heat (cal g^{-1} °C^{-1})	0.073	0.074
Oxide stoichiometry	XO_2	GeO_2
Oxide density (g cm^{-3})	4.7	4.703
Chloride stoichiometry	XCl_4	$GeCl_4$
Chloride boiling point	<100 °C	86 °C
Chloride density (g cm^{-3})	1.9	1.844

(B) Periodic trends in atomic properties

i.   *The importance of the periodic table: Mendeleev predicts Ge in 1871*

Over a century ago, Mendeleev demonstrated how useful it is to understand the periodic trends in atomic mass, size, ionization energies, and electronegativity. According to the periodic law that he formulated in 1869, 'the chemical properties of the elements are not arbitrary, but vary in a systematic way according to atomic mass.' In 1913, Henry Mosely discovered that it was actually atomic number (the number of protons and, thus, the number of electrons in the neutral atom), rather than atomic mass, that underpins the periodic law. The periodic law allowed Mendeleev to make a number of interesting predictions. For example, although element 32 (Ge) was not yet known, he successfully predicted many of its properties, as shown in Table 1.1. This example demonstrates the useful predictive power that comes with an understanding of periodicity. In the following subsections, the periodic trends in metallicity, electronegativity, and size are discussed.

ii. *Metallicity*

The property of *metallicity* can be defined as the tendency of an atom to donate electrons to metallic or ionic bonds. Metallicity increases from top to bottom and from right to left on the periodic chart. The metallicity trend can be understood according to the following line of reasoning. Metallicity increases as an atom binds its valence electrons with diminished strength. As you descend in a group on the periodic chart, the valence electron–nuclear separation is greater, and the binding force is diminished. The decrease in the binding force is a result both of the increased electron–nuclear distance and the screening of the nuclear charge by core electrons. On the other hand, as you go from right to left on the chart, the valence electron–nuclear separation is nearly the same (the electrons occupy the same principal levels), but the nuclear charge decreases. The decrease in nuclear charge is accompanied by a decrease in the electron binding force and an increase in the metallicity.

iii. *Electronegativity*

Metallicity is a good property to begin with because most people have a fairly clear idea of the difference between metals and nonmetals. However, it is far more common to describe the properties of atoms in terms of their *electronegativity*, which is the opposite of the metallicity. The electronegativity can be defined as the tendency of an atom to attract an electron. Based on this definition and the reasoning applied in the previous paragraph, you can see that the electronegativity trend is opposite to the metallicity trend. Numerous electronegativity scales have been proposed, but the most commonly used is the one originally devised by Pauling [2]. Because Pauling was an academic, he graded electronegativities on a 0 to 4.0 scale, with fluorine having the highest electronegativity of 4.0 and cesium having the lowest with 0.7. The Pauling electronegativities are shown in Fig. 1.1, and throughout this book we will use these values. In Chapter 7, more recent efforts to determine improved values will be described and alternative values will be presented.

iv. *Size and mass*

The periodic trends in *size* are the same as those for metallicity for the same reasons. Descending or moving from right to left on the chart, the atomic size increases. It is also worth remembering that cations (positive ions) are smaller than neutral atoms, while anions (negative ions) are larger. Ions always shrink with increasing positive charge and expand with increasing negative charge. *Mass*, of course, increases with atomic number.

As a closing note, it should be recognized that the periodic trends are not absolute. For example, when moving from left to right, the electronegativity does not increase continuously for every element. Note for example, that the electronegativity actually decreases to the immediate right of the noble metals (group IB). The fact that the mass of tellurium is actually greater than the mass of iodine illustrates that even the masses are not perfectly ordered. Despite these

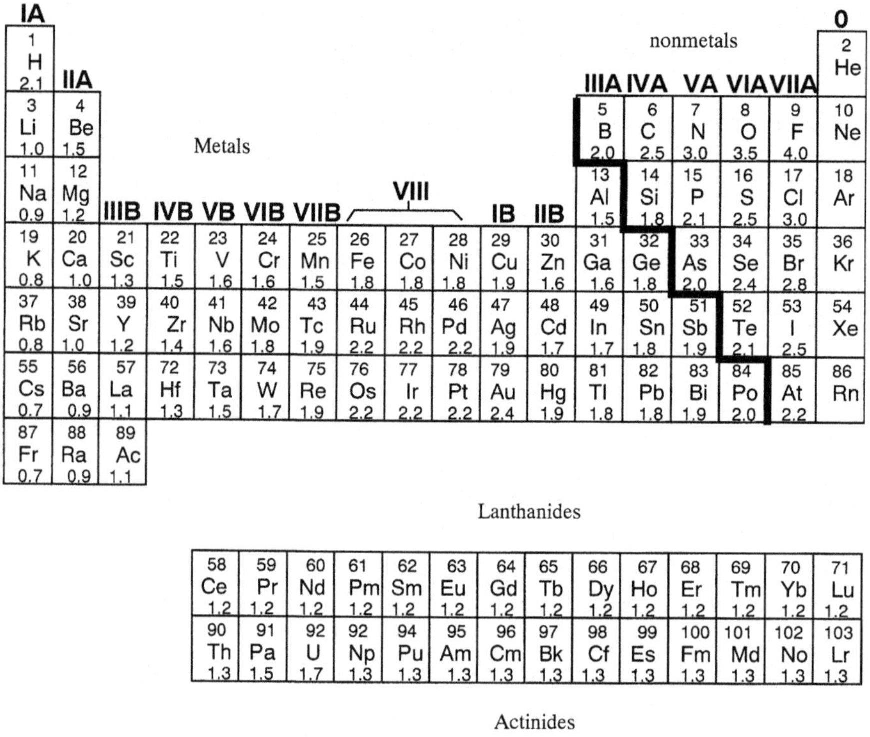

IA ... nonmetals ... **0**

Metals

IA	IIA	IIIB	IVB	VB	VIB	VIIB	VIII			IB	IIB	IIIA	IVA	VA	VIA	VIIA	0
1 H 2.1																	2 He
3 Li 1.0	4 Be 1.5											5 B 2.0	6 C 2.5	7 N 3.0	8 O 3.5	9 F 4.0	10 Ne
11 Na 0.9	12 Mg 1.2											13 Al 1.5	14 Si 1.8	15 P 2.1	16 S 2.5	17 Cl 3.0	18 Ar
19 K 0.8	20 Ca 1.0	21 Sc 1.3	22 Ti 1.5	23 V 1.6	24 Cr 1.6	25 Mn 1.5	26 Fe 1.8	27 Co 1.8	28 Ni 1.8	29 Cu 1.9	30 Zn 1.6	31 Ga 1.6	32 Ge 1.8	33 As 2.0	34 Se 2.4	35 Br 2.8	36 Kr
37 Rb 0.8	38 Sr 1.0	39 Y 1.2	40 Zr 1.4	41 Nb 1.6	42 Mo 1.8	43 Tc 1.9	44 Ru 2.2	45 Rh 2.2	46 Pd 2.2	47 Ag 1.9	48 Cd 1.7	49 In 1.7	50 Sn 1.8	51 Sb 1.9	52 Te 2.1	53 I 2.5	54 Xe
55 Cs 0.7	56 Ba 0.9	57 La 1.1	72 Hf 1.3	73 Ta 1.5	74 W 1.7	75 Re 1.9	76 Os 2.2	77 Ir 2.2	78 Pt 2.2	79 Au 2.4	80 Hg 1.9	81 Tl 1.8	82 Pb 1.8	83 Bi 1.9	84 Po 2.0	85 At 2.2	86 Rn
87 Fr 0.7	88 Ra 0.9	89 Ac 1.1															

Lanthanides

58 Ce 1.2	59 Pr 1.2	60 Nd 1.2	61 Pm 1.2	62 Sm 1.2	63 Eu 1.2	64 Gd 1.2	65 Tb 1.2	66 Dy 1.2	67 Ho 1.2	68 Er 1.2	69 Tm 1.2	70 Yb 1.2	71 Lu 1.2
90 Th 1.3	91 Pa 1.5	92 U 1.7	92 Np 1.3	94 Pu 1.3	95 Am 1.3	96 Cm 1.3	97 Bk 1.3	98 Cf 1.3	99 Es 1.3	100 Fm 1.3	101 Md 1.3	102 No 1.3	103 Lr 1.3

Actinides

Figure 1.1. A periodic chart with the Pauling electronegativities [2]. The bold line marks an arbitrary boundary between metals (to the left) and nonmetals (to the right).

exceptions, we will use the periodic trends in metallicity, electronegativity, and size (summarized in Fig. 1.2) to predict bonding types. This will, in turn, allow us to make predictions about crystal structures and properties.

C Bonding generalizations based on periodic trends in the electronegativity

i. Classification of the elements

We begin by classifying all elements as either metals or nonmetals. Because the change in properties from 'metallic' to 'nonmetallic' is continuous across the periodic table, it is not clear how to implement a binary definition. However, after some consideration, a line can be drawn, as shown on the chart in Figs. 1.1 and 1.2 (the 'bold' stepped line across the right hand side of the chart). With the elements divided up in this fashion, we establish the following rules. First, metallic elements form metallically bonded solids and metal–metal combinations form

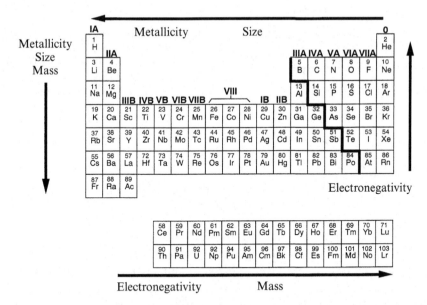

Figure 1.2. Summary of the periodic trends in atomic properties [1]. Arrows indicate the direction of increase in the property values.

metallically bonded solids. Second, nonmetallic elements and nonmetal–non-metal combinations are covalently bonded. Third, bonds between metals and nonmetals are either ionic or covalent, depending on the electronegativity difference.

These rules are fairly clear, except for the third which requires some critical electronegativity difference to separate ionic bonds from covalent bonds. We get this critical electronegativity difference from Pauling's expression for the ionicity fraction of a bond (f) [2], where

$$f = 1 - e^{-\frac{1}{4}(x_{nm} - x_m)^2} \tag{1.1}$$

and x_{nm} is the electronegativity of the nonmetallic element and x_m is the electro-negativity of the metallic element. We will assume that when $f > 0.5$ ($\Delta x > 1.7$), the bonds are ionic and that when $f \leq 0.5$ ($\Delta x \leq 1.7$), the bonds are covalent. In ternary or more complex compounds, the fractional ionicity can be determined by using stoichiometrically weighted averages for the values of x_m and/or x_{nm} in Eqn. 1.1.

It must be emphasized that the change from metallic to nonmetallic charac-ter is continuous and complex, so much so that many authors would refute the

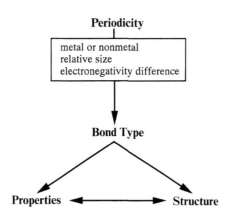

Figure 1.3. Knowledge of periodicity allows one to distinguish elements as metals or nonmetals and to gauge relative electronegativities and sizes. Based on this, it is possible to assign a bonding type. From knowledge of the bond type, characteristic structures and properties can be inferred.

apparently arbitrary binary categorization defined above. It is more common to define a third class of elements, the metalloids, which straddle the metal–non-metal boundary (for example: B, Si, Ge, As, Te, and Sb). However, with such criticism noted, a binary classification is nevertheless implemented because it has the practical advantage of leading to a simple set of rules to determine bond types. Once the bond type is defined, the type of atomic structure and properties that the solid might have can also be inferred. This relationship, upon which the following sections elaborate, is illustrated schematically in Fig. 1.3.

ii. *Simple bonding models and typical properties*

The simplest *metallic bonding model* assumes that positively charged ion cores are arranged periodically in a 'sea' of free electrons (formed by valence electrons which leave the sphere of influence of the atom). Metals include those elements from group IA and IIA where the s levels are filled (the alkali and alkaline earth metals), the B-group or transition metal series where the d levels are filled, and the lanthanide and actinide series where the f levels are filled. A number of post-transition metals are also found in the region of the chart where the p levels are filled. Materials that we would classify as metals include elemental substances such as Cu, Ag, Au, Al, Fe, Pb, intermetallic compounds such as Ni_3Al, NiAl, CuZn, $CuZn_3$, and random solid solutions or alloys, $A_x B_{1-x}$, where both A and B are metallic elements. Typical properties of metals include high reflectivity (when polished), high electronic and thermal conductivity, low to intermediate melting temperatures, and high ductility at temperatures less than half of their melting points. As exceptions, we should note that many intermetallic compounds and refractory metals have very high melting points and little ductility at room temperature.

The simplest model for the *ionic bond* assumes that charge is transferred from the more metallic (low electronegativity) atom to the less metallic (high electro-negativity) atom forming oppositely charged species, the cation (+) and anion (−). The electrostatic interaction between the two ions, F_{12}, increases with increasing charge (q_1 and q_2) and decreases with increasing separation, r_{12}, according to Coulomb's law:

$$F_{12} = \frac{kq_1q_2}{r_{12}^2}.$$ (1.2)

Crystals that we consider to be ionically bound include salts (such as NaCl and CaCl$_2$) and ceramics (such as MgO, ZrO$_2$, TiO$_2$). In each case, the electronega-tivity difference between the two atoms is greater than 1.7. Ionically bound materials are typically transparent and colorless, electronically and thermally insulating, have intermediate to high melting temperatures, are brittle at ambient temperatures, and are soluble in polar solvents or acids. Although these generalizations are well accepted, there are numerous exceptions, especially to the optical, electrical, and solubility descriptions. For example, we can compare TiO and CaO, both of which have the same crystal structure (rock salt) and, according to our definition, would be considered ionically bonded ceramics. However, while CaO is a transparent, colorless insulator, TiO is a reflective, metallic conductor that superconducts at sufficiently low (near absolute zero) temperatures.

The simple model for *covalent bonding* assumes that electrons are shared between atoms and that electron charge density accumulates between relatively positive atomic cores. Before going further, we must make an important dis-tinction between the two types of solids that contain covalent bonds. The first type includes three-dimensional covalent networks such as Si, SiC, GaAs, and BN. These crystals are composed of individual atoms, all linked by covalent bonds. In other words, there is a covalently bonded path between any two atoms in the solid. The second type includes molecular solids or polymeric solids. In these crystals, atoms within each molecule are linked by covalent bonds, but the molecules that make up the crystal are held together only by the weak interactions known collectively as intermolecular forces or secondary bonds (including van der Waals, dipolar, and hydrogen bonds). In such solids, not all atoms are connected by a path of strong covalent bonds. The difference between these two types of solids is illustrated schematically in Fig. 1.4. Examples of molecular solids include crystalline N$_2$, O$_2$, H$_2$O, C$_{60}$, and even macromolecular materials such as polyethylene. While it is easy to decide when a material will bond covalently, it is difficult to decide if it will form a three dimensional covalent network or a molecular solid. If more than two thirds of the components in a covalently bonded compound are H, C, O, N, or a

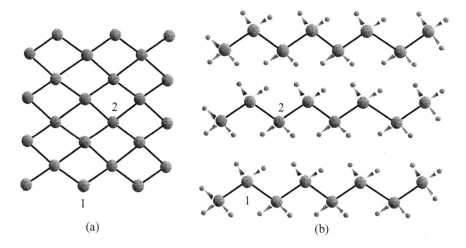

Figure 1.4. Comparison of (a) a covalently bonded three-dimensional network and (b) a molecular solid. The molecular solid has covalent bonds (dark lines) only within individual molecules. Thus, there is no covalently bonded path between the atom labeled 1 and the atom labeled 2; the molecules are bonded to one another only by weak secondary forces. In the covalently bonded network, however, there is a covalently bonded path between any two atoms.

halogen, then it is likely to be a molecular solid. However, diamond is a noteworthy example illustrating that this guideline should be applied with caution.

Covalently bonded networks typically have high melting points and are nonreflective, insulating, and brittle. On the other hand, molecular solids held together by secondary forces have low melting temperatures and are transparent, insulating, soft, and soluble. Perhaps one of the most obvious inadequacies of the simple models proposed here for assigning bond types is the inability to distinguish between these two types of solids.

iii. *Ketelaar's triangle*

Based on our discussion above, we can identify three types of primary bonds: metallic, ionic, and covalent; we will classify the weaker intermolecular forces as secondary. For simplicity, a set of rules has been defined that allow all substances to be placed in one of these three categories. However, one of the important objectives of this book is to establish the idea that these three types of bonding are limiting cases and that very few substances are well described by such an insensitive classification system. Most substances exhibit characteristics associated with more than one type of bonding and must be classified by a comparison to the limiting cases. In other words, when all of the possibilities are

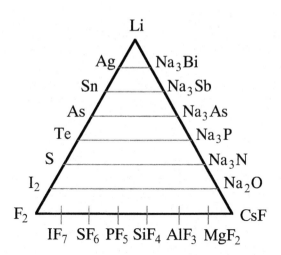

Figure 1.5. Ketelaar's triangle illustrates that there is a continuum of bonding types between the three limiting cases [3].

considered, we can say that there is a continuous transition from one type of bonding to another and that most materials are in the transition region rather than at the limits.

Ketelaar [3] expressed this idea in the simple diagram shown in Fig. 1.5. Taking the substance with the most nearly ideal metallic bond to be Li, and taking CsF and F_2 to have the most nearly ideal ionic and covalent bonds, respectively, these three substances form the vertices of the Ketelaar's triangle. All other substances fall at intermediate points; their proximity to the vertices corresponds to how well any of the three limiting cases will describe the bonding. The substances listed on the lateral edges of the triangle are merely examples chosen based on periodicity; all materials can be located on this triangle. So, when trying to understand the bonding and properties of any particular chemical compound, it is more useful to think about where it lies on Ketelaar's triangle than to try to associate it with one of the three limiting cases. In the next section, we cite some examples of how bonding is related to the properties of some real materials.

iv. *Examples of trends in bonding*

When the metal/nonmetal boundary on the periodic chart is crossed, the properties of the elements in group IV change dramatically, as is illustrated in Table 1.2. The properties of diamond are representative of a covalently bonded material and the properties of Pb are representative of a metallic material. The properties of Si and Ge are intermediate between these two limits. Note the continuous change in the melting points of these solids. To a first approximation, we can gauge relative bond strengths by melting points.

Table 1.2. *Properties of selected group IV elements.*

element	Electronic conductivity $(\Omega\text{-cm})^{-1}$	hardness	appearance	melting point/ boiling point T_m/T_b °C
C (diamond)	10^{-10}	10	transparent	3550/4827
Si	$>10^{-10}$		black	1410/2355
Ge	$<10^{-9}$		black	937/2830
Pb	10^{9}	2	reflective	327/1740

Table 1.3. *Properties of selected fourth row elements.*

element	electronic conductivity	appearance	T_m/T_b °C
Zn	conductive metal	reflective	420/907
Ga	conductive metal	reflective	30/2403
Ge	semiconductor	black	937/2830
As	insul./ photocond.	dull	817/(high press.)
Se	insul./ photocond.	dull	217/685
Br	insulator	diatomic gas	−7.2/59

The properties of elements in the fourth row of the periodic table (where the 4p shell is being filled) are shown in Table 1.3. Here, the metal–nonmetal boundary on the periodic chart is crossed in the horizontal direction. The series begins with a typical metal (Zn), goes to a three-dimensionally bonded covalent network (Ge), and finishes with a molecular solid (Br_2). Although the two atoms in a single diatomic bromine molecule are held together by a covalent bond, the molecules in the solid are held in place only by weak, secondary bonds. The difference between the melting points of solid Ge and Br_2 illustrates the difference between the properties of a three-dimensional covalent network and a molecular covalent solid.

To illustrate the changes that accompany the transition from covalent to ionic bonding, we examine the properties of isoelectronic compounds. As an example, we choose the oxides of group IV elements, which are given in Table 1.4. By examining these data, you can see that the bonding changes from ionic (ZrO_2) to a covalent network (SiO_2) and then to molecular covalent (CO_2). Note the profound difference between the behaviors of the isoelectronic compounds SiO_2 (a crystalline solid) and CO_2 (a molecular solid).

Table 1.4. *Selected properties of the oxides of group IV elements.*

compound	common name	T_m °C	Δ electroneg.
CO_2	dry ice	−57 (at 5.2 atm)	1.0
SiO_2	quartz	1610	1.7
GeO_2		1090	1.7
SnO_2	cassiterite	1630	1.7
TiO_2	rutile	1830	2.0
ZrO_2	zirconia	2700	2.1

Table 1.5. *Selected properties of three isoelectronic polar-covalent solids.*

group(s)	material	ionicity (f)	band gap	T_m °C
IV	Ge	0 %	0.7 eV	1231
III–V	GaAs	4 %	1.4 eV	1510
II–VI	ZnSe	15 %	2.6 eV	1790
I–VIII	CuBr	18 %	5.0 eV	492

Finally, the variation of melting temperature and band gap with the ionicity fraction in isoelectronic solids that exhibit partial ionic and covalent bonding is illustrated in Table 1.5. The band gap is the separation, in energy, between the highest filled electron energy level in the crystal and the lowest empty electron energy level. Radiation at energies equal to or greater than the band gap will be absorbed by the solid and promote electrons to higher energy unfilled states. Thus, the band gap is a quantitative parameter that influences the appearance of the solid. Since visible light varies in energy from 1.7 to 3.0 eV, nondefective solids with band gaps greater than 3.0 eV transmit *all* visible light and are thus transparent and colorless. Solids with band gaps less than 1.8 eV are opaque. If the band gap is much less than 1.7 eV (but greater than zero), the crystal will be black. From Table 1.5, we can see that the band gap increases with ionicity and we can infer that compounds with greater than 50% ionicity should have large band gaps and, therefore, be colorless. This is a simple explanation for why most ceramics (ionically bound materials) are colorless and most semiconductors (covalently bound materials) are black.

In conclusion, a brief survey of properties demonstrates that the periodicity of electronegativity and metallicity leads to a periodicity of bonding type. Because certain properties (electronic, optical) are also linked to bonding type, we see that there is also a periodicity of properties, as implied by Fig. 1.3.

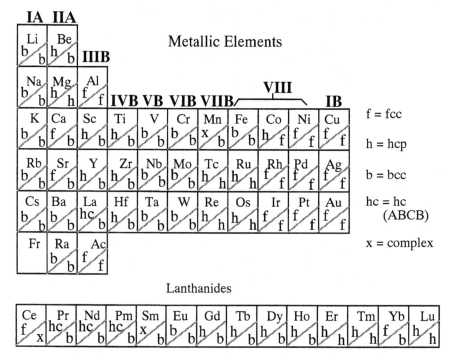

Figure 1.6. The crystal structures of the metallic elements. The symbol in the upper left refers to the room temperature crystal structure. The symbol in the lower right refers to the structure of the element just before it melts [5,6].

(D) Generalizations about crystal structures based on periodicity

i. *Close-packing in metallic solids*

The Coulombic attraction between delocalized valence electrons and positively charged cores is isotropic. Therefore, metallic bonding leads to close-packed crystal structures that maximize space filling and coordination number. The coordination number is the number of nearest neighbor atoms. Most elemental metals crystallize in the fcc (face centered cubic), hcp (hexagonal close packed), or bcc (body centered cubic) structures, as shown in Fig. 1.6. If you are not familiar with these structure types, you can read about them in Chapter 4, Section B. The closest-packed structures (fcc and hcp) have 12 nearest neighbors and the next coordination shell is 41% further away. While there are only 8 nearest-neighbors in the bcc structure, there are 6 next-nearest-neighbors only 15% further away; this gives atoms in this structure an effective coordination number of 14. The packing fraction is defined as the ratio of the sum of the atomic volumes within a representative, space-filling, structural repeat unit to the volume of that unit. In the bcc structure, the packing fraction is 68%. In the fcc and hcp structures, the packing fraction is 74%; this is the largest possible packing fraction for identical spheres.

Sm and Mn, which have comparatively complex structures, are two significant exceptions to the simple notion of close-packing driven by isotropic forces. For example, Mn has two separate bond lengths in its first coordination sphere.

Intermetallic compounds also typically have close-packed structures with high coordination numbers, but in special cases, low coordination arrangements can also occur. For example, coordination numbers of 4, 6, 8, 9, 10, 11, 12, 13, 14, 15, 16 and even 24 have been observed. The most common situation where low coordination number metallic structures are found is when one atom is much smaller than the other and fits into the octahedral or tetrahedral interstice of the close-packed structure (interstitial sites are defined in Chapter 4, Section C). Examples of such intermetallic compounds include metal nitrides and carbides such as TiC and TiN.

Finally, regarding the data in Fig. 1.6, we note that many metals are polymorphic. In other words, they take different crystal structures at different temperatures and pressures. For example, while iron is bcc at room temperature, above approximately 900 °C, it transforms to fcc (and then back to bcc before it finally melts). It should also be noted that when comparing the structures of different elements, it might not be best to pick a fixed temperature, such as room temperature. Because the variation in the melting points of the metallic elements is large (consider, for example, that Hg melts at -39 °C while W melts at 3410 °C), at any fixed temperature, the elements have very different stabilities with respect to melting. A more consistent comparison can be made by using the homologous temperature. The homologous temperature (T_h) is the absolute temperature (T), normalized by the melting point (T_m). In other words, $T_h = T/T_m$ such that all materials melt at a homologous temperature of 1.0. Note that at a homologous temperature of 0.999, 35 of the 54 elements listed in Fig. 1.6 have the bcc structure.

Example 1.1 Calculating Packing Fractions
Calculate the packing fraction for a crystal with the bcc structure.

We begin by reviewing the definition of the packing fraction: the ratio of the sum of the atomic volumes within a space-filling structural repeat unit to the volume of that unit.
1. Our first step is to choose a structural unit. The most convenient choice is the unit cell (which will be defined more precisely in Section C of the next chapter). The crystal

---

structure is composed of many identical copies of this fundamental unit. The arrangement of the bcc unit cell is shown in Fig. 1.7a.

2. In this case, the unit cell is a cube with edge length a and, therefore, the volume of
this structural unit is a^3.

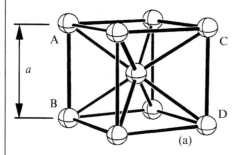 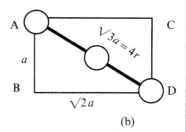

(a)   (b)

Figure 1.7. Schematic diagram for Example 1.1

3. Assuming that the atoms are spheres with radius r, we will take the volume of each
atom to be $4/3\pi r^3$ and the volume of atoms in the unit cell to be $N \times 4/3 \pi r^3$, where N
is the number of atoms in the cell. In this case, there are two atoms in the cell, one in
the center and one at the vertices (you can think of 1/8 of each of the eight atoms at
the vertices as being within the boundaries of a single cell).

4. To calculate the ratio, we have to write a in terms of r. In this cell, the nearest neighbors to the atom in the center are the atoms at the vertices. Assuming that this atom
contacts its nearest neighbors, there is a line of contact, $4r$ long, that stretches from
opposite corners of the cell, across the body diagonal. Using simple geometry, you can
see that the length of this line is $\sqrt{3}a = 4r$.

5. We can now write the ratio and compute the packing fraction:

$$\text{Packing fraction} = \frac{2 \cdot \frac{4}{3}\pi r^3}{\left(\frac{4}{\sqrt{3}}r\right)^3} = \frac{\pi\sqrt{3}}{8} = 0.68.$$

ii. *Radius ratios in ionic structures*

Because the electrostatic attractions in our simple model for the ionic bond are
isotropic, we should also expect ionically bonded solids to form close-packed
structures. However, the coordination numbers in ionically bonded structures
are influenced by steric factors, or the relative size of the cation and anion. The
relative size is quantified by the radius ratio (ρ), which is the ratio of the cation
radius (r_+) to the anion radius (r_-).

$$\rho = \frac{r_+}{r_-}. \tag{1.3}$$

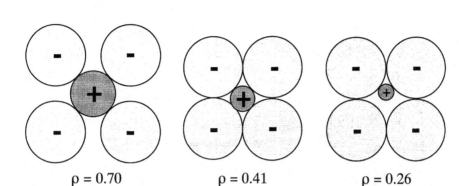

Figure 1.8. Geometric configurations with different radius ratios. (a) shows a stable configuration. The ions in (b) have the minimum radius ratio for stability in this arrangement. The radius ratio of the ions in (c) makes this an unstable configuration.

Stable and unstable configurations are illustrated in Fig. 1.8. Basically, a configuration is stable until anion–anion repulsions force longer and less stable anion–cation bond distances (as in 1.8c). The critical or minimum stable radius ratio is defined by the point when the cation contacts all of the neighboring anions, and the anions just contact one another, as shown in Fig. 1.8(b). For any given coordination number, there is a minimum stable radius ratio that can be derived through simple geometric arguments (see Example 1.2). Atoms with radius ratios as shown in 1.8(c) would be more stable in a configuration with a lower coordination number. The minimum radius ratios for selected geometries are summarized in Fig. 1.10.

Example 1.2 Calculating minimum stable radius ratios
Determine the minimum stable radius ratio for octahedral (six-fold) coordination.

1. First, assume that the cation in the center contacts the surrounding anions and that the anions just contact one another (this is the minimum stable configuration). Using Fig. 1.9, we find a plane that contains both cation–anion and anion–anion contacts. One such plane is the equatorial plane (see Fig. 1.9b).
2. Next, we note that the sides of the isosceles triangles in Fig. 1.9 have the lengths: $a=b=r_-$ and $c=r_+ +r_-$. Based on these geometric relationships, the radius ratio can be easily determined:

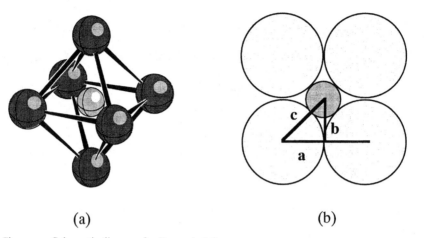

(a)   (b)

Figure 1.9. Schematic diagram for Example 1.2

$$c^2 = a^2 + b^2$$

$$r_+ + r_- = \sqrt{2r_-^2}$$

$$r_+ = r_-(\sqrt{2} - 1)$$

$$\rho = \frac{r_+}{r_-} = 0.41.$$

3. Therefore, the critical radius ratio is 0.41.

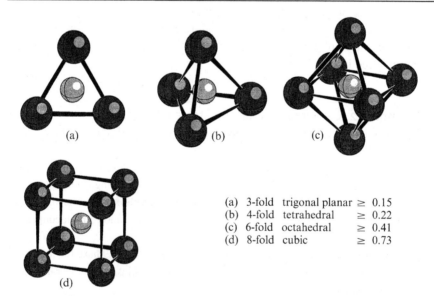

(a)	3-fold	trigonal planar	≥ 0.15
(b)	4-fold	tetrahedral	≥ 0.22
(c)	6-fold	octahedral	≥ 0.41
(d)	8-fold	cubic	≥ 0.73

Figure 1.10. Summary of the minimum radius ratios for common configurations.

Table 1.6. *A test of the radius ratio rules for binary 1:1 compounds*

compound	r_+	r_-	r_+/r_-	predicted CN	observed CN
CsCl	1.67	1.81	0.92	8	8
CsBr	1.67	1.95	0.86	8	8
RbCl	1.48	1.81	0.82	8	6
TlBr	1.15	1.95	0.59	6	8
TlI	1.15	2.16	0.53	6	8
KI	1.33	2.16	0.62	6	6
MgO	0.82	1.76	0.46	6	6
MnO	0.75	1.76	0.43	6	6
CoO	0.72	1.76	0.41	6	6
NiO	0.69	1.76	0.39	4	6
ZnO	0.88	1.76	0.50	6	4
CdO	1.14	1.76	0.65	6	6
ZnS	0.88	2.19	0.40	4	4
ZnSe	0.88	2.38	0.37	4	4

Carefully chosen examples can make the radius ratio concept look like an accurate predictive tool. However, it is often in error, particularly in complex structures and when the bonding becomes increasingly covalent (where the hard sphere model breaks down). Pauling [9] originally used the univalent radii (rather than crystal radii) to calculate the radius ratio. A table of these values, together with the crystal radii, can be found in Appendix 1A. However, the success rate of the predictions does not depend strongly on which set of radii are used, as long as both the cation and anion radii come from a set that was derived using consistent assumptions. Examples of accurate and inaccurate predictions are shown in Table 1.6 (calculated using Pauling's univalent radii). Important reasons for the inadequacy of this theory are the assumption of spherically symmetric forces and symmetric coordination, the assumption that atoms have the same size in all chemical environments, and the underestimation of 'like-atom' repulsions.

iii.   *Orbital hybridization in covalent structures*
In contrast to isotropically bound metallic and ionic systems, covalent bonding is directional. When a covalent bond is formed, electron density is increased along the line connecting two bonding atoms. Covalent structures are formed from atoms that have both s and p valence electrons (in effect, those on the right-hand side of the periodic chart with relatively high electronegativities). The formation of sp hybrid orbitals results in four equivalent sp^3 orbitals directed

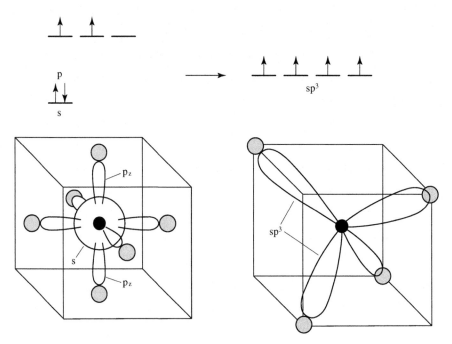

Figure 1.11. The hybridization of the inequivalent s and p orbitals of a group IV atom leads to four tetrahedrally arranged sp^3 orbitals that have the same energy. The shaded circles represent neighboring atom positions.

towards the vertices of a tetrahedron, as shown in Fig. 1.11. The geometry of these orbitals leads to the 4-fold coordination which is the signature of covalently bound structures.

Examples of covalently bonded solids include C (diamond), Si, Ge, and SiC. In all of these crystals, the atoms have tetrahedral coordination. Many III–V compounds (these are compounds formed between group III and V atoms) such as BN, BP, BAs, AlP, AlAs, AlSb, GaP, GaAs, GaSb, InP, InAs, and InSb, crystallize in the zinc blende (sphalerite) structure. In this structure, all of the atoms are tetrahedrally coordinated and the bonding is considered to be primarily covalent. Many II–VI compounds, such as ZnS, ZnSe, ZnTe, CdS, CdSe, and BeO, crystallize in the wurtzite structure in which all atoms are again situated at tetrahedral sites. Despite the increased ionicity of the bonding in these compounds, we would still consider their bonding to be mostly covalent.

The same sp^3 hybridization also influences the structure of hydrocarbon chains, such as polyethylene. The backbone of a polymer chain is formed by a string of C-C bonds. For example, a C atom in polyethylene has two C nearest

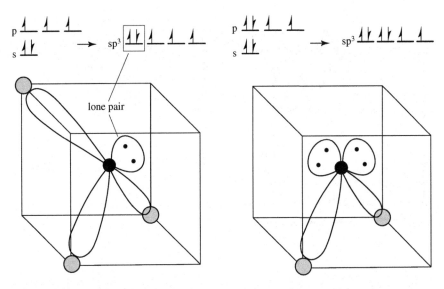

Figure 1.12. The hybridization of the inequivalent s and p orbitals of group V and VI atoms leads to four sp³ orbitals that have a tetrahedral arrangement, even though some of the ligands are nonbonding lone pairs. The shaded circles represent neighboring atom positions and the dots represent electrons in lone pair orbitals. The bond angles of the actual ligands usually vary from the ideal.

neighbors and two H nearest neighbors. While the bonding geometry is not perfectly tetrahedral, the C-C-C bond angle is 120°, so that the C backbone forms a zig-zag chain on the atomic scale. Pseudo-tetrahedral arrangements are found even when there are fewer than four nearest neighbors. In these cases, the lone pairs of electrons complete the tetrahedron, as illustrated in Fig. 1.12.

As a final example of the importance of orbital hybridization, we consider the silicates. Silicates are a wide class of silicon and oxygen containing compounds that make up much of the earth's crust. Substances such as quartz, micas, clays, and zeolites, which are commonly thought of as minerals, are also important engineered materials that can be used for piezoelectrics (quartz) and hydrocarbon cracking catalysts (clays and zeolites). The basic structural unit of a silicate is the SiO_4 tetrahedron which links in complex patterns to form a wide variety of structures. The tetrahedral coordination of the Si atom by O is a result of orbital hybridization. Furthermore, these tetrahedral SiO_4 units always link by corners, placing the O atom in two-fold coordination (in this case, O has two electropositive ligands and two nonbonding lone pairs). Figure 1.13 shows the atomic arrangement found in cristobalite, a representative silicate structure.

In conclusion, there are three established generalizations which relate the

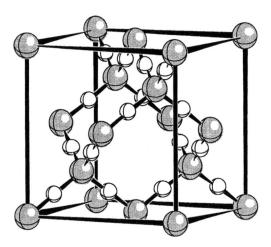

Figure 1.13. The structure of cristobalite shows the same coordination found in all silicates; the Si is in tetrahedral coordination and the O is in two-fold coordination.

type of chemical bonding in a compound to the type of crystal structure that it is likely to form. First, metals form close-packed structures with high coordination numbers. Second, structures held together by ionic bonds pack atoms as if they were hard spheres that obey radius ratio rules. Third, covalent bonds in nonmetallic structures are often formed from sp³ hybridized orbitals which leave atoms in tetrahedral coordination or a pseudo-tetrahedral arrangement that includes nonbonded pairs.

iv. Examples of periodic trends in structures

To illustrate the systematic changes in structure that occur as bonding changes, we again examine the oxides of group IV. In this case, both the ionicity of the bonding and the radius ratio of the atomic components change and there is a transformation from covalently bound, low coordination number structures to ionically bound, higher coordination number structures (see Table 1.7).

Table 1.8 shows how metallicity affects structure. In this table, we compare the structures of elements as the 3p shell is being filled. Mg and Al are examples of metallic solids that assume high coordination number structures. Si and P are examples of covalent solids that assume low coordination number structures.

In summary, we can say that the rules of periodicity give us a metal/nonmetal division, relative electronegativities, and relative atomic and ionic sizes. Based on metallicity and relative electronegativity, a bonding type can be assigned. Using established generalizations, some assumptions can be made about properties. Also, knowing the type of bonding and the relative sizes of the atoms, the type of structure that the solid is likely to take can be predicted. Although the rules

Table 1.7. *The structures of the group IV dioxides.*

compound	structure (RT)	CN	Δelectroneg.	r_+/r_-
CO_2	molecular	2	1.0	0.21
SiO_2	quartz	4	1.7	0.46
GeO_2	rutile	6	1.7	0.54
SnO_2	rutile	6	1.7	0.68
TiO_2	rutile	6	2.0	0.68
ZrO_2	zirconia	8	2.1	0.78

Table 1.8. *Structures of the elements in the 3p filling series.*

element	structure	coordination	type
Mg	hcp	12	metal
Al	fcc	12	metal
Si	diamond	4	nonmetal
P_{white}	complex	4	nonmetal
P_{black}	layered	3	nonmetal

will sometimes lead to questionable conclusions, they have the virtue of being simple, easy to apply, and, therefore, a good starting point for the development of more detailed models.

(E) The limitations of simple models

i. *Polymorphs and allotropes*
On the basis of the simple models, it is impossible to rationalize the polymorphic behavior of some materials. We define polymorphs as structurally distinct forms of the same compound and allotropes as structurally distinct forms of an elemental substance (in modern usage, the distinction between allotropes, polymorphs, and isomers is frequently lost). The range of structures taken by carbon provides one striking example. Carbon crystallizes in three different forms which can be classified as molecular (C_{60} is known as buckminster fullerene), layered (graphite), and three-dimensional (the well known diamond structure). These structures are illustrated in Fig. 1.14. The properties of this single element span the range from metallic to insulating, hard to lubricating, and transparent to reflective (see Table 1.9).

Table 1.9. *The properties of three carbon allotropes.*

polymorph	coordination	bonding type	properties	band gap, eV
diamond	4	covalent, three dimensional	hard, transparent, insulating	6
graphite	3	covalent/ van der Waals, layered	reflective, black, metal, lubricating	0
C_{60}	3	covalent/ molecular	soft, black, semicond.	~1

(a)   (b)

(c)

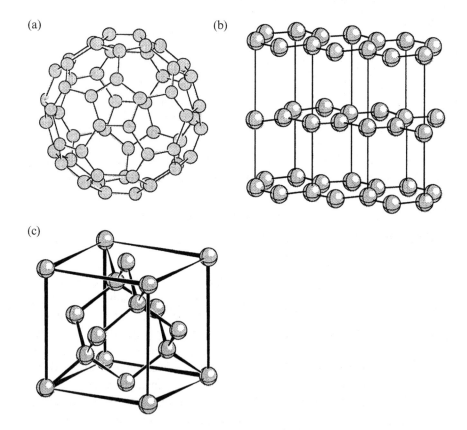

Figure 1.14. Three forms of carbon. (a) molecular C_{60}, (b) layered graphite, (c) diamond.

Note that C_{60} and graphite were omitted from Table 1.2 because their properties are not easily justified in terms of the simple models. In these cases, the simultaneous presence of both strong primary (covalent) bonds and weak secondary (van der Waals) bonds is responsible for their properties. The difference between the properties of C_{60} and diamond illustrates the difference between covalent molecular bonds and extended covalent bonds. The same phenomenon distinguishes the properties of CO_2 and SiO_2.

There are, of course, many other examples of solids that can take more than one structure. Many metallic materials (Fe, for example) can take either the fcc or bcc structure. TiO_2 is an example of an ionically bonded compound that can take different crystal structures (rutile, anatase, brookite, and TiO_2-B). Silicon carbide exhibits a special type of polymorphism that is known as polytypism. Polytypes are structures distinguished by different stacking sequences along one direction. In the more than 70 known polytypes of SiC, both Si and C are always in tetrahedral coordination, as we would expect for a three dimensional covalently bonded compound of group IV elements. The difference between the polytypes is in the long range order (the stacking sequence of its close packed layers). This phenomenon will be discussed in greater detail in Chapter 4.

ii. *Systems with mixed properties and mixed bonding*

In Section D, generalizations relating bonding type to properties were reviewed. However, one must remember that these bonding models represent limiting cases and, therefore, do not provide accurate descriptions of many materials. In most cases, bonding is best described as a mixture of these limiting cases and, because of this, the generalizations that relate bonding type to properties are often misleading.

Consider, for example, $YBa_2Cu_3O_7$ and $Y_3Fe_5O_{12}$. Both of these materials are yttrium-transition metal oxides and meet the established criterion for ionic bonding. However, the most important properties of these materials, superconductivity in the first and magnetism in the second, are often associated with metals. $YBa_2Cu_3O_7$ is a high critical temperature (high T_c) superconductor and the yttrium iron garnet (known as YIG) is ferrimagnetic and used for 'soft' magnetic applications. According to accepted bonding and property generalizations, ionically bonded compounds should not have metallic electrical properties, but many do. This example illustrates a failure of the simple generalizations and demonstrates the need for improved models.

In fact, based on the progress of superconductivity research over the past eight decades (see Fig. 1.15), one must conclude that the 'conventional wisdom' regarding the electrical properties of metallic and ionic compounds impeded the

Table 1.10. *Superconducting oxides known prior to 1985.*

compound	T_c, K	structure	reference
$SrTiO_{3-x}$	0.28	perovskite	[13]
Na_xWO_3	3	TTB*	[14]
	≈ 1	rock salt	[15]
NbO_x	≈ 1	rock salt	[16]
$Li_{1+x}Ti_{1-x}O_4$	13.7	spinel	[17]
$BaPb_{1-x}Bi_xO_3$	13	perovskite	[18]

Note:

*TTB stands for the tetragonal tungsten bronze structure type (see Table 3B.17).

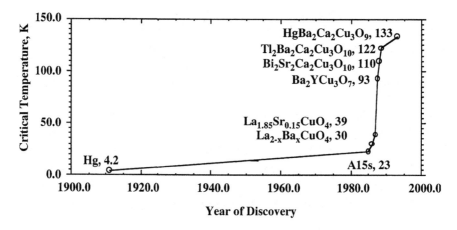

Figure 1.15. The highest T_c materials are all 'ionic' oxides. The designation A15 refers to a series of intermetallics with the so-called A15 structure (such as Nb_3Sn) which is described in Chapter 4, Section F(iv).

development of improved materials. Between the discovery of superconductivity (1911) and the discovery of the first superconducting cuprate (1985), the critical temperature was improved at a rate of 3 °C per decade. During this time, more than 1000 superconducting materials had been identified, but only six were 'ionic' metal oxides (see Table 1.10). This is mostly because few people had the foresight to check the already well known metallic oxides for superconductivity. In 1998, all of the highest T_c materials were metal oxides. What is particularly interesting is that some of the phases now recognized as high T_c superconductors were known long before their electrical properties were fully characterized and appreciated. For example, the phases that created the breakthrough in the latter half of the 1980s, $La_{2-x}Ba_xCuO_{4-y}$ (Bednorz & Müller [18]) and $La_{2-x}Sr_xCuO_{4-y}$ (Cava *et al.* [19]), had been identified more than five years before their unique electronic properties were discovered (Shaplygin *et al.* [20] and Raveau *et al.* [21]). In this particular case, Bednorz and Müller's insight was rewarded with the Nobel Prize.

Just as some ionic compounds have properties more often associated with metals, some metals have properties associated with ionic or covalent compounds. For example, many intermetallic compounds, such as NiAl, have melting temperatures greater than 1500°C and little of the ductility normally associated with metals. In fact, detailed studies of the bonding in this compound suggest that there is a small peak in the electron density along the line connecting the Ni and Al atoms, a clear signature of directional covalent bonding. The low density, good oxidation resistance, high melting temperature, high thermal conductivity, and appropriate stiffness of this intermetallic compound make it an excellent material for advanced aerospace structures and propulsion systems [22]. This interesting combination of properties has motivated research aimed at the development and understanding of several Ni- and Ti-based intermetallic alloy systems.

We conclude with an example of a polymer, Li-doped polyacetylene, $(Li_yCH)_x$, that has the optical and electrical properties of a metal combined with the chemical composition and molecular structure of a plastic. This material can potentially be used as an extremely pliable, low density, electronic conductor [23–25].

The complexities demonstrated by the C allotropes, the high T_c superconductors, intermetallic compounds, and conducting polymers motivate materials researchers to develop better descriptions of the relationships between crystal structures, bonding, and the physical properties of materials. Throughout the book, we will explore both formal, quantitative, physical models, as well as more flexible, qualitative, chemical models.

(F) Problems

(1) In Table 1.2, we see that Ge is a semiconductor and that Pb is a metal. The element Sn, omitted from the list, lies between them on the periodic table (see Fig. 1.1). In one of its allotropic forms, Sn is a semiconductor and in another, it is a metal. Thus, it bridges the gap between the two different electrical properties. The structure of 'gray tin' is diamond cubic (see Fig. 1.10) and the structure of 'white tin' is body centered tetragonal (bct), a structure that is closely related to the more common bcc structure. Which one do you think is a semiconductor and which do you think is a metal and why?

(2) Consider Ketelaar's triangle, shown in Fig. 1.5.

(i) Explain the choice of CsF as the best example of an ionic bond.

(ii) At what positions on Ketelaar's triangle would the elements in Group IVA be located?

(iii) In which region or regions of Ketelaar's triangle are molecular materials found? Can they be differentiated from the three-dimensional covalent network materials, or do they overlap?

(3) The chart in Fig. 1.6 shows the room temperature structures of metallic elements. Most metals are bcc when they melt. Using data available in the literature, determine the homologous temperature at which these elements transform to the bcc structure. Are the transition temperatures similar for each metal?

(4) Compare the packing fractions of fcc and diamond cubic crystals. Demonstrate that the packing fraction of hcp is the same as fcc. (If you are not familiar with these structure types, you can read about them in Chapter 4, Section B.)

(5) Determine the minimum stable radius ratio for three-fold (trigonal planar), tetrahedral, and eight-fold (cubohedral) coordination.

(6) Determine and compare the minimum stable radius ratios for tetrahedral and square planar coordination.

(7) Determine and compare the minimum stable radius ratios for octahedral and trigonal prismatic coordination.

(8) (i) Rank the metallicity of the following elements: Si, Na, Al, Ge, Zr, Cu.

(ii) Rank the ionicity of the following compounds: LiF, TiO, SiC, CuBr, CaO, NiAl, ZnSe, and ZnO.

(iii) Classify the materials in (i) and (ii) as likely to form either a close-packed metallic structure, a tetrahedally coordinated structure, or a crystal whose structure is governed by radius ratio rules.

(9) The band gaps of alloys with the composition $GaAs_{1-x}P_x$ vary with x.

(i) As x in the alloy above changes from 0 to 0.8, how does the ionicity fraction of the material change?

(ii) As x in the alloy above changes from 0 to 0.8, how does the band gap of the material change?

(10) GaN and AlN are important wide band gap semiconductors that have potential applications in both short wavelength optoelectronic and high power/high frequency devices. How would you describe the bonding in these materials? What structures are they likely to take? Calculate the ionicity fraction of the bonds in these compounds and compare it to the semiconductors listed in Table 1.5. Do you expect these materials to have a larger or smaller band gap than those listed in Table 1.5? If AlN is added to GaN, will the band gap increase or decrease? What if BN is added to GaN?

(11) How would you describe the bonding in SiC? Calculate the ionicity fraction of the bonds in SiC and compare it to the semiconductors listed in Table 1.5. Do you expect SiC to be transparent or opaque?

(12) Crystalline oxynitrides in the SiO_2-Si_3N_4-Al_2O_3-AlN system have been commercially developed for structural applications and are commonly referred to as 'sialon' ceramics.

> (i) Classify the bonding type of each of the four binary compounds in the sialon system.
>
> (ii) How do you expect the Si and Al atoms to be coordinated in each of the four binary compounds listed above? Check your expectations against established data.
>
> (iii) In addition to solid solutions in this system, there are also some ternary compounds including Al_3O_3N, Si_2N_2O, $Al_6Si_2O_{13}$ (mullite), and Al_7O_9N. Classify the bonding type in each of these materials.
>
> (iv) Where do you think that each of the four binary and four ternary compounds mentioned above fall on Ketelaar's triangle? (Draw a triangle and put a point for each compound on the triangle).

(G) References and sources for further study

[1] C.H. Yoder, F.H. Suydam, and F.H. Snavely, *Chemistry*, 2nd ed., (Harcourt Brace Jovanovich, New York, 1980) pp. 81–96 (Chapter 4). Table 1.1 is based on Table 4.2, p. 84. Fig. 1.2 is drawn after Fig. 4.7, p. 92.

[2] L. Pauling, *The Nature of the Chemical Bond*, (Cornell University Press, Ithaca, 1960). The data in Fig. 1.1 is from Table 3.8, p. 93. Eqn. 1.1 is found on p. 98.

[3] J.A.A. Ketelaar, *Chemical Constitution* (Elsevier Publishing Co., Amsterdam, 1953). Fig. 1.5 is drawn after Fig. 1, p. 21.

[4] L. Van Vlack, *Elements of Materials Science and Engineering*, 6th edition, (Addison-Wesley, Reading, MA, 1989) pp. 19–53, Chapter 2. In this source, you can read a summary of simple bonding models.

[5] A.F. Wells, *Structural Inorganic Chemistry*, 5th edition (Clarendon Press, Oxford, 1984) p. 1274–1326, Chapter 29. Some of the data in Fig. 1.6 is from Fig. 29.3 on p. 1281.

[6] T.B. Massalski, *Binary Alloy Phase Diagrams*, 2nd edition (ASM International, Materials Park, OH, 1990). Some of the data in Fig. 1.6 were taken from this source.

[7] *The CRC Handbook of Chemistry and Physics*, 61st edition, 1980–81, edited by R.C. Weast and M.J. Astle (CRC Press, Inc., Boca Raton, 1981). Much of the data in Tables 1.2, 1.3, and 1.4 were taken from tables in this source.

[8] W.D. Kingery, H.K. Bowen, and D.R. Uhlmann, *Introduction to Ceramics* (John Wiley & Sons, New York, 1976) pp. 25–87, Chapter 2. A summarized account of the radius ratio rules.

[9] L. Pauling, *The Nature of the Chemical Bond*, (Cornell University Press, Ithaca, 1960) p. 544.

[10] J.E. Fischer, P.A. Heiney, and A.B. Smith III, Solid State Chemistry of Fullerene-Based Materials, *Acc. Chem. Res.* **25** (1992) 112–18. Further reading on C_{60}.

[11] *MRS Bulletin*, vol. 14, no. 1 (January 1989). This issue is devoted to high T_c superconductors.

[12] C.N.R Rao and J. Gopalakrishnan, *New Directions in Solid State Chemistry* (Cambridge University Press, Cambridge, 1989) pp. 475–90, Chapter 9. Further reading on high T_c superconductors.

[13] J. Schooley, W. Hosler, and M. Cohen, *Phys. Rev. Lett.* **12** (1964) 474; *Phys. Rev. Lett.* **15** (1965) 108.

[14] A. Sweedler, Ch. Raub, and B. Matthias, *Phys. Rev. Lett.* **15** (1965) 108.

[15] J. Holm, C. Jones, R. Hein, and J. Gibson, *J. Low Temp. Phys.* **7** (1972) 291.

[16] D. Johnston, *J. Low Temp. Phys.* **25** (1976) 145.

[17] A.W. Sleight, J. Gillson, and P. Bierstedt, *Solid State Comm.* **17** (1975) 27.

[18] J.G. Bednorz and K.A. Müller, *Z. Phys. B* **64** (1986) 189.

[19] R.J. Cava, R.B. van Dover, B. Batlogg, and E.A. Rietmen, *Phys. Rev. Lett.* **58** (1987) 408.

[20] I.S. Shaplygin, B.G. Kakhan, and V.B. Lazarev, *Russ. J. Inorg. Chem.* **24** (1979) 620.

[21] N. Nguyen, J. Choisnet, M. Hervieu, and B. Raveau, *J. Solid State Chem.* **39** (1981) 120.

[22] D.B. Miracle, The Physical and Mechanical Properties of NiAl *Acta Metall. Mater.* **41** (1993) pp. 649–84. A good review about NiAl.

[23] A.G. MacDiarmid and A.J. Heeger, *Synth. Met.* **1** (1979/1980) 101. Polymeric conductors.

[24] P.J. Higrey, D. MacInnes, D.P. Mairns, and A.G. MacDiarmid, *J. Electrochem. Soc.* **128** (1981) pp. 1651–4. Polymeric conductors.

[25] *C&E News*, January 26, 1981, p. 39. Polymeric conductors.

[26] H. Morkoç and S.N. Mohammad, *Science* **267** (1995) 51. More about GaN.

[27] Y.-M. Chiang, D. Birnie III, and W.D. Kingery, *Physical Ceramics*, (John Wiley & Sons, New York, 1997) pp. 69–72. A brief description of sialons.

[28] K.H. Jack, Sialons and Related Nitrogen Ceramics, *J. Mater. Sci.* **11** (1976) 1135–58. This is a good review of sialon ceramics.

Chapter 2
Basic Structural Concepts

(A) Introduction

Crystal structures have already been mentioned and a few simple diagrams (Figs. 1.7a, 1.13, and 1.14c, for example) were presented in the last chapter. To understand crystal structures in a systematic way, one must be familiar with the formal methods of describing them. The objective of this chapter is to begin to explain how crystal structures are described and classified. The most important topics in this chapter are the definition of the lattice (both direct and reciprocal), the description of techniques for quantifying lattice components (directions and planes), and the idea that every structure can be described as a combination of a lattice and a basis. This chapter also includes a description of how crystal structures are visually represented and an introduction to polycrystallography.

(B) The Bravais lattice

The Bravais lattice is the periodic array in which the repeated units of the crystal are arranged. Because there are only 14 distinct arrangements, the Bravais lattice system provides a convenient mechanism for classifying the structural diversity that occurs in nature. However, one must remember that the Bravais lattice describes only the underlying configuration of the repeat units. It says nothing about the arrangement of atoms within the repeat unit. In this case, specificity is sacrificed for simplicity.

i. Definition

A good definition of a Bravais lattice is that it is an infinite array of discrete points with an arrangement and orientation that appears exactly the same regardless of the point from which the array is viewed. A more practical quantitative definition is that a Bravais lattice consists of all points defined by position vectors, $\vec{R}$, of the form:

$$\vec{R} = u\vec{a} + v\vec{b} + w\vec{c} \tag{2.1}$$

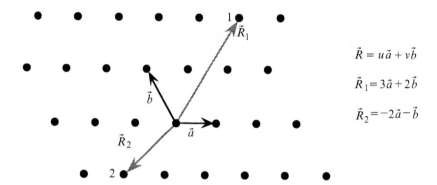

$$\vec{R} = u\vec{a} + v\vec{b}$$

$$\vec{R}_1 = 3\vec{a} + 2\vec{b}$$

$$\vec{R}_2 = -2\vec{a} - \vec{b}$$

Figure 2.1. A two-dimensional Bravais lattice, the oblique net. By definition, all points on the net are specified by linear combinations of the primitive vectors, $\vec{a}$ and $\vec{b}$, with integer coefficients.

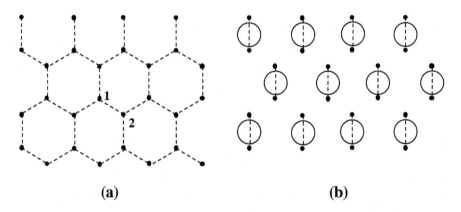

(a)   **(b)**

Figure 2.2. The honeycomb (a) is not a Bravais lattice because the environment around points 1 and 2 differ. If, however, we consider pairs of points from the honeycomb, as in (b), then we have a Bravais lattice where each pair, now called the basis, has 6 identical neighbors.

where $\vec{a}$, $\vec{b}$, and $\vec{c}$ are three noncoplanar vectors and u, v, and w may assume any positive or negative integer values, including zero. The two definitions are equivalent.

The oblique net shown in Fig. 2.1 is a two-dimensional Bravais lattice with no particular symmetry. Because it is a Bravais lattice, by definition, the locations of all the points that make up the net are specified by linear combinations of the primitive vectors, $\vec{a}$ and $\vec{b}$, with integer coefficients. On the other hand, the honeycomb shown in Fig. 2.2a is not a Bravais lattice. Notice that as you move from point 1 to point 2, the environment changes by a reflection. If, on the other hand,

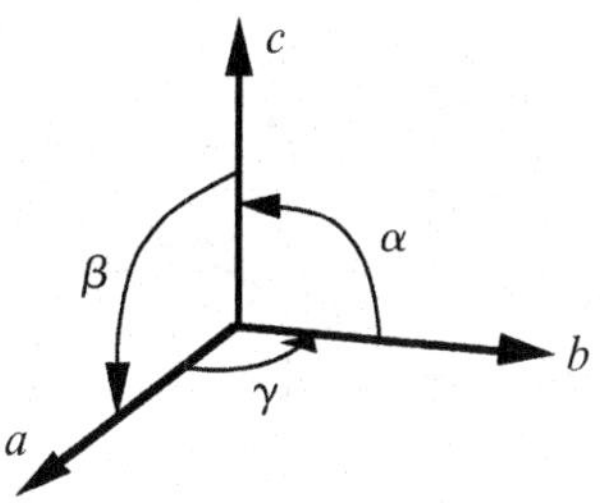

Figure 2.3. The conventional lattice parameter definitions.

you tie pairs of points together, these pairs form a two-dimensional Bravais lattice, the hexagonal net. Periodic patterns of objects can always be divided into identical groups in such a way that each group occupies the sites of a Bravais lattice. Each of the identical groups is called a basis and it is the combination of a Bravais lattice and a basis that specifies a structure. This concept will be described in greater detail in Section D.

ii. *Geometry of the 14 three-dimensional Bravais lattices*
All Bravais lattices are defined by three primitive lattice vectors, $\vec{a}$, $\vec{b}$, and $\vec{c}$. In practice, however, the lattice is conventionally specified by six scalar quantities known as lattice parameters. The dimensions of the lattice are given by a, b, and c, the magnitudes of the primitive vectors, and the relative orientation of the primitive vectors is described by three angles, α, β, and γ, that are defined in Fig. 2.3.

Special symmetrical relationships can exist among the primitive vectors which provide a means of classification. For example, a special case exists when $a = b$ or when $\vec{a}$ and $\vec{b}$ are perpendicular. The most symmetric Bravais lattices are cubic. In cubic Bravais lattices, the magnitudes of the three vectors are equal ($a = b = c$) and they are mutually perpendicular ($\alpha = \beta = \gamma = 90°$). The cubic P lattice is shown in Fig. 2.4. Any unit cell that encloses a volume $\vec{a} \cdot (\vec{b} \times \vec{c})$ and contains only one lattice point is designated as primitive and given the symbol P. Each point in the cubic P lattice has six nearest neighbors. Note that the boundaries of the cell are drawn so that the lattice points are situated at the vertices of the cube. If we consider a sphere around each lattice point, then exactly 1/8 of a sphere around each vertex lies within the boundaries of the unit cell so that there is a total of one lattice point per cell. If we shift the origin so that it is not coincident with a lattice site, it is also clear that there is only one lattice site per cell. However, by convention a lattice point is placed at the origin.

We can make a new Bravais lattice by putting a second lattice point in the center of the cube. This cubic lattice is given the symbol I and referred to henceforth as a cubic I lattice (it is helpful to remember this as *inner* centered). Notice

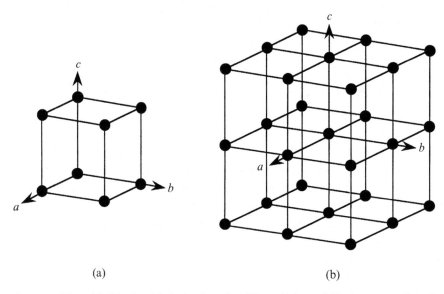

(a)   (b)

Figure 2.4. The cubic P lattice. (a) shows the unit cell boundaries and (b) shows a portion of the extended lattice.

in Fig. 2.5 that the eight-fold coordination of each point is identical and, thus, this arrangement forms a Bravais lattice. The conventional cell contains two lattice points (see Fig. 2.5).

The next distinct Bravais lattice, shown in Fig. 2.6, is the cubic F lattice (it is helpful to remember that there is a lattice point at the center of each *face*). In this case, the conventional cell contains four lattice points and each point has 12 nearest neighbors.

Readers should note that the nomenclature used for lattices is carefully chosen to avoid confusion with crystal structures. The commonly used terms simple cubic, body centered cubic, and face centered cubic are reserved here exclusively for crystal structures (specifically, the crystal structures that are created when an elemental, monoatomic basis is added to each site of the cubic P, cubic I, and cubic F lattices, respectively). This distinction is essential for clarity. For example, while both rock salt and sphalerite have cubic F lattices, it would be a mistake to call either of them face centered cubic. Atoms in a face centered cubic structure have twelve nearest neighbors, atoms in the rock salt structure have six nearest neighbors, and atoms in the sphalerite structure have four nearest neighbors. The distinction between a lattice and a crystal structure is described further in Section D of this chapter.

Having created new Bravais lattices by centering first the cube volume and

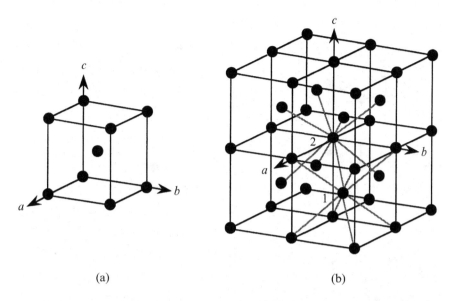

Figure 2.5. The cubic I lattice. (a) shows the repeat unit with points at coordinates (0,0,0) and (1/2,1/2,1/2). (b) The extended structure demonstrates that the cube center (1) and the vertices (2) have an identical eight-fold coordination environment. Thus, this configuration is a Bravais lattice.

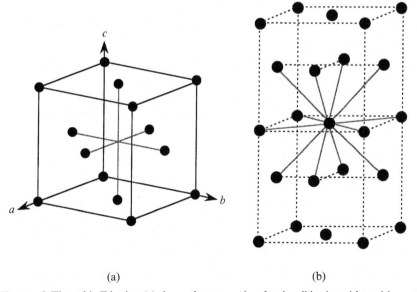

Figure 2.6. The cubic F lattice. (a) shows the conventional unit cell lattice with positions at coordinates (0,0,0), (1/2,1/2,0), (1/2,0,1/2), and (0,1/2,1/2). (b) A portion of the extended lattice shows that each lattice point has 12 nearest neighbors.

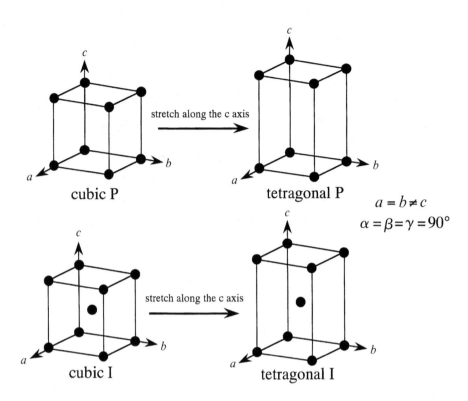

$$a = b \neq c$$
$$\alpha = \beta = \gamma = 90°$$

Figure 2.7. Axial strain on the cubic P and I lattices creates the tetragonal I and P lattices.

then the faces, it is reasonable to think that there are other possibilities. For example, what happens if we simultaneously center the cube and the faces? In this case, the environment of the central point differs from the others so this arrangement can not be a Bravais lattice. Centering only one set of faces or centering four co-planar edges creates a new Bravais lattice, the primitive tetragonal. It is, however, easiest to see the origin of the tetragonal P lattice by considering an axial distortion of the cubic P lattice, as shown in Fig. 2.7.

If the cubic P lattice is stretched along one axis, so that $a = b \neq c$ and $\alpha = \beta = \gamma = 90°$, then we have the tetragonal P cell. Similarly, stretching the cubic I lattice leads to creation of the tetragonal I lattice. What happens when we stretch the cubic F lattice along one axis? In this case, the result is identical to the tetragonal I. This is demonstrated in Fig. 2.8, which is a projection along the unique c axis. Dark points lie in the plane of the paper and lighter points are at face centered positions at $c/2$. Note that by drawing new lattice vectors rotated 45° with respect to the original vectors and shorter by a factor of $\sqrt{2}/2$, we can define a

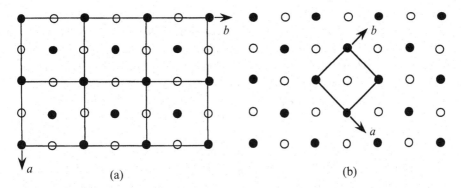

Figure 2.8. (a) A view of distorted cubic I lattice such that $a=b\neq c$, projected down the unique c-axis. Dark circles are at $c=0$ and light circles are at $c=1/2$. In (b), the identical lattice geometry is shown, but with a different assignment of lattice vectors. Because this unit has $a=b\neq c$, and a centered lattice point at $(1/2,1/2,1/2)$, it is a tetragonal I cell [1].

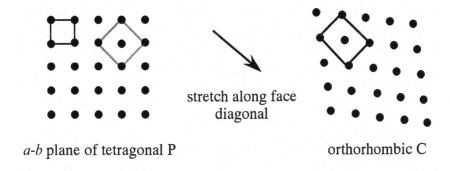

a-b plane of tetragonal P   orthorhombic C

Figure 2.9. An orthorhombic C cell is formed by stretching the tetragonal P along the face diagonal [1].

tetragonal I lattice from the same points. So, only one unique lattice, the tetragonal I, is created from axial strain of the cubic F and I lattices. An alternative way of looking at this is that cubic F is simply a special case of tetragonal I where $c=2a_I/\sqrt{2}$.

The symmetry can be further lowered by making the lengths of all three sides unequal. When $a\neq b\neq c$, and $\alpha=\beta=\gamma=90°$, the cell is said to be orthorhombic. It can easily be seen that distortion of the tetragonal P lattice along $\vec{a}$ or $\vec{b}$ will lead to an orthorhombic P lattice. A second way to distort the tetragonal P lattice is to stretch the face diagonal, as shown in Fig. 2.9. This creates a new orthorhombic Bravais lattice in which one set of opposite faces is centered. We call this an end centered lattice and the symbolic designation depends on which of the faces is centered. If the a–b plane is centered (in other words, if lattice points with coordinates (1/2,1/2,0) are added), it is orthorhombic C, if the c–b plane is centered, it is orthorhombic A, and if the a–c plane is centered, it is orthorhombic B. Axial and face diagonal distortions of the tetragonal I lattice lead to the orthorhombic I lattice and the orthorhombic F lattice, respectively.

The remaining Bravais lattices are created by distorting the angles away from 90°. Monoclinic Bravais lattices have $a \neq b \neq c$ and $\alpha = \gamma = 90°$, but β, the angle between $\vec{a}$ and $\vec{c}$, is not equal to 90°. In principle, either α or γ (rather than β) could be inclined from 90°. However, modern convention dictates that the single inclined angle is always designated as β. The monoclinic P lattice is formed by a distortion of the orthorhombic P lattice. The same distortion, applied to an end centered orthorhombic lattice, also produces the monoclinic P lattice. Distortion of both the orthorhombic I and orthorhombic F lattices leads to an end centered monoclinic lattice as shown in Fig. 2.10. Note that if the a–b plane is centered, it is called a monoclinic C lattice and if the b–c plane is centered, it is called a monoclinic A lattice. However, these two geometries are indistinguishable before the a and c axes are arbitrarily labeled and, therefore, the A and C cells are identical (as are the orthorhombic A, B, and C). By modern convention, one assigns the axes such that the a–b plane is centered and a C-type cell is formed.

The final symmetry reduction creates the triclinic lattice which has parameters such that $a \neq b \neq c$ and $\alpha \neq \beta \neq \gamma \neq 90°$. This lattice has the lowest symmetry. It is interesting to note that when atoms combine to form their minimum energy crystal structures, they usually have some symmetry greater than triclinic. In fact, in a survey of crystallographic data describing 26 000 different inorganic compounds, 27.6% have one of the cubic Bravais lattices and more than 72% have orthorhombic or higher symmetry [3]. If lattice parameters were distributed throughout the available space in a strictly random fashion, one would expect to find far more triclinic structures than anything else. The fact that so many structures are not triclinic strongly suggests that bonding electrons prefer to occupy electron energy levels that are symmetrically distributed in the solid.

The thirteenth Bravais lattice, shown in Fig. 2.11(a), is hexagonal. The hexagonal lattice has six-fold rotational symmetry about the c-axis and its cell parameters are: $a = b \neq c$, $\alpha = \beta = 90°$, and $\gamma = 120°$. The hexagonal system has only a P lattice. There is, however, a closely related trigonal Bravais lattice. The trigonal lattice can be created by stretching a cube along the body diagonal and has the following lattice parameters: $a = b = c$ and $\alpha = \beta = \gamma \neq 90°$. Some special trigonal lattices are identical to hexagonal P and an alternative primitive cell with $a = b \neq c$ and $\gamma = 120°$ can be specified. In other cases, the only primitive cell is trigonal and such groups are designated by the symbol R, which stands for rhombohedral.

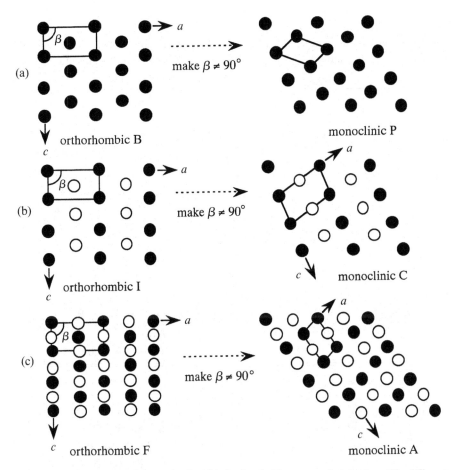

Figure 2.10. Distortions of the orthorhombic lattices lead to monoclinic lattices. The different shadings indicate lattice points at $\pm b/2$. To preserve a consistent coordinate system, the structure in (c) is labeled as an A-type cell. By convention, $\vec{a}$ and $\vec{c}$ should be transposed to form a C-type cell.

It is useful to remember that a rhombohedral lattice can always be expressed as a nonprimitive hexagonal with three points per cell. In fact, because the hexagonal cell is easier to visualize than the trigonal cell, rhombohedral lattices are almost always represented in terms of a nonprimitive hexagonal unit cell. Therefore, it is practical to think of the trigonal lattice as a special case of hexagonal. The relative coordinates of the three lattice points in a nonprimitive R cell are (0,0,0), (2/3,1/3,1/3), and (1/3,2/3,2/3). The 14 Bravais lattices are shown in Fig. 2.12 and their important parameters are specified in Table 2.1.

Table 2.1. *Parameters for the Bravais lattices.*

System	Number of lattices	Lattice symbols	Restrictions on conventional cell axes and angles
Cubic	3	P, I, F	$a=b=c$ $\alpha=\beta=\gamma=90°$
Tetragonal	2	P, I	$a=b\neq c$ $\alpha=\beta=\gamma=90°$
Orthorhombic	4	P, C, I, F	$a\neq b\neq c$ $\alpha=\beta=\gamma=90°$
Monoclinic	2	P, C	$a\neq b\neq c$ $\alpha=\gamma=90°\neq\beta$
Triclinic	1	P	$a\neq b\neq c$ $\alpha\neq\beta\neq\gamma$
Trigonal	1	R	$a=b=c$ $\alpha=\beta=\gamma<120°,\neq90°$
Hexagonal	1	P	$a=b\neq c$ $\alpha=\beta=90°,\ \gamma=120°$

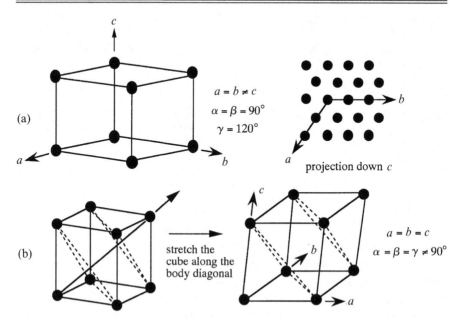

Figure 2.11. (a) The hexagonal lattice. (b) The trigonal lattice.

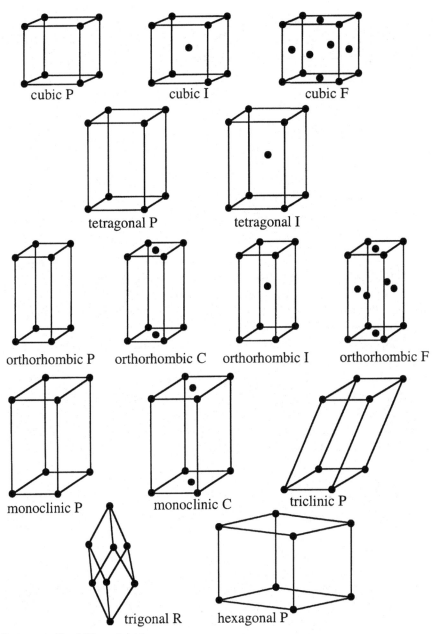

Figure 2.12. The 14 Bravais lattices.

Table 2.2. *Relative coordinates of points in the different cell types*

Cell type	Coordinates
P	(0,0,0)
I	(0,0,0) (1/2,1/2,1/2)
F	(0,0,0) (1/2,1/2,0) (1/2,0,1/2) (0,1/2,1/2)
A	(0,0,0) (0,1/2,1/2)
B	(0,0,0) (1/2,0,1/2)
C	(0,0,0) (1/2,1/2,0)
R	(0,0,0) (2/3,1/3,1/3) (1/3,2/3,2/3)

Example 2.1 Specifying nearest neighbor lattice points

We normally define the 'nearest neighbors' of a lattice point as being the set of closest equidistant points. The second nearest neighbors are the next set of equidistant lattice points just beyond the first nearest neighbors. For example, by inspection of Fig. 2.4b, you can see that points on the cubic P lattice have six nearest neighbors (at a distance of a) and 12 second nearest neighbors at a distance of √2 times a. How many 10th nearest neighbors does the cubic P lattice have and how far away are they?

1. First, you should not attempt to do this by inspection. Instead, rely on the regularity of the lattice, which is completely specified by the position vectors, $\vec{R}$:

$$\vec{R} = u\vec{a} + v\vec{b} + w\vec{c}.$$

2. Since the lattice is cubic P, we can simplify the position vectors:

$$\vec{R} = ua\hat{x} + va\hat{y} + wa\hat{z}.$$

Recall that the indices, u, v, and w, are integers. Thus, all positions are specified by the set of three indices (u,v,w) and the length of each vector is $a\sqrt{u^2 + v^2 + w^2}$.

3. The first nearest neighbor is at (1,0,0), which is a distance of a away. There are five additional first nearest neighbors, at the same distance away, specified by the indices $(-1,0,0)$, (0,1,0), $(0,-1,0)$, (0,0,1), $(0,0,-1)$. The second nearest neighbors are specified by the indices (1,1,0) and 11 additional related permutations. These 12 points are at a distance of $\sqrt{2}a$ from the origin.

4. The scheme for finding distant nearest neighbors is to systematically increment the indices. The third are at (1,1,1), the fourth at (2,0,0), the fifth at (2,1,0), the sixth at (2,1,1), the seventh at (2,2,0), the eighth at (2,2,1) and (3,0,0), which are equidistant, the ninth at (3,1,0), and the tenth are at (3,1,1).

5. Using the distance formula, the length of this position vector is $a\sqrt{11}$.

6. By permuting the indices, we can see that there are 24 distinct lattice points at this same distance:

(3,1,1) (−3,1,1) (3,−1,1) (3,1,−1) (−3,−1,1) (−3,1,−1) (3,−1,−1) (−3,−1,−1)
(1,3,1) (−1,3,1) (1,−3,1) (1,3,−1) (−1,−3,1) (−1,3,−1) (1,−3,−1) (−1,−3,−1)
(1,1,3) (−1,1,3) (1,−1,3) (1,1,−3) (−1,−1,3) (−1,1,−3) (1,−1,−3) (−1,−1,−3)

7. In conclusion, there are 24 tenth nearest neighbors at a distance of $a\sqrt{11}$.

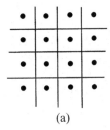
(a)

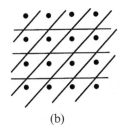
(b)

Figure 2.13. The choice of a primitive cell is not unique. The cells in (a) and (b) both have the same volume and contain exactly one point per cell and are, therefore, primitive cells.

(C) The unit cell

i. The primitive cell

The primitive cell is the volume that, when translated through all of the vectors of the Bravais lattice, just fills all space without overlapping itself or leaving voids. The primitive cell contains exactly one lattice point. Although the volume of the primitive cell is unique (it is the inverse of the lattice point density), the shape of the cell and the origin of the lattice vectors are not. Both of the primitive cells in Fig. 2.13 have the same area, but they have different shapes. Although the primitive cell might seem to be a logical choice, it often obscures the symmetry of the lattice, as is the case for the cubic F lattice shown in Fig. 2.15. If you see only the primitive cell, indicated by the solid lines, it is not easy to recognize that the lattice is cubic.

ii. The non-primitive conventional cell

Unit cells which contain more than one lattice point are called non-primitive. The volume of a non-primitive cell must be an integer multiple of the primitive cell volume. Eight of the 14 Bravais lattices are described in terms of non-primitive cells also known as conventional cells. For example, Fig. 2.14 compares the primitive cubic I cell with the conventional cell which has two lattice points and twice the volume. The lattice vectors for the primitive cell, written in terms of three orthogonal unit vectors, $\hat{x}$, $\hat{y}$, and $\hat{z}$, are:

$$\vec{a}=\frac{a}{2}(-\hat{x}+\hat{y}+\hat{z}) \quad \vec{b}=\frac{a}{2}(\hat{x}-\hat{y}+\hat{z}) \quad \vec{c}=\frac{a}{2}(\hat{x}+\hat{y}-\hat{z}) \tag{2.2}$$

and the conventional lattice vectors are:

$$\vec{a}=a\hat{x} \quad \vec{b}=a\hat{y} \quad \vec{c}=a\hat{z} \tag{2.3}$$

Conventional cells can also be translated to exactly fill space with no overlaps or voids. These non-primitive cells are convenient to use and chosen to acknowledge the symmetry of the lattice. However, because conventional cells contain an integer multiple of primitive cells, the conventional lattice vectors terminate at only a fraction of the lattice positions. For example, if the lattice vectors in Eqn. 2.3 are used to build a lattice using the rule expressed by Eqn. 2.1, a cubic P lattice results with half the lattice point density of the true cubic I lattice. Therefore, it is important to remember that when describing a lattice, you must either use the primitive vectors or add a set of basis vectors to the conventional vectors. For the case of cubic I, the basis vectors (0,0,0) and (1/2,1/2,1/2) produce the two lattice points in each conventional cell. Basis vectors are described more completely in the next section.

The vectors that describe the primitive and conventional cell of the cubic F lattice are given in Eqns. 2.4 and 2.5, respectively.

$$\vec{a}=\frac{a}{2}(\hat{y}+\hat{z}) \quad \vec{b}=\frac{a}{2}(\hat{z}+\hat{x}) \quad \vec{c}=\frac{a}{2}(\hat{x}+\hat{y}) \tag{2.4}$$

$$\vec{a}=a\hat{x} \quad \vec{b}=a\hat{y} \quad \vec{c}=a\hat{z} \tag{2.5}$$

The volume of the conventional cell for the cubic F lattice is four times the volume of the primitive cell. Because it contains four lattice points, four basis vectors are needed to specify the configuration of lattice points: (0,0,0), (1/2,1/2,0), (1/2,0,1/2), and (0,1/2,1/2).

iii. *The Wigner–Seitz cell*

The Wigner–Seitz cell is a special primitive cell which is relevant because it is used in some models for cohesion. It is defined as the volume of space about a Bravais lattice point that is closer to that point than to any other point of the lattice. The recipe for construction is to connect a Bravais lattice points with all other lattice points and to bisect each of these lines with a perpendicular plane. The smallest

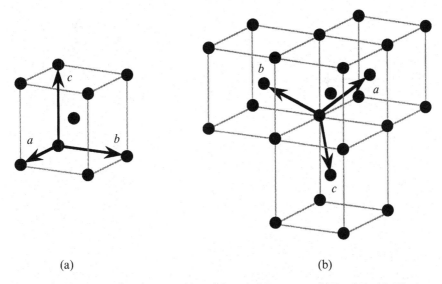

Figure 2.14. The conventional vectors (a) and the primitive vectors (b) for the cubic I lattice are compared. Linear combinations of the primitive vectors point to every site in the lattice while the same combinations of the conventional vectors point to only one half of the sites.

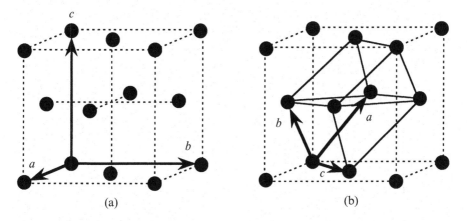

Figure 2.15. The conventional vectors (a) and the primitive vectors (b) for the cubic F lattice are compared. Linear combinations of the primitive vectors point to every site in the lattice while the same combinations of the conventional vectors point to only a quarter of the sites. While the cubic symmetry of the cubic F lattice is clearly reflected in the shape of the conventional unit cell, it is obscured by the primitive cell (solid lines in b).

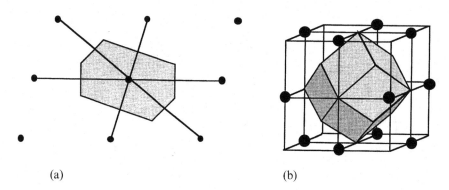

(a)   (b)

Figure 2.16. (a) The Wigner–Seitz cell construction in two dimensions. (b) The Wigner–Seitz cell for the cubic F lattice [1].

polyhedron bounded by these planes and containing the original lattice point is the Wigner–Seitz cell. Two Wigner–Seitz cells are shown in Fig. 2.16.

(D) The crystal structure. A Bravais lattice plus a basis

Until now, we've described only lattices, which are simply three dimensional arrays of points. What we are actually interested in is the crystal structure or the arrangement of atoms in space. It is important to recognize the distinction between the lattice and the crystal structure. A crystal structure is obtained when identical copies of a basis are located at all of the points of a Bravais lattice. A basis is a set of one or more atoms that has the stoichiometry of the bulk material. So, a crystal structure is always made up of a lattice and a basis. As an example, consider the atomic arrangements illustrated in Fig 2.17. Figure 2.17a shows the structure of Cr and Fig. 2.17b shows the structure of CsCl. The chromium crystal structure has identical copies of the basis, a single Cr atom, at each point on a cubic I lattice. We can refer to this structure as body centered cubic, while the lattice is cubic I. In contrast to this elemental structure, the structures of all compound materials necessarily have a multiatomic basis. For example, the basis for the CsCl structure is one formula unit of CsCl. The two atoms in this basis are located at the relative positions (0,0,0) and (1/2,1/2,1/2), and when the basis is placed at all the vertices of a cubic P lattice, the CsCl structure is formed. It is useful to note that while the Cr and the CsCl crystal structures have similar geometric arrangements, neither the lattice, the structure, nor the basis are the same. The rock salt and diamond structures, illustrated in Fig. 2.18a and c, both have a cubic F lattice, but neither have the face centered cubic structure. It is for this reason that we have been careful in our taxonomy for the Bravais lattices.

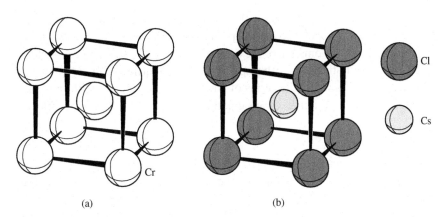

(a)   (b)

Figure 2.17. (a) The structure of Cr: a cubic I lattice with a basis of a single Cr atom. (b) The structure of CsCl: a cubic P lattice with a diatomic basis containing one Cs and one Cl atom.

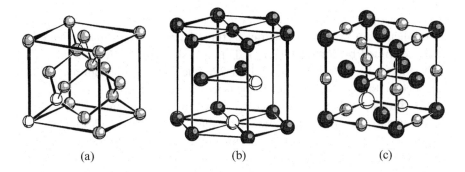

(a)   (b)   (c)

Figure 2.18. (a) The diamond structure has a cubic F lattice with two atoms in the basis related by a displacement of (1/4,1/4,1/4). (b) The hexagonal close-packed structure has a hexagonal P lattice with two atoms in the basis related by the displacement of (1/3,2/3,1/2). (c) The NaCl structure has a cubic F lattice and a two atom NaCl basis with the atoms related by a (1/2,0,0) displacement. In each case, the basis atoms are white.

When trying to understand the distinction between a lattice and a basis, it is useful to remember that the basis of a structure always has a chemical composition that is either the same as the formula unit, or an integer multiple of the formula unit. Many crystal structures are very complicated (the basis can be composed of hundreds of atoms), so it is important to have a systematic way of thinking about them; all structures consist of one of the 14 well-known Bravais lattices plus a collection of atoms, the basis, that decorates each lattice point. Additional examples of some simple crystal structures are illustrated in Fig. 2.18.

Just as Eqn. 2.1 provides us with a systematic description of the Bravais lattice, we can use a set of vectors to specify the basis. We will say that for any basis composed of J atoms, there are a set of J vectors, $\vec{r}_j$, that specify the location of each atom in the basis with respect to a Bravais lattice point at $(0,0,0)$. The position vector, $\vec{r}_j$, is written in terms of the three lattice vectors with fractional coordinates (x_j, y_j, z_j) that are always greater than or equal to 0 and less than 1:

$$\vec{r}_j = x_j \vec{a} + y_j \vec{b} + z_j \vec{c}. \tag{2.6}$$

Since this basis is identically repeated at each lattice site, every atom position in the crystal can be specified by the sum of a Bravais lattice vector $(\vec{R})$ and a basis vector $(\vec{r}_j)$.

E Specifying locations, planes and directions in a crystal

i. Locations

Specific positions within the unit cell are identified by coefficients of location. The coefficients of location are written as three fractional coordinates, in parentheses, separated by commas: (x,y,z). This is a shorthand way of writing the coefficients of the vector $\vec{r}_j$ defined in Eqn. 2.6. The coefficients are always greater than or equal to zero, but less than 1; coefficients outside this range specify equivalent positions in adjacent unit cells. An out-of-range coordinate can always be transformed to an equivalent position in the reference cell by the addition or subtraction of a Bravais lattice vector. Examples of locations and their specifications are shown in Fig. 2.19. Remember that the coordinates always refer to the lattice vectors and that these vectors are not necessarily orthogonal.

ii. Directions

Directions in the unit cell are specified by three integer indices, u, v, and w, in square brackets with no commas (for example, $[u\,v\,w]$). These integer indices are simply the components of the Bravais lattice vector (Eqn. 2.1) that specify the direction of interest. A less systematic way to find the indices is to specify the smallest integers proportional to the unit cell intercepts of a line whose origin is at $(0,0,0)$. Examples are shown in Fig. 2.20. To specify the direction of a line that doesn't begin at $(0,0,0)$, translate the line to a parallel direction that does. Figure 2.20c shows a direction from point 1 at $(0,1/2,0)$ to point 2 at $(1,1,1)$. The translation is carried out by subtracting the coefficients of point 1 from point 2: $(1-0, 1-1/2, 1-0) = (1,1/2,1)$. Therefore, the direction is $[2\,1\,2]$. The opposite direction, found by subtracting point 2 from point 1, is indicated by $[\bar{2}\,\bar{1}\,\bar{2}]$, where the 'bar' represents an opposite or negative direction. The directions $[2\,1\,2]$ and $[\bar{2}\,\bar{1}\,\bar{2}]$ are indistinguishable except for the choice of origin. Because of this, we say that they belong to the same family of directions. The family is specified by

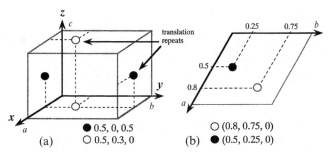

Figure 2.19. Coefficients of location are illustrated. The coordinates refer to displacements parallel to the lattice vectors, even when the lattice vectors are not perpendicular, as in the *a–b* plane of the hexagonal cell, illustrated in (b).

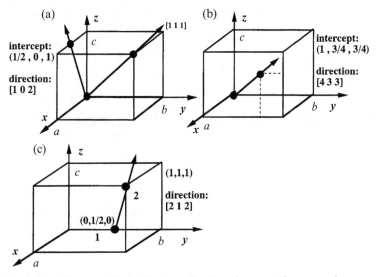

Figure 2.20. Examples illustrating how directions in a crystal are named.

angled brackets; [2 1 2] and [$\bar{2}$ $\bar{1}$ $\bar{2}$] belong to the ⟨212⟩ family of directions. The number of equivalent directions in a family depends on the Bravais lattice symmetry and the values of the indices, *u*, *v*, and *w*.

iii.   *Indices for planes*

Indices for planes, usually called Miller indices, are the reciprocals of axis intercepts, cleared of common fractions and multipliers. When a plane does not intersect an axis, the index is set equal to zero. Miller indices are always written in parentheses, with no commas; examples are shown in Fig. 2.21. To specify the name of a plane that intersects an axis at all points, as shown in Fig. 2.22(b), translate it by one lattice unit normal to the plane. This will always produce an identical plane that can be named according to the rule given above.

47

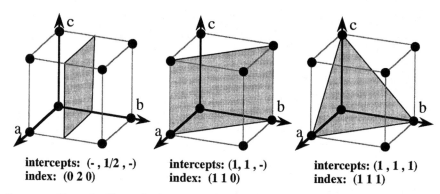

intercepts: (-, 1/2, -)   intercepts: (1, 1, -)   intercepts: (1, 1, 1)
index: (0 2 0)   index: (1 1 0)   index: (1 1 1)

Figure 2.21. Examples illustrating how some common planes in a cubic P crystal are named.

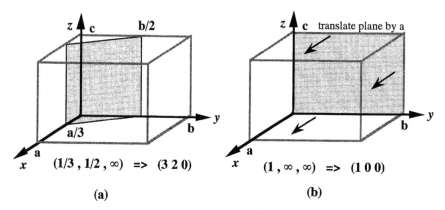

(1/3, 1/2, ∞) => (3 2 0)   (1, ∞, ∞) => (1 0 0)

(a)   (b)

Figure 2.22. (a) Another example of naming a plane. (b) If a plane intersects an axis at all points, translate it by one lattice unit normal to the plane (this is an identical position) to name it.

Sets of identically spaced, indistinguishable planes belong to the same family. For example, if we consider the cubic rock salt structure illustrated in Fig. 2.18(c), the (001) and (100) planes both contain the same configuration of Na and Cl atoms and both have an interplanar spacing of a. In fact, they can be distinguished only when the axes are labeled. Families of identical planes are specified in 'curly' brackets $\{hkl\}$. For example, in a cubic structure, the (100), (010), (001), ($\bar{1}$00), (0$\bar{1}$0), and (00$\bar{1}$) planes all belong to the $\{100\}$ family.

Directions and planes in hexagonal systems are usually specified by a set of four indices, $[uvtw]$ and $(hkil)$. The third indices (t and i) refer to a third, redundant, axis in the basal plane, as illustrated in Fig. 2.23. Frank [6] pointed out that these indices can be thought of as four-dimensional vectors confined to a

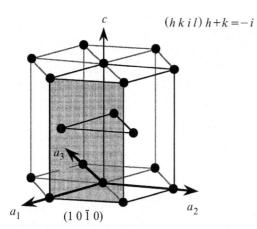

$(h\,k\,i\,l)\;h+k=-i$

Figure 2.23. Indexing a lattice plane in the hexagonal system.

three-dimensional section of space by the rule that the sum of the first three indices is zero. The direction in a hexagonal coordinate system with indices $[uvtw]$ is the same as the four-dimensional vector in Cartesian space with components $[u, v, t, \lambda w]$. The normal to the plane indexed as $(hkil)$ is the vector in Cartesian space with components $[h, k, i, l/\lambda]$. In both of these cases, $\lambda = (2/3)^{1/2}(c/a)$. With these vectors, one can then use the well known equations for vector algebra in three dimensions that are used to compute lengths, dihedral angles, and interplanar spacings (this topic is reviewed in Section G of this chapter).

It is sometimes necessary to convert between four-index and three-index notation. For a hexagonal unit cell defined by the following basis vectors:

$$\vec{a} = \frac{a}{\sqrt{6}}(2,\bar{1},\bar{1},0)$$

$$\vec{b} = \frac{a}{\sqrt{6}}(\bar{1},2,\bar{1},0)$$

$$\vec{c} = c(0,0,0,1) = \frac{a}{\sqrt{6}}(0,0,0,3\lambda), \tag{2.7}$$

the direction $[uvwt]$ is $[(u\text{-}t)\,(v\text{-}t)\,w]$ in three-index notation and the direction $[uvw]$ is $[(2u-v), (-u+2v), -u-v), 3w]$ in four-index notation. The plane $(hkil)$ is (hkl) in three-index notation and the plane (hkl) is $(h\,k\,(-h-k)\,l)$ in four-index notation.

(F) The reciprocal lattice

In this section, lattices in reciprocal space are described. Every real space Bravais lattice has an analogous lattice in reciprocal space and understanding the relationship between the two is the key to understanding how diffraction patterns are related to crystal structures. Even though the utility of the reciprocal lattice might not be entirely clear until our discussion of diffraction (Chapter 5), it is appropriate to introduce it now, while we are concentrating on understanding the properties of lattices.

i. Formal description

We begin a mathematical description of the reciprocal lattice by considering the plane wave, Ψ:

$$\Psi = e^{i\vec{k}\cdot\vec{r}}. \tag{2.8}$$

In the expression for the plane wave, $\vec{k}$ is the wave vector and $\vec{r}$ is the position vector. The direction of the wave vector specifies the direction in which the wave propagates and the magnitude of the wave vector is proportional to the reciprocal of its wavelength or the frequency:

$$|k| = \frac{2\pi}{\lambda}. \tag{2.9}$$

The name plane wave implies that the amplitude is constant in any plane perpendicular to $\vec{k}$. Furthermore, along the direction of propagation parallel to $\vec{k}$, $e^{i\vec{k}\cdot\vec{r}}$ is periodic with a wavelength of λ. These characteristics of the plane wave are illustrated schematically in Fig. 2.24.

We define the reciprocal lattice by posing the following question. If we wish to represent all of the possible sets of parallel planes in the Bravais lattice by a set of plane waves, what values of $\vec{k}$ must be used? The answer is that there will be one wave vector for each set of parallel planes and that each wave vector will have a direction normal to its corresponding real space lattice plane (with index hkl) and a wavelength equal to the interplanar spacing, d_{hkl}. The set of discrete wave vectors with these properties will be labeled $\vec{G}_{hkl}$. Note that there is a one-to-one correspondence between the planes in the direct lattice and the reciprocal lattice vectors, $\vec{G}_{hkl}$.

The plane waves that have the same periodicity as the Bravais lattice are specified in the following way. If $\vec{r}$ is the position vector that ranges over the volume of a single unit cell (Eqn. 2.6), and $\vec{R}$ is a Bravais lattice vector (Eqn. 2.1), then any vector $\vec{R}+\vec{r}$ points to a position that is indistinguishable from the

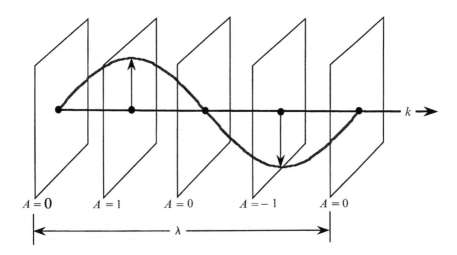

Figure 2.24. Schematic illustration of a plane wave. Planes of constant amplitude (A) are shown perpendicular to the direction of propagation. The amplitude repeats with a wavelength of λ. The sinusoidal line represents the variation of the amplitude with k.

position specified by $\vec{r}$. For a plane wave to have the same periodicity as the Bravais lattice,

$$e^{i\vec{G}_{hkl}\cdot(\vec{r}+\vec{R})} = e^{i\vec{G}_{hkl}\cdot\vec{r}}. \tag{2.10}$$

Following from Eqn. 2.10,

$$e^{i\vec{G}_{hkl}\cdot\vec{R}} = 1 = e^{in2\pi} \tag{2.11}$$

$$\vec{G}_{hkl}\cdot\vec{R} = 2\pi n \tag{2.12}$$

where n is any integer. Inspection of Eqn. 2.12 tells us that the units of $\vec{G}_{hkl}$ must be reciprocal length. Therefore, we have a new set of wave vectors, $\vec{G}_{hkl}$, in reciprocal space that defines the points of a reciprocal lattice. Just as every point in the Bravais lattice is specified by a vector, $\vec{R}$, every point in the reciprocal lattice is specified by a reciprocal lattice vector, $\vec{G}_{hkl}$. Another way of stating this is to say that in the reciprocal lattice, every real space plane is represented by a perpendicular vector, $\vec{G}_{hkl}$.

The reciprocal lattice vectors can be expressed as linear combinations of a set of non-coplanar primitive vectors (just as $\vec{R}$ was expressed in terms of the primitive vectors, $\vec{a}$, $\vec{b}$, and $\vec{c}$). The reciprocal lattice vectors are usually named $\vec{a}^*$, $\vec{b}^*$, and $\vec{c}^*$, and are defined with respect to the real space lattice. According to the definitions above, the vector $\vec{a}*$ ($\vec{G}_{100}$) is perpendicular to the (100) plane, which contains $\vec{b}$ and $\vec{c}$. Therefore, the direction of $\vec{a}*$ is $\vec{b} \times \vec{c}$. For $\vec{G}_{100}$ to have the proper magnitude, we must divide by a factor of $\vec{a} \cdot (\vec{b} \times \vec{c})$. Thus, the three basis vectors of the reciprocal lattice can be determined from the real space basis vectors using the relations:

$$\vec{a}* = \frac{2\pi \vec{b} \times \vec{c}}{\vec{a} \cdot (\vec{b} \times \vec{c})}, \quad \vec{b}* = \frac{2\pi \vec{c} \times \vec{a}}{\vec{a} \cdot (\vec{b} \times \vec{c})}, \quad \vec{c}* = \frac{2\pi \vec{a} \times \vec{b}}{\vec{a} \cdot (\vec{b} \times \vec{c})} \tag{2.13}$$

This definition guarantees that the reciprocal lattice vectors will always have the properties described above, which also can be expressed in the following way:

$$\begin{aligned} \vec{a}* \cdot \vec{a} &= 2\pi & \vec{b}* \cdot \vec{a} &= 0 & \vec{c}* \cdot \vec{a} &= 0 \\ \vec{a}* \cdot \vec{b} &= 0 & \vec{b}* \cdot \vec{b} &= 2\pi & \vec{c}* \cdot \vec{b} &= 0 \\ \vec{a}* \cdot \vec{c} &= 0 & \vec{b}* \cdot \vec{c} &= 0 & \vec{c}* \cdot \vec{c} &= 2\pi. \end{aligned}$$

In summary, we define the reciprocal lattice as the set of points specified by vectors of the form:

$$\vec{G}_{hkl} = h\vec{a}* + k\vec{b}* + l\vec{c}*, \tag{2.14}$$

where h, k, and l are any positive or negative integers, including zero, and $\vec{a}*$, $\vec{b}*$, and $\vec{c}*$ are defined in Eqn. 2.13. For every direct lattice in real space, there is a reciprocal lattice in reciprocal space and for every set of parallel planes in the primitive direct lattice, specified by the Miller indices h, k, and l, there is a point on the reciprocal lattice. It is important to note that if the reciprocal lattice vectors are defined with respect to conventional rather than primitive vectors, some values of h, k and l will produce reciprocal lattice vectors that do not correspond to points on the reciprocal lattice. These so-called systematic absences are an artifact of the conventional cell choice and will be discussed more completely in Chapter 5.

As the book progresses, we shall see that the reciprocal lattice concept is a useful framework for the study of several topics, especially X-ray diffraction. In particular, we will show that an X-ray diffraction pattern gives a view of a crystal's reciprocal lattice.

ii.   *The relationship between the direct lattice and the reciprocal lattice*
Without physical justification, the definition of the reciprocal lattice seems obscure. To get an intuitive feel for the reciprocal lattice, it is instructive to

graphically examine the direct lattice/reciprocal lattice relationship, without the mathematics [9].

Consider a monoclinic P direct lattice with the unique b-axis oriented normal to the page which contains the a–c plane. A projected unit cell of this lattice is shown in Fig. 2.25a and the extended lattice is shown in Fig. 2.25c. Two planes, (100) and (001), are highlighted in the figure. In the reciprocal lattice, each direct lattice plane is represented by a single point. These points are at a distance $2\pi/d_{hkl}$ from the origin (labeled O) of the reciprocal lattice and oriented in a direction normal to the plane with the same index. This is equivalent to the statement that $|\vec{G}_{hkl}| = 2\pi/d_{hkl}$ and that the direction of $\vec{G}_{hkl}$ is normal to (hkl). Notice in this example that because $\beta \neq 90°$, $\vec{a}^*$ and $\vec{c}^*$ are not parallel to $\vec{a}$ and $\vec{c}$. However, because $\alpha = \gamma = 90°$, $\vec{b}^*$ is parallel to $\vec{b}$.

Each direct lattice plane gives rise to a reciprocal lattice point, and from this relationship the entire reciprocal lattice can be constructed. Alternatively, we can find the entire reciprocal lattice by finding only the points specified by the planes (100), (001), and (010). For example, the primitive reciprocal lattice vectors, $\vec{a}^*$ and $\vec{c}^*$, shown in Fig. 2.25b, specify the (100) and (001) points. Linear combinations of these primitive vectors can be used to construct the lattice shown in Fig. 2.25d. Because spots on diffraction patterns correspond to reciprocal lattice points, one can use the diffraction pattern as a basis for determining the orientation of a crystal.

Although the picture in Fig. 2.25d shows only a slice (one plane) of the reciprocal lattice, it extends infinitely in three dimensions above and below the plane of the paper. Here we see only points from planes having (h0l) indices. We will call this the zero layer. The first layer would show points from planes having (h1l) or (h$\bar{1}$l) indices. In conclusion, the reciprocal lattice/direct lattice relationship can be summarized by the following statements: $\vec{a}^*$ is perpendicular to the b–c plane, $\vec{b}^*$ is perpendicular to the a–c plane, $\vec{c}^*$ is perpendicular to the a–b plane. For the special case when $\alpha = \beta = \gamma = 90°$, then $\alpha^* = \beta^* = \gamma^* = 90°$, $\vec{a}$ is parallel to $\vec{a}^*$, $\vec{b}$ is parallel to $\vec{b}^*$, and $\vec{c}$ is parallel to $\vec{c}^*$. For the general case, $\alpha + \alpha^* = 180°$, $\beta + \beta^* = 180°$, and $\gamma + \gamma^* = 180°$.

iii. The first Brillouin zone
While on the subject of the reciprocal lattice, it is worthwhile defining the first Brillouin zone. Simply put, the first Brillouin zone is the Wigner–Seitz cell of the reciprocal lattice and is found in the same way. Because the reciprocal lattice of a cubic I lattice has the cubic F arrangement, the Wigner–Seitz cell of a cubic F direct lattice has the same geometry as the first Brillouin zone of a cubic I reciprocal lattice. Also, the cubic F direct lattice has a reciprocal lattice with a cubic I arrangement. Therefore, the first Brillouin zone of a cubic F has the same geometry as the Wigner–Seitz cell of a cubic I lattice.

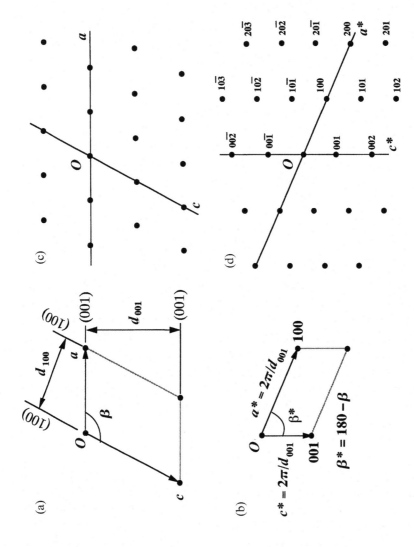

Figure 2.25. The relationship between a direct lattice and the reciprocal lattice. See the text for a complete description [9].

Example 2.2 Constructing a reciprocal lattice

Specify the primitive reciprocal lattice vectors for the cubic I lattice. Using these vectors, draw a projection of the reciprocal lattice, as you would see it if you were looking down the c axis of the conventional cell. On a new diagram, showing the reciprocal lattice projected in the same orientation, indicate the orientation and length of the conventional reciprocal lattice vectors and re-label the reciprocal lattice points based on these conventional vectors.

1. Beginning with the primitive vectors of the direct lattice (illustrated in Fig. 2.14 and specified in Eqn. 2.2), we can use Eqn. 2.13 to find the primitive vectors of the reciprocal lattice:

$$\vec{a}* = \frac{2\pi}{a}(\hat{y}+\hat{z}), \qquad \vec{b}* = \frac{2\pi}{a}(\hat{x}+\hat{z}), \qquad \vec{c}* = \frac{2\pi}{a}(\hat{x}+\hat{y}).$$

2. Substituting into Eqn. 2.14, the reciprocal lattice vectors are written in the following way:

$$\vec{G} = h\vec{a}* + k\vec{b}* + l\vec{c}* = \frac{2\pi}{a}[(k+l)\hat{x}+(h+l)\hat{y}+(h+k)\hat{z}].$$

3. Now we can draw a projection along [001]. It is easiest to begin by finding the reciprocal lattice points in the plane of the paper. These points have no z component and, therefore, must have indices such that $h+k=0$: (000), (001), (00$\bar{1}$), (1$\bar{1}$0), ($\bar{1}$10), ($\bar{1}$1$\bar{1}$), ($\bar{1}$11), (1$\bar{1}$$\bar{1}$), ($\bar{1}1\bar{1}$), (002), (00$\bar{2}$). The positions of these points are indicated on Fig. 2.26 by solid dots.

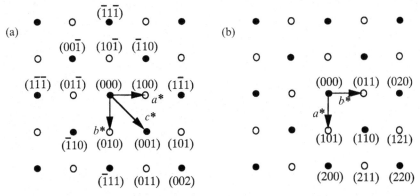

Figure 2.26. The reciprocal lattice of a cubic I direct lattice, projected on the *x-y* plane. Dark points are at $z=0$. Open points are at $z=1$. In (a), the points are labeled according to the primitive vectors. In (b), the conventional vectors are used. Note that in (a), the vectors $\vec{a}*$ and $\vec{b}*$ are inclined to the plane of the paper by 45° and in (b), they lie in the plane and do not terminate at reciprocal lattice points.

4. Next, we find points with a z component of 1 $(h+k=1)$: (100), (010), (011), (101). These points are plotted in Fig. 2.26 as open dots. The x and y coordinates of those points with $z=2$ are identical to those with $z=0$ and the x and y coordinates of those points with $z=3$ are identical to those with $z=1$. Therefore, the first two layers represent the entire lattice.

5. The reciprocal lattice vectors derived from the conventional cell are parallel to the corresponding direct lattice vectors and have a magnitude of $2\pi/a$. They are illustrated in Fig. 2.26 and the reciprocal lattice points are re-labeled. Note that the conventional cell is the arrangement of the cubic F direct lattice, where the cell edge length is $2a^*$.

6. This example illustrates the idea that the reciprocal lattice must be independent of the choice of basis vectors; only the labeling scheme is altered. Note also that when conventional, non-primitive vectors are used, not all linear combinations point to positions where reciprocal lattice points exist. For example, the vectors with indices (100), (010), (001), and (111) do not terminate at reciprocal lattice points. Thus, only a fraction of all the possible indices are used to label the points. In this case, only the indices which sum to an even number $(h+k+l=2n)$ give vectors that terminate at reciprocal lattice points. Because the conventional direct lattice vectors define a cell that has twice the volume of the primitive cell, the conventional reciprocal lattice vectors define a reciprocal cell that has half the volume of the primitive reciprocal cell. Therefore, only one half of all the smaller conventional reciprocal lattice vectors are needed to define all of the reciprocal lattice points.

(G) **Quantitative calculations involving the geometry of the lattice**

When computing distances, angles, and interplanar spacings in lattices, it is important to remember that Bravais lattice basis vectors are not always mutually orthogonal. In this section, generalized rules for computing the geometric characteristics of lattices, applicable to all seven crystal classes, are presented [10–12].

i. *Distances, angles, and volumes in the direct lattice*
The length (L) of any vector $(\vec{R})$ with components $x\vec{a}$, $y\vec{b}$, and $z\vec{c}$ is:

$$L=(\vec{R}\bullet\vec{R})^{1/2}. \tag{2.15}$$

In matrix notation, the scalar product can be written in the following way:

$$L^2=[x\ y\ z]\begin{bmatrix} \vec{a}\bullet\vec{a} & \vec{a}\bullet\vec{b} & \vec{a}\bullet\vec{c} \\ \vec{b}\bullet\vec{a} & \vec{b}\bullet\vec{b} & \vec{b}\bullet\vec{c} \\ \vec{c}\bullet\vec{a} & \vec{c}\bullet\vec{b} & \vec{c}\bullet\vec{c} \end{bmatrix}\begin{bmatrix} x \\ y \\ z \end{bmatrix}. \tag{2.16}$$

With reference to the unit cell parameters $(\alpha, \beta, \gamma, a, b, c)$, Eqn. 2.16 can be re-written:

$$L^2 = [x \ y \ z] \begin{bmatrix} a^2 & ab\cos\gamma & ac\cos\beta \\ ba\cos\gamma & b^2 & bc\cos\alpha \\ ca\cos\beta & cb\cos\alpha & c^2 \end{bmatrix} \begin{bmatrix} x \\ y \\ x \end{bmatrix} = [x \ y \ z][\Gamma] \begin{bmatrix} x \\ y \\ x \end{bmatrix}. \quad (2.17)$$

The 3×3 matrix in Eqn. 2.17 is known as the metric tensor, Γ. We will use this tensor for all of our geometric calculations. Note that Eqn. 2.17 expresses the metric tensor for the most general (triclinic) case; for most crystal systems, Γ reduces to a simpler form. Metric tensors for specific crystal systems are listed in Appendix 2A. Equation 2.17 can also be used to compute the distance between two points, (x_1, y_1, z_1) and (x_2, y_2, z_2). In this case, the components of the vector $\vec{R}$ are $\Delta x \vec{a} = (x_2 - x_1)\vec{a}$, $\Delta y \vec{b} = (y_2 - y_1)\vec{b}$, and $\Delta z \vec{c} = (z_2 - z_1)\vec{c}$, and:

$$L^2 = [\Delta x \ \Delta y \ \Delta z][\Gamma] \begin{bmatrix} \Delta x \\ \Delta y \\ \Delta z \end{bmatrix}. \quad (2.18)$$

The angle, θ, between two vectors, $\vec{R}_1$ and $\vec{R}_2$, can be calculated by using the definition of the scalar product:

$$\vec{R}_1 \cdot \vec{R}_2 = R_1 R_2 \cos\theta, \quad (2.19)$$

where both the vector magnitudes (R_1, R_2) and the scalar product $(\vec{R}_1 \cdot \vec{R}_2)$ are calculated according to Eqn. 2.17.

The volume (V) of the unit cell is always given by $\vec{a} \cdot (\vec{b} \times \vec{c})$. In terms of the lattice parameters, we have:

$$V = \vec{a} \cdot (\vec{b} \times \vec{c}) = abc\sqrt{1 - \cos^2\alpha - \cos^2\beta - \cos^2\gamma + 2\cos\alpha\cos\beta\cos\gamma}. \quad (2.20)$$

Equation 2.20 gives the volume in the most general case. For more symmetric crystal systems, the expression is simpler (for example, the volume of a cubic unit cell is a^3). Crystal system specific equations are listed in Appendix 2B.

ii. *Interplanar spacings and the angles between planes*
The reciprocal lattice is used to determine the geometric characteristics of lattice planes. By definition, the magnitudes of the reciprocal lattice vectors, $\vec{G}_{hkl}$, are proportional to the inverse of the separation of the planes (hkl), or d_{hkl}. Analogous to the way we computed the length of real space vectors using the

metric tensor, Γ (see Eqn. 2.17), we compute the length of reciprocal lattice vectors $(2\pi/d_{hkl})$ using the reciprocal metric tensor, Γ^*. The values of d_{hkl}, often referred to as d-spacings, are essential for the interpretation of diffraction data.

$$\frac{1}{d_{hkl}^2} = \frac{\vec{G}_{hkl} \cdot \vec{G}_{hkl}}{(2\pi)^2} = \frac{1}{(2\pi)^2} [h\ k\ l] \begin{bmatrix} \vec{a}^* \cdot \vec{a}^* & \vec{a}^* \cdot \vec{b}^* & \vec{a}^* \cdot \vec{c}^* \\ \vec{b}^* \cdot \vec{a}^* & \vec{b}^* \cdot \vec{b}^* & \vec{b}^* \cdot \vec{c}^* \\ \vec{c}^* \cdot \vec{a}^* & \vec{c}^* \cdot \vec{b}^* & \vec{c}^* \cdot \vec{c}^* \end{bmatrix} \begin{bmatrix} h \\ k \\ l \end{bmatrix}$$

$$\frac{1}{d_{hkl}^2} = \frac{1}{(2\pi)^2} [h\ k\ l] \begin{bmatrix} a^{*2} & a^* b^* \cos \gamma^* & a^* c^* \cos \beta^* \\ b^* a^* \cos \gamma^* & b^{*2} & b^* c^* \cos \alpha^* \\ c^* a^* \cos \beta^* & c^* b^* \cos \alpha^* & c^{*2} \end{bmatrix} \begin{bmatrix} h \\ k \\ l \end{bmatrix}. \tag{2.21}$$

The matrix expression in Eqn. 2.21 is completely general and works in all crystal systems. The expanded forms of this equation are greatly simplified in most crystal systems; for example, for a cubic crystal:

$$\frac{1}{d_{hkl}^2} = \frac{h^2 + k^2 + l^2}{a^2}. \tag{2.22}$$

Simplified expressions relevant to the other crystal systems are provided in Appendix 2C.

Since the angle between two planes (hkl) and $(h'k'l')$ is equal to the angle between their normals, we can simply compute the angle between $\vec{G}_{hkl}$ and $\vec{G}_{h'k'l'}$ in a manner analogous to Eqn. 2.19:

$$\frac{\vec{G}_{hkl} \cdot \vec{G}_{hkl}}{G_{hkl} G_{h'k'l'}} = \cos \theta. \tag{2.23}$$

The vector lengths and dot product are computed using the reciprocal metric tensor, Γ^*, as in Eqn. 2.21.

iii. Zones

The crystallographic direction, $[u\ v\ w]$, that is common to two or more planes, $(h_1 k_1 l_1)$ and $(h_2 k_2 l_2)$, is called a zone. The geometry of a zone axis and the planes that belong to the zone are illustrated in Fig. 2.27. For example, in a cubic system, if the zone axis in Fig. 2.27 is [001], then the planes belonging to this zone all intersect along the c-axis and must have indices $(hk0)$. The line of intersection of two planes (their zone axis) is given by the cross product of their normals (reciprocal lattice vectors).

$$[u\ v\ w] = \vec{G}_{h_1 k_1 l_1} \times \vec{G}_{h_2 k_2 l_2}. \tag{2.24}$$

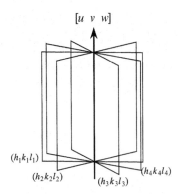

[u v w]

$(h_1k_1l_1)$

$(h_2k_2l_2)$

$(h_3k_3l_3)$

$(h_4k_4l_4)$

Figure 2.27. The zone axis of the planes with indices $(h_i k_i l_i)$ is the direction [u v w], their line of mutual intersection.

Since directions are indexed according to relative integer values, we can use the following simplified form of Eqn. 2.24 to determine the indices for the direction.

$$u = k_1 l_2 - l_1 k_2$$
$$v = l_1 h_2 - h_1 l_2$$
$$w = h_1 k_2 - k_1 h_2. \tag{2.25}$$

If the plane (*hkl*) belongs to the zone [u v w], then the zone axis lies in the plane and is perpendicular to the reciprocal lattice vector, $\vec{G}_{hkl}$, associated with that plane. So, a plane belongs to a zone if the scalar product of the zone axis and the reciprocal lattice vector is equal to zero. In practice, this amounts to the condition that:

$$hu + kv + lw = 0. \tag{2.26}$$

Determining which planes belong to a given zone is important for interpreting diffraction data, and determining the line along which planes intersect is important in the study of extended defects and surfaces.

(H) Visual representations of crystal structures

Although the most systematic way of specifying a crystal structure is to enumerate all of the atomic coordinates, this is often not the most efficient way of communicating its most important features. The importance of being able to visualize atomic geometries can not be underestimated. Thus, it is essential that we have efficient graphic methods for the representations of crystal structures. For the simplest of structures, such as tetrahedrally bound semiconductors and close packed metals, simple ball and stick or hard sphere pictures usually suffice. However, the majority of solid materials have more complex structures that are difficult to visualize without the help of a good drawing. Because drawings with many atoms become increasingly difficult to use (not to mention sketch), simplified methods of representing crystal structures have been devised. The basic scheme is to divide the crystal into simple units, each of which corresponds to a frequently occurring group of atoms.

i. *Polyhedral models*

One common way of representing crystals is through polyhedral models. The concept originated with Pauling, who proposed that structures could be viewed as coordinated polyhedra of anions about cations. This method has been used in the fields of solid state chemistry and mineralogy for decades and has recently moved into the mainstream of solid state science due to the growing interest in materials with more complex structures such as the high critical temperature superconducting cuprates. Each solid polyhedron in a structure model represents an electropositive atom and its coordinating ligands (the electronegative nearest neighbors). The metal atom is at the center of the polyhedron and its ligands are at the vertices. Common polyhedra from which complex structures are built are shown in Fig. 2.28: the tetrahedron, the octahedron, the square pyramid, the trigonal prism, and the cube.

The most common use of the polyhedral representation is to simplify a complex structure. This is demonstrated in Fig. 2.29, where a hard sphere model and a polyhedral model of the same structure are compared. There are, however, several other advantages to this form of representation. One is that it emphasizes the coordination number of the metal cations and, thus, gives some information about the bonding. Another is that it emphasizes connectivity, which has important consequences with regard to structural stability. A third advantage of this system is that it emphasizes the locations of interstitial positions. We conclude this section by noting that although coordination polyhedra are often distorted, they may still be represented by the ideal shapes.

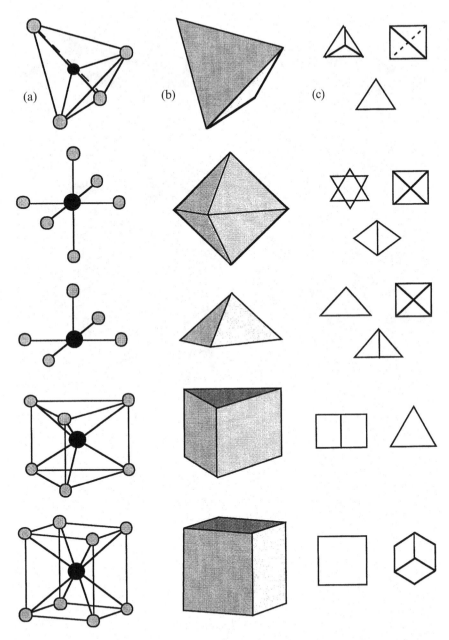

(a)   (b)   (c)

Figure 2.28. Common polyhedral units. (a) shows the metal atom and its ligands. (b) shows the solid polyhedra that represent these groups of atoms. (c) Representations of the polyhedra viewed in projection along high symmetry axes. From top to bottom, the tetrahedron, the octahedron, the square pyramid, the trigonal prism, and the cube are represented.

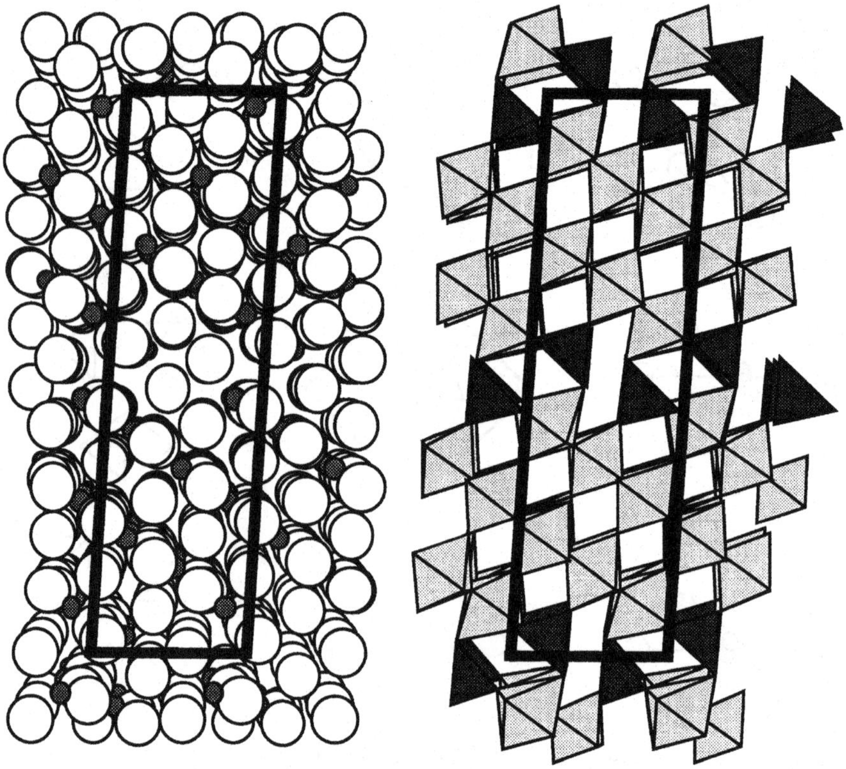

Figure 2.29. Sketches of the structure of Mo_4O_{11}, projected along [010]. (a) A hard sphere model. (b) A polyhedral model. In the hard sphere model, the smaller circles are the Mo. The larger circles represent O. The polyhedral model shows that the structure can be thought of as layers of connected octahedral groups (MoO_6), which have a lighter shading, separated by layers of tetrahedral (MoO_4) groups, which are darker [13].

Example 2.3 Using Polyhedral Models

Consider the polyhedral representation of a binary metal oxide compound in Fig. 2.30, below. What is the coordination of the metal atoms in this structure? The O atoms? What is the stoichiometry (metal-to-oxygen ratio)?

1. Based on Fig. 2.30, the structure is built from corner-sharing octahedra. The octahedron is an eight-sided solid with six vertices. Therefore, the metal atoms have six nearest neighbor oxygen atoms.

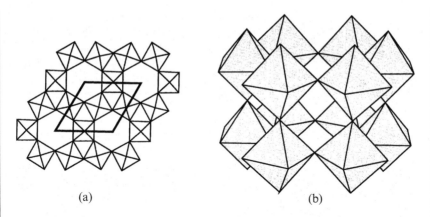

(a)   (b)

Figure 2.30. (a) An [0001] projection and (b) a perspective view of a binary M-O structure.

2. The corner-sharing indicates that the O has two nearest-neighbor metal atoms.
3. Six O surround each metal and each O is shared between two M atoms. Therefore, there is one M per 6 x 1/2 O = one M per three O. Therefore, the stoichiometry is MO_3.

ii.   *Representations of layered structures*

Complex crystals are sometimes visualized as an arrangement of structural units or 'blocks' that are larger than a single polyhedron, but smaller than the unit cell. Typically, one would choose blocks or sections of the structure that resemble more well known structure types. Such representations are often used to visualize layered compounds (compounds that have a well defined two-dimensional structural unit that is repeated in the third dimension). However, since all structures can ultimately be visualized as stacks of planes, layer models can also be used to represent three-dimensional compounds. One example of a block representation is shown in Fig. 2.31, where La_2CuO_4 is represented as repeating layers of the more well known structures, rock salt (NaCl, Chapter 4, Section F.vii) and perovskite ($CaTiO_3$, Chapter 4, Section F.xiii). Lanthanum cuprate can be considered the prototype compound for the high T_c cuprate superconductors. Because the superconducting cuprates have complex structures, this form of representation is common.

   Thinking of complex structures as being built of blocks of simpler structures can also be used to describe families of related crystal structures and compounds. For example, the so-called Ruddlesden–Popper Phases [14] in the Sr–Ti–O system consist of n perovskite layers separated by a rock salt layer. Distinct phases in this ternary system have the composition $Sr_{n+1}Ti_nO_{3n+1}$, where n is a small integer. The

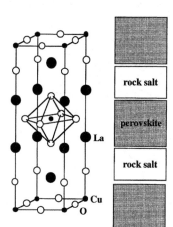

Figure 2.31. On the left, a ball and stick representation of the La_2CuO_4 structure is shown. The structure can be viewed as being made up of alternating perovskite and rock salt layers. A block representation is shown on the right.

$n=1$ phase has the structure depicted in Fig. 2.31. The $n=2$ structure, $Sr_3Ti_2O_7$, has two perovskite layers and the $n=3$ structure, $Sr_4Ti_3O_{10}$, has three perovskite layers. Note that the $n=\infty$ phase is $SrTiO_3$. In practice, entropic effects make it difficult to synthesize phases with large (finite) values of n using conventional preparation techniques. Recent advances in thin film synthesis, however, have made it possible to synthesize long period structures in this system [15]. There are other families of intergrowth structures that can be described by similar principles. These include the Aurivillius phases [16, 17], the hexagonal barium ferrites [18], and the intergrowth tungsten bronzes [19].

Figure 2.32 shows block representations of selected clay minerals. Many clay minerals can be viewed as if they were built from layers of AlO_6 octahedra, layers of SiO_4 tetrahedra, and interlayer materials (usually water or alkali cations) [20, 21]. Thus, these regions are easily represented schematically as rectangular blocks (for the Al^{3+} octahedra) and as blocks with angled ends (for the Si^{4+} tetrahedra).

Layered compounds that undergo intercalation reactions, such as graphite and TiS_2, are also typically represented in a schematic form as shown in Fig. 2.33 [22]. An intercalation reaction is one in which electropositive (or electronegative) species with relatively small ionization energies (or electron affinities) diffuse into the interstitial or interlayer spaces of a layered compound and react so that electrons are transferred to the host framework (or removed from it). In most cases, the host structure is composed of layers that are held together only by weak van der Waals forces. Because changes in electrical and optical properties often accompany intercalation, these materials are used in a variety of electrochemical systems. In graphite, it has been found that the distribution of intercalated

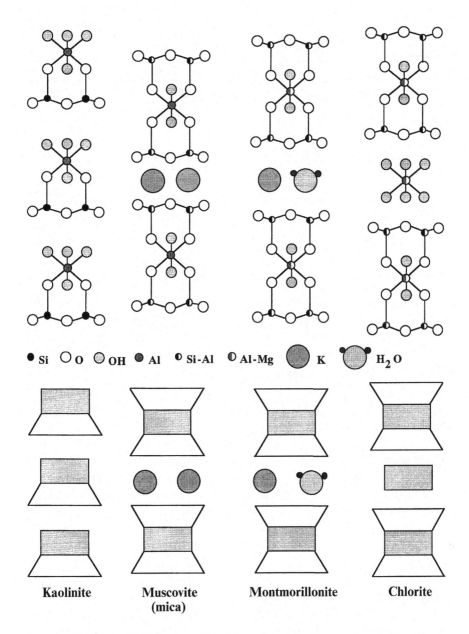

Figure 2.32. Ball and stick diagrams and block representations of four common clay minerals.

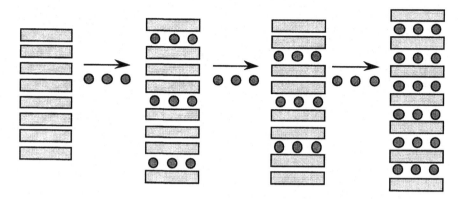

Figure 2.33. The intercalation of graphite can lead to a variety of compounds with different repeat units. From right to left are a stage one compound, a stage two compound, and a stage three compound.

species between the layers depends on their concentration and that the repeat unit changes with concentration. This process, called 'staging', is illustrated in Fig. 2.33.

Other layered structures conveniently represented in this way include the β-aluminas [23, 24], the magnetoplumbites [25], molybdate bronzes [26], and vanadate bronzes [27]. Finally, zeolites such as faujasite are broken up into three-dimensional units or 'cages' that can be joined to form a wide variety of three-dimensional, nanoporous solids [29].

iii.   *The stereographic projection*

While the representations of crystal structures discussed in the previous sections emphasized atomic arrangements, analysis of diffraction experiments requires knowledge of the relationships among the planes in a crystal. The stereographic projection is an alternative representation of a crystal that emphasizes the angular relationships among the planes. It is, thus, ideal for the interpretation of X-ray diffraction patterns.

In the stereographic projection [29], a plane is represented by its normal, which we call its pole. If we consider a crystal at the center of a sphere (see Fig. 2.34), then a pole pointing away from a specific crystal plane will intersect the sphere at a location that is characteristic of the orientation of the pole and, therefore, the plane it specifies. Figure 2.34a shows the points where the poles of the {100} family of planes in a cubic crystal intersect the sphere. The important thing about this idea is that the angle between two planes, α, can be determined by the relative positions at which the poles of the two planes intersect this

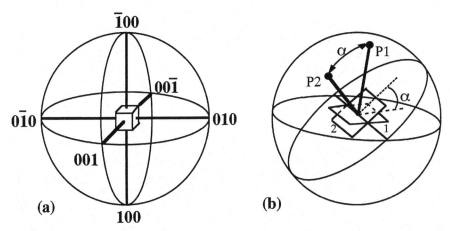

Figure 2.34. Poles or normals to crystallographic planes intersect the surface at a characteristic point. The orientations of the planes can be determined from the relative positions at which the poles intersect the sphere [29].

sphere. For example, the poles from plane 1 and plane 2 in Fig. 2.34b intersect the sphere at points P1 and P2, respectively. The relative positions of P1 and P2 can be used to determine the angle between planes 1 and 2. We note that the intersection of the sphere with any plane that contains the center of the sphere is a circle of maximum diameter known as a great circle. The intersection of the sphere with a plane that does not contain the center is a smaller circle called a small circle.

We would like to represent all of the points where poles intersect the sphere (such as P1 and P2 in Fig. 2.34a) on a two-dimensional drawing. Thus, we have to project the sphere onto a flat surface. One way of doing this is to project the points onto a tangent plane from a point on the other side of the sphere. It is easiest to understand this by consider only a two-dimensional slice of the sphere. The drawing in Fig. 2.35a shows the orientations of three poles that all lie in a single plane. The tangent line on the left represents a plane (the plane of projection) that is perpendicular to the paper and intersects the sphere at point P'_{001}. The points at which the poles intersect the sphere (P_{hkl}) are represented on the two-dimensional projection by the points at which rays starting at point B (which is opposite the point at which the projection plane intersects the sphere) and passing through P_{hkl} intersect the plane of the projection. These points are labeled P'_{hkl}. One way of visualizing this is to imagine that the points labeled P'_{hkl} are shadows of the points P_{hkl} which are formed by a light source at B. A simplified stereographic projection for a cubic crystal is shown in Fig. 2.36. The

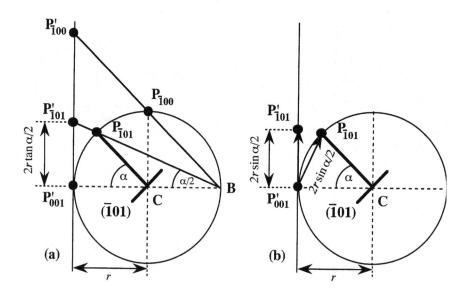

Figure 2.35. The (a) stereographic and (b) equal area projections are formed by projecting the points at which poles of planes intersect a sphere onto a tangent plane. See the text for more details.

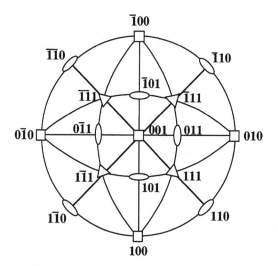

Figure 2.36. A simplified stereographic projection for a cubic structure [29].

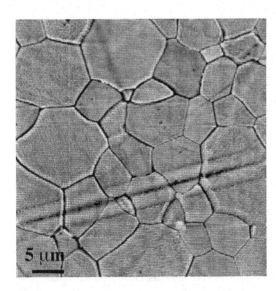

Figure 2.37. Contact atomic force microscope image of an Al_2O_3 polycrystal, thermally etched at 1400 °C.

vertical line on this projection is the line of projection in Fig. 2.35a. An alternative projection method, the equal area projection, is illustrated in Fig. 2.35b. Here, the position of the pole on the projection is determined by the length of the chord connecting pole's position on the sphere and the projection plane's point of tangency with the sphere.

ⓘ Polycrystallography

Until this point, we have exclusively dealt with structural descriptions of single crystals. The majority of useful materials, however, are polycrystalline in nature. In other words, the macroscopic solid is actually composed of many small single crystals (called grains) with different orientations, joined at interfaces called grain boundaries. As an example, consider the micrograph in Figure 2.37, which shows the surface of an Al_2O_3 polycrystal. The sample was polished and thermally etched to groove the grain boundary regions, which appear as dark contrast. The single crystal regions in this example are approximately 5 μm in diameter.

Microstructures have both geometric and crystallographic characteristics that influence their properties. Among the geometric characteristics are the distribution of grain sizes and aspect ratios; we will not consider these characteristics further. The two important crystallographic characteristics of a microstructure that we will consider are orientation of the crystallites with respect to the sample reference frame (the external, macroscopic surfaces of the specimen) and the

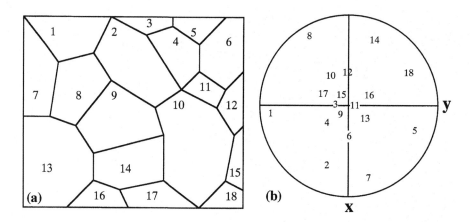

Figure 2.38. A schematic microstructure (a) with each of the grains labeled. The (100) pole figure shows the positions of the (100) poles of each grain. In this case, the grain's numerical label is used to designate the location of its pole. The center of the circle represents the sample normal (out of the plane of the paper). This type of pole figure could be constructed from the results of a microdiffraction experiment (in an SEM or TEM) in which the probe can be diffracted from specific points.

orientation of the crystallites with respect to one another (the misorientation). The orientation of the grains with respect to an external sample surface is typically shown on a special stereographic projection known as a pole figure. For example, Fig 2.38b shows a (100) pole figure for the schematic microstructure in Fig. 2.38a. The stereographic projection is oriented with its center parallel to the sample's surface normal and each point on the projection shows the relative position of the (100) pole of a specific crystallite with respect the sample's reference axes. If the distribution of the poles is not random (for example, if the majority of the crystallites have their (100) pole oriented within a few degrees of the sample normal, which is at the center of the circle in Fig. 2.38b), the sample is said to have some texture. Geologists call this fabric.

The situation in Fig. 2.38 is somewhat artificial in that specific poles are not associated with specific grains on conventional pole figures. Instead, pole figures usually give a statistical distribution of poles from a very large number of grains. The pole figures in Fig. 2.39 show results from an Al alloy with a cube texture. In this case, a rolling and annealing process caused the $\langle 100 \rangle$ axes to be preferentially aligned along the rolling, normal, and transverse directions. The relative densities of the labeled poles with respect to a random orientation are represented by contours. Pole figures of the type illustrated in Fig. 2.39 are the most

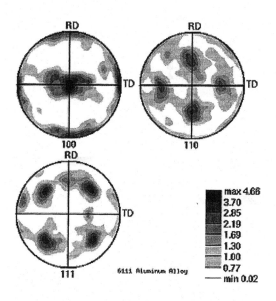

Figure 2.39. Pole figures for rolled and recrystallized Al sheet showing the (100), (110), and (111) poles with respect to the sample normal. The sample exhibits a strong cube texture with the ⟨100⟩ axes aligned along the rolling, normal and transverse directions. In this case, RD (the rolling direction) is equivalent to X and TD (the transverse direction) is equivalent to Y[30].

common and are most frequently obtained by X-ray diffraction. However, we also note that pole figures of the type illustrated in both Fig. 2.38 and Fig. 2.39 can be constructed based on grain specific orientation data accumulated by electron diffraction in an SEM or TEM.

i. Orientation

The relative orientation of a grain with respect to an external reference frame can be specified by Euler angles, by a transformation matrix, by Miller indices, or by a misorientation axis–angle pair. Each of these methods is briefly described below.

Euler angles are an ordered set of rotations, about specific axes, that can be used to rotate a crystal into coincidence with a frame of reference. We will use the Euler angles defined by Bunge, (ϕ_1, Φ, ϕ_2) [31]. Consider the primed (crystal) and unprimed (reference) axes, initially parallel in 2.40a. The first Euler angle refers to a positive rotation of ϕ_1 about the Z' axis. The second angle refers to a positive rotation of Φ about the X' axis, in its new orientation. The final angle refers to a positive rotation of ϕ_2 around the Z' axis in its new orientation. The angles ϕ_1 and ϕ_2 can take any values between 0 and 2π, while Φ lies between 0 and π. Note that using these definitions, Φ is the angle between the two Z axes and ϕ_1 and ϕ_2 indicate the angles between the two X axes and the line of intersection of the two X–Y planes.

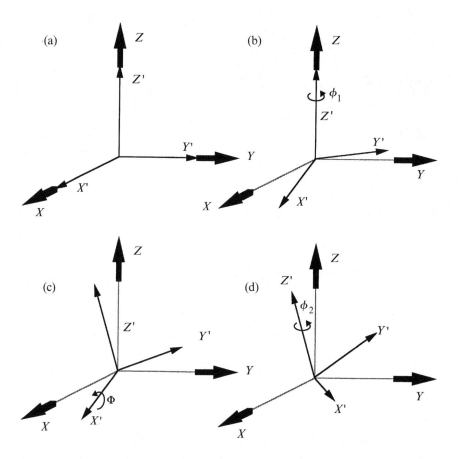

Figure 2.40. Definition of the Euler angles, (ϕ_1,Φ,ϕ_2). (a) The coincident crystal axes. (b) After the first rotation, ϕ_1, about the Z' axis. (c) After the second rotation, Φ, about the new X' axis. (d) After the third rotation, ϕ_2, about the new Z' axis [31].

Each of the three rotations can be expressed in matrix form:

$$g_{\phi_1}^{Z'} = \begin{bmatrix} \cos\phi_1 & \sin\phi_1 & 0 \\ -\sin\phi_1 & \cos\phi_1 & 0 \\ 0 & 0 & 1 \end{bmatrix} \tag{2.27}$$

$$g_{\Phi}^{X'} = \begin{bmatrix} 1 & 0 & 0 \\ 0 & \cos\Phi & \sin\Phi \\ 0 & -\sin\Phi & \cos\Phi \end{bmatrix} \tag{2.28}$$

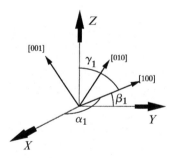

Figure 2.41. Definition of the angles used to compute the orientation matrix in Eqn 2.31. α_1 is the angle between [100] in the crystal and the X axis in the reference frame. β_1 is the angle between [100] in the crystal and the Y axis in the reference frame. γ_1 is the angle between [100] in the crystal and the Z axis in the reference frame. α_2, β_2, and γ_2 are the angles between [010] and the X, Y, and Z axes, respectively, and α_3, β_3, and γ_3 are the angles between [001] and the X, Y, and Z axes, respectively. For clarity, only the first three angles are shown in the figure [32].

$$g_{\phi_2}^{Z'} = \begin{bmatrix} \cos\phi_2 & \sin\phi_2 & 0 \\ -\sin\phi_2 & \cos\phi_2 & 0 \\ 0 & 0 & 1 \end{bmatrix} \tag{2.29}$$

and the multiplication of these matrices leads to the transformation matrix which can be used to transform vectors from one reference frame to another.

$g(\phi_1 \Phi \phi_2) =$

$$\begin{bmatrix} \cos\phi_1\cos\phi_2 - \sin\phi_1\sin\phi_2\cos\Phi & \sin\phi_1\cos\phi_2 + \cos\phi_1\sin\phi_2\cos\Phi & \sin\phi_2\sin\Phi \\ -\cos\phi_1\sin\phi_2 - \sin\phi_1\cos\phi_2\cos\Phi & -\sin\phi_1\sin\phi_2 + \cos\phi_1\cos\phi_2\cos\Phi & \cos\phi_2\sin\Phi \\ \sin\phi_1\sin\Phi & -\cos\phi_1\sin\Phi & \cos\Phi \end{bmatrix}$$

$$\tag{2.30}$$

An alternative way to express the orientation of a crystal with respect to a reference frame is a matrix of direction cosines [32]. The elements of the orientation matrix are the direction cosines of the angles between each of the crystal axes and each of the axes in the reference frame. Therefore, there are nine angles which are defined in Fig. 2.41. The orientation matrix, $[a_{ij}]$, is given by Eqn. 2.31, which is equivalent to the transformation specified by Eqn. 2.30.

$$[a_{ij}] = \begin{bmatrix} \cos\alpha_1 & \cos\beta_1 & \cos\gamma_1 \\ \cos\alpha_2 & \cos\beta_2 & \cos\gamma_2 \\ \cos\alpha_3 & \cos\beta_3 & \cos\gamma_3 \end{bmatrix}. \tag{2.31}$$

For any coordinate (x,y,z) in the reference frame, the identical point in the crystal frame (x',y',z') is given by Eqns. 2.32 to 2.34.

$$x' = a_{11}x + a_{12}y + a_{13}z \tag{2.32}$$

$$y' = a_{21}x + a_{22}y + a_{23}z \tag{2.33}$$

$$z' = a_{31}x + a_{32}y + a_{33}z. \tag{2.34}$$

Similarly, it can be shown that any point in the crystal frame (x',y',z') can be transformed to the equivalent coordinates in the reference frame (x,y,z) according to the following equations:

$$x = a_{11}x' + a_{21}y' + a_{31}z' \tag{2.35}$$

$$y = a_{12}x' + a_{22}y' + a_{32}z' \tag{2.36}$$

$$z = a_{13}x' + a_{23}y' + a_{33}z'. \tag{2.37}$$

While the orientation matrix representation has nine parameters (six more than necessary), there are certain advantages. For example, the elements of the orientation matrix are easily related to Miller indices that can also be used to express orientation information. If we write the orientation of a crystal as $(hkl)[uvw]$, where (hkl) is the index of the plane normal to the specimen surface and $[uvw]$ is the direction parallel to the X axis of the reference frame, then

$$a_{13} = \frac{h}{\sqrt{h^2 + k^2 + l^2}}, \ a_{23} = \frac{k}{\sqrt{h^2 + k^2 + l^2}}, \ a_{33} = \frac{l}{\sqrt{h^2 + k^2 + l^2}},$$

$$a_{11} = \frac{u}{\sqrt{u^2 + v^2 + w^2}}, \ a_{21} = \frac{v}{\sqrt{u^2 + v^2 + w^2}}, \ a_{31} = \frac{w}{\sqrt{u^2 + v^2 + w^2}}. \tag{2.38}$$

Furthermore, since the rows of the orientation matrix are the direction cosines of each of the crystal axes with respect to the reference frame, they can be used to construct pole figures. For example, the elements of the first row of the matrix in Eqn. 2.31 are the direction cosines of the [100] pole with respect to the sample reference frame; these data, from each grain in Fig. 2.38a, could be used to construct the pole figure in Fig. 2.38b.

Another way to describe relative orientations is to specify a common crystallographic axis, $\langle uvw \rangle$, and the rotation angle (ω) about that axis that brings the crystal into coincidence with the reference frame. The axis–angle pair concept is illustrated in Fig. 2.42b for two misoriented crystals. The direction cosines of the rotation axis (d_x, d_y, d_z), which contain two independent variables, and the rotation angle can be written in term of the elements of the orientation matrix.

$$d_1 = a_{23} - a_{32}$$

$$d_2 = a_{31} - a_{13}$$

$$d_3 = a_{12} - a_{21}$$

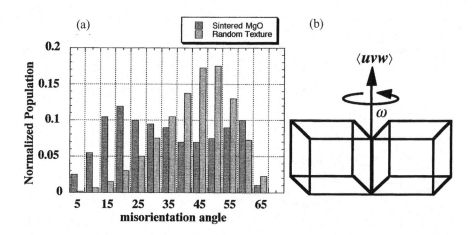

Figure 2.42. (a) Distribution of misorientation angles for 200 grain boundaries in hot pressed MgO, compared to a predicted random distribution. (b) Illustration of a common axis between two misoriented grains.

$$\cos\omega = \frac{1}{2}(a_{11} + a_{22} + a_{33} - 1).$$
(2.39)

The graphical representation of orientation and misorientation data from many grains is difficult, since the minimum number of parameters that must be used is three. One simplified way to do this is to disregard the two parameters that specify the rotation axis and consider only the misorientation angle. For example, in Fig. 2.42a, the number of observed grain pairs with misorientation angles within a certain range is plotted against the misorientation angle and compared to a random distribution [33]. This particular sample has a (111) texture and a high fraction of low misorientation angle grain boundaries.

ii. *Misorientation*
The misorientation between two crystallites can be expressed using any of the means described above, with the sample reference frame being replaced by that of a second crystallite. Similarly, if the orientation parameters with respect to an external reference frame have been measured for more than one crystal, then the misorientation between the two crystallites can be determined. The Rodrigues–Frank parameterization [34] is the favored method of representing misorientation data. In this case, each orientation is represented by a vector in a three-dimensional space, $\vec{R}_m = \tan(\omega/2)[d_1, d_2, d_3]$. The direction of the vector

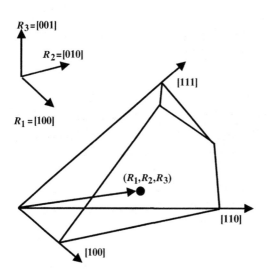

Figure 2.43. The irreducible wedge of Rodrigues–Frank space for cubic symmetry, with reference to the rotation axes parallel to the low index crystallographic directions.

is determined by the misorientation axis $\langle d_1\ d_2\ d_3 \rangle$ ordered in such a way that $d_1 > d_2 > d_3 > 0$. Note that the crystal symmetry must be accounted for in order to perform this ordering. The magnitude of the vector is proportional to the misorientation angle, ω.

$$|R_m| = \tan\left(\frac{\omega}{2}\right). \tag{2.40}$$

The shape of Rodrigues–Frank space for cubic materials is illustrated in Fig. 2.43. One of the advantages of this representation method is that it is easy to see where a misorientation axis lies with respect to low index crystallographic directions. Another is that each distinguishable orientation appears only once in the space shown in Fig. 2.43, hence the space is known as the fundamental zone. In general, there might be several rotations about an axis that bring the crystals to an indistinguishable state. Rodrigues–Frank space contains only the smallest rotation necessary to achieve any particular misorientation. By convention, this minimum misorientation is distinguished from equivalent misorientations by the term disorientation. Finally, grains with common misorientation axes lie along lines in Rodrigues–Frank space.

iii. *Grain boundary character and coincident site lattices*
Until now, we've concentrated on the misorientation between a grain and a reference frame; three parameters suffice to describe this situation. If we consider the boundary between two grains in a polycrystal, there are three degrees of freedom associated with the misorientation between the crystallites, and an

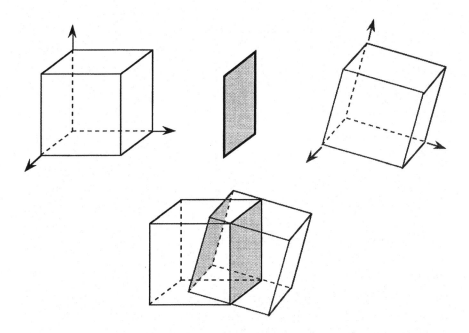

Figure 2.44. Schematic representation of the five macroscopic degrees of freedom associated with a grain boundary. Above, the misorientation between the crystals on the left and right can be specified by three degrees of freedom. Choice of the plane requires two more. Below, two boundary plane choices (shaded) are illustrated.

additional two degrees of freedom required to specify the plane of the grain boundary. Thus, a grain boundary has five macroscopic degrees of freedom (while translations potentially add three additional microscopic degrees of freedom, we will not consider these further). The five degrees of freedom associated with the grain boundary are illustrated schematically in Fig. 2.44.

Given a fixed misorientation, there are two limiting cases for the placement of the boundary plane. If the misorientation axis lies in the boundary plane, the boundary is said to be of the tilt type. If the boundary plane is perpendicular to the misorientation axis, it is a twist boundary. Historically, experimental studies of bicrystals have concentrated on symmetric tilt or twist boundaries with low index misorientation axes (for example, see Fig. 2.45). In nature, most boundaries have mixed character and are neither pure tilt nor pure twist. The potential number of distinct grain boundaries is quite large. If the resolution is Δ, then the number of distinguishable interfaces for a cubic material is approximately $\pi^5/288\Delta^5$. Therefore, at 2° of angular resolution, there are approximately 10^7 possible distinct grain boundaries for cubic materials. While it is currently not

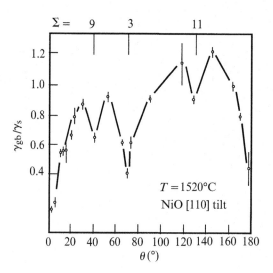

Figure 2.45. The experimentally measured relative energy of symmetric tilt boundaries in NiO, as a function of misorientation angle [35].

known if all of these possibilities are realized, it is known that the properties (energy, mobility) of grain boundaries are anisotropic. For example, Fig. 2.45 shows the relative energy of symmetric NiO tilt boundaries as a function of their misorientation angle [35]. Note that the energy is relatively lower at specific misorientation angles corresponding to special configurations where a coincident site lattice is formed between the two grains.

Special boundaries (known as coincident site lattice boundaries) occur when a relatively large number of sites in two adjacent crystals are coincident. For example, consider two square lattices, superimposed, as shown in Fig. 2.46. We can begin by assuming that both lattices are aligned in the same orientation so that all of the points overlap (there is 100% coincidence). Positive rotations of the second lattice around the axis normal to the paper lead to orientations with partial site coincidence. To be most accurate, we should recognize that all possible rotations lead to some level of coincidence in the infinite superimposed lattices. In practice, however, we are only interested in those orientations where a significant number of the sites (say >2%) are coincident. Two coincidence lattices with a high degree of coincidence are illustrated in Fig. 2.46.

We can find the angles at which a coincident site lattice (CSL) occurs as follows [36]. If θ is a positive angle between the y-axis and a vector in the first lattice with components u and v, then the existence of the mirror plane whose normal is along the x-axis guarantees that there is a vector of the same length in the second lattice which is oriented at an angle of $-\theta$ with respect to the positive y axis. Therefore, a positive rotation of $2\theta = \omega$ will bring a point from lattice

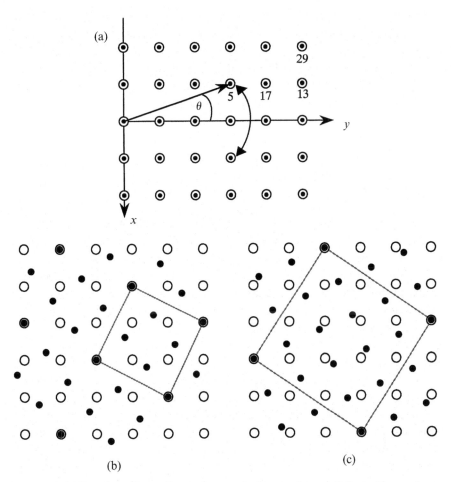

Figure 2.46. (a) Rotations of superimposed square lattices produce partial coincidence when a site in one lattice from below the y-axis at $-\theta$ rotates to the mirror equivalent position above the axis at θ. The points leading to $\Sigma5$, $\Sigma17$, $\Sigma13$, and $\Sigma29$ lattices are marked. (b) The $\Sigma5$ CSL ($\omega = 36.86°$). (c) The $\Sigma13$ CSL ($\omega = 22.62°$).

2 into coincidence with the point from lattice 1. Given components u and v, the angle can, therefore, be determined using the following equation:

$$\frac{u}{v} = \tan\frac{\omega}{2}. \tag{2.41}$$

While any integer values for the components u and v will lead to an angle for a CSL, all distinct CSLs are found within a limited angular range determined by the symmetry of the lattice. Furthermore, since we are only interested in those lattices with a reasonable degree of coincidence, we consider only relatively small values of u and v.

Each CSL is named based on its degree of coincidence. Since any integer values of u and v lead to a CSL, there is at least one site of coincidence in every u by v unit. The number of normal sites in such a unit is Σ, where:

$$\Sigma = \frac{1}{\alpha}(u^2 + v^2), \tag{2.42}$$

for $u^2 + v^2$ odd, $\alpha = 1$,
for $u^2 + v^2$ even, $\alpha = 2$.

Therefore, Σ is the number of normal lattice sites per coincidence lattice site. The $\Sigma 1$ lattice has perfect coincidence, every third lattice site is coincident in the $\Sigma 3$ lattice, and every fifth site in the $\Sigma 5$ lattice. The factor of α is included because if $u^2 + v^2$ is even, there is always a smaller CSL with half the repeat distance and double the coincidence. For large choices of u and v, some care must be exercised since some points can belong to more than one CSL. CSLs with $\Sigma > 50$ are not expected to have any physical significance.

The existence of special CSL boundaries, where atoms can occupy undistorted lattice sites, can explain the observation of cusps in the relative grain boundary free energy. For example, in the data shown in Fig. 2.45, the lowest energies occur at the $\Sigma 1$, $\Sigma 9$, $\Sigma 3$, and $\Sigma 11$ orientations. It is assumed that because one out of Σ atoms on these boundaries can occupy ideal positions, that the energy is more nearly equal to that of the bulk crystal. Table 2.3 lists the CSLs for boundaries in cubic materials with $\Sigma < 30$. The suffixes a and b are added to the name of the CSL when geometrically distinct lattices with the same degree of coincidence must be distinguished. Without some level of generalization, the CSL orientations account for only a very small number of the possible boundaries. The concept has, therefore, been extended to account for orientations that are near special boundaries. For low angle grain boundaries that are near coincidence ($\Sigma 1$), it was proposed by Read and Shockley [38] that such boundaries could be viewed as defective versions of $\Sigma 1$ with the small misorientation accounted for by a set of dislocations. Using this idea, it is possible to imagine that low angle tilt boundaries are made from edge dislocations and low angle twist boundaries result from screw dislocations. Brandon [39] extended this idea to other near-CSL orientations and proposed a condition (which has become known as the Brandon criterion) which can be used to separate special boundaries from general boundaries. For a grain boundary near a CSL orientation, if the misorientation angle between the actual

Table 2.3 *CSL boundaries with $\Sigma < 30$, for cubic–cubic misorientations [37]*

Σ	ϕ_1	Φ	ϕ_2	$\langle uvw \rangle$	ω
3	45.00	70.53	45.00	111	60.00
5	0.00	90.00	36.86	100	36.86
7	26.56	73.40	63.44	111	38.21
9	26.56	83.62	26.56	110	38.94
11	33.68	79.53	33.68	110	50.47
13a	0.00	90.00	22.62	100	22.62
13b	18.43	76.66	71.57	111	27.79
15	19.65	82.33	42.27	210	48.19
17a	0.00	90.00	28.07	100	28.07
17b	45.00	86.63	45.00	221	60.92
19a	18.44	86.98	18.44	110	26.53
19b	33.69	71.59	56.31	111	46.83
21a	14.03	79.02	75.97	111	21.78
21b	22.83	79.02	50.91	211	44.41
23	15.25	82.51	52.13	311	40.45
25a	0.00	90.00	16.26	100	16.26
25b	36.87	90.00	36.87	331	51.68
27a	21.80	85.75	21.80	110	31.59
27b	15.07	85.75	31.33	210	35.43
29a	0.00	90.00	43.60	100	43.60
29b	33.69	84.06	56.31	221	46.40

misorientation and the particular CSL misorientation, ω, is within the following range:

$$\omega < \frac{15°}{\sqrt{\Sigma}}, \tag{2.43}$$

then the boundary can be consider as a CSL or a special boundary. Otherwise, it is known as a general boundary.

J Problems

(1) (i) Draw and identify the Bravais lattice that is formed when additional lattice points are placed at the centers of two opposite faces of a cubic P cell (for example, at (1/2,1/2,0)).

(ii) Draw and identify the Bravais lattice that is formed when additional lattice points are placed at the centers of four parallel edges of a cubic P cell (for example, at (0,0,1/2)).

(iii) Among the fourteen Bravais lattices, there is a monoclinic C and an orthorhombic C. Explain why there is no tetragonal C.

(iv) To find the 14 Bravais lattices, we considered distortions along all of the axes and the body diagonal of the cubic cell. What happens when the cube is deformed along the face diagonal?

(2) The unit cell of an oxide crystal with a cubic lattice parameter of 4.00 Å contains the following atoms (the ordered sets are the coefficients of location that define $\vec{r}_j$):

> B at (0,0,0)
> A at (1/2,1/2,1/2)
> O at (1/2,0,0)
> O at (0,1/2,0)
> O at (0,0,1/2).

(i) Sketch the unit cell, including all atoms that are within or partially within the cell.

(ii) What is the formula unit of this compound?

(iii) Specify the lattice and the basis of this crystal.

(iv) What are the coordination numbers of A and B?

(v) Determine the A–O and B–O bond lengths.

(vi) If A and B are not the same size, which is smaller?

(vii) Specify the sequence of atoms found along the $\langle 111 \rangle$ and $\langle 110 \rangle$ directions.

(viii) Specify the contents of the (200) plane and the contents of the (110) plane.

(3) In the cubic fluorite structure, Ca atoms are at the cube vertices and the face centers (in effect, the positions of a cubic F lattice). The fluorine atoms are at the eight $(\pm 1/4, \pm 1/4, \pm 1/4)$ positions.

(i) Specify the lattice and basis of this compound.

(ii) Sketch a diagram of the atoms in the (002) plane.

(iii) Sketch a diagram of the atoms in the (400) plane.

(iv) Sketch a diagram of the atoms in the (110) plane.

(v) If a planar defect occurs along the (111) plane and intersects the (110) plane at the surface, along what direction does the intersection occur?

(vi) What is the maximum packing efficiency of this structure and how does it compare to that of close-packed fcc and bcc metals?

(4) Draw a [001]* projection of the reciprocal lattice (with h and k ranging from -2 to 2) of a simple orthorhombic direct lattice assuming that $a = 2b$ and $c = 3a$. Indicate the reciprocal lattice vectors and the unit cell of this reciprocal lattice. What happens to the volume of this reciprocal lattice when the direct lattice unit cell expands so that b remains the same, but $a = 4b$?

(5) (i) Specify the primitive reciprocal lattice vectors for the cubic I lattice.

(ii) Using these vectors, draw a projection of the reciprocal lattice, as you would see it if you were looking down the [110] axis of the conventional cell.

(6) (i) Specify the primitive reciprocal lattice vectors for the cubic F lattice.

(ii) Using these vectors, draw a projection of the reciprocal lattice, as you would see it if you were looking down the c-axis of the conventional cell.

(iii) On a new diagram, showing the reciprocal lattice projected in the same orientation, indicate the orientation and length of the conventional reciprocal lattice vectors and re-label the reciprocal lattice points based on these conventional vectors.

(iv) What fraction of the hkl indices (possible values of h,k,l) specify vectors that terminate at reciprocal lattice points. Are there systematic absences in the indices?

(7) A monoclinic P lattice has parameters $a = 4$ Å, $b = 8$ Å, $c = 6$ Å, and $\beta = 110°$.

(i) Specify the primitive reciprocal lattice vectors.

(ii) Sketch a projection of the reciprocal lattice, down the b-axis. (Make the drawing to scale.)

(8) TiO_2 has a tetragonal crystal structure with $a = 4.59$ Å and $c = 2.96$ Å.

(i) Along what direction does the (121) plane intersect the (110) surface?

(ii) What angle does this direction make with the [001] axis?

(9) (i) In a cubic crystal, what is the angle between [111] and [110]?

(ii) What is this angle in a tetragonal crystal with $a = 5$ Å and $c = 3$ Å?

(iii) What is this angle in an orthorhombic crystal with $a = 5$ Å, $b = 4$ Å, and $c = 3$ Å?

(iv) What is this angle in a monoclinic crystal with $a = 5$ Å, $b = 4$ Å, $c = 3$ Å, and $\beta = 95°$?

(10) NiAs has a hexagonal structure with $a = 3.6$ Å and $c = 5.0$ Å. There are Ni atoms at (0,0,0) and (0,1,0) and an As atom at (1/3,2/3,1/4).

(i) What is the Ni–As bond distance?

(ii) What is the Ni–As–Ni bond angle?

(11) In a monoclinic crystal with $a = 5$ Å, $b = 4$ Å, $c = 3$ Å, and $\beta = 95°$, there is a metal (M) atom at (0,0,0), another at (0,0,1), and an O at (1/4,1/4,1/2).

(i) What are the M–O bond distances?

(ii) What is the M–O–M bond angle?

(12) Show that the c/a ratio of the ideal hexagonal close packed structure is 1.633.

(13) The compound $TiAl_3$ has a tetragonal I lattice with $a = 3.85$ Å and $c = 8.596$ Å. In this structure, there is a Ti atom at coordinates $(1/2,1/2,1/2)$ that is bonded to two Al at $(0,1/2,1/4)$ and $(0,1/2,3/4)$.

  (i) What is the Ti–Al bond distance?

  (ii) What is the Al–Ti–Al bond angle?

  (iii) Specify the reciprocal lattice vectors.

  (iv) Make a sketch comparing the direct and reciprocal lattices.

  (v) What is the angle between the 101 reciprocal lattice vector and the $\langle 101 \rangle$ direction?

  (vi) What is the angle between the (102) and (010) plane?

  (vii) For rotations about [001], specify the three lowest Σ coincident site lattice orientations.

(14) Let N_n be the number of nth nearest neighbors of a given Bravais lattice point (for example, in cubic P, $N_1 = 6$, $N_2 = 12$, etc.). Let r_n be the distance to the nth nearest neighbor expressed as a multiple of the nearest neighbor distance (for example, in simple cubic, $r_1 = 1$, $r_2 = 1.414$, etc.) [1].

  (i) Make a table of N_n and r_n for $n = 1$ to 6 for cubic I and F Bravais lattices.

  (ii) Repeat for the tetragonal P and I Bravais lattices, assuming that $c/a = 1.3$.

(15) Consider the polyhedral representation of a binary metal oxide compound in Fig. 2.47, opposite. What is the coordination of the metal atoms in this structure? The O atoms? What is the stoichiometry (metal-to-oxygen ratio)?

(16) Resisting conventions, some authors name monoclinic P and monoclinic I as the two unique Bravais lattices with monoclinic symmetry. Is there a difference between monoclinic I and monoclinic C? How will the indexing of planes be changed by the unconventional choice?

(17) Consider the pole normal to a plane (hkl) that makes angles ρ, θ, and ϕ with the coordinate axes a, b, and c. Show that for a cubic system, the Miller indices are in the same ratios as the direction cosines of the pole.

(K) References and sources for further study

[1] N.W. Ashcroft and N. D. Mermin, *Solid State Physics* (Holt Rinehart and Winston, New York, 1976) Chapters 4 and 7. Fig. 2.8 is drawn after 7.4 on p. 117. Fig. 2.9 is drawn after Fig. 7.5 on p. 117. Fig. 2.16 is drawn after Figs. 4.14 and 4.15. Problem 14 is based on problem 7 on p. 83. Definitions of Bravais lattices and unit cells can be found in this source.

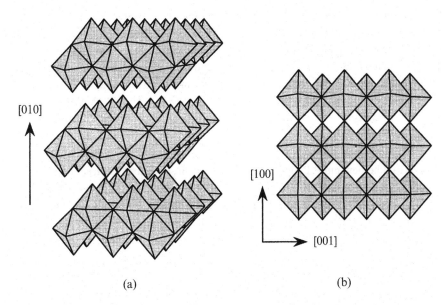

[010]

[100]

[001]

(a)   (b)

Figure 2.47. Two projections showing the polyhedral structure of a binary metal oxide. (a) is a perspective drawing and (b) is a [010] axial projection of a single layer.

[2] C. Kittel, *Introduction to Solid State Physics*, 5th edition (J. Wiley & Sons, New York, 1976) Chapter 1. Definitions of Bravais lattices and unit cells can be found in this source.

[3] A.D. Mighell and J.R. Rodgers, *Acta Cryst.* **A36** (1980) 321. This paper discusses the distribution of known crystal structures among the Bravais lattice types.

[4] L. Van Vlack, *Elements of Materials Science and Engineering* (Addison-Wesley, Reading, MA, 1989) pp. 81–95. On specifying locations, planes, and directions in crystals.

[5] C. Kittel, *Introduction to Solid State Physics*, 5th edition (J. Wiley & Sons, New York, 1976) pp. 18–19. On specifying locations, planes, and directions in crystals.

[6] F.C. Frank, On Miller-Bravais Indices and Four Dimensional Vectors, *Acta Cryst.* **18** (1965) 862. A detailed consideration of four-index notation in hexagonal and other systems.

[7] N.W. Ashcroft and N.D. Mermin, *Solid State Physics* (Holt Rinehart and Winston, New York, 1976) Chapter 5. Definition of the reciprocal lattice.

[8] C. Kittel, *Introduction to Solid State Physics*, 5th edition (J. Wiley & Sons, New York, 1976) p. 47. Definition of the reciprocal lattice.

[9] A.R. West, *Solid State Chemistry and its Applications* (J. Wiley & Sons, Chichester, 1984) 695–9. Fig. 2.25 is drawn after Fig. A7.1 on p. 696. This source provides an excellent intuitive description of the reciprocal lattice.

[10] J.B. Fraleigh, *Calculus with Analytic Geometry* (Addison-Wesley Publishing Co., Reading, 1980) Chapter 14. This source provides a clear (noncrystallographic) description of points, planes, directions (vectors), and relationships among them.

[11] B.D. Cullity, *Elements of X-ray Diffraction*, 2nd edition (Addison-Wesley, Reading, MA, 1978) pp. 501–3. Some useful geometric equations can be found here.

[12] M. DeGraef and M. McHenry, *Crystallography, Symmetry, and Diffraction*, to be published by Cambridge University Press, 2002. A description of the metric tensor and its uses.

[13] L. Kihlborg, *Ark. Kemi* **21** (1963) 365. The structure of Mo_4O_{11}.

[14] S.N. Ruddlesden and P. Popper, The Compound $Sr_3Ti_2O_7$ and its Structure, *Acta Cryst.* **11** (1958) 54–5.

[15] J.H. Haeni, C.D. Theis, D.G. Schlom, W. Tian, X.Q. Pan, H. Chang, I. Takeuchi, and X.-D. Xiang, Epitaxial Growth of the First Five Members of the $Sr_{n+1}Ti_nO_{3n+1}$ Ruddlesden–Popper Homologous Series, *App. Phys. Lett.*, submitted 1999.

[16] B. Aurivillius, Mixed Oxides with Layer Lattices, *Ark. Kemi* **2** (1950) 519–27.

[17] J. Gopalakrishnan, A. Ramanan, C.N.R. Rao, D.A. Jefferson, and D.J. Smith, A Homologous Series of Recurrent Intergrowth Structures of the Type $Bi_4A_{m+n-2}B_{m+n}O_{3(m+n)+6}$ Formed by Oxides of the Aurivillius Family, *J. Solid State Chem.* **55** (1984) 101–5.

[18] J.S. Anderson and J.L. Hutchison, The Study of Long Range Order in Hexagonal Barium Ferrite Structures, *Cont. Phys.* **16** (1975) 443–67.

[19] A. Hussaun and L. Kihlborg, Intergrowth Tungsten Bronzes, *Acta Cryst.* A**32** (1976) 551–7.

[20] W.D. Kingery, H.K. Bowen, and D.R. Uhlmann, *Introduction to Ceramics* (John Wiley & Sons, New York, 1976) pp. 77–80. Summary of clay structures.

[21] H. van Olphen, *An Introduction to Clay Colloid Chemistry* (John Wiley & Sons, New York, 1977) Chapter 5. A detailed description of the structures of clay minerals.

[22] M.S. Whittingham and A.J. Jacobson, eds., *Intercalation Chemistry* (Academic Press, New York, 1982).

[23] C.R. Peters, M. Bettman, J.W. Moore, and M.D. Glick, Refinement of the Structure of Sodium β-alumina, *Acta Cryst.* B**27** (1971) 1826.

[24] S. Sattar, B. Ghoshal, M.L. Underwood, H. Mertwoy, M.A. Saltzberg, W.S. Frydrych, G.S. Rohrer, and G.C. Farrington, Synthesis of Di- and Trivalent ß″-Aluminas by Ion Exchange, *J. Solid State Chem.* **65** (1986) 231.

[25] A.F. Wells, *Structural Inorganic Chemistry*, 5th edition (Clarendon Press, Oxford, 1984) p. 599.

[26] M. Greenblatt, Molybdenum Oxide Bronzes with Quasi-Low-Dimensional Properties, *Chem. Rev.* **88** (1988) 31–53.

[27] C. Delmas, H. Cognac-Auradou, J.M. Cocciantelli, M. Menetrier, and J.P. Doumerc, The Li$_x$V$_2$O$_5$ System: An Overview of the Structure Modifications Induced by the Lithium Intercalation, *Solid State Ionics* **69** (1994) 257–64.

[28] R.M. Barrer, *Zeolites and Clay Minerals as Sorbants and Molecular Sieves* (Academic Press, New York, 1978).

[29] B.D. Cullity, *Elements of X-ray Diffraction*, 2nd edition (Addison-Wesley, Reading, MA, 1978) pp. 63–78. Fig. 2.34 is drawn after Figs. 2.25 and 2.26 on pp. 63 and 64. Fig. 2.36 is drawn after Fig. 2.36a, p. 76. A description of the stereographic projection.

[30] A.D. Rollett, Private communication.

[31] H.-J. Bunge, *Texture Analysis in Materials Science*, translated by P.R. Morris (Butterworths, London, 1982). Figure 2.40 is drawn after Fig. 2.2 on p. 5.

[32] V. Randle. *Microtexture Determination and its Applications* (The Institute of Materials, London, 1992). Fig. 2.41 is drawn after Fig. 4.1 on p. 75.

[33] A. Morawiec, *J. Appl. Cryst.* **28** (1995) 289.

[34] F.C. Frank, Orientation Mapping, *Met. Trans.* **19A** [3] 403–08 (1988).

[35] G. Dhalenne, M. Dechamps, and A. Revcolevschi, *Advances in Ceramics*, Vol. 6, *Character of Grain Boundaries* (American Ceramic Society, Columbus, OH, 1983) pp. 139–50.

[36] W. Bollmann, *Crystal Defects and Crystalline Interfaces* (Springer-Verlag, New York, 1970) pp. 143–9.

[37] B.L. Adams, J. Zhao, and H. Grimmer, Discussion of the Representation of Intercrystalline Misorientations in Cubic Materials, *Acta Cryst.* **A46** [7] (1990) 620–2.

[38] W.T. Read, *Dislocations in Crystals* (McGraw-Hill, New York, 1953).

[39] D.G. Brandon, The Structure of High Angle Grain Boundaries, *Acta Met.* **14** (1966) 1479.

Chapter 3
Symmetry in Crystal Structures

(A) Introduction

In Chapter 2, we noted that all crystal structures could be specified by a set of Bravais lattice vectors ($\vec{R}$) and a set of vectors describing the positions of the basis atoms ($\vec{r}_j$). In practice, the number of basis atoms can be quite large and simply listing them is both cumbersome and unenlightening. By acknowledging the symmetry of the atomic configuration, it is usually possible to distill the description of the basis down to a small number of parameters. For this reason, crystal structure data are always presented with reference to the space group or underlying symmetry of the structure. It is, therefore, important that we understand the formal mechanisms for the description of a crystal's symmetry. In the next sections of this chapter, the symmetry operators and groups of operators that are used to describe the long range configuration of atoms in a crystal are defined. The objective is to provide you with the information necessary for the interpretation of conventional crystal structure and diffraction data. It will then be demonstrated that from such data it is possible to construct a model of the crystal structure and specify the geometry of the bonding. Although it is not considered further in this book, knowledge of symmetry is also valuable for understanding the thermal, mechanical, optical, magnetic, and electrical properties of solids.

(B) Symmetry operators

Symmetry can be defined as the spatial relationships between objects in a pattern [1]. A symmetry operator describes an imaginary action that can be used to develop a pattern. Operators change the position and/or orientation of an object in space. This is analogous to the action of a mathematical function that changes the value of a variable. The seven symmetry operators are translation, rotation, reflection, inversion (center of symmetry), roto-inversion (inversion axis), glide (translation + reflection), and screw (rotation + translation). Symmetry elements are imaginary objects that perform the symmetry operation. The primary function of the symmetry element is to specify the reference point about which an action occurs. The first five symmetry elements that we consider are translation

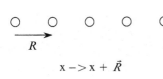

Figure 3.1. The lattice translation vector moves the object to an equivalent position in a different unit cell.

$x \rightarrow x + \vec{R}$

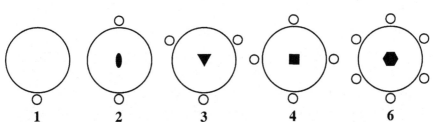

Figure 3.2. The five rotation operators that are consistent with translational symmetry. The large circles are lines of construction to guide the eye. The solid object in the center shows the position of the rotation axis and the small circle is the object which is repeated to form the pattern. The 2 axis is referred to as a diad, the 3 axis as a triad, the 4 axis as a tetrad, and the 6 axis as a hexad.

vectors, rotation axes, mirror planes, centers of symmetry (inversion points), and inversion axes.

i. *Translation*

Translation is the replication of an object at a new spatial coordinate. If the operator is the vector, $\vec{R}$, and there is an object located at a position specified by the vector $\vec{r} = x\vec{a} + y\vec{b} + z\vec{c}$, then we know that an identical object will be located at $\vec{r} + \vec{R}$. In other words, this operator is the same as a Bravais lattice vector, as defined in Chapter 2 (see Eqn. 2.1). Translation is used to build a crystal structure by replicating an object (the basis) at each of the Bravais lattice points. The generation of a pattern in one-dimension by translation is illustrated in Fig. 3.1.

ii. *Rotation*

Rotation is motion through an angle about an axis. Since repeated operations must eventually place the object in its original position, the possible angles are constrained by the condition that $n\alpha = 2\pi$, where n is an integer number of rotations and α is the angle of each rotation, in radians. For reasons that will be specified later, we will only consider rotations of $n = 1, 2, 3, 4,$ and 6; the notation and the patterns generated by these rotations are illustrated in Fig. 3.2. In a diagram or picture of a pattern, a polygon with the symmetry of the rotation specifies the position of the axis (the symmetry element) when it is normal to the plane of the paper. In the written notation, the numeral n is used; this is the

Figure 3.3. The reflection operator. The normal of the mirror is parallel to x and the positive x direction is to the right. The left-handed replica is specified by a comma.

number of repeated operations required to bring the object back to its original position.

Analytically, the coordinates of the objects in a pattern formed by rotation can be generated using a rotation matrix, g. The matrices used to perform rotations about the $x, y,$ and z axes were already given in Eqns. 2.27, 2.28, and 2.29, respectively. Assuming that the rotation axis is parallel to z, the rotations are carried out in the x–y plane. For a tetrad axis, the relevant angles are $\pi/2$, π, and $3\pi/2$. Using these angles in Eqn. 2.29, we can generate the coordinates of the objects in the pattern:

$$[x\ y\ z]\begin{bmatrix} 0 & 1 & 0 \\ -1 & 0 & 0 \\ 0 & 0 & 1 \end{bmatrix} = \begin{bmatrix} y \\ -x \\ z \end{bmatrix}; \quad [x\ y\ z]\begin{bmatrix} -1 & 0 & 0 \\ 0 & -1 & 0 \\ 0 & 0 & 1 \end{bmatrix} = \begin{bmatrix} -x \\ -y \\ z \end{bmatrix};$$

$$[x\ y\ z]\begin{bmatrix} 0 & -1 & 0 \\ 1 & 0 & 0 \\ 0 & 0 & 1 \end{bmatrix} = \begin{bmatrix} -y \\ x \\ z \end{bmatrix} \tag{3.1}$$

According to Eqn. 3.1, the coordinates of the four objects are (x, y, z), $(-y, x, z)$, $(-x, -y, z)$, and $(y, -x, z)$. The matrix that creates a clockwise rotation of θ about the axis $\langle n_1\ n_2\ n_3 \rangle$ (a vector with unit magnitude) has the following elements:

$$g_{ij} = \delta_{ij}\cos\theta - \varepsilon_{ijk}n_k \sin\theta + (1 - \cos\theta)n_i n_j. \tag{3.2}$$

In Eqn. 3.2, δ_{ij} is the Kronecker delta ($\delta_{ij} = 1$ for $i = j$ and 0 for $i \neq j$) and ε_{ijk} is the permutation tensor which takes the value of $+1$ for even permutations of 1, 2 and 3, -1 for odd permutations, and is zero otherwise.

iii. *Reflection*
Reflection describes the operation of a mirror, as shown in Fig. 3.3. In this case, the symmetry element, the mirror plane, is specified by an m. The axis of a mirror refers to its normal. On diagrammatic representations of patterns, the positions

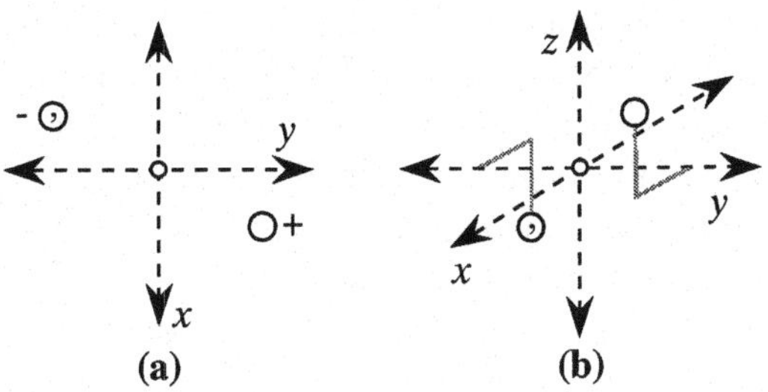

Figure 3.4. (a) Projection (along z) of the pattern formed by an inversion center. The plus and minus sign indicate small vertical displacements above and below the plane of the paper, respectively. (b) An oblique projection of the same pattern.

(a)

(b)

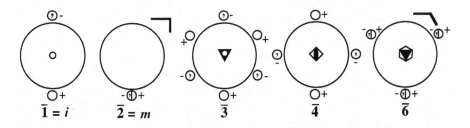

Figure 3.5. The roto-inversion operators. Note that the mirror in the plane of the paper is indicated by the bold lines in the upper right-hand corner.

of mirror planes are specified by bold lines. A mirror plane at $x=0$, with its axis parallel to x, reflects an object at (x, y, z) to $(-x, y, z)$. It is important to note that reflection converts a right-handed object into a left-handed or enantiomorphous replica. In most drawings of patterns, left-handed objects are specified by a comma.

iv. Inversion
The inversion operation occurs through the element called a center of symmetry. A center of symmetry at the origin transforms an object at (x, y, z) to the position $(-x, -y, -z)$, an action illustrated in Fig. 3.4. The written symbol is an i or $\bar{1}$ and on diagrams it is usually indicated by a small, open circle. Like the mirror, inversion also creates a left-handed replica.

v. Roto-inversion
The roto-inversion operator rotates an object in a pattern about its axis and then inverts the object through a center of symmetry on the axis. The roto-inversion axes (shown in Fig. 3.5) produce only one pattern that could not be produced by other operators used alone or in combination. The diagrammatic symbols are shown in Fig. 3.5 and the written symbol is the same as the n-fold rotational axis,

but with a 'bar' above. In Fig. 3.5, the vertically divided circles represent two objects with the same x and y coordinates, but different values of z. One of the objects has a positive z coordinate, indicated by the plus sign, and the other has a negative z coordinate, indicated by the minus sign. Note that the inversion diad, $\bar{2}$, is always written as 'm'. You can also see by comparing Fig. 3.2 and 3.5 that the combination of a rotation triad and a perpendicular mirror produces the $\bar{6}$ pattern. The combination of the triad and the perpendicular mirror is written as $3/m$, indicating that the mirror's normal and the triad axis are parallel.

Point symmetry groups are operators or combinations of operators that leave at least one point unchanged. We have now defined a total of ten unique point symmetry groups that have the symbols: 1, 2, 3, 4, 6, $\bar{1}$, m, $\bar{3}$, $\bar{4}$, and $\bar{6}$.

vi. *The distinction between proper and improper operations*
Operators fall into one of two classes, those producing a 'right-handed' replica (proper) and those producing a 'left-handed' or mirror image replica (improper). Left-handed and right-handed objects can not be superimposed by any combination of rotation or translation.

Ⓒ The 32 distinct crystallographic point groups

Two or more operators can be combined to form a symmetry group. The term 'group' has a very specific mathematical implication and, if you are curious, you can find many textbooks dedicated to the theory of groups [2]. Although group theory is not described in this course, we will apply some of the results that are relevant to crystallography. For example, we will see that the symmetry of any crystal can be described as a combination of a point symmetry group and a set of lattice translation vectors (the Bravais lattice vectors defined in Chapter 2). The operators in a point symmetry group leave at least one point in the pattern unchanged. The groups obtained by combining a point symmetry group with a set of lattice translation vectors are known as space groups and will be discussed in Section D of this chapter.

i. *Compatibility of rotations and translations*
In principle, there are a tremendous number of point symmetry groups that can be constructed from the operators defined in the last section. However, we are only interested in the ones that can be combined with Bravais lattice vectors to build a crystal. Since the Bravais lattice translations themselves already have some symmetry, this limits the possibilities. For example, tetragonal crystals have a rotation tetrad parallel to [001]. Therefore, this operator must be part of any

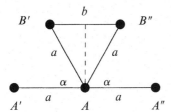

Figure 3.6. See text for description.

point symmetry group used in conjunction with tetragonal Bravais lattice vectors. There are 32 (distinct) groups that are compatible with one of the 14 Bravais lattice translations and they are known as the 32 crystallographic point groups.

We can determine which rotation operators are compatible with translational periodicity in the following way [3]. Referring to Fig. 3.6, we see that A is an axis of rotation that is replicated by translations of a to form A' and A''. Symmetry around A requires the A–A'' translation be rotated through α to produce B' and the A–A' translation to be rotated through $-\alpha$ to form B'. Since B' and B'' must be related by a translation, the distance between them is $b = pa$, where p is an integer. Furthermore, from the geometry in Fig. 3.6, $a\cos\alpha = b/2$, so $\cos\alpha = (b/2a) = (pa/ba) = p/2$. The values of $\cos\alpha$ that satisfy this equation are 0, 1/2, and 1. Thus, the possible values of α are 60°, 90°, 120°, 180°, or 360° and the allowed rotation axes are 6, 4, 3, 2, and 1. Other rotation axes will not generate space filling patterns.

ii. *Combinations of proper rotations*

Six distinct groups can be formed by the combination of the five proper rotations. One of the defining characteristics of a group is that the effect of any two operators can be reproduced by a third operation in the group [4, 5]. To demonstrate this, consider three operations on a cube, A, B, and C, where A is a rotation triad about a body diagonal, B is a rotation triad about another body diagonal, and C is a rotation diad about a line through the cube's face center. These operations are illustrated in Fig. 3.7. Note that because the effect of AB is the same as the effect of C, we can say that $AB = C$. An alternative way to test the compatibility of rotation operators is to use Euler's equation; this geometric formula relates the angles between rotation axes (A, B, C) to the angles of rotation (α, β, γ).

$$\cos(A \wedge B) = \frac{\cos(\gamma/2) + \cos(\alpha/2)\cos(\beta/2)}{\sin(\alpha/2)\sin(\beta/2)}$$

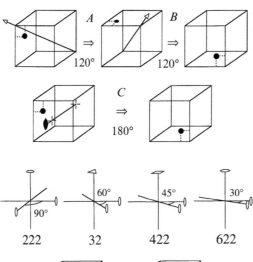

Figure 3.7. Applying the operators A and B in sequence has the same effect as the operator C.

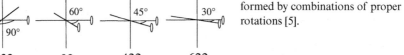

Figure 3.8. The five point groups formed by combinations of proper rotations [5].

$$\cos(A \wedge C) = \frac{\cos(\beta/2) + \cos(\alpha/2)\cos(\gamma/2)}{\sin(\alpha/2)\sin(\gamma/2)}$$

$$\cos(B \wedge C) = \frac{\cos(\alpha/2) + \cos(\beta/2)\cos(\gamma/2)}{\sin(\beta/2)\sin(\gamma/2)}. \tag{3.3}$$

If substituting α, β, and γ into Eqn 3.3 yields acceptable values for the angles between the axes, then the three rotation operators form a group and meet the condition that $AB = C$. For example, for the combination illustrated in Fig. 3.7, we find that $\cos(A \wedge B) = 1/3$ and that the angle is 70.53°. Operators that form a group and intersect at an unchanged point are written as products. In this case, the group formed by the combination of A, B, and C is named 'ABC'. In this illustration, the unchanged point is found at the center of the cube.

The six distinct groups formed by the combination of the proper rotation axes are: 222, 223, 224, 226, 233, and 234. The conventional names of these groups, illustrated in Fig. 3.8, are 222, 32, 422, 622, 23, 432. The rules that govern this conventional notation will be clarified later in this section. One point that you might notice is that in some cases only two symbols are used. While this notation could be applied to all groups, the third rotation axis is often preserved to clarify the crystal system to which the point group belongs and the relative positions of the elements.

iii.   *Groups containing improper operators*

Improper operators can be combined to form groups under the condition that the final object must be right-handed. The 16 point groups that contain improper operators derive from the groups containing proper rotations by assuming that any of the proper axes could be $\bar{n}$ or could have a parallel $\bar{n}$. To name these groups, we note that the combination of two parallel operators is written as a fraction. For example, the combination of an *n*-fold rotation axis and a parallel $\bar{n}$ axis would be written as $n/\bar{n}$. Furthermore, recalling that the *m* notation represents a mirror operator, when n is an even integer, $n/\bar{n} = n/m$; when *n* is an odd integer, $n/\bar{n} = \bar{n}$. The operator $\bar{2}$ is always written as *m*. As an example, consider the group of proper rotations, 222. If two of the three diads are inversion diads, the group is $\bar{2}\bar{2}2$, which has the conventional name *mm2*. The conventional names for the 16 distinct point groups containing improper operators are:

> 2/*m*
> *mm2*,   *mmm*
> 4/*m*, $\bar{4}2m$,   4*mm*,   4/*m mm*
> $\bar{3}m$,   3*m*
> 6/*m*,   6*mm*, $\bar{6}m2$,   6/*m mm*
> $m\bar{3}$, $\bar{4}3m$, $m\bar{3}m$

This completes the total of 32 point groups.

iv.   **The 32 point groups classified by Bravais lattice**

It is the combination of the crystallographic point groups with the translational symmetry of the Bravais lattices that leads to the formation of the space groups which describe the full symmetry of crystals. Because the rotational symmetry must be consistent with the translational symmetry, only a certain subset of the point symmetry groups can be combined with each Bravais lattice type. The allowed combinations of point group symmetries and Bravais lattices are shown on Table 3.1.

v.   *Point group nomenclature*

The rules for the nomenclature of the point groups are enumerated below [6]. By understanding these rules, you can easily interpret the meaning of the symbols when you encounter them.

Table 3.1. *The crystallographic point groups and the lattice types [6].*

crystal system	Schoenflies symbol	Hermann–Mauguin symbol	order of the group	Laue Group
Triclinic	C_1	1	1	$\bar{1}$
	C_i	$\bar{1}$	2	
Monoclinic	C_2	2	2	$2/m$
	C_s	m	2	
	C_{2h}	$2/m$	4	
Orthorhombic	D_2	222	4	mmm
	C_{2v}	$mm2$	4	
	D_{2h}	mmm	8	
Tetragonal	C_4	4	4	$4/m$
	S_4	$\bar{4}$	4	
	C_{4h}	$4/m$	8	
	D_4	422	8	$4/m\ mm$
	C_{4v}	$4mm$	8	
	D_{2d}	$\bar{4}2m$	8	
	D_{4h}	$4/m\ mm$	16	
Trigonal	C_3	3	3	$\bar{3}$
	C_{3i}	$\bar{3}$	6	
	D_3	32	6	$\bar{3}m$
	C_{3v}	$3m$	6	
	D_{3d}	$\bar{3}m$	12	
Hexagonal	C_6	6	6	$6/m$
	C_{3h}	$\bar{6}$	6	
	C_{6h}	$6/m$	12	
	D_6	622	12	$6/m\ mm$
	C_{6v}	$6mm$	12	
	D_{3h}	$\bar{6}m2$	12	
	D_{6h}	$6/m\ mm$	24	
Cubic	T	23	12	$m\bar{3}$
	T_h	$m\bar{3}$	24	
	O	432	24	$m\bar{3}m$
	T_d	$4\bar{3}m$	24	
	O_h	$m\bar{3}m$	48	

(1) Each component in the name refers to a different direction. For example, the symbol for the orthorhombic group, 222, refers to the symmetry around the x, y, and z axes, respectively.

(2) The position of the symbol m indicates the direction perpendicular to the mirror plane.

(3) Fractional symbols mean that the axes of the operators in the numerator and denominator are parallel. For example, $2/m$ means that there is a mirror plane perpendicular to a rotation diad.

(4) For the orthorhombic system, the three symbols refer to the three mutually perpendicular x, y, and z axes, in that order.

(5) All tetragonal groups have a 4 or $\bar{4}$ rotation axis in the z-direction and this is listed first. The second component refers to the symmetry around the mutually perpendicular x and y axes and the third component refers to the directions in the x–y plane that bisect the x and y axes.

(6) In the trigonal systems (which always have a 3 or $\bar{3}$ axis first) and hexagonal systems (which always have a 6 or $\bar{6}$ axis first), the second symbol describes the symmetry around the equivalent directions (either 120° or 60° apart) in the plane perpendicular to the 3, $\bar{3}$, 6, or $\bar{6}$ axis.

(7) A third component in the hexagonal system refers to directions that bisect the angles between the axes specified by the second symbol.

(8) If there is a 3 in the second position, it is a cubic point group. The 3 refers to rotation triads along the four body diagonals of the cube. The first symbol refers to the cube axis and the third to the face diagonals.

vi.   *The importance of an inversion center*
One final classification is based on the presence or absence of a center of symmetry (an inversion center). Only 11 of the point groups have an inversion center; the other 21 do not. These 11 groups are indicated on Table 3.1 as Laue groups. If a center of symmetry is added to any of the members of a Laue group, it becomes identical to the point group for which the Laue group is named. For example, if a center of symmetry is added to 222 or $mm2$, they are the same as mmm. The presence of a center of symmetry can have a profound effect on some of the physical properties of crystals. For example, the absence of a center of symmetry is an essential requirement for piezoelectricity. Among these piezoelectric crystals, those which polarize spontaneously below a critical temperature are called pyroelectric. If the polarization is reversible, the crystal is ferroelectric.

vii.   *Two-dimensional projections of point groups*
The crystallographic point groups provide a set of instructions (well defined symmetry operations) that allow a pattern to be generated based on a single

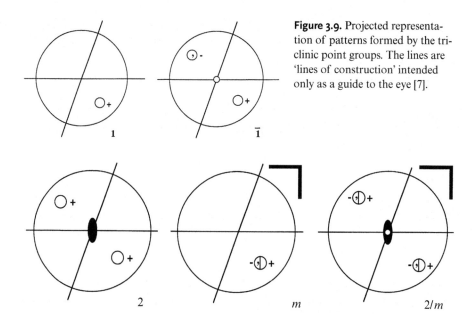

Figure 3.9. Projected representation of patterns formed by the triclinic point groups. The lines are 'lines of construction' intended only as a guide to the eye [7].

Figure 3.10. Projected representation of patterns formed by the monoclinic groups [7].

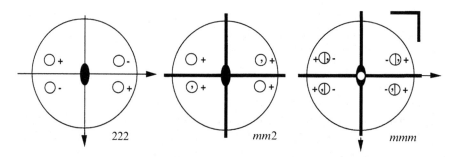

Figure 3.11. Projected representation of patterns formed by the orthorhombic point groups [7].

general coordinate, (x, y, z). In other words, the point group can be thought of as a concise way to list the coordinates of the objects in a pattern. By translating this pattern, or basis, to all of the points of a Bravais lattice, the crystal structure is formed.

The projections shown in Figs. 3.9 through 3.14 are helpful in understanding the symmetry of a group because they show the pattern created by the operators and the relative positions of the different elements [7, 8]. The replicas in the pattern indicate the equivalent positions in a pattern. The outer circle is a

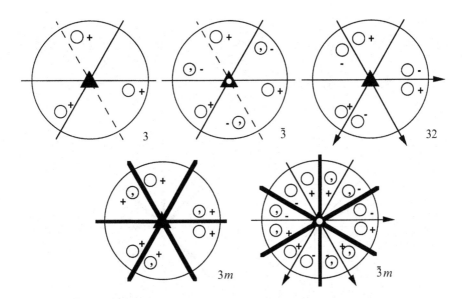

Figure 3.12. Projected representation of patterns formed by the trigonal point groups [7].

reference for construction and the smaller circles are the objects used to form the pattern. Following the convention described earlier, objects divided in half represent two different positions that overlap in projection along the z direction. The plus and minus signs indicate a small positive or negative displacement with respect to the plane of the paper and a comma indicates that it is a left-handed replica. Bold lines indicate a mirror plane and two bold lines, intersecting one another, in the upper right-hand part of the diagram, indicate a mirror in the plane of the paper. Because the 27 noncubic point groups have only one axis with greater than two-fold symmetry, the projection is normally produced along the highest symmetry axis and the other operations are carried out in the plane of the paper (in which case they are represented by arrows for diads and bold lines for mirrors). Cubic point groups have axes inclined to the plane of the paper and are more difficult to depict in two dimensions; two are shown in Fig. 3.15.

viii.   *The cubic point groups*
In the first 27 point groups, the axis with the highest rotational symmetry is unique. However, the five cubic point groups have multiple triad and tetrad axes and it is a challenge to visualize the patterns in two dimensions. Point group 23 has the minimum number of symmetry operations for a cubic group; the group's three mutually perpendicular diads form a convenient set of axes for reference. In order to find the triads, it is helpful to visualize the three diad axes as joining

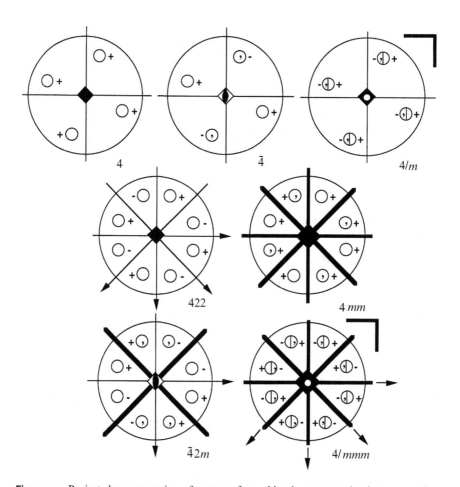

Figure 3.13. Projected representation of patterns formed by the tetragonal point groups [7].

the midpoints of opposite edges of a tetrahedron, as shown in Fig. 3.15. The triads are directed from the four vertices of the tetrahedron to the centers of the opposite faces. The cubic group with maximum symmetry is $m\bar{3}m$. It is useful to remember that in the conventional name, the operators are ordered to describe the symmetry around the cube edge, body diagonal, and face diagonal, respectively. Also, the symbols for the cubic point groups are easily identified because there is always a 3 or a $\bar{3}$ in the second position [9].

ix.   *General and special positions in point groups*
Based on our knowledge of the symmetry elements and their relative orientation, it is possible to generate a list of coordinates of the symmetrically related objects

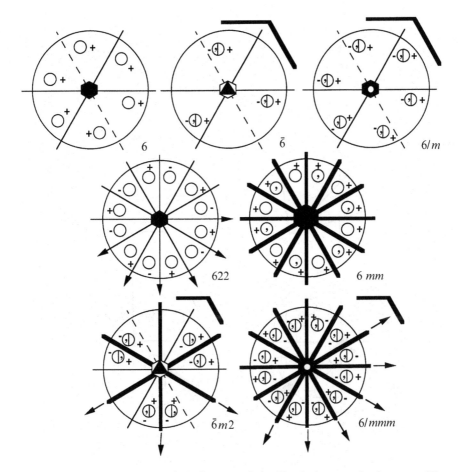

Figure 3.14. Projected representation of patterns formed by the hexagonal point groups [7].

that are consistent with each crystallographic point group. For example, consider a pattern that has point group symmetry 222 (see Fig. 3.11 for reference). The general coordinates for the objects in this pattern are deduced by assuming that one object is initially placed at an arbitrary position given by the coordinates (x, y, z) which does not lie on a symmetry element. In this case, x, y, and z are any fractional coordinates that are greater than zero and less than one. The diad along the z-axis must produce a replica of this object at $(-x, -y, z)$. The diad along the x-axis creates a replica of the initial point at $(x, -y, -z)$ and of the second point at $(-x, y, -z)$. The third diad, along y, is redundant and creates no additional points. So, the general coordinates of the object forming a pattern with point symmetry 222 are: (x, y, z), $(-x, -y, z)$, $(x, -y, -z)$, $(-x, y, -z)$.

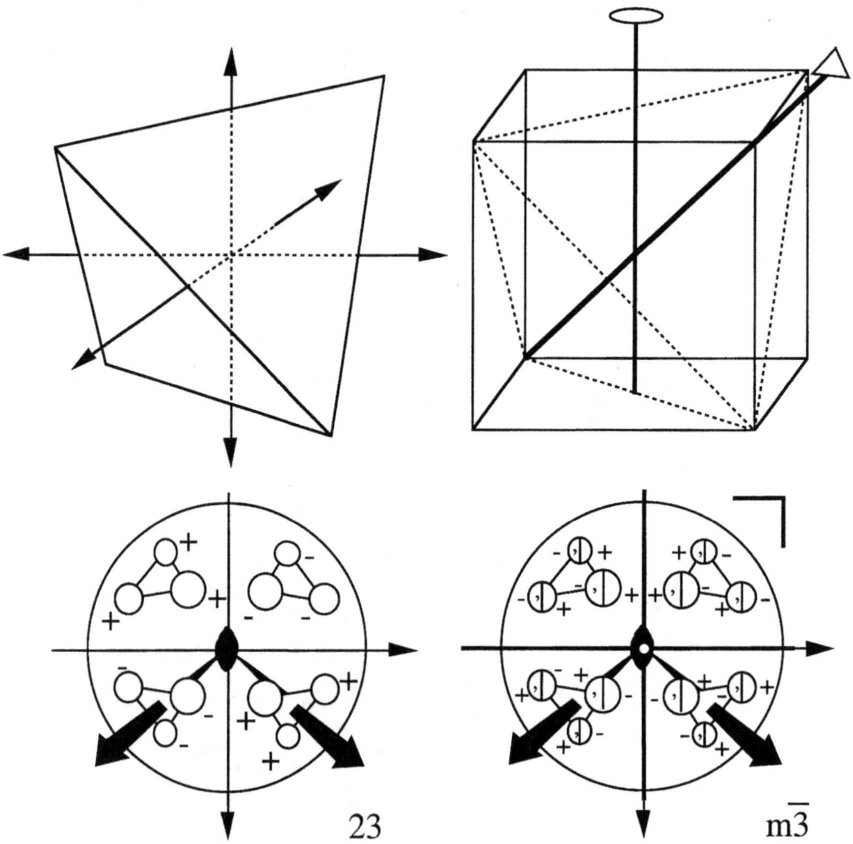

Figure 3.15. The positions of the rotation axes in the cubic point groups and two projections showing equivalent positions. For clarity, only two of the triads are shown in each projection.

Because there are four general coordinates, we say that the order of the group is four. Note that once the point group is specified, only three numerical values are needed to completely determine the coordinates of all of the objects in the pattern; this is the importance of symmetry.

It was noted above that to produce the general positions of the objects in the pattern, the initial coordinate must not lie on a symmetry element. If it does, then a more restrictive list of special coordinates is generated. For example, in group 222, if we assume that the initial coordinate is $(0, y, 0)$, instead of (x, y, z), then the initial point lies on the diad along y. While the diads along x and z produce the same replica at $(0, -y, 0)$, the third diad does not produce a replica.

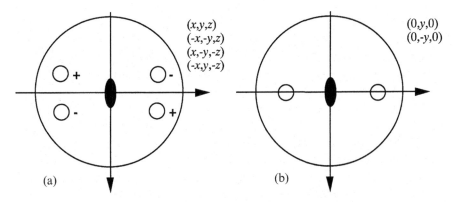

(a)

(x,y,z)
(-x,-y,z)
(x,-y,-z)
(-x,y,-z)

(b)

(0,y,0)
(0,-y,0)

Figure 3.16. Illustration of (a) the general equivalent positions in point group 222 and (b) a set of special equivalent positions.

Therefore, this pattern has only two points instead of four. The coordinates (0, y, 0) and (0, $-y$, 0) represent special positions in the group 222. To summarize, the general equivalent positions of a group have the maximum multiplicity and do not lie on a symmetry element. Special equivalent positions, however, lie on symmetry elements, have restricted coordinates (containing zeros, fractional, or duplicated values), and a reduced multiplicity with respect to the general coordinates. General and special coordinates in group 222 are illustrated in Fig. 3.16.

x. *Determining the crystallographic point group of an object*
When examining a pattern, it is possible to specify the point group that describes its symmetry by using the following procedure outlined by Sands [10]. First, check for the highest or lowest symmetry groups. If no rotational symmetry axes exist, the group is 1, $\bar{1}$, or m. On the other hand, if the four triad axes characteristic of the cubic point groups are present (these are usually the easiest to recognize), then the object has the symmetry of one of the cubic point groups. To determine which of the five cubic point groups it is, systematically search for the additional elements. If it is neither cubic nor triclinic, then search for the axis of highest rotational symmetry. If it has a rotation hexad, it is hexagonal. If it has a single triad or tetrad axis, then it is trigonal or tetragonal, respectively. Remember, the simultaneous occurrence of these two elements would imply a cubic group. If the object's highest symmetry element is a diad and there are two mutually perpendicular elements, it is orthorhombic. If not, it is monoclinic. In each case, once the crystal system has been identified, the presence or absence of perpendicular diads and mirrors tells you which group it is.

Example 3.1

Specify the coordinates of the general equivalent positions in the point group 422. What point group is created if a center of symmetry is added to 422? Specify the general equivalents in this group.

1. The general positions in a group are generated by carrying out all of the symmetry operations on the coordinates (x, y, z). From the symbol, we know that this is a tetragonal group and that a tetrad axis exists along z, diads exist along x and y, and two more diads exist along the directions that bisect the x and y axes, $\langle 110 \rangle$.

2. Using the rotation operator specified by Eqn. 3.1, and the angles $\pi/2$, π, and $3\pi/2$, it has already been shown in Eqn. 3.2 that point 1 (x, y, z) generates points 2 $(-y, x, z)$, 3$(-x, -y, z)$, and 4 $(y, -x, z)$.

3. Next, we rotate each of these by π radians around the x axis using g_x:

$$
g_x = \begin{bmatrix} 1 & 0 & 0 \\ 0 & \cos\theta & \sin\theta \\ 0 & -\sin\theta & \cos\theta \end{bmatrix}; \quad g_x(\pi) = \begin{bmatrix} 1 & 0 & 0 \\ 0 & -1 & 0 \\ 0 & 0 & -1 \end{bmatrix}
$$

$$
[x\, y\, z]\begin{bmatrix} 1 & 0 & 0 \\ 0 & -1 & 0 \\ 0 & 0 & -1 \end{bmatrix} = \begin{bmatrix} x \\ -y \\ -z \end{bmatrix}; \quad [-y\, x\, z]\begin{bmatrix} 1 & 0 & 0 \\ 0 & -1 & 0 \\ 0 & 0 & -1 \end{bmatrix} = \begin{bmatrix} -y \\ -x \\ -z \end{bmatrix}; \quad (3.4)
$$

$$
[-x\, -y\, z]\begin{bmatrix} 1 & 0 & 0 \\ 0 & -1 & 0 \\ 0 & 0 & -1 \end{bmatrix} = \begin{bmatrix} -x \\ y \\ -z \end{bmatrix}; \quad [y\, -x\, z]\begin{bmatrix} 1 & 0 & 0 \\ 0 & -1 & 0 \\ 0 & 0 & -1 \end{bmatrix} = \begin{bmatrix} y \\ x \\ -z \end{bmatrix}.
$$

4. The third operator is redundant and generates no new points. So, the coordinates of the eight general equivalents are:

$1 (x, y, z); 2 (-y, x, z); 3 (-x, -y, z); 4 (y, -x, z)$
$5 (x, -y, -z); 6 (-y, -x, -z); 7 (-x, y, -z); 8 (y, x, -z)$

The pattern formed by these coordinates is shown in Fig. 3.17(a).

5. Beginning with the pattern in 3.17(a), we can add a center of symmetry by inverting each point through the origin to get the pattern in 3.17(b), which we recognize as 4/*mmm*.

6. To specify these coordinates, we simply carry out the inversion operation by changing the signs of all of the coordinates. For example, this operation takes (x, y, z) to

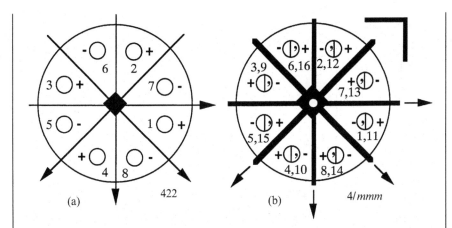

Figure 3.17. Locations of the general positions in the patterns generated by (a) 422 and (b) $422 + i = 4/mmm$.

$(-x, -y, -z)$. Each of the eight coordinates above creates a new set of coordinates, making a total of 16. The additional coordinates are:

$$9\,(-x, -y, -z);\ 10\,(y, -x, -z);\ 11\,(x, y, -z);\ 12\,(-y, x, -z);$$
$$13\,(-x, y, z);\ 14\,(y, x, z);\ 15\,(x, -y, z);\ 16(-y, -x, z).$$

(D) The 230 space groups

The 32 crystallographic point groups, whose operations leave at least one point unchanged, are sufficient for the description of finite objects. However, since ideal crystals extend indefinitely in all directions, we must also include translations in our description of symmetry. The combination of a point group and translational symmetry operators (Bravais lattice vectors) leads to the formation of a space group. The combination of point symmetries and translations also leads to two additional operators known as screw and glide. The screw and glide operators are described below.

i.   The screw operator

The screw operation is a combination of a rotation and a translation parallel to the rotation axis [11]. As for simple rotations, only diad, triad, tetrad, and hexad axes that are consistent with the Bravais lattice translation vectors can be used for a screw operator. In addition, the translation on each rotation, $\vec{r}$, must be a rational fraction of the entire translation, $\vec{R}$. The symbol for the screw operator is N_q, where N is the rotational operator, q is rN, and $q < N$. Figs. 3.18 and 3.19 show the 11 possible screw axes in oblique and plane projection, respectively.

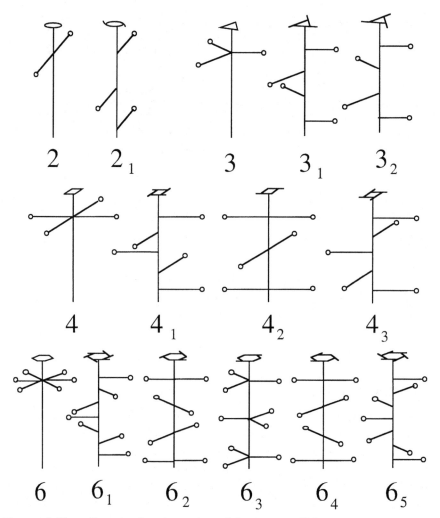

Figure 3.18. Three-dimensional representations of the screw axes [11].

Note that the screw operation can be clockwise or counterclockwise and this results in a handedness or chirality to the pattern. For example, there is no combination of rotations or translations that can transform the pattern produced by 4_1 to the pattern produced by 4_3.

ii. *The glide operator*
Glide is the combination of a translation and a mirror, as shown in Fig. 3.20. In this figure, the dashed, bold line labeled '*g*' is the mirror component of the glide

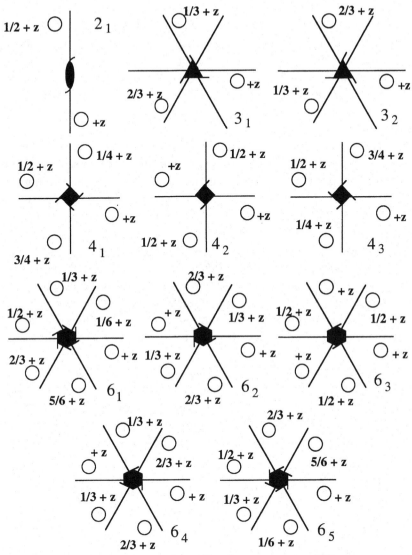

Figure 3.19. Projections of the patterns formed by the screw operators [12].

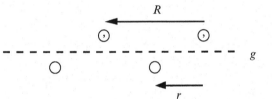

Figure 3.20. The action of the glide operator. In this case, $1/2R = r = 1/2a$.

operator. Glide must, of course, be compatible with the translations of the Bravais lattice. Thus, the translation components of glide operators must be rational fractions of lattice vectors. In practice, the translation components of a glide operation are always 1/2 or 1/4 of the magnitude of the translation vectors.

If the translation is parallel to a lattice vector, it is called axial glide. Glide planes with translations $a/2$, $b/2$, or $c/2$ are designated with the symbols a, b, and c, respectively. Another type of glide, called diagonal glide, is indicated by the symbol n and has translation components of $a/2 + b/2$, $b/2 + c/2$, or $a/2 + c/2$. For any particular glide plane, the appropriate translation is the one parallel to the plane of the mirror. In the tetragonal, rhombohedral, and cubic systems, the translation component of the n glide operator is $a/2 + b/2 + c/2$ when the mirror is along a $\langle 110 \rangle$-type direction. Finally, we consider the relatively more rare diamond glide. Diamond glide planes, designated by the symbol d, are found in only two orthorhombic groups, five tetragonal groups and five cubic groups. In the orthorhombic groups, the translation components for diamond glide are $a/4 + b/4$, $a/4 + c/4$, or $b/4 + c/4$, depending on the orientation of the mirror (the translation is always parallel to the plane of the mirror). In tetragonal and cubic groups, the translation component is $a/4 + b/4 + c/4$. Groups containing diamond glide are always F or I centered. While there are only two equivalent positions generated by axial and diagonal glide, diamond glide generates four equivalent positions [12].

iii *The origin of the space groups and their symbols*

The 230 symmetry groups, called space groups, are used to describe three-dimensional, infinite crystals. Each space group contains both a point group (one of the 32 that we discussed earlier) and translational symmetry in the form of Bravais lattice translation vectors and, in many cases, screw axes and/or glide planes.

Instead of attempting to derive the 230 space groups, we simply outline the reasoning [15]. The first step is to consider one of 32 point groups, for example $2/m$, and combine it with all the possible consistent Bravais lattice types. In this case, we know that $2/m$ is a monoclinic point group and that there are two monoclinic Bravais lattices, P and C. From this combination, we get 2 space groups, P2/m and C2/m. Next, we consider the possibility that each rotation axis can be replaced by a screw (with the same rotation component) and that each mirror plane can be replaced by a glide plane. In this process, many duplications are generated and must be eliminated. For example, P $2/m$ creates P $2/a$ and P $2/c$ which turn out to be identical, except for the naming of the axes. Eliminating the duplications and choosing the conventional names are the challenging parts of this process. The 13 unique monoclinic space groups that are derived from the three monoclinic point groups and the two monoclinic Bravais lattices are illustrated

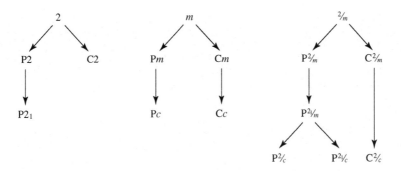

Figure 3.21. Derivation of the monoclinic space groups from the monoclinic point groups.

in Fig. 3.21. When this process is repeated for the other six crystal classes, the 230 space groups listed in Appendix 3A are derived.

Combining the Bravais lattice translations can be more complex than presented above. Consider point group *mm2*. This is an orthorhombic point group, so there are four possible Bravais lattices that lead to four space groups, P*mm2*, I*mm2*, F*mm2*, and C*mm2*. However, C*mm2* should be considered further. 'C' specifies that one face is centered and that it is the face perpendicular to the *z* axis. The fact that there is a 2 in the last position tells you that there is a diad parallel to *z*, normal to the centered face. Therefore, this is distinct from the *A* and *B* groups so you must also consider the possibility that the diad lies in the plane of the centered face. This would give the space group the name A*mm2*. So, the point group *mm2* creates five distinct space groups.

iv. *The space group symbols*
Before continuing, it is important to make a few clarifying statements regarding the space group symbols that we have been using (for example, P*mm2*). First, these symbols are the standard, internationally recognized Hermann–Mauguin symbols. Although they are the best ones for a solid state scientist to know, they are not the only ones. Other common symbols are the extended forms and the Schoenflies forms. Second, although the symbol does not show all existing symmetry operators, it shows a sufficient number to allow all of the equivalent positions to be specified. Third, the space group symbols always begin with a capital letter indicating the type of lattice centering. The following characters, when subscripts are dropped from the screw axes and glide designations replaced by simple mirrors, are the symbol for the point group from which the space group was derived. The conventional names of the 230 space groups are given in Tables 3A.1 through 3A.7. By looking at any of these names, you should be able to immediately identify the Bravais lattice type and the point symmetry group.

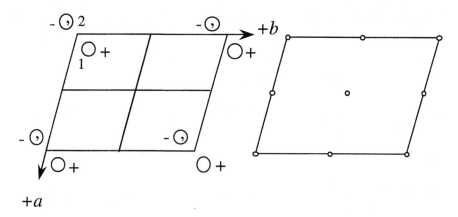

Figure 3.22. Equivalent positions and symmetry elements in the group P1̄. The equivalent positions are (x, y, z) and $(\bar{x}, \bar{y}, \bar{z})$ [17].

ν *Illustrations of space groups*

The space group is the central component of any conventional crystal structure description and it is absolutely essential to understand the notations and conventions used in this description. Graphic representations which show the relative orientation of the symmetry elements and the general positions of the group are useful for this purpose. Projections of all of the space groups can be found in the *International Tables for Crystallography*, Vol. 4, *Space Group Symmetry* [14]. Graphic representations of selected examples of space groups are presented here so that the range of possibilities and the different conventional symbols can be illustrated [15–17]. In the final section of this chapter, we will use this information to interpret crystal structure data.

vi. *Triclinic groups*

Diagrams of space groups include two parallelograms with the origin in the top left corner and the z axis normal to the paper, as shown in Fig. 3.22, a representation of space group no. 2, P1̄. The parallelogram on the left shows the equivalent positions and the one on the right shows the relative positions of the symmetry elements. The primary symmetry element of the group P1̄ is the inversion center at the origin. The other inversion centers at the edge, face, and body centered positions are a consequence of the translational symmetry.

The general equivalent positions in a group are found by carrying out the symmetry operations in a group's name on a point (x, y, z), where x, y, and z are small positive numbers less than one. For example, point 2 in Fig. 3.22 is generated by applying the inversion operator to point 1. As with our diagrams for the point groups, the minus sign at point 2 indicates a small negative value for its z coordinate and the comma indicates that the object is enantiomorphous or left-handed relative to point 1. Unit cell translations of point 1 and point 2 generate the additional coordinates. Conventionally, only equivalent positions in the same unit cell are listed. In this case, there are only two equivalent positions. The co-ordinates of the first point are (x, y, z). The point labeled 2, at $(-x, -y, -z)$, can be translated within the unit cell boundaries by adding the vector $\vec{a}+\vec{b}+\vec{c}$. The new coordinate is $(1-x, 1-y, 1-z)$ which, by convention, is written as $(\bar{x},\bar{y},\bar{z})$. So, the two general equivalent positions in this space group are (x,y,z) and $(\bar{x},\bar{y},\bar{z})$. There are also eight special positions, each of which has a multiplicity of one. Recall that special positions lie on a symmetry element. In this case, the only symmetry elements are the inversion centers and they are located at: $(0,0,0)$, $(1/2,0,0)$, $(0,1/2,0)$, $(0,0,1/2)$, $(1/2,1/2,1/2)$, $(0,1/2,1/2)$, $(1/2,0,1/2)$, $(1/2,1/2,0)$.

vii   Monoclinic groups

Space group no. 5, C2, is an end-centered monoclinic cell. Remember, the desig-nation 'C' means that there is a lattice position in the a–b plane at $(1/2,1/2,0)$. It is an unfortunate convention that the unique axis in the monoclinic system is b, unlike the tetragonal, trigonal, and hexagonal systems where the unique axis is c. Thus, the conventional c-axis projection shows a rectangle, as in the upper part of Fig. 3.23. Although older versions of the International Tables show only the c-axis projection, the newest version includes multiple projections. The b-axis projection of C2 is illustrated in the lower portion of Fig. 3.23.

To generate the equivalent positions, we start with the general position (x, y, z), labeled 1. The diad coincident with the b-axis generates the point labeled 2. When the C-centering operation $(x+1/2, y+1/2, 0)$ is applied to points 1 and 2, points 3 and 4 are generated.

It is noteworthy that the C-centering and the diad create new symmetry ele-ments not specified in the name of the group. This is common. In this case, there is a diad at $x-1/2$, $z=0$, and parallel to y and there are screw diads at $x=1/4$ and $x=3/4$, $z=0$ (note the symbols used in Fig. 3.23). Recognition of these addi-tional symmetry elements is challenging without some practice and experience. However, remember that you will always be able to generate all of the equivalent positions using only the symmetry operators in the space group name. Thus, for the generation of positions, the recognition of the additional elements is not nec-essary. The recognition of the additional symmetry elements can, however, become important for understanding the systematic absences in diffraction pat-terns (see Chapter 5).

The general equivalent positions in C2 are: 1 (x, y, z), 2 $(-x, y, -z)$, which is translated to $(1-x, y, 1-z)$ and equals $(\bar{x}, y, \bar{z})$, 3 $(x+1/2, y+1/2, z)$, and

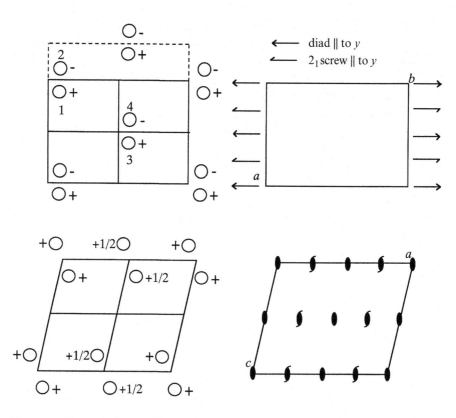

Figure 3.23. The equivalent positions and symmetry elements in the group C 2. The upper part of the diagram shows the projection onto the *a–b* plane while the lower part shows the projection down the *b*-axis. The dashed lines are intended to indicate the location at which the top face of the cell (at $z=1$) projects onto the *a–b* ($z=0$) plane.

4 ($1/2 - x$, $y + 1/2$, $-z$), which is translated to ($1/2 - x$, $y + 1/2$, $1 - z$), and equals ($1/2 - x$, $y + 1/2$,). By convention, it is customary to list only points 1 and 2, (x, y, z) and ($\bar{x}$, y, $\bar{z}$). The other two points are implied by the C operation. This convention holds for all groups with lattice centering; it is assumed that you know the lattice translation vectors implied by the symbols A, B, C, I, F, and R (see Table 2.2). Special positions, (0, y, 1/2) and (0, y, 0), exist on the diads at $x = z = 0$ and at $x = 0$, $z = 1/2$. There are additional special positions at (1/2, $y + 1/2$, 1/2) and (1/2, $y + 1/2$, 0) which are created by the C operation. Finally, we note that this group illustrates a general rule that applies to all groups that have no symmetry greater than two-fold. For every symmetry plane or diad, there is another element of the same type, in the same direction, midway between the first and its translation repeat.

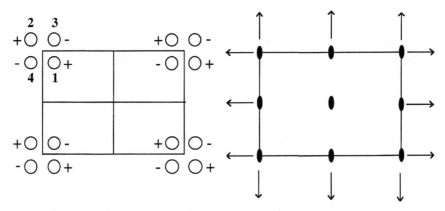

Figure 3.24. The equivalent positions and symmetry elements in the group P222, where the three diads intersect.

viii. *Orthorhombic groups*

In orthorhombic space groups, there is an ambiguity regarding whether or not all three symmetry axes intersect. Some groups possess intersecting axes, while others do not. The ambiguity can be resolved either by checking the International Tables [14] or by considering the positions generated by the specified axes. Since any two of the axes generate the third, it is not too difficult to explore their spatial relationship. For example, consider P222, space group no. 16, a primitive orthorhombic group shown in Fig. 3.24. We can use the two intersecting in-plane diads to generate points 2, 3 and 4 from 1. From the generation of these four equivalent points, it is obvious that it is entirely consistent to make the final diad intersect the origin. So, internal consistency shows that the three diads intersect at a single point in this space group.

The same is not true for the related group, P222$_1$, shown in Fig. 3.25. When the screw axis is present, the x and y diads do not intersect at $z=0$. The diad along y is displaced by $z-1/4$ along the z axis. Using the screw diad, point 2 is generated from point 1. Using the diad along the a-axis (at $z=0$), point 1 generates point 3 and point 2 generates point 4. From this pattern, we can see that the diad along the b-axis must be at $z=1/4$. Thus, the three diads do not intersect. The general positions are (x, y, z), $(x, \bar{y}, \bar{z})$, $(\bar{x}, \bar{y}, 1/2+z)$, $(\bar{x}, y, 1/2-z)$.

There are a few anomalous cases where the space group symbols are ambiguous without knowledge of a convention [18]. Consider, for example, P222, the space group mentioned earlier that has intersecting diads. Performing the body centering operation to get I222, no. 23, has the usual effect of producing additional symmetry elements parallel to the named elements. In this case, a set of three mutually perpendicular screw diads are created. We must then ask, why

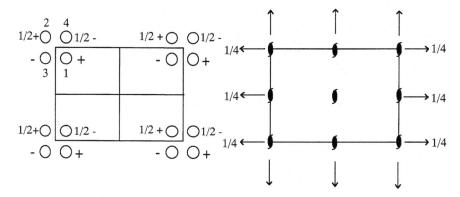

Figure 3.25. The equivalent positions and symmetry elements in the group P222$_1$. In this case, the three diads do not intersect [17].

should the group labeled I2$_1$2$_1$2$_1$, which also contains three mutually perpendicular screw diads, be any different from I222? It turns out that the screw axes in the group labeled I2$_1$2$_1$2$_1$ generate nonintersecting diads. The conventional names were assigned arbitrarily so that I2$_1$2$_1$2$_1$ has nonintersecting rotation diads and I222 has intersecting rotation diads. A similar situation occurs for the cubic space groups I23 and I2$_1$3.

Space group no. 33, Pna2$_1$, has two types of glide operators. By the positions of the operators in the space group symbol, the orientation of the mirrors can be determined. The normal of the mirror associated with the diagonal glide plane is along the a axis. The translation is parallel to this plane and has the magnitude $b/2 + c/2$. The axial glide in this group has a mirror whose normal is along b and a translation of $a/2$. Note in Fig. 3.26 that the screw diad does not intersect the glide planes. The general positions for this group are (x, y, z), $(\bar{x}, \bar{y}, \bar{z} + 1/2)$, $(x + 1/2, \bar{y} + 1/2, z)$, and $(\bar{x} + 1/2, y + 1/2, z + 1/2)$.

Centering operations create additional symmetry operators. For example, in Fdd2 (see Fig. 3.27), there are numerous rotation diads, screw diads, and diamond glide planes that generate 16 general equivalent positions. The general equivalents are (x, y, z), $(\bar{x}, \bar{y}, z)$, $(x + 1/4, \bar{y} + 1/4, z + 1/4)$, and $(\bar{x} + 1/4, y + 1/4, z + 1/4)$; by convention, the remaining 12 general equivalent positions, implied by the F in the space group symbol, are obtained by adding $(1/2, 1/2, 0)$, $(1/2, 0, 1/2)$, and $(0, 1/2, 1/2)$ to each of the first four.

ix. Tetragonal groups

The tetragonal space group I41, number 80, is illustrated in Fig. 3.28. Note that the principal axis of rotation, the screw tetrad, is not placed at the unit

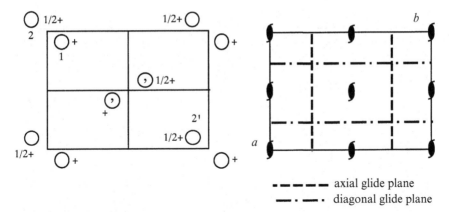

Figure 3.26. The equivalent positions and symmetry elements in space group number 33, P$na2_1$. Note the two different kinds of glide planes.

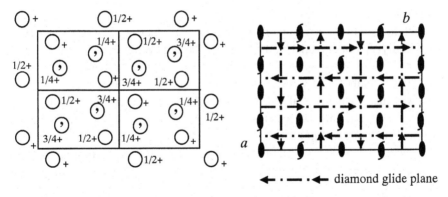

Figure 3.27. The equivalent positions and symmetry elements in space group number 43, F$dd2$. The centering operations create many additional symmetry operators.

cell origin. A new cell can be defined, with axes rotated by 45° with respect to the *a* and *b* axes shown in Fig. 3.28, that has screw tetrads on the corners. In this case, we see that another screw tetrad is generated in the center of the new square and some kind of axes (tetrad or diad) are generated at the midpoints of the edges. In this case, note that the central screw tetrad rotates in the opposite direction from the screw tetrads at the vertices; this is a 4_3 axis. The general equivalent positions in this group are (x, y, z), $(\bar{x} + 1/2, \bar{y} + 1/2, z + 1/2)$, $(\bar{x}, x + 1/2, z + 1/4)$, and $(y + 1/2, \bar{x}, z + 3/4)$. Four additional positions are generated by adding the translation implied by the I symbol, $(1/2, 1/2, 1/2)$, to each point.

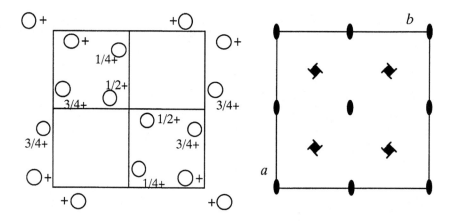

Figure 3.28. The equivalent positions and symmetry elements in the group I4₁ [17].

x. Trigonal and hexagonal groups

The relationship between trigonal and hexagonal cells is worthy of special consideration [19]. For crystals that have one of the 25 trigonal space group symmetries, it is always possible to choose a hexagonal cell ($a=b\neq c$ and $\alpha=\beta=90°$, $\gamma=120°$) instead of a trigonal cell ($a=b=c$ and $\alpha=\beta=\gamma\neq90°$). By convention, we always use the hexagonal cell. This is for the simple practical reason that the hexagonal cell has a unique axis normal to the a–b plane and is, therefore, easier to visualize than the trigonal cell.

There are two types of trigonal space groups. Eighteen groups are primitive (P) and, in these cases, a primitive hexagonal cell is used. The remaining seven are rhombohedral (R) and, in these cases, the smallest hexagonal cell that can be assigned has a volume that is three times larger than that of the primitive trigonal cell. The relationship between the primitive trigonal lattice vectors (with the T subscripts) and the rhombohedral lattice vectors (with the R subscripts) is illustrated in Fig. 3.29.

In cases where the crystal structure is described in terms of a trigonal cell, the rhombohedral vectors can be found using Eqn. 3.5.

$$\vec{a}_R = \vec{a}_T - \vec{b}_T$$
$$\vec{b}_R = \vec{b}_T - \vec{c}_T$$
$$\vec{c}_R = \vec{a}_T + \vec{b}_T + \vec{c}_T. \tag{3.5}$$

Also, there are three times as many points in the R cell than in the primitive cell. For any point in the primitive trigonal cell specified by the position vector $\vec{P}_T$,

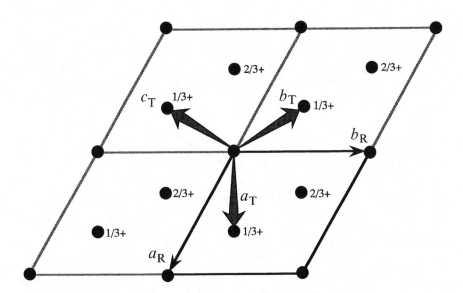

Figure 3.29. The relationship between the trigonal P cell and the hexagonal cell of the rhombohedral lattice. This view shows a projection down the c axis of the hexagonal cell and the (111) axis of the trigonal cell. There are three lattice points in each R cell, one at the origin, one at (2/3,1/3,1/3) and another at (1/3,2/3,2/3) [19].

$$\vec{P}_T = x\vec{a}_T + y\vec{b}_T + z\vec{c}_T, \tag{3.6}$$

there are three positions in the rhombohedral cell given by the three vectors $\vec{P}_{Ri}$:

$$\vec{P}_{R1} = \left(\frac{2x}{3} - \frac{y}{3} - \frac{z}{3}\right)\vec{a}_R + \left(\frac{x}{3} + \frac{y}{3} - \frac{2z}{3}\right)\vec{b}_R + \left(\frac{x}{3} + \frac{y}{3} + \frac{z}{3}\right)\vec{c}_R$$

$$\vec{P}_{R2} = \vec{P}_{R1} + \left(\frac{2}{3}\right)\vec{a}_R + \left(\frac{1}{3}\right)\vec{b}_R + \left(\frac{1}{3}\right)\vec{c}_R$$

$$\vec{P}_{R3} = \vec{P}_{R1} + \left(\frac{1}{3}\right)\vec{a}_R + \left(\frac{2}{3}\right)\vec{b}_R + \left(\frac{2}{3}\right)\vec{c}_R. \tag{3.7}$$

A primitive trigonal group, P3m1 (no. 156), is represented in Fig. 3.30. Its six general equivalent positions, generated by the rotation triad and mirror, are (x, y, z), $(\bar{y}, x-y, z)$, $(\bar{x}+y, \bar{x}, z)$, $(\bar{y}, \bar{x}, z)$, $(\bar{x}+y, y, z)$, and $(x, x-y, z)$.

A rhombohedral group, R3m (no. 160), is shown in Fig. 3.31. For the rhombohedral groups, it is conventional to show the projection down the c-axis of the hexagonal cell together with the (111) projection of the trigonal cell. This group has 18 general equivalent positions. The six associated with the (0, 0, 0) point are

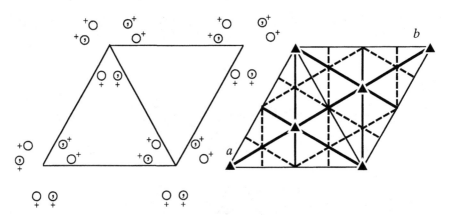

Figure 3.30. Equivalent positions and symmetry elements in space group P3m1. By convention, the axes of the primitive hexagonal cell are shown instead of the trigonal axes.

(x, y, z), $(\bar{y}, x-y, z)$, $(\bar{x}+y, \bar{x}, z)$, $(\bar{y}, \bar{x}, z)$, $(\bar{x}+y, y, z)$, and $(x, x-y, z)$. The remaining twelve are found by adding the translations (2/3, 1/3, 1/3) and (1/3, 2/3, 2/3) to these coordinates.

An example of a hexagonal space group (P6, no. 168) is shown in Fig. 3.32. A rule for hexagonal groups is that if there are hexad axes at the corners of the unit cell, a triad is generated at the center of each equilateral triangle and rotation diads are found on the triangles' edges, midway between the hexads [15]. The general equivalent positions in this group are (x, y, z), $(\bar{y}, x-y, z)$, $(\bar{x}+y, \bar{x}, z)$, $(\bar{x}, \bar{y}, z)$, $(y, \bar{x}+y, z)$, and $(x-y, x, z)$.

xi. *Choice of origin*

In order to correctly interpret and use conventional crystal structure data, it is essential to identify the origin with respect to the symmetry elements. In centrosymmetric space groups, it is customary to place the origin at a center of symmetry. This sometimes means that other important symmetry elements are shifted from the origin. When a number of reasonable choices exist, the *International Tables* [14] show them separately as different settings.

xii. *Information in the* International Tables

While it is always possible to deduce the relative positions of symmetry operators and lists of general and special positions from the space group symbol, differences in the choice of origin, labels for the orthogonal axes, and labels for the different sets of special positions can sometimes cause confusion when examining literature data or using crystallographic software. For this reason, it is always useful to check the data in the *International Tables*. Two particularly

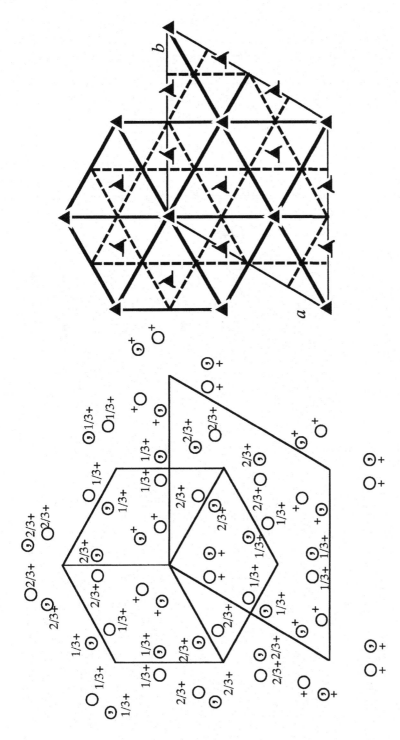

Figure 3.31. The positions of the symmetry elements and general positions in space group no. 160, R3*m*.

Table 3.2. *All positions in space group no. 17, P222₁.*

multiplicity	Wyckoff letter	point symmetry	coordinates	
4	e	1	x, y, z	$\bar{x}, \bar{y}, z + 1/2$
			$\bar{x}, y, \bar{z} + 1/2$	$x, \bar{y}, \bar{z}$
2	d	2	$1/2, y, 1/4$	$1/2, \bar{y}, 3/4$
2	c	2	$0, y, 1/4$	$0, \bar{y}, 3/4$
2	b	2	$x, 1/2, 0$	$\bar{x}, 1/2, 1/2$
2	a	2	$x, 0, 0$	$\bar{x}, 0, 1/2$

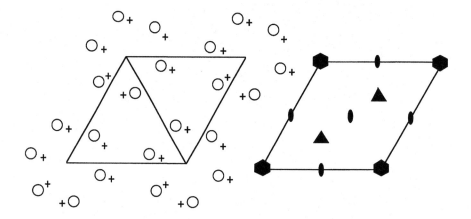

Figure 3.32. The equivalent positions and symmetry elements in the group P6.

useful pieces of information are projections of the operators in different orientations and tabulated lists of all the general and special positions.

For example, the possible positions in space group no. 17, P222₁, are reproduced in Table 3.2. The first column lists the multiplicity of the positions. The general positions always have the highest multiplicity and are listed first. The special positions have reduced multiplicities and follow. The Wyckoff letter is used to name the different possible sites in each group. The letter 'a' is used for the most symmetric site. The least symmetric sites (the general positions) are given the last letter of the alphabetic sequence (in this case, 'e'). When describing an atomic position in a crystal structure, one typically refers to the multiplicity and the Wyckoff letter. In most cases, atoms occupy only a subset of the possible sites. Next, the point symmetry of each site is given, and then the coordinates. Note that when 0 or a fractional coordinate value is used, the value is

Table 3.3 *The crystal structure of La$_2$CuO$_4$ (Lanthanum Cuprate) [20].*

Formula unit:	La$_2$CuO$_4$, K$_2$NiF$_4$-type
Space group:	I4/*mmm*
Cell dimensions:	$a = 3.7873$ Å, $c = 13.2883$ Å
Cell contents:	2 formula units
Atomic positions:	La in (4e): $(0, 0, z)$; $(0, 0, \bar{z})$; $z = 0.3606$
	Cu in (2a): $(0, 0, 0)$
	O(1) in (4c): $(0, 1/2, 0)$; $(1/2, 0, 0)$
	O(2) in (4e): $(0, 0, z)$; $(0, 0, \bar{z})$; $z = 0.1828$

exact and fixed by the symmetry of the group. When the variables x, y, or z are used, the coordinate is free to take any value between zero and one and must be experimentally determined. Note that the first set of coordinates for the general positions is always (x, y, z) and that the special positions always have restricted coordinates that contain zeros, fractions, or constraints (for example, the set of coordinate (x, x, z) constrains y to be equal to x).

(E) The interpretation of conventional crystal structure data

Figure 2.31 shows two possible representations of the La$_2$CuO$_4$ structure. While such structural pictures are informative in a qualitative way, they are not systematic and it is impossible to derive quantitative information (for example, bond lengths) from such pictures. Table 3.3, on the other hand, shows how this crystal structure would be represented in the archival literature. These data, combined with a model or drawing, tell all that you need to know about a structure. In this section, we will demonstrate, using two illustrative examples, how data in the standard format (such as Table 3.3) are used to produce a model and how the atomic structure is related to the space group symmetry.

Before beginning, we should point out that crystal structure data are available from a variety of sources, most of which can be found in any technical library. For example, if a structure has been known since the early 1960s, it can probably be found in the set of reference books entitled 'Crystal Structures' by Ralph W.G. Wyckoff, volumes 1 through 6, (John Wiley & Sons, New York, 1964) or one of the periodic releases of Structure Reports. For metallic structures, W.B. Pearson's 'A Handbook of Lattice Spacings and Structures of Metals and Alloys', (Pergamon, Oxford, 1967) is very useful. If the crystal structure has been determined more recently, it will have been published in one of several periodical journals. For example, many crystal structures are reported in *Acta Crystallographica*.

To illustrate how conventional crystal structure data are interpreted, we consider the structure of lanthanum cuprate as described in Table 3.3. Based on the space group symbol, I4/*mmm*, we know that this structure has an I centered tetragonal Bravais lattice and that it is derived from the point symmetry group 4/*mmm*. Because it is assumed that the reader recognizes this symbol and knows the geometry of the tetragonal lattice ($a = b \neq c$, $\alpha = \beta = \gamma = 90°$), only the a and c parameters are listed in the table. The basis is always made up of an integer number of formula units of the compound and, in this case, the table tells us that the basis contains two La_2CuO_4 units. Therefore, we can compute the density of this compound by dividing the masses of 4 La, 2 Cu, and 8 O atoms by the volume of the cell, a^2c.

Table 3.3 tells us that the atoms occupy 2a, 4c and 4e sites. These are Wyckoff designations for three different special positions in the space group. The coordinates are usually given in their most general form, using variable parameters such as x, y, and z together with fixed fractions. In this case, only the coordinate z is needed for the La and O(2) site. The experimentally determined values of the variable parameters for each atom type are then listed as decimal numbers. Note that for sites with a multiplicity of 4, only two coordinates are listed, and for sites with a multiplicity of 2, only a single site is listed. By convention, positions generated by centering are not listed; it is assumed that the reader understands that the symbol I implies that (1/2, 1/2, 1/2) must be added to each coordinate to generate the second half of the occupied sites.

Note that there are 14 atoms in the unit cell. Each of these potentially has three variable fractional coordinates. These 42 coordinates, combined with the six lattice parameters, make 48 potential parameters that must be determined to specify this structure. However, knowledge of the symmetry allows us to reduce this to only four parameters: two lattice constants and two position parameters. Thus, the reward for understanding symmetry is that it allows a complex configuration of atoms to be described in a simple, yet exact, manner. The variable parameters are determined using X-ray diffraction and the remaining parameters are fixed by the conditions of the space group.

While the information in Table 3.3 completely specifies the structure, it is only possible to understand the atomic arrangement by producing a sketch. The first step in sketching a model of the structure is to determine the atomic coordinates for all the atoms in the cell and on the borders of the cell. The four La atoms are at (0, 0, 0.3606), (0, 0, 0.6394), (1/2, 1/2, 0.8606), and (1/2, 1/2, 0.1394). For the purposes of making a sketch, it is often practical to approximate the exact coordinates as simple fractional values. For example, we might say that the

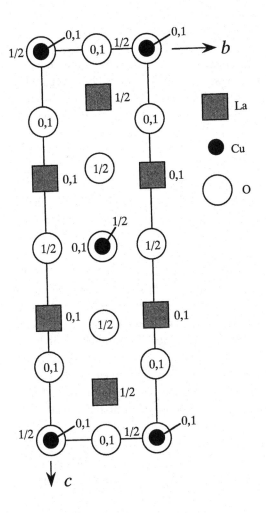

Figure 3.33. (a) Sketch of a projection of the La_2CuO_4 structure along the [100] direction (pointing into the plane of the paper).

La coordinates are (0, 0, 1/3), (0, 0, 2/3), (1/2, 1/2, 5/6), and (1/2, 1/2, 1/6). At this level of approximation, the sketch has only a very small distortion, but is much easier to draw. Note that such approximations are not appropriate when computing bond distances or other quantitative data. Using a similar level of approximation, the O(2) coordinates can be enumerated as (0, 0, 0.2), (0, 0, 0.8), (1/2, 1/2, 0.7), and (1/2, 1/2, 0.3). The remaining coordinates are trivial.

When producing a sketch, it is always easiest and most informative to begin with a projection. In this case, we project down the [100] axis so that the longest dimension of the unit cell is in the plane of the paper. Following convention, the origin is placed in the upper left-hand corner. Different symbols are chosen for each atom and the vertical coordinates are shown nearby, as illustrated in Fig. 3.33.

When identical atoms are superimposed in the projection, the vertical coordinates are written nearby, separated by commas.

When developing the model, it is important to remember to include the translation repeats. For example, the Cu atom at $(0,0,0)$ should be replicated at all eight vertices of the unit cell. When including translation repeats, it is helpful to remember the following rules: atoms on cell vertices have seven additional translation repeats, atoms on edges have three, and atoms on faces have one.

The coordination environment of each atom can be determined by a careful examination of the diagram and calculations of the bond lengths. If there is any question about whether or not two atoms are 'bonded', simply calculate the distance between the two coordinates and compare it to the expected bonding distance. You can predict an expected bonding distance either by summing the ionic radii (see Tables 1A.1 or 7B.1 in the appendix) or by using bond valence theory (see Chapter 10, Section C(ii)). As a coarse rule of thumb, it is useful to remember that few bonds in crystals are shorter than 1.5 Å or longer than 3 Å (although bonds formed by the smallest or largest atoms sometimes fall outside these limits). If the calculated and expected distances are comparable, you can consider the atoms to be bonded. In practice, bonded coordination polyhedra seldom consist of identical equidistant atoms and you must use some judgment in the assignment of 'nearest neighbors'. As a guideline, you can build polyhedra from nearest neighbors found within ± 30% of the radius sum. Note that it is frequently possible to justify more than one description of the coordination environment.

In the case of La_2CuO_4, we note that all of the Cu have four neighboring O in the a–b plane at a distance of $0.5a = 1.89$ Å. Based on the data in Table 7B.1, the ideal distance is 2.09 Å, so it is appropriate to call these atoms nearest neighbors. There are two additional O atoms displaced along the c-axis by $0.1828c = 2.43$ Å. While this separation is significantly larger than the Cu–O separation in the a–b plane, it is close enough to the Cu–O radii sum to include these two slightly further away O atoms in the coordination polyhedron, which is a distorted octahedron. Nine O atoms can be found that are reasonably close to the ideal La–O distance of 2.46 Å. The easiest approach to producing a polyhedral picture of this structure is to begin by sketching the Cu–O octahedra. After doing this, we note that all of the O atoms have been accounted for at the vertices of these octahedra. Attempting to draw polyhedra around the La (these 13-sided polyhedra share faces with four of the CuO_6 octahedra and a vertex with another) complicates rather than clarifies the picture, so we simply indicate the La as circles. A [100] projection is illustrated in Fig. 3.34a and an oblique projection is shown in Fig. 3.34b. The structure is composed of corner-sharing layers

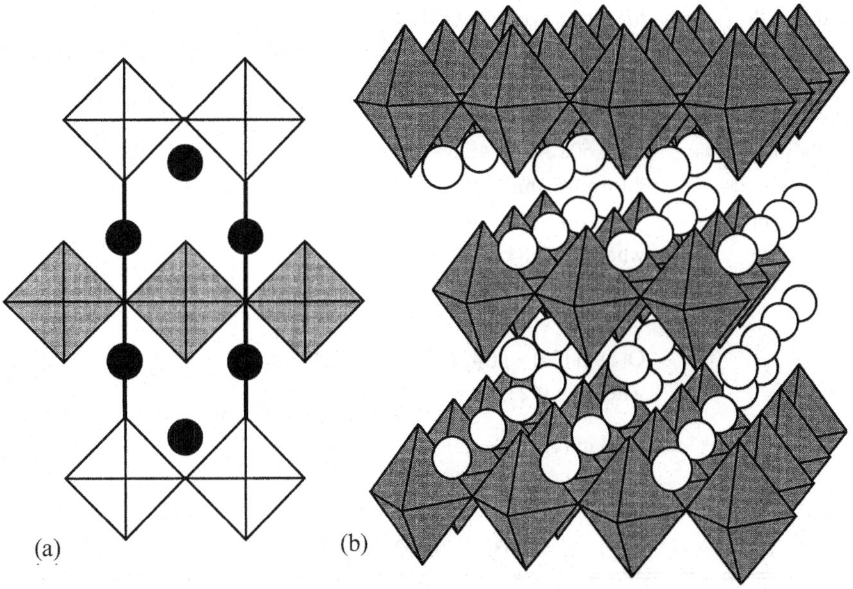

Figure 3.34. Polyhedral representations of the La_2CuO_4 structure. (a) [100] projection. The shaded polyhedra are set behind the open ones. (b) Oblique view showing the layers.

of CuO_6 octahedra stacked along the c-axis and interleaved with La. The centers of adjacent layers are offset by $1/2a$ and $1/2b$, so that they are stacked in an A-B-A-B pattern along [001].

Example 3.2

The crystal structure of $VOMoO_4$ is described in Table 3.4. Using these data, answer the following questions.
(i) Are the general equivalent positions in this group occupied and if so, by which atoms? Are special positions occupied and if so, by which atoms?
(ii) Draw a diagram of the structure projected onto the (001) plane. Show all of the atoms in one unit cell, labeling the atom type and the vertical position.
(iii) Where is the n operator in this structure?
(iv) Describe the coordination around each atom.
(v) Compute the Mo–O and V–O bond distances.
(vi) Draw a polyhedral representation of this structure, also projected on the (001) plane.

Table 3.4. *The crystal structure of VOMoO$_4$ [21]*

Formula unit	VOMoO$_4$
Space group:	P4/n (no. 85)
Cell dimensions:	$a = 6.608$ Å, $c = 4.265$ Å
Cell contents:	2 formula units
Atomic positions:	V in (2c) (1/4, 1/4, z) and (3/4, 3/4, $\bar{z}$)
	where $z = 0.84$
	Mo in (2b) (1/4, 3/4, 1/2) and (3/4, 1/4, 1/2)
	O(1) in (2c) (1/4, 1/4, z) and (3/4, 3/4, $\bar{z}$)
	where $z = 0.233$
	O(2) in (8g) (x,y,z); ($1/2-x,1/2-y,z$); ($1/2-y, x,z$); ($y,1/2-x,z$); ($\bar{x}, \bar{y}, \bar{z}$); ($1/2+x,1/2+y, \bar{z}$); ($1/2+y, \bar{x}, \bar{z}$); ($\bar{y}, 1/2+x, \bar{z}$)
	where $x = 0.703$, $y = 0.462$, $z = 0.26$

1. We recall that the general positions always have the lowest symmetry and the highest multiplicity. In the group P4/n, it is clear that the tetrad axis along z creates four points and that the diagonal glide plane in the x–y plane creates four more for a total of eight. Therefore, the 8g site has the correct multiplicity of the general positions. Also note that the general positions must always include the coordinate x, y, z. The remainder of the atoms are in special positions.

2. Since this is a tetragonal structure, the (001) projection should be a square and we use this to guide the sketch. Next, we use the information in Table 3.4 to list the coordinates:

V	(1/4,1/4,0.84)	(3/4,3/4,0.16)		
Mo	(1/4,3/4,1/2)	(3/4,1/4,1/2)		
O1	(1/4,1/4,0.233)	(3/4,3/4,0.767)		
O2	(0.703,0.462,0.26)	(0.797,0.038,0.26)	(0.038,0.703,0.26)	(0.462,0.797,0.26)
	(0.297,0.538,0.74)	(0.203,0.962,0.74)	(0.962,0.297,0.74)	(0.536,0.203,0.74)

3. Plotting these points produces the diagram shown in Fig. 3.35. As symbols, choose shapes that are simple to draw and easily distinguished.

4. From the symbol, we know that the n-glide plane lies parallel to the x–y plane. The operation takes (x, y, z) to $(x + 1/2, y + 1/2, \bar{z})$. Based on the atomic positions illustrated in Fig. 3.36, the glide plane must be situated at $z = 0$ and $z = 1/2$.

5. In this structure, we expect to find polyhedral groups that have central metal atoms coordinated by O at the vertices. Each of the Mo atoms is surrounded by four

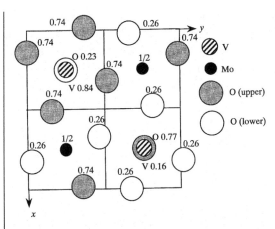

Figure 3.35. Projection of the VOMoO$_4$ structure down [001].

equidistant O at the vertices of a tetrahedron. Using the information in Chapter 2, we find that each Mo–O bond distance is 1.76 Å.

6. The V atom coordination might be described as distorted octahedral (six nearest neighbors), or as square prismatic (five nearest neighbors). The bond distances to the four equatorial O (in the *a–b* plane) are 1.978 Å. Along the *c*-axis there is one very short bond (1.68 Å) and one very long one (2.59 Å). For our drawing, we shall interpret this as octahedral coordination (even though it is very distorted).

7. Using the assumption described above, all of the polyhedra are corner-sharing and each of the O is connected to two metal atoms.

8. Based on these conclusions, the polyhedral representation in Fig. 3.36 can be drawn.

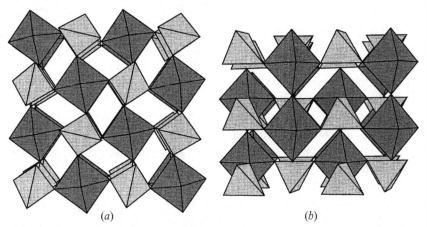

(a)   *(b)*

Figure 3.36. Polyhedral representations of the VOMoO$_4$ structure, projected along [001] (a) and [100] (b).

In closing this chapter, we note that a number of useful computer programs are available commercially that can be used to draw projections and three-dimensional representations of crystal structures. In this text, the structures that were not drawn by hand were produced either by 'Ca.R.Ine Crystallography, 3.0' (Divergent SA, Compiegne France) or 'Ball and Stick, 3.5.1' (Cherwell Scientific, Oxford, UK); the former is particularly useful because it relies on conventional crystallographic practice and notation.

(F) Problems

(1) Using sketches, determine which point groups result from adding a center of symmetry to the following groups: 1, m, $mm2$, $4mm$, 6, $\bar{6}$, $6mm$. [22]

(2) Using sketches, determine which point groups result from adding a center of symmetry to the following groups: 2, $\bar{4}$, 4, 222, 3, 422, $4/mmm$, 622. [22]

(3) Use a diagram to determine what point group results from the combination of:

>   (i) two mirror planes oriented at 90° with respect to one another and a diad perpendicular to both,
>   (ii) a tetrad axis with a perpendicular diad,
>   (iii) a $\bar{4}$ axis with a mirror oriented with its normal perpendicular to the roto-inversion axis,
>   (iv) a triad axis with a perpendicular diad,
>   (v) a hexad axis with a perpendicular mirror.

(4) What point groups result from the combination of two mirror planes oriented at 90° with respect to one another? Repeat for 60°, 45°, and 30°. [22]

(5) What point groups result from the combination of two intersecting two-fold axes oriented at 90° with respect to one another? Repeat for 60°, 45°, and 30°. [22]

(6) Does point group $\bar{3}m$ contain symmetry elements not present in the name? If so, indicate them on a diagram. Repeat for group 23.

(7) Explain why we only use crystallographic point groups and space groups with 1, 2, 3, 4, and 6-fold rotational axes.

(8) Specify the point group and Bravais lattices of the crystals that have the following space groups:

> $P\bar{4}2_1c$ $Ccc2$ $I4_1cd$ $P6_3mc$ $Fdd2$

(9) Specify the point group of crystals that have the following space groups:

> $I4_122$ $P4bm$ $R3c$ $P6_3mc$ $Aba2$

(10) In a molecule of $MoCl_5$, the Mo is at the center of a trigonal bipyramid. What is the point group of this molecule? Do any of the atoms occupy special

positions? If so, show this on a diagram. Repeat for Mo_2Cl_{10}. (You can think of this molecule as two $MoCl_6$ octahedra that share a common edge.) [23]

(11) Explain, in a brief paragraph or two, how the 230 space groups are related to the 32 crystallographic point groups.

(12) Al_2O_3 crystallizes in a structure with trigonal symmetry described by space group $R\bar{3}c$. The parameters of its primitive trigonal cell are $a = 5.128$ Å and $\alpha = 55°20'$. The Al atoms are at $4c$ sites which have coordinates $\pm(x,x,x) + I$, where $x = 0.352$ and the O atoms are in $6e$ sites with the coordinates $\pm(x,1/2 - x,1/4; 1/2 - x,1/4,x; 1/4,x,1/2 - x)$ where $x = 0.556$. Specify the lattice parameters of the conventional rhombohedral (hexagonal) cell and the positions of the atoms in this cell.

(13) The structure of α-quartz is specified by the data in Table 3B.1. Answer the following questions. [23, 24]

(i) Draw a diagram of the structure projected onto the (0001) plane. Show all of the atoms in one unit cell, labeling the atom type and the vertical position.

(ii) Describe the coordination around each atom. How are the polyhedra linked?

(iii) Are the atoms on special positions or general positions? What is the point symmetry of each position?

(iv) Calculate the Si–O bond distances.

(14) Tridymite is the high temperature form of SiO_2. Its structure is described in Table 3B.2. Answer the following questions. [23]

(i) Draw a diagram of the structure projected onto the (0001) plane. Show all of the atoms in one unit cell, labeling the atom type and the vertical position.

(ii) Describe the coordination around each atom. How are the polyhedra linked?

(iii) Are the atoms on special positions or general positions? What is the point symmetry of each position?

(iv) Calculate the Si O bond distances.

(15) One common structure for compounds with the ABO_4 stoichiometry is scheelite, the structure of $CaWO_4$. This structure is described in Table 3B.3.

(i) What is the Bravais lattice of this crystal?

(ii) Sketch a projection down [100].

(iii) Describe the coordination of the W atoms (include the bond distances).

(iv) Are the general positions of this group occupied?

(v) What is the point symmetry of the site occupied by W?

(vi) With respect to the atoms in the unit cell, where is the glide plane?

(16) The structure of PdS is specified in Table 3B.4. Use this information to answer the following questions. [23]

(i) Specify the general positions of space group P4$_2$/m.

(ii) Do any of the atoms in PdS occupy general positions? If so, which ones?

(iii) Sketch a projection of this cell along [001].

(iv) What is the coordination of the Pd atoms? What is the coordination of the S?

(17) The structure of α-MoO$_3$ [25] is given in Table 3B.5. Use this information to answer the following questions.

(i) Are the 4c sites general or special positions in this group?

(ii) The original papers on the structure of MoO$_3$ cite the space group as Pbnm. This is identical to Pnma, except for the way that the axes are labeled. In space group Pbnm, how would you specify the lattice parameters?

(iii) Sketch a projection of this structure, viewed along [001].

(iv) Specify the coordination of the Mo and O atoms in this structure and sketch polyhedral projections along all three axes.

(18) At high pressures, Rh$_2$O$_3$ transforms to the orthorhombic structure described in Table 3B.6, which has been named the Rh$_2$O$_3$ II structure [26]. Use the information in the table to answer the following questions.

(i) Are any of the occupied sites in this structure general positions?

(ii) Sketch an [010] projection of the structure.

(iii) What are the lengths of the Rh–O bonds?

(iv) Describe the Rh coordination in this structure. How do the polyhedra connect?

(19) There are 1121 known binary AB phases that are nearly stoichiometric and ordered. More than 1000 of these phases are isostructural with one of 21 structure types, listed in Table 3.5.

(i) What is the Bravais lattice of each of these structure types?

(ii) What percentage of the AB compounds that take one of these 21 structures are cubic? What percentage are tetragonal? Orthorhombic? Hexagonal?

(20) The polymorph of NiS stable at room temperature has the Millerite structure [28], described in Table 3B.7. Use the information in the table to answer the following questions.

(i) The cell described in Table 3B.7 is hexagonal. Is it possible to choose a smaller, primitive cell for this structure? If so, what are the characteristics of this cell?

(ii) Sketch a projection of this structure (using the hexagonal cell specified in Table 3B.7) along the [001] axis. Use different symbols for the different atoms and indicate the z coordinate of each one.

(iii) Calculate the Ni–S bond lengths in this structure.

(iv) How would you describe the coordination around the Ni atoms?

Table 3.5 AB *crystal structure types with at least six known representative compounds [27].*

prototype name	structure type	space group	number of compounds
LiAs	—	$P2_1/c$	6
AuCd	B19	$Pmma$	12
FeB	B27	$Pnma$	74
GeS	B16	$Pnma$	7
MnP	B31	$Pnma$	30
ErAl	—	$Pmma$	12
CrB	B_f (B33)	$Cmcm$	117
CuAu	L10	$P4/mmm$	32
γ-TiCu	B11	$P4/nmm$	7
NaPb	—	$I4_1/acd$	9
WC	B_h	$P\bar{6}m2$	9
ZnS	B4	$P6_3mc$	21
HgS	B9	$P3_221$	5
TiAs (γ'-MoC)	B_i	$P6_3/mmc$	6
NiAs	$B8_1$	$P6_3/mmc$	24
CsCl	B2	$Pm\bar{3}m$	298
FeSi	B20	$P2_13$	18
KGe	—	$P\bar{4}3n$	6
ZnS	B3	$F\bar{4}3n$	33
NaCl	B1	$Fm\bar{3}m$	282
NaTl	B32	$Fd\bar{3}m$	14

(v) Sketch a polyhedral representation of this structure, also in the [001] projection.

(21) The most highly oxidized form of vanadium is V_2O_5. Its structure is described in Table 3B.8.

(i) What is the Bravais lattice of this crystal?

(ii) Draw a diagram that specifies the coordinates of the general positions in this group.

(iii) Sketch a projection down [001].

(iv) Describe the coordination of each atom.

(v) Draw a polyhedral representation of this structure.

(22) The crystal structure of $YBaCuFeO_5$ is described in Table 3B.9 and the crystal structure of $YBa_2Cu_3O_7$ is described in Table 3B.10. Using these data, answer the following questions:

(i) What are the Bravais lattices of these crystals?

(ii) Draw a diagram that specifies the coordinates of the general positions in each of these groups. List the coordinates of the general positions.

(iii) Sketch a projection of each cell, viewed along [100]. Label each atom and its position along the a-axis.

(iv) The space group name tells us that among the symmetry elements in these groups, there are mirrors and tetrad axes in $YBaCuFeO_5$ and only mirrors in $YBa_2Cu_3O_7$. Specify the locations of these elements relative to the atoms in this structure.

(v) There is also a diad in the $YBaCuFeO_5$ structure. Specify its location.

(vi) Describe the coordination environment of the Cu and Fe atoms in $YBaCuFeO_5$ and the Cu atoms in $YBa_2Cu_3O_7$.

(vii) Sketch a polyhedral representation of each structure, projected on the (100) plane, including only Cu–O and Fe–O polyhedra. Indicate the Y and Ba positions with circles. Describe the relationship between these two structures.

(23) The structure of graphite is described in Table 3B.35. Use these data to answer the questions that follow.

(i) What is the Bravais lattice of this crystal structure?

(ii) What is the point group from which this space group derives?

(iii) Does the point group have a center of symmetry?

(iv) Define each component of the symbol '$P6_3mc$'.

(v) On a drawing, sketch the positions of the symmetry elements with respect to the lattice vectors.

(vi) Sketch the atomic structure, projected along the c-axis.

(vii) What is the C–C distance in the a–b plane? What is the C–C distance along the c-axis?

(24) The tetragonal tungsten bronze structure is described in Table 3B.17. Use these data to answer the questions that follow.

(i) What is the Bravais lattice of this crystal structure?

(ii) What is the point group from which this space group derives?

(iii) Make a sketch illustrating all of the equivalent positions in this point group.

(iv) List the coordinates of the equivalent positions in this point group.

(v) Does this point group have a center of symmetry?

(vi) Define the symmetry operation represented by each of the components in the space group symbol.

(vii) Sketch the atomic structure, projected along the c-axis.

(viii) Show the positions of the symmetry elements in the unit cell (with respect to the lattice vectors and the atom positions).

(ix) Are the general positions in this space group occupied?

(x) Describe the coordination of the tungsten atoms in this structure.

(xi) Sketch a polyhedral representation of this structure, projected along c.

(G) References and sources for further study

[1] M.J. Burger, *Contemporary Crystallography* (McGraw-Hill, New York, 1970), Chapters 1 and 2. A good introduction to symmetry operators and point groups. Much of the information in Sections A through C is derived from this source.

[2] F.A. Cotton, *The Chemical Applications of Group Theory*, 2nd edn. (Wiley-Interscience, New York, 1971). A text on the application of group theory.

[3] M.J. Burger, *Contemporary Crystallography* (McGraw-Hill, New York, 1970), p. 21. The compatibility of point groups and translations.

[4] H. Megaw, *Crystal Structures: A Working Approach* (W.B. Saunders, Philadelphia, 1973), Chapter 6.

[5] M.J. Burger, *Contemporary Crystallography* (McGraw-Hill, New York, 1970), pp. 24–9. Fig. 3.8 is drawn after Fig. 13, Chapter 2, p. 25.

[6] D.E. Sands, *Introduction to Crystallography* (W.A. Benjamin, New York, 1969), Section 3.5, pp. 55–6. Table 3.1 is drawn after Table 3.1, p. 51.

[7] L.S. Dent Glasser, *Crystallography and its Applications* (Van Nostrand Reinhold Co., New York, 1977), Chapter 1. Figures 3.9 to 3.14 are drawn after Fig. 1.9, p. 10 & 11.

[8] A.R. West, *Solid State Chemistry and its Applications* (J. Wiley & Sons, Chichester, 1984), Section 6.1.

[9] H. Megaw, *Crystal Structures: A Working Approach* (W.B. Saunders, Philadelphia, 1973), Section 6.9, p. 129.

[10] D.E. Sands, *Introduction to Crystallography* (W.A. Benjamin, New York, 1969), Sections 2.4–2.5, p. 43.

[11] M.J. Burger, *Contemporary Crystallography* (McGraw-Hill, New York, 1970), Chapter 2, p. 37. Fig. 3.18 is drawn after Fig. 18 on p. 39.

[12] H. Megaw, *Crystal Structures: A Working Approach* (W.B. Saunders, Philadelphia, 1973). Fig. 3.19 is drawn after Fig. 8.2 on p. 150.

[13] D.E. Sands, *Introduction to Crystallography* (W.A. Benjamin, New York, 1969), Section 4.4, pp. 71–2.

[14] T. Hahn (editor), *International Tables for Crystallography*, Vol. 4, *Space Group Symmetry*, 2nd edition (International Union for Crystallography, Kluwer Academic Publishers, Dordrecht, 1989).

[15] H. Megaw, *Crystal Structures: A Working Approach* (W.B. Saunders, Philadelphia, 1973), Chapter 8.

[16] D.E. Sands, *Introduction to Crystallography* (W.A. Benjamin, New York, 1969), Chapter 4.

[17] A.R. West, *Solid State Chemistry and its Applications* (J. Wiley & Sons, Chichester, 1984), Chapter 6. Fig. 3.22 is drawn after Fig. 6.8, p. 196. Fig. 3.25 is drawn after Fig. 6.12, p. 201. Fig. 3.28 is drawn after Fig. 6.14, p. 204.

[18] H. Megaw, *Crystal Structures: A Working Approach* (W.B. Saunders, Philadelphia, 1973), Section 8.9, pp. 158–9.

[19] D.E. Sands, *Introduction to Crystallography* (W.A. Benjamin, New York, 1969) Section 3.7, p. 59. Fig. 3.29 is drawn after Fig. 3.8, p. 62.

[20] J.M. Longo and P.M. Raccah, The Structure of La_2CuO_4 and $LaSrVO_4$, *J. Solid State Chem.* **6** (1973) 526–31; J.D. Jorgensen *et al., Phys. Rev. Lett.* **58** (1987) 1024.

[21] H.A. Eick and L. Kihlborg, The Crystal Structure of $VOMoO_4$, *Acta Chem. Scand.* **20** (1966) 722.

[22] A.R. West, *Solid State Chemistry and its Applications* (J. Wiley & Sons, Chichester, 1984), Chapter 6, p. 210. Problems 1 and 2 are similar to problem 6.1. Problem 4 is 6.2 and problem 5 is 6.3.

[23] D.E. Sands, *Introduction to Crystallography* (W.A. Benjamin, New York, 1969). Problem 10 is from problem 2.12 on p. 46. For problem 13, see p. 81. Problem 14 is after problem 4.7 on p. 83. For problem 16, see p. 77.

[24] R.W.G. Wyckoff, *Crystal Structures*, Volume 1 (John Wiley & Sons, New York, 1964), tridymite, p. 315, α-quartz, p. 312. The structure of SiO_2.

[25] R.W.G. Wyckoff, *Crystal Structures*, Volume 1 (John Wiley & Sons, New York, 1964), p. 81. The structure of MoO_3.

[26] R.D. Shannon and C.T. Prewitt, Synthesis and Structure of a New High-Pressure Form of Rh_2O_3, *J. Solid State Chemistry* **2** (1970) 134–6.

[27] P. Villars, *J. of Less-Common Metals*, **92** (1983) 215–38. The structures of AB compounds.

[28] R.W.G. Wyckoff, *Crystal Structures*, Volume 1 (John Wiley & Sons, New York, 1964), p. 122. The structure of Millerite (NiS).

[29] R.W.G. Wyckoff, *Crystal Structures*, Volume 2 (John Wiley & Sons, New York, 1964), p. 185. The structure of V_2O_5 (vanadium pentoxide).

[30] J.T. Vaughey and K.R. Poeppelmeier, *Structural Diversity in Oxygen Deficient Perovskites*, NIST Special Publication 804, Chemistry of Electronic Ceramic Materials, *Proceedings of the International Conference, Jackson, WY, Aug. 17–22, 1990*, (1991). The crystal structure of $YBaCuFeO_5$.

[31] J.D. Jorgensen, B.W. Veal, A.P. Paulikas, L.J. Nowicki, G.W. Crabtree, H. Claus, and W.K. Kwok, *Phys. Rev. B* **41** (1990) 1863. The crystal structure of $YBa_2Cu_3O_7$.

Chapter 4
Crystal Structures

(A) Introduction

The most unambiguous way to describe the structure of a crystal is to specify the chemical composition, the space group, the cell dimensions, the number of formula units per cell, and the coordinates of the crystallographically distinct atoms. This is the symmetry based method described in Chapter 3 and many structures are described this way in Appendix 3B. While we will continue to use this method, we should also recognize its shortcomings. First, based on tabulated data, it is difficult to understand the geometric arrangement of the atoms without taking the time to draw a sketch. Second, the conventional information does not provide the context necessary for the comparison of different structures.

The primary objective of this chapter is to describe an alternative and less rigid method for the description of crystal structures. The basic idea is to assume that all of the atoms in the cell are situated either at the sites of a prototypical close packed structure or at one of a few well defined interstitial sites. Structures are differentiated by the pattern in which these sites are occupied. While this method lacks specificity, it provides an immediate understanding of the geometric arrangement of atoms and a context for the comparison of different structures. The advantage of this method is that by understanding a few geometric principles, it is possible to describe a vast number of crystal structures. A secondary objective of this chapter is to survey important prototype crystal structures. By the end of the chapter, you will see that even very complex structures can be described by comparing them to one of a few simple prototypes.

(B) Close packed arrangements

i. Types of close packing

Arrangements in which atoms contact at least six other atoms will be referred to as *close packed* [1]. Thus, the simple cubic structure, with a coordination number of six, is the least densely packed of all the close packed structures. The packing fractions of the different close packed arrangements are given in Table 4.1.

The only element that crystallizes in the simple cubic structure is α-Po. However, As, Sb, Bi, and P_{black} all crystallize in structures that can be thought of

Table 4.1. *Ideal packing fractions for close packed structures*

Lattice arrangement	Coordination number	Packing fraction
simple cubic	6	0.5236
simple hexagonal	8	0.6046
body centered cubic	8	0.6802
body centered tetragonal	10	0.6981
tetragonal close packing	11	0.7187
closest packings (ccp, hcp)	12	0.7405

as distorted versions of the simple cubic arrangement, where three nearest neighbors are close and three are further away. These structures are most accurately described as $3 + 3$ arrangements.

No elements assume the eight-coordinate simple hexagonal structure. The body centered cubic (bcc) is the more common arrangement with eight nearest neighbors. It is important to note that in addition to the eight nearest neighbors, there are six next nearest neighbors (NNN) only 15% further away. So, this nearly 14 coordinate structure could also be described as having $8 + 6$ coordination. The fact that the six NNN are so close suggests that tetragonal distortions could lead to the formation of 10 and 12 coordinate structures. For example, if we compress the structure along c, the atoms along the vertical axis become nearest neighbors when the c to a ratio is $\sqrt{2}/\sqrt{3}$. This arrangement has a coordination number of 10 and is known as body centered tetragonal. It is also possible to shrink a and b with respect to c so that the in-plane or equatorial atoms become nearest neighbors. This 12-coordinate arrangement occurs when $c/a = \sqrt{2}$. However, since the height of the cell is now equal to the diagonal of the square base, the repeat unit is now identical to the familiar cubic F arrangement. We refer to the structure that has this arrangement as cubic close packed (ccp). From this perspective, the differences between the ccp and bcc structures are subtle. See Fig. 4.1.

Only Pa crystallizes in the ten coordinate body centered tetragonal structure and the measured c/a ratio is 0.825, close to the ideal. Finally, the recently discovered tetragonal close packed arrangement where atoms have 11 nearest neighbors is described in problem 4.1 at the conclusion of this chapter [2].

It had been accepted, but not proven in the mathematical sense, that the ccp and hcp arrangements represent the most efficient packing of equal sized spheres in three dimensions; for this reason, they are referred to as *closest packed*. In the sixteenth century, Kepler could find no better arrangement, but was not able to prove that this was the most efficient. Ever since, the idea that these configurations really are the closest packed has been known as the Kepler conjecture. In

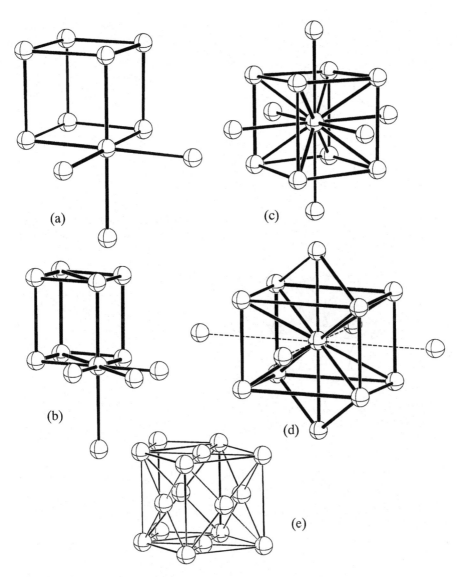

Figure 4.1. Types of close packing: (a) simple cubic, (b) simple hexagonal, (c) body centered cubic, (d) body centered tetragonal, (e) cubic close packed (ccp) [1].

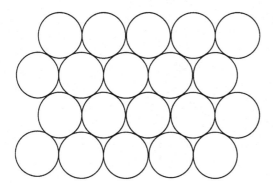

Figure 4.2. A layer of closest packed spheres.

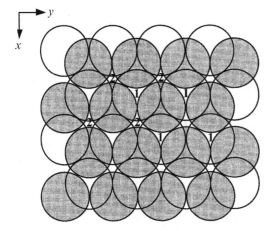

Figure 4.3. The stacking of closest packed layers. The shaded spheres fit on top of the unshaded layer. There is only one unique way to stack two closest packed layers, but the third layer can be added in either one of the two positions labeled 1 and 2.

1998, after four centuries of study by mathematicians, a proof of this conjecture has been claimed by Hales [3].

ii. *The ccp and hcp arrangements*

In this book, most crystal structures are described by comparing them to one of the two 12-coordinate arrangements (ccp or hcp). It is easiest to view these structures as stacked, closest packed layers [4]. The structure of a single closest packed layer is shown in Fig. 4.2. The three dimensional structure is formed by stacking these layers upon one another. Atoms in the second layer fit into the valleys formed by three atoms of the first layer, as illustrated in Fig. 4.3. There are two possible positions for the third layer. If the atoms take the positions labeled '1', then they are directly above the atoms in the first layer and the third layer reproduces the first. If this pattern continues, we have the ABAB stacking that characterizes the hcp structure. If, on the other hand, the atoms in the

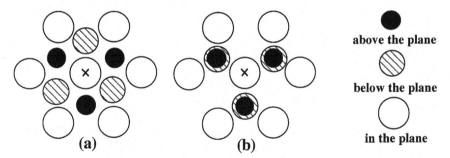

Figure 4.4. Local coordination environments in the (a) ccp and (b) hcp structures.

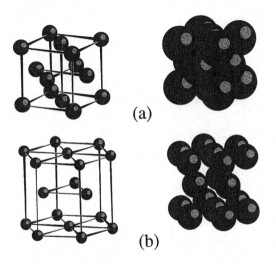

Figure 4.5. The conventional unit cells for the ccp (a) and hcp (b) structures.

third layer occupy the positions labeled '2', then this layer is distinct from the first and second. The fourth layer must then repeat either the first or second. If it repeats the first, then we have the ABCABC stacking sequence of the ccp structure.

The similarity of the packing in these two structures is noteworthy. Both are 12 coordinate structures and have identical densities. However, as shown in Fig. 4.4, the configurations of the nearest neighbor coordination shells differ. While these are by far the most common stacking sequences, more complicated sequences can also occur. For example, La, Pr, Nd, and Pm have an ABCBABCB sequence. More complex stacking patterns are discussed in Section I.

The conventional ccp and hcp unit cells are shown in Fig. 4.5. Despite the similarity in the packing, the cells' appearances differ. In the ccp structure, it is the {111} planes that are closest packed and they are stacked along the body

Table 4.2. *The cubic close packed structure, Cu, A1.*

Formula unit	Cu, *copper*
Space group:	$Fm\bar{3}m$ (no. 225)
Cell dimensions:	$a = 3.6147$ Å
Cell contents:	4 formula units
Atomic positions:	Cu in 4(a) $(0, 0, 0 + F)$

Table 4.3. *The body centered cubic structure, W, A2.*

Formula unit	W, *tungsten*
Space group:	$Im\bar{3}m$ (no. 229)
Cell dimensions:	$a = 3.1652$ Å
Cell contents:	2 formula units
Atomic positions:	W in 2(a) $(0, 0, 0 + I)$

Table 4.4. *The hexagonal close packed structure, Mg, A3.*

Formula unit	Mg, *magnesium*
Space group:	$P\,6_3/m\,mc$ (no. 194)
Cell dimensions:	$a = 3.2094$ Å; $c = 5.2105$ Å
Cell contents:	2 formula units
Atomic positions*:	Mg in 2(c) (1/3, 2/3, 1/4); (2/3, 1/3, 3/4)

Note:
* This cell is easier to visualize by shifting the origin such that the coordinates of the 2(c) sites are (0, 0, 0) and (1/3, 2/3, 1/2).

diagonal $\langle 111 \rangle$ directions. In the hcp structure, it is the basal planes perpendicular to [0001] that are closest packed and stacked along c. The conventional crystallographic data for the ccp, hcp, and bcc structure are given in Tables 4.2, 4.3, and 4.4, respectively, and a selection of elemental metals that have these structures at room temperature is listed in Table 4.5.

Ⓒ The interstitial sites

In the spaces between the sites of the closest packed lattices, there are a number of well defined interstitial positions, as shown in Fig. 4.6 [7]. The ccp structure

Table 4.5. *Structures and lattice constants of selected elements at room temp.* *[6].*

ccp	a, Å	hcp	a, Å	c, Å	bcc	a, Å
Cu	3.6147	Be	2.2856	3.5832	Fe	2.8664
Ag	4.0857	Mg	3.2094	5.2105	Cr	2.8846
Au	4.0783	Zn	2.6649	4.9468	Mo	3.1469
Al	4.0495	Cd	2.9788	5.6167	W	3.1650
Ni	3.5240	Ti	2.506	4.6788	Ta	3.3026
Pd	3.8907	Zr	3.312	5.1477	Ba	5.019
Pt	3.9239	Ru	2.7058	4.2816	Li	3.5101
Pb	4.9502	Os	2.7353	4.3191	Na	4.2908
Ce	5.1603	Re	2.760	4.458	K	5.247
Yb	5.4864	Sc	3.3091	5.2735	Rb	5.70
Rh	3.8045	Y	3.6475	5.7308	Cs	6.0797
Ir	3.8390	Hf	3.1947	5.0513	V	3.0232
Ca	5.5886	Co	2.507	4.070	Nb	3.3067
Sr	6.0851	Gd	3.6361	5.7828	Eu	4.5822
		Tb	3.6011	5.6938		
		Dy	3.5904	5.6477		
		Ho	3.5774	5.6160		
		Er	3.5589	5.5876		
		Tm	3.5376	5.5548		
		Lu	3.5032	5.5511		
		Tc	2.735	4.388		

has four octahedral interstitial sites per cell; one site is at the cell center and the rest are at the midpoints of all of the cell edges. There are also eight tetrahedral, four-coordinate sites per unit cell at the ($\pm 1/4, \pm 1/4, \pm 1/4$) positions. Although similar sites occur in the bcc lattice, they do not possess ideal tetrahedral or octahedral symmetry.

It is important to understand how the sites are configured with respect to the closest packed layers, and this is illustrated in Fig. 4.7. Of the six nearest neighbors to the octahedral site in the ccp structure, three are in one close packed layer and the remaining three are in the adjacent layer. The most symmetric octahedral position is, therefore, midway between the two close packed planes. The situation is slightly different for the tetrahedral position. Three of the atoms that make up the tetrahedron lie in the same close packed plane and the fourth (the

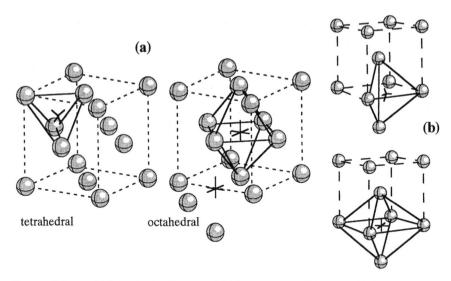

(a)

tetrahedral   octahedral

(b)

Figure 4.6. Interstitial sites in the (a) ccp and (b) bcc structures. The interstitial sites in the bcc structure do not have the ideal tetrahedral or octahedral symmetry.

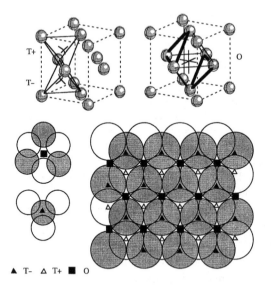

T+

T−

O

▲ T− △ T+ ■ O

Figure 4.7. Details of the interstitial sites in a close packed structure. The T+ and T− sites are not equivalent [7].

apex of the tetrahedron) lies in another. We expect the interstitial atom to occupy a position equidistant between the four ligands. Because this position is not halfway between the two close packed planes, as the octahedral site is, the sites above (T_+) and below (T_-) a central reference plane are distinct. The assignment of these sites is illustrated in Fig. 4.7.

Although it is practical to visualize the octahedral site as being sandwiched between two close packed layers, it is often depicted as shown in the upper part of Fig. 4.7. Thus, the four coplanar atoms are usually called the in-plane or equatorial ligands, while the top and bottom atoms are the axial or apical ligands. The representation is convenient since it emphasizes the arrangement of the p_x, p_y, and p_z orbitals that form chemical bonds. It is useful to remember that for each close packed site in the hcp and ccp structures, there is one six-coordinate octahedral site and two four-coordinate tetrahedral sites (one T_+ site and one T_- site). In the bcc structure, there are six equivalent pseudo-tetrahedral sites and three pseudo-octahedral sites for each close packed site.

(D) Naming crystal structures

We will define a *structure type* as a specific configuration of atoms. Two materials have the same structure type if the atoms in the crystal structure have the same configuration and connectivity. In general, the lattice parameters will differ because of the differences in atomic volumes. In some cases, we will group compounds with different symmetries in the same structure type, assuming that the distortions needed to transform from one to the other are small.

There are two prevailing methods of naming a structure type. The first is to use the name of the prototype for the structure, usually the first compound known to have that structure. The prototype name can be either the mineral name or the chemical name. For example, we have already encountered the structure of NaCl which is called the rock salt structure or the sodium chloride structure. The second common way is to use the *Strukturbericht* names which were assigned according rules of nomenclature. Although they don't work universally, the rules for interpreting the Strukturbericht names are that A is for an element, B is for a binary compound with 1:1 stoichiometry, C is a binary compound with 1:2 stoichiometry, D signifies a compound with some other stoichiometry, E is for a ternary compound or more complex composition, L is for an alloy, and K is for a complex alloy. For example, the Strukturbericht name for the NaCl structure is B1. Selected prototype names and the Strukturbericht names are compared in Table 4.6. The majority of the crystal structure data in this chapter were taken from Refs. [8] and [9].

Table 4.6. *A comparison of the Structurebericht names and the prototype names* [8].

Struc.	Proto.	Struc.	Proto.	Struc.	Proto.	Struc.	Proto.
A1	Cu	B_f	ζ-CrB	DO_9	ReO_3	$D8_1$	Fe_3Zn_{10}
A2	W	B_g	MoB	DO_{11}	Fe_3C	$D8_2$	Cu_5Zn_8
A3	Mg	B_h	WC	DO_{18}	Na_3As	$D8_3$	Al_4Cu_9
A4	C	B_i	γ'-MoC	DO_{19}	Ni_3Sn	$D8_4$	$Cr_{23}C_6$
A5	Sn	C1	CaF_2	DO_{20}	Al_3Ni	$D8_5$	Fe_7W_6
A6	In	$C1_b$	AgAsMg	DO_{21}	Cu_3P	$D8_6$	$Cu_{15}Si_4$
A7	As	C2	FeS_2	DO_{22}	Al_3Ti	$D8_8$	Mn_5Si_3
A8	Se	C3	Cu_2O	DO_{23}	Al_3Zr	$D8_9$	Co_9S_8
A10	Hg	C4	TiO_2	DO_{24}	Ni_3Ti	$D8_{10}$	Al_8Cr_5
A11	Ga	C6	CdI_2	DO_c	SiU_3	$D8_{11}$	Al_5Co_2
A12	α-Mn	C7	MoS_2	DO_e	Ni_3P	$D8_a$	$Mn_{23}Th_6$
A13	β-Mn	$C11_a$	CaC_2	$D1_3$	Al_4Ba	$D8_b$	σ phase
A15	W_3O	$C11_b$	$MoSi_2$	$D1_a$	$MoNi_4$	$D8_f$	Ge_7Ir_3
A20	α-U	C12	$CaSi_2$	$D1_b$	Al_4U	$D8_i$	Mo_2B_5
B1	NaCl	C14	$MgZn_2$	$D1_c$	$PtSn_4$	$D8_h$	W_2B_5
B2	CsCl	C15	$MgCu_2$	$D1_e$	ThB_4	$D8_l$	Cr_5B_3
B3	ZnS cub	$C15_b$	$AuBe_5$	$D1_f$	Mn_4B	$D8_m$	Si_3W_5
B4	ZnS hex	C16	Al_2Cu	$D2_1$	CaB_6	$D10_1$	Cr_7C_3
$B8_1$	NiAs	C18	FeS_2	$D2_3$	$NaZn_{13}$	$D10_2$	Fe_3Th_7
$B8_2$	$InNi_2$	C19	$CdCl_2$	$D2_b$	$Mn_{12}Th$	$E0_1$	PbFCl
B9	HgS	C22	Fe_2P	$D2_c$	MnU_6	$E1_1$	$CuFeS_2$
B10	PbO	C23	$PbCl_2$	$D2_d$	$CaCa_5$	$E2_1$	$CaTiO_3$
B11	γ-CuTi	C32	AlB_2	$D2_f$	UB_{12}	E3	Al_2CdS_4
B13	NiS	C33	Bi_2STe_2	$D2_h$	Al_6Mn	$E9_3$	CFe_3W_3
B16	GeS	C34	$AuTe_2$	$D5_1$	α-Al_2O_3	$E9_a$	Al_7Cu_2Fe
B17	PtS	C36	$MgNi_2$	$D5_2$	La_2O_3	$E9_b$	$AlLi_3N_2$
B18	CuS	C38	Cu_2Sb	$D5_3$	Mn_2O_3	$F0_1$	NiSbS
B19	AuCd	C40	$CrSi_2$	$D5_8$	Sb_2S_3	$F5_1$	$NaCrS_2$
B20	FeSi	C44	GeS_2	$D5_9$	P_2Zn_3	$H1_1$	Al_2MgO_4
B27	FeB	C46	$AuTe_2$	$D5_{10}$	C_2Cr_3	$H2_4$	Cu_3VS_4
B31	MnP	C49	$ZrSi_2$	$D5_{13}$	Al_3Ni_2	$L1_0$	AuCu
B32	NaTl	C54	$TiSi_2$	$D5_a$	Si_2U_3	$L1_2$	$AuCu_3$
B34	PdS	C_c	$ThSi_2$	$D5_c$	Pu_2C_3	$L2_1$	$AlCu_2Mn$
B35	CoSn	C_e	$CoGe_2$	$D7_1$	Al_4C_3	$L'2_b$	ThH_2
B37	TlSe	DO_2	$CoAs_3$	$D7_3$	Th_3P_4	L'_3	Fe_2N
B_e	CdSb	DO_3	BiF_3	$D7_b$	Ta_3B_4	$L6_0$	$CuTi_3$

(E) Classifying crystal structures

Based on the principles described in the previous sections, it is now possible to describe and classify a large number of crystal structures using a small number of common characteristics. The three important characteristics are the packing, the compositional ordering, and the filling of interstitial sites.

We begin by identifying the atoms that belong to a close packed framework with either the bcc, ccp, or hcp arrangement. It is important to note that the framework atoms we choose need not occupy the ideal close packed sites (we will overlook small displacements), nor must they be in contact. This expanded definition of close packed is especially important for ionic compounds, where we expect the anions and the cations to be in contact, while species of like charge should repel one another. When atoms are configured in close packed positions, but not in contact, it is called a eutactic arrangement; the roots of eutactic mean 'well arranged' [10]. To avoid implying the existence of contacts between the framework atoms, we will describe the pseudo-close packed frameworks as eutactic arrangements.

After identifying a eutactic framework and the packing type, we will specify how the framework sites are occupied. In the simplest case, only a single type of atom occupies the eutactic sites. If more than one type of atom occupies a single type of eutactic site in a periodic arrangement, we say that the structure has compositional order and this can lead to larger unit cells and/or lower symmetries than the elemental prototypes. Such arrangements are sometimes referred to as *superlattice* structures. With respect to the prototype close packed structures or the compositionally disordered forms, the diffraction patterns of the superlattice structures have more peaks. This topic is addressed more completely in Example 5.3 of the next chapter.

The final aspect of our classification scheme is to specify how the interstitial (tetrahedral and octahedral) sites are occupied. In this case, the sites can be empty, partially occupied, or completely occupied. Furthermore, it is possible to have compositional order on the interstitial sites. In Table 4.7, a number of structures are classified based on the packing configuration and the interstitial site occupancy [11]. In some cases, you might note that the distinction between framework and interstitial atoms is somewhat arbitrary. In these cases, we typically choose the more electronegative atom as the framework atom. In the section that follows, a number of the important prototype structures included in this table are discussed and compared.

Table 4.7 *Classification of crystal structures based on packing.*

Packing type	Comp.	Framework occupation; ordering	Interstitial occupation		Examples
			T	O	
bcc	A	A	—	—	A2, W, Cr
	AB	1/2A, 1/2B	—	—	B2, CsCl, NiAl
	AB_2	1/3A, 2/3B; $c \approx 3a$	—	—	$C11_b$, $AlCr_2$
	AB_3	A	B: 1/2T	—	A15, W_3O, V_3Si
ccp	A	A	—	—	A1, Cu, Au
	A	1/2A	1/2A: T_+	—	A4, diamond
	AB	1/2A, 1/2 B; alt. (002)		—	$L1_0$, CuAu
	AB	1/2A, 1/2B; alt. (111)	—	—	$L1_1$, CuPt
	AB_3	1/4A, 3/4B; A at 0, 0, 0	—	—	$L1_2$, $AuCu_3$
	AB_3	1/4A, 3/4B; $c \approx 2a$	—	—	$D0_{22}$, $TiAl_3$
	AB_3	A	B: T_+, T_-	B	$D0_3$, BiF_3
	A_2BC	C	A: T_+, T_-	B	$L2_1$, Cu_2MnAl
	AB	A	—	B	B1, NaCl, rock salt
	AB	A	B: T_+	—	B3, ZnS, sphalerite
	AB_2C_4	C	A: $1/8T_+$, $1/8T_-$	1/2B	$MgAl_2O_4$, spinel
	AB_2	B	—	1/2A	C19, $CdCl_2$
	ABC_2	C	A and B: T_+	—	$CuFeS_2$
	AB_3	B	—	1/3A	$CrCl_3$
	A_2B	B	A: T_+, T_-	—	K_2O, anti-fluorite
	ABC_3	1/4A, 3/4C		1/4B	$CaTiO_3$, perovskite
hcp	A	A	—	—	A3, Mg, Zn
	A_3B	1/4A, 3/4B	—	—	$D0_{19}$, Ni_3Sn
	A_3B	1/4A, 3/4B; ch	—	—	$D0_{24}$, Ni_3Ti
	AB	B	—	A	$B8_1$, NiAs
	A_2BC	1/2B, 1/2C; ch	—	A	Ti_2CS
	AB_2	B	—	1/2A	C6, CdI_2
	AB_2	B	—	1/2A	C4, TiO_2
	A_2B_3	B	—	2/3A	$D5_1$, Al_2O_3
	AB_2C_4	C	A: $1/8T_+$, $1/8T_-$	1/2B	Mg_2SiO_4, olivine
	A_3BC_4	C	A/B: T_+	—	β-Li_3PO_4
	A_3BC_4	C	A/B: $1/2T_+$, $1/2T_-$	—	γ-Li_3PO_4

(F) Important prototype structures

i.   *The B2 or CsCl structure*
We begin by examining some structures that have simple packing arrangements with compositional ordering. Many intermetallic compounds and a few salts crystallize in the CsCl or B2 structure. The B2 structure has a primitive cell with a two atom basis and each atom is eight coordinate. B2 compounds have an equal number of A and B atoms that alternately occupy the cube vertices and centers. In terms of our close packing description, we would say that the A and B atoms are ordered on the sites of the bcc structure. In this case, the A and B atoms occupy alternate sites along $\langle 111 \rangle$. It is important to emphasize that because of this ordering, the B2 structure is cubic P, not cubic I. It is, however, easiest to describe it by referring to the bcc structure. More 1:1 binary compounds crystallize in the B2 structure than any other type.

ii.   *The L1$_0$ or CuAu structure*
A number of other AB intermetallic compounds crystallize in the tetragonal L1$_0$ structure. In this structure, the A and B atoms are ordered on the sites of a ccp structure, in layers perpendicular to [001], as shown in Fig 4.8(b). Note that because of the ordering, the cell is no longer cubic, nor is it face centered. As long as there is a size difference between the two component atoms, the cell will be tetragonal. Nonetheless, it is easiest to describe it by referring to the ccp structure.

iii.   *The L1$_2$ structure*
The L1$_2$ structure is another example of an atomic arrangement that is most easily described as an ordered ccp cell. In this case, the stoichiometry of the compound is AB$_3$ and the A atoms occupy the cell vertices while the B atoms occupy the face centers, as illustrated in Fig. 4.9. Again, because the A and B atoms occupy the sites in an ordered fashion, this is not a cubic F lattice, but a cubic P lattice with a four atom basis.

iv.   *The Do$_3$ and L2$_1$ structures*
We next consider two structures that have occupied interstitial sites. The BiF$_3$ (D0$_3$) and L2$_1$ structures can be described as ccp structures with all of the octahedral and tetrahedral interstices filled. In the D0$_3$ structure, the Bi atoms occupy the positions of the ccp structure and the F atoms fill all of the octahedral and tetrahedral voids so that there are 16 atoms per cell. See Fig. 4.10.

The L2$_1$ structure, also known as the Heusler alloy, is very similar to the D0$_3$ structure. In the prototype, Cu$_2$AlMn, the Al atoms occupy the ccp sites, the

Table 4.8. *The CsCl structure, cesium chloride, B2*

Formula unit	CsCl, *cesium chloride*
Space group:	$Pm\bar{3}m$ (no. 221)
Cell dimensions:	$a = 4.123$ Å
Cell contents:	1 formula unit
Atomic positions:	Cs in (1a) $m\bar{3}m$ (0, 0, 0)
	Cl in (1b) $m\bar{3}m$ (1/2, 1/2, 1/2)

Examples:

compound	a (Å)	compound	a (Å)	compound	a (Å)
CsCl	4.123	AlNd	3.73	FeTi	2.976
CsBr	4.286	AgCd	3.33	InNi	3.099
CsI	4.5667	AgCe	3.731	LiAg	3.168
CsCN	4.25	AgLa	3.760	LiHg	3.287
NH_4Cl	3.8756	AgMg	3.28	MgCe	3.898
NH_4Br	4.0594	AuMg	3.259	MgHg	3.44
TlCl	3.8340	AuZn	3.19	MgLa	3.965
TlBr	3.97	AgZn	3.156	MgSr	3.900
TlI	4.198	CoTi	2.986	MgTl	3.628
AlNi	2.881	CuZn	2.945	NiTi	2.972
AlCo	2.862	CuPd	2.988	MgRh	3.099

Table 4.9. *The CuAu I structure, $L1_0$*

Formula unit	CuAu, *copper gold*
Space group:	$P4/mmm$ (no. 123)
Cell dimensions:	$a = 3.960$ Å $c = 3.670$ Å
Cell contents:	2 formula units
Atomic positions:	Cu(1) in (1a) $4/mmm$ (0, 0, 0)
	Cu(2) in (1c) $4/mmm$ (1/2, 1/2, 0)
	Au in 2(e) mmm (0, 1/2, 1/2); (1/2, 0, 1/2)

Examples:

compound	a (Å)	c (Å)	compound	a (Å)	c (Å)
AgTi	4.104	4.077	FePt	3.852	3.723
AlTi	3.976	4.049	MgPd	3.03	3.42
CoPt	2.682	3.675	MnNi	2.6317	3.5295
CrPd	3.879	3.802	β'-MnPt	2.83	3.66
FePd	3.852	3.723	PtZn	4.0265	3.474

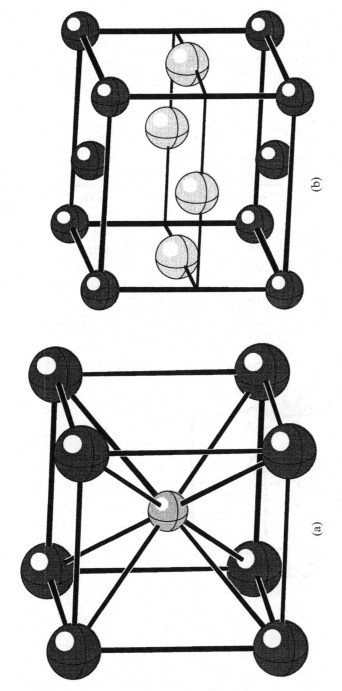

Figure 4.8. Two AB structures for binary compounds. (a) The CsCl or **B2** structure can be thought of as an ordered bcc and (b) the $L1_0$ can be thought of as an ordered ccp.

(a)

(b)

Table 4.10. *The AuCu₃ I structure, L1₂.*

Formula unit	AuCu₃ I
Space group:	$Pm\bar{3}m$ (no. 221)
Cell dimensions:	$a = 3.7493$ Å
Cell contents:	1 formula unit
Atomic positions:	Au in (1a) $m\bar{3}m$ (0, 0, 0)
	Cu in (3c) 4/*mmm* (0, 1/2, 1/2); (1/2, 0, 1/2); (1/2, 1/2, 0)

Examples:

compound	a (Å)	compound	a (Å)	compound	a (Å)
PtAg₃	3.895	SmPt₃	4.0015	ErAl₃	4.215
AgPt₃	3.90	TiPt₃	3.094	NpAl₃	4.262
AlCo₃	3.658	TiZn₃	3.9322	TmAl₃	4.200
AlNi₃	3.572	ZnPt₃	3.893	UAl₃	4.287
AlZr₃	4.372	AlLa₃	5.093	MgPt₃	3.906
FeNi₃	3.545	AlPr₃	5.007	NaPb₃	4.884
FePt₃	3.858	AlPt₃	3.876	YPd₃	4.074
PbPt₃	4.05	AlSm₃	4.901		

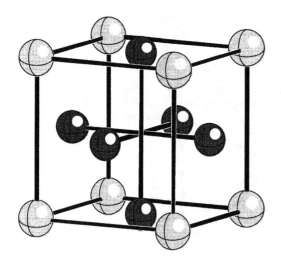

Figure 4.9. In the L1₂ structure, the atoms order on sites in the ccp structure.

Table 4.11. *The BiF$_3$ structure, bismuth trifluoride, D0$_3$.*

Formula unit	BiF$_3$, *bismuth trifluoride*
Space group:	F$m\overline{3}m$ (no. 225)
Cell dimensions:	$a = 5.865$ Å
Cell contents:	4 formula units
Atomic positions:	Bi in (4a) $m\overline{3}m$ $(0, 0, 0) + F$
	F(1) in (4b) $m\overline{3}m$ $(1/2, 1/2, 1/2) + F$
	F(2) in (8c) $\overline{4}3m$ $(1/4, 1/4, 1/4); (1/4, 1/4, 3/4) + F$
Examples:	

compound	a (Å)	compound	a (Å)
BiF$_3$	5.865	SiFe$_3$	5.662
BiLi$_3$	6.722	PrMg$_3$	7.430
PrCd$_3$	7.200	AlFe$_3$	5.78
CeMg$_3$	7.438	SiMn$_3$	5.722
SbCu$_3$	6.01	β-SbLi$_3$	6.572

Table 4.12. *The L2$_1$ structure, Heusler alloy.*

Formula unit	Cu$_2$MnAl
Space group:	F$m\overline{3}m$ (no. 225)
Cell dimensions:	$a = 5.949$ Å
Cell contents:	4 formula units
Atomic positions:	Mn in (4b) $m\overline{3}m$ $(1/2, 1/2, 1/2) + F$
	Al in (4a) $m\overline{3}m$ $(0, 0, 0) + F$
	Cu in (8c) $\overline{4}3m$ $(1/4, 1/4, 1/4); (3/4, 3/4, 3/4) + F$
Examples:	

compound	a (Å)	compound	a (Å)	compound	a (Å)
Cu$_2$MnAl	5.949	Ni$_2$TiAl	5.850	Co$_2$AlZr	6.081
Cu$_2$MnIn	6.1865	Ni$_2$MnSb	5.985	Co$_2$AlNb	5.946
Cu$_2$MnSb	6.097	Ni$_2$MnSn	6.048	Co$_2$AlHf	6.009
Cu$_2$FeSn	5.93	Co$_2$MnSn	5.989	Co$_2$AlTa	5.927
Cu$_2$AlHf	6.172			Co$_2$AlTi	5.847

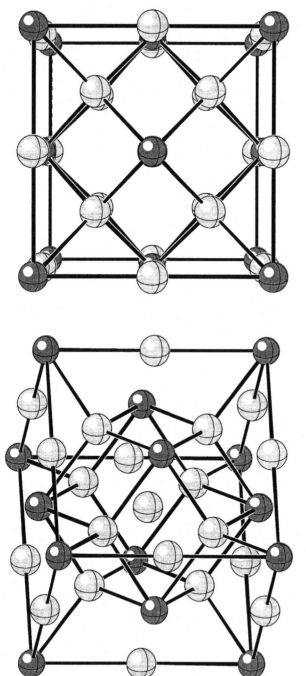

Figure 4.10. In the DO$_3$ structure the Bi atoms occupy fcc positions and the F atoms fill all of the interstitial sites.

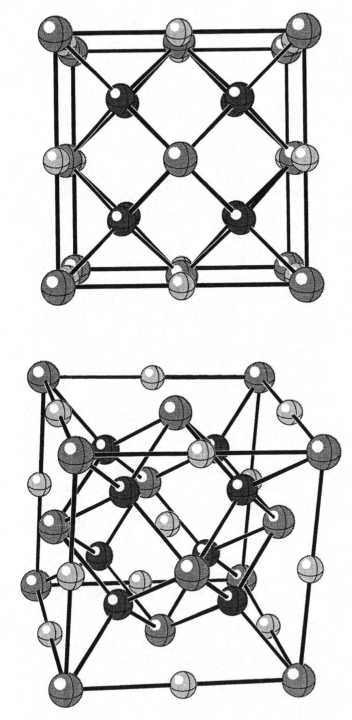

Figure 4.11. The L2$_1$ structure. Al occupy the ccp sites, Mn the octahedral sites, and Cu the tetrahedral sites.

Mn occupy the octahedral sites, and the Cu occupy all of the tetrahedral sites. One might think of this as a ternary version of the BiF_3 structure. See Fig. 4.11.

v.   *The Do_{19} and Do_{24} structures*

The previous examples have all been based on cubic packing. Here, examples of structures formed by the ordered occupation of the sites in the hcp structure are presented. For example, the $D0_{19}$ structure, described in Table 4.13 and shown in Fig. 4.12, has the AB_3 stoichiometry, with the A atoms in one quarter of the close packed sites and the B atoms in the remaining three quarters of the close packed sites. Therefore, by stoichiometry and site filling, it is the hexagonal analog of the $L1_2$ structure.

The $D0_{24}$ structure is built from close packed layers, each of which has the Ni_3Ti stoichiometry, see Fig. 4.13. However, there is a four-layer stacking sequence. For the layers at $z = 0$, 1/4, 1/2, and 3/4, the sequence is ABACABAC. This and more complex stacking sequences are discussed in Section I(ii) of this chapter. It is not unusual that multiple descriptions can be developed for the same structure. For example, in this case we might also say that the layers at $z = 1/4$ and 3/4 make up an ABAB or hcp stacking sequence and that three quarters of the octahedral interstices at $z = 0$ and 1/2 are filled by Ni and the remaining quarter are filled by Ti. Furthermore, the Ni and Ti are ordered in the octahedral interstices. While the best description is frequently a matter of choice, it is important to keep in mind the spirit of the close packing definition. Adjacent framework atoms should be separated by less than two atomic radii and the close packed units should not be distorted by more than 25 % of the appropriate cubic or hexagonal aspect ratios.

The ordered intermetallic structures described in the previous subsections often occur as precipitates in useful ferrous, aluminum, titanium, and nickel-based alloys. In this case, they usually serve to strengthen the matrix phase. Ordered intermetallics are also commonly used as high strength, high temperature materials and as magnetic materials.

vi.   *The A15 structure*

The A15 structure is an example a structure with bcc packing and filled interstices. The A15 structure is described in Table 4.15 and a projection is shown in Fig. 4.14. Note that the B (O) atoms occupy the sites of the bcc structure and that the A (W) atoms occupy one half of the pseudo-tetrahedral interstices. A number of compounds with the A15 structure (including Nb_3Ge and V_3Si) are superconductors at low temperature. In Table 4.15, there are question marks next to phases whose existence is disputed.

Table 4.13. *The DO₁₉ structure.*

Formula unit	Ni₃Sn
Space group:	P6₃ / *mmc* (no. 194)
Cell dimensions:	$a = 5.29$ Å, $c = 4.24$ Å
Cell contents:	2 formula units
Atomic positions:	Sn in (2c) 6̄m2 (1/3, 2/3, 1/4); (2/3, 1/3, 3/4); (x, 2x, 1/4);
	(2$\bar{x}$, $\bar{x}$, 1/4); (x, $\bar{x}$, 1/4)
	Ni in (6h) *mm2* ($\bar{x}$, 2$\bar{x}$, 3/4); (2x, x, 3/4); ($\bar{x}$, x, 3/4)
	$x = 0.833$

Examples

compound	a (Å)	c (Å)	compound	a (Å)	c (Å)
Ni₃In	4.1735	5.1215	Co₃Mo	5.125	4.113
Ni₃Sn	5.29	4.24	Co₃W	5.12	4.12
Fe₃Sn	5.458	4.361	Cd₃Mg	6.2334	5.0449
Ti₃Al	5.77	4.62	Mg₃Cd	6.3063	5.0803

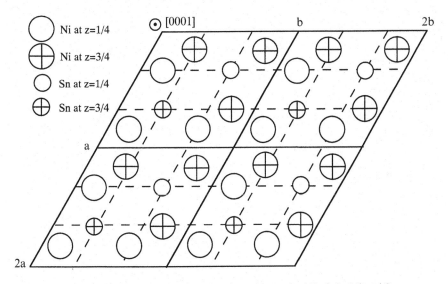

Figure 4.12. A projection of the DO₁₉ structure along the *c*-axis. All of the Ni and Sn atoms occupy hcp lattice sites in an ordered configuration.

Table 4.14. *The DO$_{24}$ structure.*

Formula unit	Ni$_3$Ti		
Space group:	P6$_3$/mmc (no. 194)		
Cell dimensions:	$a = 5.093$ Å, $c = 8.32$ Å		
Cell contents:	4 formula units		
Atomic positions:	Ti(1) in (2a)	$\bar{3}m$	(0, 0, 0); (0, 0, 1/2)
	Ti(2) in (2c)	$\bar{6}m2$	(1/3, 2/3, 1/4); (2/3, 1/3, 3/4).
	Ni in (6g)	2/m	(1/2, 0, 0); (0, 1/2, 0); (1/2, 1/2, 0); (1/2, 0, 1/2); (0, 1/2, 1/2).
	Ni in (6h)	mm2	(1/2, 1/2, 1/2); $(x, 2x, 1/4)$; $(x, \bar{x}, 1/4)$; $(\bar{x}, 2\bar{x}, 3/4)$; $(2x, x, 3/4)$; $(\bar{x}, x, 3/4)$; $x = 0.833$

Examples

compound	a (Å)	c (Å)	compound	a (Å)	c (Å)
Ni$_3$Ti	5.093	8.32	Pd$_3$Hf	5.595	9.192
Co$_3$Ti	2.55	8.24	Pt$_3$Hf	5.636	9.208
Pd$_3$Zr	5.612	9.235	Pt$_3$Zr	5.645	9.235
Al$_3$Dy	6.097	9.534	Pd$_3$Ti	5.489	8.964
Al$_3$Ru	4.81	7.84	Pd$_3$Th	5.856	9.826

Table 4.15. *The A15 structure*

Formula unit	W$_3$O	
Space group:	Pm$\bar{3}$m (no. 223)	
Cell dimensions:	$a = 5.04$ Å	
Cell contents:	2 formula units	
Atomic positions:	O in (2a) $m\bar{3}$	(0, 0, 0); (1/2, 1/2, 1/2)
	W in (6c) $\bar{4}m2$	(1/4, 0, 1/2); (3/4, 0, 1/2); (1/2, 1/4, 0); (0, 1/2, 1/4); (0, 1/2, 3/4); (1/2, 3/4, 0)

Examples

compound	a (Å)	compound	a (Å)	compound	a (Å)	compound	a (Å)
W$_3$O	5.04	Cr$_3$Si	4.558	V$_3$Co	4.6813	Ti$_3$Au	5.0974
W$_3$Si	4.91	Mo$_3$O (?)	5.019	V$_3$Si	4.7243	Nb$_3$Ir	5.134
Cr$_3$O (?)	4.544	V$_3$Al	4.829	V$_3$Pb	4.937	Nb$_3$Bi	5.320

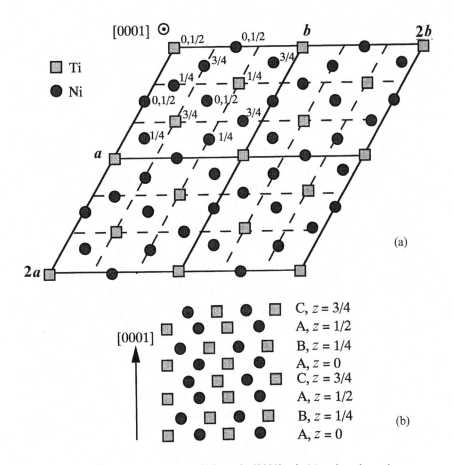

Figure 4.13. The DO_{24} structure, projected along the [0001] axis (a) and a schematic representation of the stacking sequence (b).

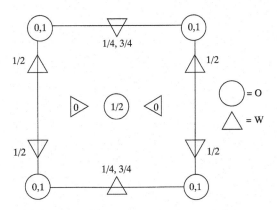

Figure 4.14. A projection of the A15 structure. The atomic arrangement can be described as bcc with filled tetrahedral interstices.

Example 4.1: Applying the close packed description to binary compounds

Table 4.16. *The TiAl$_3$ structure, D0$_{22}$.*

Formula unit	TiAl$_3$, *titanium aluminide*				
Space group:	I4/*mmm*				
Cell dimensions:	$a = 3.85$ Å, $c = 8.596$ Å				
Cell contents:	2 formula units				
Atomic positions:	Ti in (2a)	4/*mmm*	$(0, 0, 0) + I$		
	Al(1) in (2b)	4/*mmm*	$(0, 0, 1/2) + I$		
	Al(2) in (4d)	$\bar{4}m2$	$(0, 1/2, 1/4); (1/2, 0, 1/4) + I$		

Examples:

compound	a (Å)	c (Å)	compound	a (Å)	c (Å)
Al$_3$Nb	3.845	8.601	Ga$_3$Ti	3.789	8.734
Al$_3$Ta	3.842	8.553	Ni$_3$V	3.5424	7.1731
Al$_3$V	3.780	8.321	Pd$_3$Ta	3.880	7.978
Ga$_3$Hf	3.881	9.032	Pd$_3$V	3.847	7.753
Ga$_3$Nb	3.78	8.71			

The D0$_{22}$ crystal structure is specified in Table 4.16. Describe the packing in this structure. In other words, if close packed sites are occupied, specify which atoms occupy them and what the packing arrangement is (bcc, ccp, or hcp). If interstitial sites are occupied, specify the type of sites (O or T), the fraction that are occupied, and which atoms occupy them.

1. The first step in all problems of this sort is to produce a simple, but reasonably accurate sketch, as we did in Chapter 3. It is always easiest to draw the projection on the shortest axis. In this case, inspection of the lattice parameters leads to the conclusion that $a = b \approx 1/2c$, so we project the structure along [010]. The sketch is shown in Fig. 4.15.

2. When developing a packing description, you must look for the signatures of the different packing arrangements: eight coordinate atoms for the bcc structure, face centered cubes for ccp, and alternating stacked close packed layers for hcp and ccp.

3. Here, we note that the unit cell can be approximated as two cubes stacked along c. In each cube, the ccp sites are occupied. Because of the arrangement of Al and Ti, the repeat unit includes two cubes and this makes the c-axis twice the length of a.

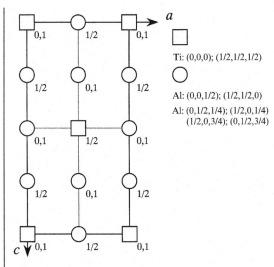

Figure 4.15. A projection of the $D0_{22}$ structure.

Ti: (0,0,0); (1/2,1/2,1/2)

Al: (0,0,1/2); (1/2,1/2,0)
Al: (0,1/2,1/4); (1/2,0,1/4)
    (1/2,0,3/4); (0,1/2,3/4)

4. Therefore, we describe the structure as having a ccp eutactic structure where three quarters of the ccp sites are filled by Al and one quarter are filled by Ti. None of the interstitials are occupied.

5. Note that if the Al and Ti positions in the (002) plane were switched, this would be identical to the $L1_2$ structure. Therefore, it would be appropriate to say that the $D0_{22}$ structure is a superlattice that derives from the $L1_2$ structure.

vii.   *The rock salt and NiAs structures*

It is interesting to note that there are many ccp/hcp analog structures. In other words, structures that differ only in the packing. For example, compare the rock salt and NiAs structures, shown in Figs. 4.16 and 4.17, respectively. In both cases, the cations fill all of the octahedral interstices. However, the anions are ccp in rock salt and hcp in NiAs. While all of the atoms in both of these structures occupy six coordinate sites, not all are octahedral. The hcp packing in NiAs situates the As in trigonal prismatic coordination. Therefore, $NiAs_6$ octahedra share faces along the *c*-axis; this configuration puts the Ni atoms very near one another (the Ni–Ni distance is 1.16 times the Ni–As distance) and suggests that metal–metal bonds are formed. This is why many of the compounds in Table 4.18 exhibit metallic properties.

The fact that the Na and Cl in rock salt have the same coordination means that our choice of which atom is in the eutactic position and which is in the interstitial is arbitrary. In other words, a new structure is not formed by interchanging

Table 4.17. *The rock salt structure, sodium chloride, B1.*

Formula unit	NaCl, *sodium chloride*		
Space group:	$Fm\bar{3}m$ (no. 225)		
Cell dimensions:	$a = 5.6402$ Å		
Cell contents:	4 formula units		
Atomic positions:	Na in (4b) $m\bar{3}m$ $(0, 0, 0) + F$		
	Cl in (4a) $m\bar{3}m$ $(1/2, 1/2, 1/2) + F$		

Examples:

compound	a (Å)	compound	a (Å)	compound	a (Å)	compound	a (Å)
MgO	4.213	MgS	5.200	LiF	4.0270	KF	5.347
CaO	4.8105	CaS	5.6948	LiCl	5.1396	KCl	6.2931
SrO	5.160	SrS	6.020	LiBr	5.5013	KBr	6.5966
BaO	5.539	BaS	6.386	LiI	6.00	KI	7.0655
TiO	4.177	αMnS	5.224	LiH	4.083	RbF	5.6516
MnO	4.445	MgSe	5.462	NaF	4.64	RbCl	6.5810
FeO	4.307	CaSe	5.924	NaCl	5.6402	RbBr	6.889
CoO	4.260	SrSe	6.246	NaBr	5.9772	RbI	7.342
NiO	4.1769	BaSe	6.600	NaI	6.473	AgF	4.92
CdO	4.6953	CaTe	6.356	NaH	4.890	AgCl	5.549
SnAs	5.7248	SrTe	6.660	ScN	4.44	AgBr	5.7745
TiC	4.3285	BaTe	7.00	TiN	4.240	CsF	6.014
UC	4.955	LaN	5.30	UN	4.890	LuSb	6.0555

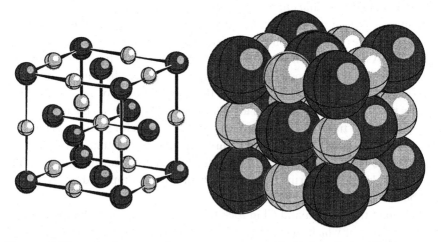

Figure 4.16. The rock salt structure is a ccp arrangement with all of the O sites filled.

Table 4.18. *The nickel arsenide structure, B8₁.*

Formula unit	NiAs, *nickel arsenide*
Space group:	P6₃/mmc (no. 194)
Cell dimensions:	$a = 3.602$ Å, $c = 5.009$ Å
Cell contents:	2 formula units
Atomic positions:	Ni in (2a) $\bar{3}m$ (0, 0, 0); (0, 0, 1/2)
	As in (2c) $\bar{6}m2$ (1/3, 2/3, 1/4); (2/3, 1/3, 3/4)

Examples

compound	a(Å)	c(Å)	c/a	compound	a(Å)	c(Å)	c/a
NiS	3.4392	5.3484	1.555	CoS	3.367	5.160	1.533
NiAs	3.602	5.009	1.391	CoSe	3.6294	5.3006	1.460
NiSb	3.94	5.14	1.305	CoTe	3.886	5.360	1.379
NiSe	3.6613	5.3562	1.463	CoSb	3.866	5.188	1.342
NiSn	4.048	5.123	1.266	CrSe	3.684	6.019	1.634
NiTe	3.957	5.354	1.353	CrTe	3.981	6.211	1.560
FeS	3.438	5.880	1.710	CrSb	4.108	5.440	1.324
FeSe	3.637	5.958	1.638	MnTe	4.1429	6.7031	1.618
FeTe	3.800	5.651	1.487	MnAs	3.710	5.691	1.534
FeSb	4.06	5.13	1.264	MnSb	4.120	5.784	1.404
δ'-NbN*	2.968	5.549	1.870	MnBi	4.30	6.12	1.423
PtB*	3.358	4.058	1.208	PtSb	4.130	5.472	1.325
PtSn	4.103	5.428	1.323	PtBi	4.315	5.490	1.272

Note:

*Anti-NiAs structure

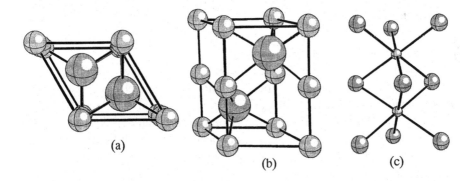

(a)

(b)

(c)

Figure 4.17. The NiAs structure. (a) shows a projection down the *c*-axis. (b) Another view of the unit cell that emphasizes the trigonal prismatic coordination of the As atoms. (c) The Ni atoms are in octahedral coordination and the octahedra share faces.

the atoms. However, in NiAs, the Ni site is octahedral and the As site is trigonal prismatic. If the components are interchanged the so-called 'anti-NiAs' structure is formed. Distinct coordination geometries are a requirement for the existence of an anti-structure. The rock salt structure is one of the most common binary prototypes. The MnP and FeB structures, which are related to NiAs, are described in Tables 3B.11 and 3B.12, in Appendix 3B.

viii.   *The rutile and CdI$_2$ structures*

One must exercise caution when examining structures with the partial occupation of interstitials. Note that one of the limitations of the simple descriptive methods that are employed in this chapter is that the pattern of occupation is not specified. For example, using the methods described in this chapter, TiO$_2$ and CdI$_2$ have the same structural description; both hcp arrays of anions with one half of the octahedral sites filled. However, they are actually quite different because of the way in which the occupied octahedral sites are arranged. In the CdI$_2$ structure, the Cd ions fill all of the octahedral sites in every other layer. Thus, the structure has an anisotropic two-dimensional character. Although there are strong bonds between the Cd and I in the occupied layers, the bonding is very weak between the layers where the I atoms meet, and the crystal is micaceous. In TiO$_2$, on the other hand, one half of the octahedral sites in every layer are filled and, as a result, the structure has strong three-dimensional bonds and resists cleavage.

ix.   *The CdCl$_2$ structure*

The CdCl$_2$ structure, shown in Fig. 4.20, is analogous to CdI$_2$ in the same way that NaCl is analogous to NiAs. Cations occupy the octahedral sites in CdCl$_2$ and CdI$_2$ in exactly the same way (they fill all of the O sites in every other layer), but the anions in CdCl$_2$ are cubic close packed and in CdI$_2$ they are hexagonal close packed.

x.   *The spinel and olivine structures*

The relationship between the olivine (Mg$_2$SiO$_4$) and the spinel (MgAl$_2$O$_4$) structures is similar to that between rock salt and NiAs. Spinel has a cubic close packed anion arrangement and olivine has a hexagonal close packed anion arrangement. The distribution of cations, on the other hand, is equivalent. The chemically diverse spinel compounds (see Table 4.22) can all be thought of as AB$_2$O$_4$, where all of the B atoms occupy octahedral sites and all of the A atoms occupy tetrahedral sites [13].

It is interesting to note that all the details of the rather complex spinel structure are specified entirely by only two parameters, x and a. The oxygen parameter,

Table 4.19. *The rutile structure, titanium dioxide, C4.*

Formula unit	TiO_2, *titanium dioxide*
Space group:	$P4_2/mnm$ (no. 136)
Cell dimensions:	$a = 4.594$Å, $c = 2.958$Å
Cell contents:	2 formula units
Atomic positions:	Ti in (2a) *mmm* $(0, 0, 0)$; $(1/2, 1/2, 1/2)$
	O in (4f) *m2m* $(x, x, 0)$; $(\bar{x}, \bar{x}, 0)$
	$(1/2 + x, 1/2 - x, 1/2)$, $(1/2 - x, 1/2 + x, 1/2)$
	$x = 0.3$

Examples

compound	a(Å)	c(Å)	x	compound	a(Å)	c(Å)	x
TiO_2	4.5937	2.9581	0.305	$CrSbO_4$	4.57	3.042	
CrO_2	4.41	2.91		$CrTaO_4$	4.626	3.009	
GeO_2	4.395	2.859	0.307	$FeNbO_4$	4.68	3.05	
IrO_2	4.49	3.14		$FeSbO_4$	6.623	3.011	
β-MnO_2	4.396	2.871	0.302	$FeTaO_4$	4.672	3.042	
MoO_2	4.86	2.79		$GaSbO_4$	4.59	3.03	
NbO_2	4.77	2.96		$RhSbO_4$	4.601	3.100	
OsO_2	4.51	3.19		$RhVO_4$	6.607	2.923	
PbO_2	4.946	3.379		CoF_2	4.6951	3.1796	0.306
RuO_2	4.51	3.11		FeF_2	4.6966	3.3091	0.300
SnO_2	4.7373	3.1864	0.307	MgF_2	4.623	3.052	0.303
TaO_2	4.709	3.065		MnF_2	4.8734	3.3099	0.305
WO_2	4.86	2.77		NiF_2	4.6506	3.0836	0.302
$AlSbO_4$	4.510	2.961	0.305	PdF_2	4.931	3.367	
$CrNbO_4$	4.635	3.005		ZnF_2	4.7034	3.1335	0.303

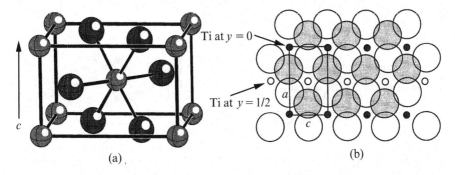

Ti at $y = 0$

Ti at $y = 1/2$

(a)   (b)

Figure 4.18. The rutile structure. (a) shows an oblique projection, (b) shows a projection down [010]. The Ti ions occupy half of the octahedral sites by filling every other row along [100]. The filled rows are staggered in every other layer [12].

Table 4.20. *The CdI$_2$ structure, cadmium iodide, C6.*

Formula unit	CdI$_2$, *cadmium iodide*
Space group:	P$\bar{3}$m1 (no. 164)
Cell dimensions:	$a = 4.24$Å, $c = 6.84$Å
Cell contents:	1 formula unit per cell
Atomic positions:	Cd in (1a) $\bar{3}$m (0, 0, 0)
	I in (2d) 3m (1/3, 2/3, z); (2/3, 1/3, z); z = 1/4

Examples

compound	a (Å)	c (Å)	compound	a (Å)	c (Å)
CdI$_2$	4.24	6.84	TiCl$_2$	3.561	5.875
CaI$_2$	4.48	6.96	VCl$_2$	3.601	5.835
CoI$_2$	3.96	6.65	Mg(OH)$_2$	3.147	4.769
FeI$_2$	4.04	6.75	Ca(OH)$_2$	3.584	4.896
MgI$_2$	4.14	6.88	Fe(OH)$_2$	3.258	4.605
MnI$_2$	4.16	6.82	Co(OH)$_2$	3.173	4.640
PbI$_2$	4.555	6.977	Ni(OH)$_2$	3.117	4.595
ThI$_2$	4.13	7.02	Cd(OH)$_2$	3.48	4.67
TiI$_2$	4.110	6.820	TiS$_2$	3.408	5.701
TmI$_2$	4.520	6.967	TaS$_2$	3.36	5.90
VI$_2$	4.000	6.670	ZrS$_2$	3.662	5.813
YbI$_2$	4.503	6.972	TiSe$_2$	3.535	6.004
ZnI$_2$(I)	4.25	6.54	VSe$_2$	3.353	6.101
VBr$_2$	3.768	6.180	ZrSe$_2$	3.77	6.14
TiBr$_2$	3.629	6.492	SnSe$_2$	3.811	6.137
MnBr$_2$	3.82	6.19	TiTe$_2$	3.766	6.491
FeBr$_2$	3.74	6.17	VTe$_2$	3.689	6.405
CoBr$_2$	3.68	6.12	ZrTe$_2$	3.95	6.63

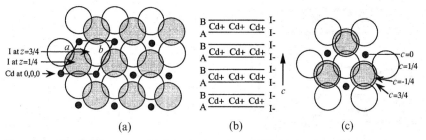

Figure 4.19. The CdI$_2$ structure. (a) A projection along [001]. (b) A schematic projection normal to [001]. Horizontal lines represent layers of I atoms in close packed positions. Cd atoms occupy half of the octahedral positions by occupying all of the sites in every second layer along [001]. (c) Another projection down [001] showing the atoms in three close packed layers [12].

Table 4.21. *The CdCl₂ structure, cadmium chloride, C19.*

Formula unit	$CdCl_2$, *cadmium chloride*			
Space group:	$R\bar{3}m$ (no. 166)			
Cell dimensions:	$a = 3.85\text{Å}$, $c = 17.46\text{Å}$			
Cell contents:	3 formula units per cell (hex)			
Atomic positions:	Cd in (1a) $3m$ $(0, 0, 0) + \text{rh}$			
	Cl in (2c) $3m$ $(0, 0, z); (0, 0, \bar{z}) + \text{rh}$, $z = 1/4$			

Examples

compound	a (Å)	c (Å)	compound	a (Å)	c (Å)
$CdCl_2$	3.845	17.457	$NiCl_2$	3.543	17.335
$CdBr_2$	3.95	18.67	$NiBr_2$	3.708	18.300
$CoCl_2$	3.544	17.43	NiI_2	3.892	19.634
$FeCl_2$	3.579	17.536	$ZnBr_2$	3.92	18.73
$MgCl_2$	3.596	17.589	ZnI_2	4.25	21.5
$MnCl_2$	3.6786	17.470	Cs_2O^*	4.256	18.99

Note:

*Anti-$CdCl_2$ structure

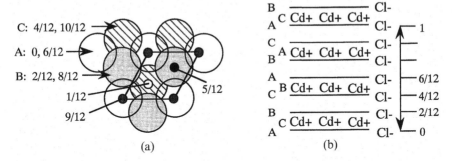

Figure 4.20. The structure of $CdCl_2$ is similar to CdI_2; cations fill the octahedral sites in every other layer. However, the stacking sequence of the anion layer is cubic close packed, as shown in the [001] projection (a) and the schematic (b) [12].

x, is very close to 3/8. It varies from structure to structure and must be determined experimentally. We should also mention that there are *inverse-spinels* where all of the A and half of the B atoms (in AB_2O_4) occupy the octahedral sites and the other half of the B atoms occupy the tetrahedral sites. It is important to realize that interstitial site occupation is frequently less than ideal. For example, the so-called normal and inverse cases of the spinel are actually limiting conditions that are not

Table 4.22. *The spinel structure, magnesium aluminate, H1$_1$ [13].*

Formula unit	$MgAl_2O_4$, *magnesium aluminate*
Space group:	$Fd\bar{3}m$ (no. 227)
Cell dimensions:	$a = 8.086$
Cell contents:	8 formula units
Atomic positions:	Mg in (8a) $\bar{4}3m$ (0, 0, 0); (1/4, 1/4, 1/4) + F

Al in (16c) $\bar{3}m$ (5/8, 5/8, 5/8); (5/8, 7/8, 7/8); (7/8, 5/8, 7/8); (7/8, 7/8, 5/8) + F

O in (32e) $3m$ $(x, x, x); (x, \bar{x}, \bar{x}); (1/4-x, 1/4-x, 1/4-x); (1/4-x, x+1/4, x+1/4); (\bar{x}, \bar{x}, x); (\bar{x}, x, \bar{x}); (x+1/4, 1/4-x, x+1/4); (x+1/4, x+1/4, 1/4-x) + F; x = 3/8$

Examples

compound	a (Å)	x	compound	a (Å)	x
$MgAl_2O_4$	8.086	0.387	$CdMn_2O_4$	8.22	
$MgTi_2O_4$	8.474		$CdFe_2O_4$	8.69	
MgV_2O_4	8.413	0.385	$CdGe_2O_4$	8.39	
$MgCr_2O_4$	8.333	0.835	$CdRh_2O_4$	8.781	
$MgMn_2O_4$	8.07	0.385	$MgYb_2S_4$	10.957	
$MgRh_2O_4$	8.530		$CaIn_2S_4$	10.774	<0.393
$MnTi_2O_4$	8.600		$MnCr_2S_4$	10.129	
MnV_2O_4	8.522	0.388	$FeCr_2S_4$	9.998	
$MnCr_2O_4$	8.437		$CoCr_2S_4$	9.934	
Mn_3O_4	8.13		$CoRh_2S_4$	9.71	
$MnRh_2O_4$	8.613		$CuTi_2S_4$	9.880	0.382
$FeCr_2O_4$	8.377		CuV_2S_4	9.824	0.384
$CoAl_2O_4$	8.105	0.390	$CuCr_2S_4$	9.629	0.381
CoV_2O_4	8.407		$CuRh_2S_4$	9.72	
$CoCr_2O_4$	8.332		$ZnAl_2S_4$	9.988	0.384
$CoMn_2O_4$	8.1		$ZnCr_2S_4$	9.983	
Co_3O_4	8.083		$CdCr_2S_4$	10.207	0.375
$CoRh_2O_4$	8.495		$CdIn_2S_4$	10.797	0.386
$NiCr_2O_4$	8.248		$HgCr_2S_4$	10.206	0.392
$NiRh_2O_4$	8.36		$HgIn_2S_4$	10.812	<0.403
$CuCr_2O_4$	8.532		$CuCr_2Se_4$	10.365	0.380
$CuMn_2O_4$	8.33	0.390	$ZnCr_2Se_4$	10.443	0.378
$CuRh_2O_4$	8.702		$CdCr_2Se_4$	10.721	0.383
$ZnAl_2O_4$	8.086		$CuCr_2Te_4$	11.049	0.379

Table 4.22. (*cont.*)

compound	a (Å)	x	compound	a (Å)	x
Examples					
ZnV_2O_4	8.414		$MoNa_2O_4$	8.99	
$ZnCr_2O_4$	8.327		WNa_2O_4	8.99	
$ZnMn_2O_4$	8.087		$MoAg_2O_4$	9.26	0.364
$ZnFe_2O_4$	8.416	0.380	$GeMg_2O_4$	8.225	
$ZnCo_2O_4$	8.047		$GeFe_2O_4$	8.411	0.375
$ZnGa_2O_4$	8.37		$GeCo_2O_4$	8.317	0.375
$ZnRh_2O_4$	8.54		$GeNi_2O_4$	8.221	0.375
$CdCr_2O_4$	8.567	0.385	$ZnK_2(CN)_4$	12.54	0.37

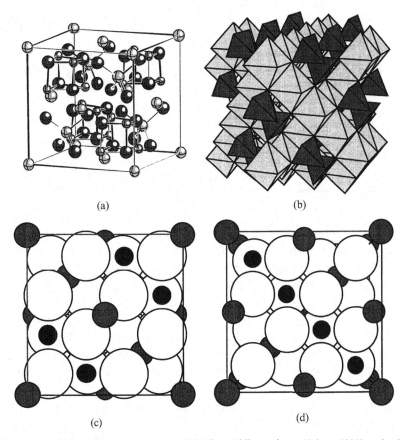

(a)　　　　　　　　　　　　　(b)

(c)　　　　　　　　　　　　　(d)

Figure 4.21. The spinel structure. (a) and (b) show oblique views. (c) is an [001] projection of the upper half of the cell and (d) is an [001] projection of the lower half of the cell.

Table 4.23. *Inverse and partially inverse spinels.*

compound	λ	a (Å)	x	compound	λ	a (Å)	x
$MgIn_2O_4$	0.5	8.81	0.372	$SnMg_2O_4$	0.5	8.60	
Fe_3O_4	0.5	8.394	0.379	$TiMn_2O_4$	0.5	8.67	
$FeCo_2O_4$	0.5	8.254		$SnMn_2O_4$	0.5	8.865	
$FeGa_2O_4$	0.5	8.360		$TiFe_2O_4$	0.5	8.50	0.390
$CoFe_2O_4$	0.5	8.390		$TiCo_2O_4$	0.5	8.465	
$NiMn_2O_4$	0.5	8.390	0.383	VCo_2O_4	0.5	8.379	
$NiFe_2O_4$	0.5	8.325	0.381	$SnCo_2O_4$	0.5	8.644	0.375
$NiCo_2O_4$	0.5	8.121		$TiZn_2O_4$	0.5	8.445	0.380
$NiGa_2O_4$	0.5	8.258	0.387	VZn_2O_4	0.5	8.38	
$CuFe_2O_4$	0.5	8.445	0.380	$SnZn_2O_4$	0.5	8.665	0.390
$MgIn_2S_4$	0.5	10.687	<0.387	$MgFe_2O_4$	0.45	8.389	0.382
$CrAl_2S_4$	0.5	9.914	0.384	$MgGa_2O_4$	0.33	8.280	0.379
$CrIn_2S_4$	0.5	10.59	0.386	$MnAl_2O_4$	0.15	8.242	
$FeIn_2S_4$	0.5	10.598	<0.387	$MnFe_2O_4$	0.10	8.507	0.835
$CoIn_2S_4$	0.5	10.559	0.384	$MnGa_2O_4$	0.10	8.435	
$NiIn_2S_4$	0.5	10.464	0.384	$FeMn_2O_4$	0.33	8.31	
$NiLi_2F_4$	0.5	8.31	0.381	$CoGa_2O_4$	0.45	8.307	
$TiMg_2O_4$	0.5	8.445	0.390	$NiAl_2O_4$	0.375	8.046	0.831
VMg_2O_4	0.5	8.39	0.386	$CuAl_2O_4$	0.20	8.086	
$SnMg_2O_4$	0.5	8.60		$MnIn_2S_4$	0.33	10.694	<0.390
				VMn_2O_4	0.40	8.575	0.382

always realized; it is also possible for the A and B cations to be completely disordered over the octahedral and tetrahedral sites. The exact distribution of the metal atoms is influenced by a number of crystal chemical factors including ionic size, charge, and the crystal field stabilization energy. The degree of disorder in the cation distribution is represented by the factor λ, which is the fraction of B cations at tetrahedral sites. $\lambda = 0$ specifies an ideal normal spinel. $\lambda = 1/2$ specifies an ideal inverse spinel. Intermediate values indicate some degree of disorder. When $\lambda = 1/3$, the cations are distributed at random. Examples of inverse and disordered spinels are given below. The manner in which the coordination polyhedra are linked is illustrated in Fig. 4.21(b).

A number of the transition metal ferrite spinels have useful magnetic properties. For example, magnetite (Fe_3O_4) is the naturally occurring magnetic ore that was known as lodestone. When the formula is written with respect to the

Table 4.24. *The olivine structure, magnesium silicate.*

Formula unit	Mg_2SiO_4, *magnesium silicate*
Space group:	*Pnma* (no. 62)
Cell dimensions:	$a = 10.26$Å, $b = 6.00$Å, $c = 4.77$Å
Cell contents:	4 formula units
Atomic positions:	Mg(1) in (4a) $\bar{1}$ $(0, 0, 0); (0, 1/2, 0); (1/2, 0, 1/2); (1/2, 1/2, 1/2)$
	Mg(2) in (4c) m $\pm(x, 1/4, z); (x + 1/2, 1/4, 1/2 - z)$
	$x = 0.277$ $z = 0.99$
	Si in (4c) m $x = 0.094$ $z = 0.427$
	O(1) in (4c) m $x = 0.092$ $z = 0.776$
	O(2) in (4c) m $x = 0.448$ $z = 0.22$
	O(3) in (8d) 1 $\pm(x, y, z); (x + 1/2, y - 1/2, 1/2 - z);$
	$(x, 1/2 - y, z); (x + 1/2, y, 1/2 - z)$
	$x = 0.163$ $y = 0.034$ $z = 0.278$

Examples

compound	a(Å)	b(Å)	c(Å)
Mg_2SiO_4	10.26	5.00	4.77
Fe_2SiO_4	10.49	6.10	4.83
γ-Ca_2SiO_4	11.371	6.782	5.091
Sr_2SiO_4	9.66	7.262	5.59
Ba_2SiO_4	10.17	7.56	5.76
Mg_2GeO_4	10.30	6.02	4.92
Ca_2GeO_4	11.13	6.57	5.20
$CaMgSiO_4$	11.08	6.37	4.815

dual valence of the iron ($Fe^{+2}Fe_2^{+3}O_4$), the manner in which this conforms to the inverse spinel structure is clarified. The so-called *defect-spinels* such as γ-Al_2O_3 and γ-Fe_2O_3 are also noteworthy. γ-alumina is a metastable form of alumina formed at (relatively) low temperature by the thermal decomposition of aluminum hydroxide, $Al(OH)_3$, also known as gibbsite. The gamma forms of Al_2O_3 and Fe_2O_3 have essentially the same framework as the spinel, but some of the octahedral sites that would normally be occupied are vacant. In this case, there are 8/3 Al vacancies per unit cell, with respect to the ideal spinel. The formula for the defect spinel can be alternatively represented as $(Al_8^{+3})_T[Al_{40/3}V_{8/3}]_OO_{32}$, where the V stands for a vacancy.

   Olivine is the name for the mineral $(Mg,Fe)_2SiO_4$. The magnesium rich version of the mineral, Mg_2SiO_4, is most accurately called forsterite. As mentioned earlier, this is the hexagonally packed analog of the spinel structure. Note that although the anions occupy the eutactic positions of the hcp arrangement, the unit cell is not hexagonal. The olivine structure is also sometimes referred to as the chrysoberyl structure, for the isostructural mineral Al_2BeO_4.

xi. *The sphalerite and wurtzite structures*

Periodic trends are frequently evident in the selection of a structure type. As an example, we can consider the ccp/hcp analog structures sphalerite (zinc blende or ZnS) and wurtzite. These structure are closely related to the diamond structure. The atoms in the diamond structure occupy all of the ccp sites and all of the T+ sites (Table 3B.13). The binary sphalerite structure is a compositionally ordered version of the diamond structure. It has a ccp arrangement of S atoms, with the Zn filling all of the T+ sites. The Zn atoms also fill the T+ sites in ZnO. However, the O atoms have an hcp arrangement. The details of these two structures are given in Tables 4.25 and 4.26 and the structures are compared in Fig. 4.22. Note that in the sphalerite structure, the pseudo-close packed layers are stacked along the [111] direction. In the wurtzite structure, they are stacked along [0001]. Most of the elemental group IV materials, including C, Si, Ge, and Sn, assume the diamond structure. The periodicity is evident when one notes that more ionic compounds (group I–VII and II–VI combinations) are more likely to form the wurtzite structure and more covalent compounds (group III–V combinations) are more likely to form the sphalerite structure. The wurtzite and sphalerite structures are important because of the vast majority of useful electronic materials (III–V and II–VI compounds and their ternary combinations) that either take one of these structures or a closely related superlattice structure. The more complex structures are described in Section I(i) of this chapter.

xii. *The fluorite structure*

It is important to note that the anions are not always in the eutactic positions. For example, the fluorite (CaF_2) structure is illustrated in Fig. 4.23. In this structure, the Ca ions occupy the ccp eutactic positions and the F ions fill the tetrahedral sites. In the anti-fluorite structure, the occupations are reversed and the anions form a eutactic array, while the cations fill the tetrahedral voids.

xiii. *The perovskite structure*

The perovskite or $CaTiO_3$ structure provides us with an important example of the fact that cations and anions can mix on the eutactic sites. In this case, it is a combination of the O and the Ca that form the cubic close packed framework. The O atoms occupy three quarters of the ccp sites and the Ca atoms occupy the

Table 4.25. *The sphalerite (zinc blende) structure*, zinc sulfide, *B3*.

Formula unit	ZnS, *zinc sulfide*
Space group:	F$\bar{4}$3m (no. 216)
Cell dimensions:	$a = 5.4060$Å
Cell contents:	4 formula units per cell
Atomic positions:	S in (4a) $\bar{4}$3m (0, 0, 0) + F
	Zn in (4c) $\bar{4}$3m (1/4, 1/4, 1/4) + F

Examples

compound	a (Å)	compound	a (Å)	compound	a (Å)
CuF	4.255	BeS	4.8624	β-CdS	5.818
CuCl	5.416	BeSe	5.07	CdSe	6.077
γ-CuBr	5.6905	BeTe	5.54	CdTe	6.481
γ-CuI	6.051	β-ZnS	5.4060	HgS	5.8517
γ-AgI	6.495	ZnSe	5.667	HgSe	6.085
β-MnS, red	5.600	ZnTe	6.1026	HgTe	6.453
β-MnSe	5.88	β-SiC	4.358	AlAs	5.662
BN	3.616	GaP	5.448	InAs	6.058
BP	4.538	GaAs	5.6534	AlSb	6.1347
BAs	4.777	GaSb	6.095	InSb	6.4782
AlP	5.451	InP	5.869		

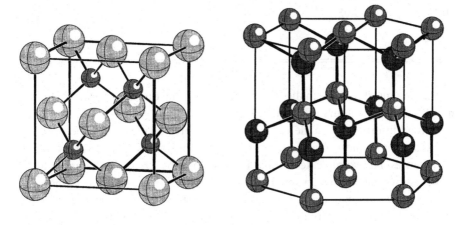

Figure 4.22. A comparison of the (a) sphalerite and (b) wurtzite structures.

Table 4.26. *The wurtzite structure, zinc oxide, B4.*

Formula unit	ZnO, *zinc oxide*
Space group:	$P6_3mc$ (no. 186)
Cell dimensions:	$a = 3.2495\text{Å}, c = 5.2069\text{Å}$
Cell contents:	2 formula units per cell
Atomic positions:	Zn in (2b) 3m (1/3, 2/3, z); (2/3, 1/3, 1/2 + z) z = 0
	O in (2b) 3m (1/3, 2/3, z); (2/3, 1/3, 1/2 + z) z = 3/8

Examples

compound	a (Å)	c (Å)	u	c/a
ZnO	3.2495	5.2069	0.345	1.602
ZnS	3.811	6.234		1.636
ZnSe	3.98	6.53		1.641
ZnTe	4.27	6.99		1.637
BeO	2.698	4.380	0.378	1.623
CdS	4.1348	6.7490		1.632
CdSe	4.30	7.02		1.633
MnS	3.976	6.432		1.618
MnSe	4.12	6.72		1.631
AgI	4.580	7.494		1.636
AlN	3.111	4.978	0.385	1.600
GaN	3.180	5.166		1.625
InN	3.533	5.693		1.611
TaN	3.05	4.94		1.620
NH_4F	4.39	7.02	0.365	1.600
SiC	3.076	5.048		1.641

remaining quarter. The smaller Ti atoms occupy one quarter of the octahedral positions. The structure is shown in Fig. 4.24.

   In addition to all of the compounds that form as ideal cubic perovskites, there are many more that have perovskite arrangements, but distorted lattice symmetries. For example, tetragonal $BaTiO_3$ and $PbTiO_3$ are both based on the cubic perovskite, but have a small dilation along the c-axis that reduces the symmetry. It is common to classify compounds with the same arrangement of atoms and bonding connectivity as the same structure type, even if they do not have exactly the same symmetry. It is interesting to note that compounds with the perovskite structure have a wide range of interesting and important properties. Perovskite structure materials are known for their superconducting properties

Table 4.27. *The fluorite structure, calcium fluoride, C1.*

Formula unit	CaF_2, *calcium fluoride*
Space group:	$Fm\bar{3}m$ (no. 225)
Cell dimensions:	$a = 5.4626$Å
Cell contents:	4 formula units per cell
Atomic positions:	Ca in (4a) $m\bar{3}m$ $(0, 0, 0) + F$
	F in (8c) $\bar{4}3m$ (1/4, 1/4, 1/4); (1/4, 1/4, 3/4) + F

Examples

	fluorite				anti-fluorite		
compound	a (Å)	compound	a (Å)	compound	a (Å)	compound	a (Å)
CaF_2	5.4626	PbO_2	5.349	Li_2O	4.6114	K_2O	6.449
SrF_2	5.800	CeO_2	5.411	Li_2S	5.710	K_2S	7.406
$SrCl_2$	6.9769	PrO_2	5.392	Li_2Se	6.002	K_2Se	7.692
BaF_2	6.200	ThO_2	5.600	Li_2Te	6.517	K_2Te	8.168
$BaCl_2$	7.311	PaO_2	—	Na_2O	5.55	Rb_2O	6.74
CdF_2	5.3895	UO_2	5.372	Na_2S	6.539	Rb_2S	7.65
HgF_2	5.5373	NpO_2	5.433	Na_2Se	6.823		
EuF_2	5.836	PuO_2	5.386	Na_2Te	7.329		
β-PbF_2	5.940	AmO_2	5.376				
		CmO_2	5.3598				

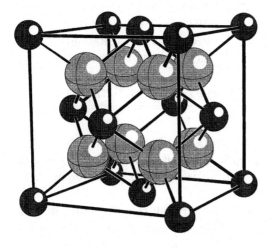

Figure 4.23. In the fluorite (CaF_2) structure, the cations occupy the ccp positions and the anions fill the tetrahedral sites.

Table 4.28. *The perovskite structure, calcium titanate, E2$_1$.*

Formula unit	$CaTiO_3$, *calcium titanate*		
Space group:	$Pm\bar{3}m$ (no. 221)		
Cell dimensions:	$a = 3.84$Å		
Cell contents:	1 formula unit per cell		
Atomic positions:	Ti in (1a) $m\bar{3}m$ (0, 0, 0)		
	Ca in (1b) $m\bar{3}m$ (1/2, 1/2, 1/2)		
	O in (3d) $4/mmm$ (0, 0, 1/2); (0, 1/2, 0); (1/2, 0, 0)		

Examples

compound	a (Å)	compound	a (Å)	compound	a (Å)	compound	a (Å)
$BaCeO_3$	4.397	$KNbO_3$	4.007	$PrAlO_3$	3.757	$AgZnF_3$	3.98
$BaTiO_3$	4.012	$KTaO_3$	3.9858	$PrCrO_3$	3.852	$CsCaF_3$	4.552
$BaMoO_3$	4.0404	$LaAlO_3$	3.778	$PrFeO_3$	3.887	$CsCdBr_3$	5.33
$BaPbO_3$	4.273	$LaCrO_3$	3.874	$PrGaO_3$	3.863	$CsCdCl_3$	5.20
$BaPrO_3$	4.354	$LaFeO_3$	3.920	$PrMnO_3$	3.82	$CsHgBr_3$	5.77
$BaTiO_3$	4.0118	$LaGaO_3$	3.874	$PrVO_3$	3.89	$CsPbCl_3$	5.605
$BaZrO_3$	4.1929	$LaRhO_3$	3.94	$SmAlO_3$	3.734	$CsPbBr_3$	5.874
$CaTiO_3$	3.84	$LaTiO_3$	3.92	$SmCoO_3$	3.75	$KCdF_3$	4.293
$CaVO_3$	3.76	$LaVO_3$	3.99	$SmCrO_3$	3.812	$KCoF_3$	4.069
$CeAlO_3$	3.772	Li_xWO_3	3.72	$SmFeO_3$	3.845	$KFeF_3$	4.122
$DyMnO_3$	3.70	$NaAlO_3$	3.73	$SmVO_3$	3.89	$KMgF_3$	3.973
$EuAlO_3$	3.725	$NaTaO_3$	3.881	$SrFeO_3$	3.869	$KMnF_3$	4.190
$EuCrO_3$	3.803	$NaWO_3$	3.8622	$SrMoO_3$	3.9751	$KNiF_3$	4.012
$EuFeO_3$	3.836	$NdAlO_3$	3.752	$SrTiO_3$	3.9051	$KZnF_3$	4.055
$EuTiO_3$	3.897	$NdCoO_3$	3.777	$SrZrO_3$	4.101	$RbCoF_3$	4.062
$GdAlO_3$	3.71	$NdCrO_3$	3.835	$YAlO_3$	3.68	$RbCaF_3$	4.452
$GdCrO_3$	3.795	$NdFeO_3$	3.870	$YCrO_3$	3.768	$RbMnF_3$	4.250
$GdFeO_3$	3.820	$NdMnO_3$	3.80	$YFeO_3$	3.785	$TlCoF_3$	4.138

[(K,Ba)BiO$_3$], piezoelectric properties [Pb(Zr,Ti)O$_3$], relaxor ferroelectric properties [Pb(Nb,Mg)O$_3$], dielectric properties [BaTiO$_3$], electro-optic properties [(Pb,La)(Zr,Ti)O$_3$], magneto-resistive properties [LaMnO$_3$], catalytic properties [LaCrO$_3$], and protonic conductivity [BaCeO$_3$].

In addition to all of the metal oxides that crystallize in the perovskite structure, it is interesting to note that many metal carbides and nitrides, such as AlCFe$_3$, AlCMn$_3$, AlCTi$_3$, TlCTi$_3$, InCTi$_3$, ZnCMn$_3$, Fe$_4$N, PdNCu$_3$, and TaThN$_3$, also take this structure.

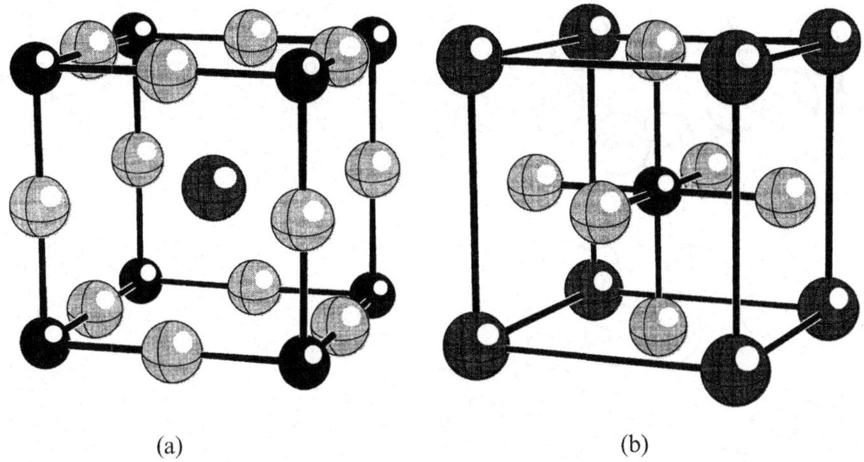

(a)   (b)

Figure 4.24. Two representations of the perovskite structure. In (a), the Ti is at the cell vertices and the Ca in the center site. In (b), the Ti is at the center.

Example 4.2: Applying the close packed description to more complex crystals

(i) ABO_4 compounds sometimes take the scheelite structure, specified in Table 3B.3. Describe the scheelite structure in terms of its packing. In other words, if close packed sites are occupied, specify which atoms occupy them and what the packing arrangement is (bcc, ccp, or hcp). If interstitial sites are occupied, specify the type of sites (O or T), the fraction that are occupied, and which atoms occupy them.

(ii) Compare this structure to one of the prototypes described earlier.

(iii) Ternary ABO_4 compounds can also take the rutile structure (see Table 4.19) where the A and B atoms are disordered on the 2a site. Specify a criterion that can be used to predict which of the two structures an ABO_4 compound will assume.

1. The first step is always to produce a sketch of the structure showing the atomic positions. In this case, we choose a [100] projection that will show the stacking along the longer [001] axis. This sketch is shown in Fig. 4.25.

From the sketch, we can see that the metal atoms are compositionally ordered on the sites of a ccp eutactic structure. Based on this arrangement, we can say that the O atoms fill all of the tetrahedral interstices, noting that in each case the O is actually displaced from the center of the tetrahedron so that it is quite close to three metals, and much further from the fourth. The octahedral interstices are empty. In this arrangement, the A cations have eight O nearest neighbors and the B cations have four.

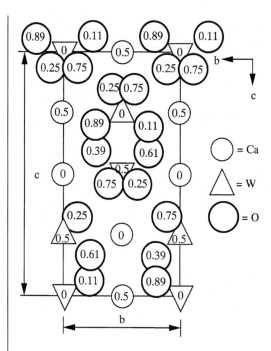

Figure 4.25. A sketch of the atomic positions in the scheelite structure. The projection is shown along the [100] direction.

2. We note the similarity between this structure and the fluorite structure, based on the ccp packing of metal ions and the placement of the anions in the tetrahedral inter-stices. In this case, however, the compositional ordering of the two types of metal cations on the ccp sites makes the repeat unit along the c-axis approximately twice the length of a and b. Therefore, it is similar to two fluorite cells stacked upon one another.

3. In the rutile structure, all of the metal cations are in octahedral coordination. In the scheelite structure, the A cations are eight-coordinate and the B cations are four-coordinate. This implies that the A and B cations in rutile are approximately the same size, while those in scheelite are different; the A cation should be larger than the B cation. Therefore, based on ratios of the A and B radii, it should be possible to predict which structure each compound will form. Begin by looking up the cation radii of all of the examples listed in Table 3B.3 and 4.19. For a consistent comparison, use the octahedral radii. A plot of r_B v. r_A, shown in Fig. 4.26, clearly illustrates that the schee-lite structure is formed when r_A is sufficiently larger than r_B. Based on this, we can say that if $r_A/r_B < 1.3$, then the rutile structure is likely and if $r_A/r_B > 1.3$, then the scheelite structure is likely. It is an interesting exercise to extend this comparison to other com-pounds with the ABO_4 stoichiometry.

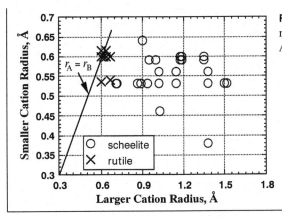

Figure 4.26. Radius ratios for rutile and scheelite structured ABO_4 compounds

(G) Interstitial compounds

Crystal structures of compounds formed when a metal is reacted with smaller nonmetallic atoms such as H, C, B, or N can also be described as eutactic arrangements with the smaller atoms in the interstitial sites. For this reason, these compounds are widely known as *interstitial compounds* [14–16]. While the name has persisted, several observations indicate that it is not wholly accurate to think of most of these compounds as ordered, interstitial solid solutions. First, metal borides stand apart from the hydrides, carbides, and nitrides because B–B bonding influences the structure. Second, there are a number of dicarbides (for example, CaC_2, LaC_2, CeC_2) where C–C bonding leads to the formation of a C_2^{2-} anion which occupies a greatly distorted octahedral position. Finally, we should not imagine that the interstitial atom simply dissolves into the metal with a negligible perturbation of the structure; the arrangement of the metal atoms in most transition metal hydrides, carbides, and nitrides is usually different from that of the pure metal. For example, pure Ti metal takes the hcp structure, but TiH has the sphalerite structure where the Ti occupy ccp sites. The ccp Ti arrangement is maintained in TiH_2, which has the fluorite structure. Similarly, while the Cr atoms in pure Cr metal have the bcc arrangement, in CrH they have the hcp arrangement (anti-NiAs), and in CrH_2 they have a ccp arrangement (fluorite).

Alkali metal hydrides (LiH, NaH, KH, RbH, CsH) and alkaline earth metal hydrides (MgH_2, CaH_2, SrH_2, and BaH_2) are called the 'salt-like' hydrides because they are colorless and conduct ions rather than electrons. The alkali metal hydrides take the rock salt structure, MgH_2 takes the rutile structure, and the other group II hydrides take the fluorite structure. The transition metal

Table 4.29. *The structures of transition metal carbides and nitrides [14].*

metal	metal structure	carbide structure	nitride structure
Sc	ccp, hcp	—	NaCl
La	ccp, hcp	—	NaCl
Ce	ccp, hcp	—	NaCl
Pr	hcp	—	NaCl
Nd	hcp	—	NaCl
Ti	bcc, hcp	NaCl	NaCl
Zr	bcc, hcp	NaCl	NaCl
Hf	bcc, hcp	NaCl	NaCl (?)
Th	ccp	NaCl	NaCl
V	bcc	NaCl	NaCl
Nb	bcc	NaCl	NaCl
Ta	bcc	NaCl	TaN
Cr	bcc	Hex?	NaCl
Mo	bcc, hcp	WC	WC
W	bcc	WC	WC
U(γ)	bcc	NaCl	NaCl

hydrides are typically metallic or semiconducting in nature and are often non-stoichiometric. Furthermore, while the metal–metal bond distances in the salt-like hydrides are contracted with respect to the pure metal, they are extended in the transition metal hydrides. Ti, Zr, and Hf form dihydrides with the fluorite structure (although it is distorted at room temperature). V, Nb, and Ta each form a bcc solid solution at low H concentration, and later transform to a distorted tetragonal version of the solution phase.

Most transition metals dissolve small amounts of C or N before compound phases form. Most of the M_2X compounds, including Fe_2N, Cr_2N, Mn_2N, Nb_2N, Ta_2N, V_2N, Ti_2N, W_2C, Mo_2C, Ta_2C, V_2C, and Nb_2C, take structures with the metal atoms in hcp sites and the nonmetallic atom in an interstitial site. Mo_2N and W_2N are exceptions; in these compounds, the metal atoms have a ccp arrangement. Compounds with the MX stoichiometry are known for their high hardness, high melting points (typically greater than 3000 K), and electrical conductivity. Most of these compounds adopt the NaCl structure (see Table 4.29) where the metal atoms are in ccp sites and the nonmetal atoms have six nearest neighbors. Although many of these compounds are nonstoichiometric, exceptions such as UC and UN exist.

Table 4.30. *The C14 structure.*

Formula unit	$MgZn_2$
Space group:	$P6_3/mmc$ (no. 194)
Cell dimensions:	$a = 5.18$Å, $c = 8.52$Å
Cell contents:	4 formula units
Atomic positions:	Mg in (4f) $3m$ $(1/3, 2/3, z)$; $(2/3, 1/3, \bar{z})$
	$(2/3, 1/3, 1/2+z)$
	$(1/3, 2/3, 1/2-z)$
	$z = 0.062$
	Zn in (2a) $\bar{3}m$ $(0, 0, 0)$; $(0, 0, 1/2)$;
	$(x, 2x, 1/4)$; $(2\bar{x}, \bar{x}, 1/4)$; $(x, \bar{x}, 1/4)$
	Zn in (6h) $mm2$ $(\bar{x}, 2\bar{x}, 3/4)$; $(2x, x, 3/4)$; $(\bar{x}, x, 3/4)$
	$x = 0.833$

Examples

compound	a (Å)	c (Å)	compound	a (Å)	c (Å)
$MgZn_2$	5.18	8.52	$CaCd_2$	5.993	9.654
$TiZn_2$	5.064	8.210	$CaMg_2$	6.2386	10.146
$TiFe_2$	4.785	7.799	$CdCu_2$	4.96	7.98
$ZrAl_2$	5.275	8.736	$TaFe_2$	4.816	7.868
$MoBe_2$	4.434	7.275	WFe_2	4.727	7.704
$MoFe_2$	4.73	7.72	$SmOs_2$	5.336	8.879

The existence and structures of these compounds were systematized by Hägg, who also proposed a radius ratio rule to predict the formation of interstitial compounds. If the ratio of the radii of the nonmetallic element and the metallic element is less than or equal to 0.59, then a simple structure such as rock salt, sphalerite, rutile, or fluorite is formed. If, on the other hand, the radius ratio is larger than 0.59, then the structure is more complex, such as cementite, Fe_3C.

(H) Laves phases

There are a number of intermetallic structures that, while densely packed, do not easily fit into our classification scheme. For example, the structure types usually known as Laves phases include the cubic $MgCu_2$ (C15) structure, the hexagonal $MgZn_2$ (C14) structure, and the hexagonal $MgNi_2$ (C36) structure (the C14 and C15 crystal structures are described in Tables 4.30 and 4.31). If one is liberal with definitions, it is possible to develop descriptions based on

Table 4.31. *The C15 structure.*

Formula unit	Cu_2Mg
Space group:	$Fd\bar{3}m$ (no. 227)
Cell dimensions:	$a = 7.05$Å
Cell contents:	8 formula units
Atomic positions:	Mg in (8a) $\bar{4}3m$ (0, 0, 0); (3/4, 1/4, 3/4) + F
	Cu in (16d) $\bar{3}m$ (5/8, 5/8, 5/8); (3/8, 7/8, 1/8);
	(7/8, 1/8, 3/8); (1/8, 3/8, 7/8) + F

Examples

compound	a (Å)	compound	a (Å)	compound	a (Å)
Cu_2Mg	7.05	Co_2U	6.992	Fe_2Dy	7.325
Al_2Gd	7.900	Co_2Zr	6.929	Fe_2Y	7.357
Al_2Ca	8.038	Cr_2Ti	6.943	Mg_2Gd	8.55
Al_2U	7.766	Fe_2Zr	7.070	Mn_2Gd	7.724
W_2Zr	7.63	Fe_2U	7.062	Mo_2Zr	7.596

eutactic sites and interstitial positions. However, these structures are usually visualized as being built from four-atom tetrahedral units that link by corners. For example, in the $MgCu_2$ structure, the Cu atoms form the tetrahedral units and the Mg atoms occupy the spaces between the tetrahedral clusters. This arrangement leads to 12-coordinate Cu and 16-coordinate Mg.

Laves proposed a radius ratio rule to predict the existence of intermetallic compounds with these characteristic structures. Assuming that the ideal structures are formed from hard spheres, he proposed that these AB_2 compounds should have a radius ratio (r_A/r_B) equal to 1.225. In fact, the compounds that crystallize in these structures have radius ratios in the 1.05 to 1.67 range, and for the majority of the compounds, the radius ratio is greater than 1.225. Furthermore, many AB_2 combinations that we might expect to assume one of these phases do not. For example, of the 45 AB_2 compounds that form between group IIIB and group IB elements, none have a Laves phase crystal structure. In general, we can say that arguments based on fixed atomic volumes and hard sphere atoms have limited quantitative accuracy.

In the cubic $MgCu_2$ structure, the B atoms form tetrahedral units that link at corners. The larger A atom, which occupies a space between the tetrahedral building blocks, has 16 nearest neighbors, four A atoms and 12 B. The B atom

Table 4.32. *Selected Laves phase compounds [17].*

MgZn$_2$ (C14)		MgCu$_2$ (C15)		MgNi$_2$ (C36)
BaMg$_2$	TaMn$_2$	AgBe$_2$	NaAu$_2$	HfCr$_2$
CaAg$_2$	TiCr$_2$	BiAu$_2$	NbCo$_2$	HfMn$_2$
CaCd$_2$	TiFe$_2$	CaAl$_2$	NbCr$_2$	HfMo$_2$
CaLi$_2$	TiMn$_2$	CeAl$_2$	PbAu$_2$	NbZn$_2$
CaMg$_2$	TiZn$_2$	CeCo$_2$	PrNi$_2$	ThMg$_2$
CdCu$_2$	UNi$_2$	CeFe$_2$	TaCo$_2$	UPt$_2$
CrBe$_2$	VBe$_2$	CeMg$_2$	TaCr$_2$	
FeBe$_2$	WBe$_2$	CeNi$_2$	TiBe$_2$	
HfFe$_2$	WFe$_2$	GdAl$_2$	TiCo$_2$	
KNa$_2$	ZrAl$_2$	GdFe$_2$	TiCr$_2$	
MnBe$_2$	ZrCr$_2$	GdMg$_2$	UAl$_2$	
MoBe$_2$	ZrIr$_2$	GdMn$_2$	UCo$_2$	
MoFe$_2$	ZrMn$_2$	HfCo$_2$	UFe$_2$	
NbFe$_2$	ZrRe$_2$	HfFe$_2$	UMn$_2$	
NbMn$_2$	ZrRu$_2$	HfMo$_2$	YFe$_2$	
ReBe$_2$	ZrOs$_2$	HfV$_2$	ZrCo$_2$	
SrMg$_2$	ZrV$_2$	HfW$_2$	ZrCr$_2$	
TaCo$_2$		KBi$_2$	ZrFe$_2$	
TaCr$_2$		LaAl$_2$	ZrMo$_2$	
TaFe$_2$		LaMg$_2$	ZrV$_2$	
		LaNi$_2$	ZrW$_2$	

has 12 nearest neighbors, six A and six B. As a fraction of the cubic lattice constant a, the A–A separation is $0.433a$, the B–B separation is $0.354a$, and the A–B separation is $0.414a$. This means that in any single coordination polyhedron (an A atom surrounded by A and B or a B atom surrounded by B and A), the like-atom bond distances are always shorter than the A–B bond distances. This is the opposite of what happens in ionic structures.

There are more than 200 binary Laves phases and about 90% of them contain a transition metal element. Approximately 70% of these phases assume the MgCu$_2$ structure. Most of the others assume the MgZn$_2$ structure. Only a few take the MgNi$_2$ structure. A number of the Laves phases are listed in Table 4.32. The atomic structures of the Laves phases, which are formed from the packing of tetrahedral units, illustrate the tendency of atoms in metallic crystals to fill space as efficiently as possible (they should maximize their density), to form symmetric arrangements, and to have the highest possible connectivity (coordination and dimensionality).

We close this section by noting a few important points about the structures of metallic alloys. First, many metals are polymorphic (at different temperatures, they assume different crystal structures). Second, many intermetallic structures tolerate significant levels of nonstoichiometry. In other words, the elemental ratio might not take an exact integer value. For example, the actual composition of a hypothetical B2 compound, called 'AB', might be $A_{1-x}B_{1+x}$ where x is a small but measurable quantity greater than zero. When stoichiometric disorder occurs, the lattice parameters typically are linearly dependent on the value of x (Vegard's law).

(i) Superlattice structures and complex stacking sequences

Some crystals have structures that are similar to one of the common prototype materials, but with a longer periodicity. For example, in Example 4.1, we noted that the $D0_{22}$ structure is closely related to the $L1_2$ structure. However, compositional ordering doubles the length of the c-axis with respect to the a-axis in the $D0_{22}$ structure. While $D0_{22}$ is distinguished as a separate structure type, many superlattice compounds that result from compositional order are most easily described by reference to the simple prototype structure.

Compositional order most frequently arises in alloy systems where there is a difference in atomic size or charge. For example, consider a material with the perovskite structure described in the preceding section. Normally, the stoichiometry would be ABX_3. If we form an equimolar alloy phase of ABX_3 and $AB'X_3$, we can express the compound as $A(B_{0.5}B'_{0.5})X_3$. If B and B' substituted randomly in the octahedral site, the cell would be a garden variety perovskite with a cubic lattice parameter of a. However, if the B and B' cations (at (1/2,1/2,1/2) type sites) form an ordered arrangement such that alternate (002) planes along [001] are occupied completely by B or B', the repeat distance along this direction would be $2a$. We would say that the cell is a perovskite, doubled along c with dimensions $a \times a \times 2a$. In this case, it might be appropriate to write the formula as $A_2BB'X_6$. Many other ordering patterns can occur and, in general, should be suspected when components with different sizes or charges are combined in rational fractions.

In many cases, multicomponent mixtures such as the one described above can exist as a random solid solution phase at elevated temperature or as an ordered superlattice phase at lower temperature. The order/disorder transformation is easily detected in diffraction experiments. If the unit cell increases its size, the reciprocal lattice shrinks and extra reflections are observed in the diffraction pattern. Even in cases where the unit cell volume remains more or less constant, ordering lowers the symmetry and, therefore, the ordered structure usually exhibits fewer systematic absences and its diffraction pattern has more peaks. For example, in the case above of the perovskite compound with ordering on the B site, the cell would change from a cubic symmetry, in the disordered state, to a tetragonal symmetry in the ordered state.

i. *Superlattice structures based on sphalerite and wurtzite*

Ternary combinations of III–V and II–VI compounds frequently crystallize in structures that are closely related to the sphalerite and wurtzite prototypes. These related structures are formed when atoms order on one of the two sub-lattices. For example, in the $InGaAs_2$ structure, the two types of metal atoms order so that alternate planes parallel to (001) are composed entirely of In or Ga. In the chalcopyrite structure, $CuFeS_2$, the metal atoms order so that they occupy alternate (201) planes. Some of the more common ordered variants of the sphalerite structure are shown in Fig. 4.27.

There are also a number of similar ternary compounds with structures closely related to ZnO by ordering. Again, the ordered occupation of the T+ sites by multiple cation types leads to larger unit cells. Some of these are shown schematically in Fig. 4.28. These are important examples of how complex structures can be described in terms of more simple ones.

The examples above were chosen to illustrate an important general principle; complex ternary and quaternary structures can usually be described in terms of a simpler binary prototype structure (see also Chapter 2, H.ii). Depending on the structure, the binary prototypes might be any of those discussed earlier in the chapter.

Example 4.3: Visualizing a structure as a close packed arrangement

The $LiGaO_2$ structure is specified in Table 3B.14 in Appendix 3B. First, sketch the structure, projected along [001]. Second, describe the coordination of the Li and Ga atoms. Based on this description, sketch a polyhedral representation of this structure projected on (001). Third, describe the $LiGaO_2$ structure in terms of its packing. In other words, if close packed sites are occupied, specify which atoms occupy them and what the packing arrangement is (bcc, ccp, or hcp). If interstitial sites are occupied, specify the type of sites (O or T), the fraction that are occupied, and which atoms occupy them. Finally,

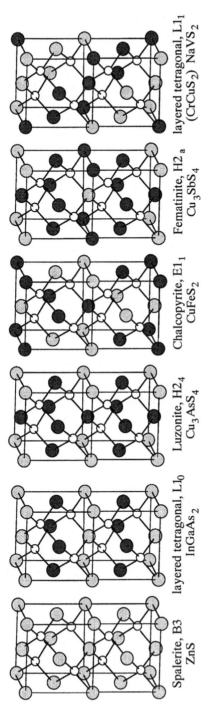

Figure 4-27. Ordered structures based on sphalerite [20].

Spalerite, B3
ZnS

layered tetragonal, L1$_0$
InGaAs$_2$

Luzonite, H2$_4$
Cu$_3$AsS$_4$

Chalcopyrite, E1$_1$
CuFeS$_2$

Fematinite, H2$_a$
Cu$_3$SbS$_4$

layered tetragonal, L1$_1$
(CrCuS$_2$) NaVS$_2$

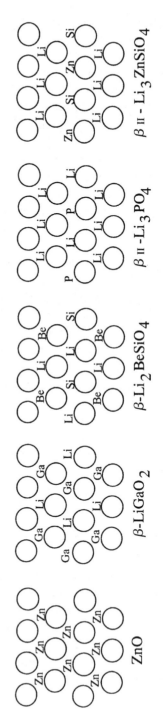

Figure 4-28. Ordered structures based on wurtzite [21].

ZnO

β-LiGaO$_2$

β-Li$_2$BeSiO$_4$

β_{II}-Li$_3$PO$_4$

β_{II} - Li$_3$ZnSiO$_4$

compare this structure to one of the common binary prototype structures that we have described in this chapter.

1. You should always begin a problem like this by sketching the unit cell boundaries of the projection with the proper aspect ratio, as we have in Fig. 4.29. In this picture, we have projected a two-unit cell by two-unit cell area, to emphasize the periodic pattern. 2. Next, we generate a list of coordinates and plot the atomic positions in the cell, as we practiced in Chapter 3. This task is simplified by using the following approximate coordinates:

atom	x	y	z
Ga	0.1	1/8	0
Li	0.4	1/8	1/2
O(1)	0.4	1/8	7/8
O(2)	0.1	1/8	3/8

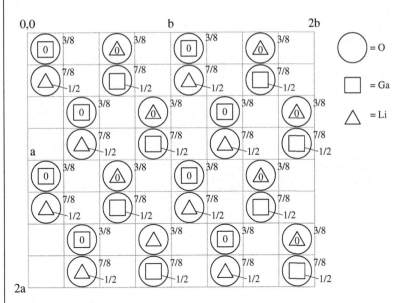

Figure. 4.29. See text for description.

3. Based on inspection of Fig. 4.29 and a number of trial bond length calculations, we see that both the Li and the Ga are surrounded by four O atoms. The Ga–O bond distances are all between 1.835 and 1.860 Å. The Li–O bond distances are between 1.95 and 2.00 Å. In both cases, the O are arranged about the metal atoms in a tetrahedron.

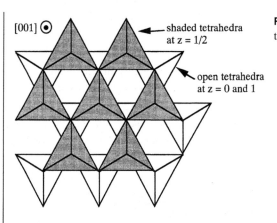

[001] ⊙
shaded tetrahedra
at z = 1/2

open tetrahedra
at z = 0 and 1

Figure 4.30. A polyhedral projection of $LiGaO_2$ along [001].

4. Based on such a description, we can construct the projection of the polyhedral structure shown in Fig. 4.30.

5. Looking along the [001] direction, we see that at $z = 3/8$ and at $z = 7/8$, the O ions
occupy eutactic sites in each layer. Since the two layers repeat in an ABAB sequence,
this is hcp packing. All of the metal ions are tetrahedrally coordinated and, therefore,
occupy the T sites. Only one half of the available T sites are filled in each layer and
they are all the same type (T_+). In summary, we can say that in the $LiGaO_2$ structure,
the O occupy the hcp eutactic positions and the Li and Ga fill all of the T_+ sites in an
ordered configuration.

6. This arrangement is essentially identical to the ZnO, wurtzite, or B4 structure where
O occupy all of the hcp eutactic sites and Zn occupies all of the T_+ sites. Note that
$LiGaO_2$ and ZnO have the same overall metal-to-oxygen ratio. The difference between
the binary prototype and the ternary structure considered here is that the metal atoms
are compositionally ordered in the T_+ sites so that the repeat unit is orthorhombic.
Further, the Ga and Li layering leads to a non-ideal O packing.

ii. *More complex stacking sequences and polytypic disorder*

While the two most common stacking sequences are hcp (ABABAB) and ccp
(ABCABCABC), many others are possible and found in nature. In this section,
the notation for alternate stacking sequences is described and the stacking polytypes of SiC are described.

When the stacking of closest packed layers was described in Section B(ii) of
this chapter, different layers in each sequence were labeled A, B, and C. With
reference to hexagonal coordinates, the B layer is displaced in the (001) plane by
(1/3, 2/3) with respect to the A layer and the C layer is displaced by (2/3, 1/3).

Table 4.33 *SiC polytypes.*

Name	Stacking	a (Å)	c (Å)	η (%)	notes
3C	c	3.073	7.57	0	sphalerite, $E_g = 2.39$ eV
2H	h	3.076	5.048	100	wurtzite, $E_g = 3.330$ eV
4H	ch	3.073	10.053	50	$E_g = 3.265$ eV
6H	cch	3.073	15.08	33	$E_g = 3.023$ eV
9R	chh			67	
$8H_1$	$ccch$	3.079	20.147	25	
$8H_2$	$chhh$	3.079	20.147	75	
12R	$cchh$			50	
5T	$ccchh$			40	
$10H_1$	$cccch$	3.079	25.083	20	
$10H_2$	$cchhh$	3.079	25.083	60	
$10H_3$	$chchh$	3.079	25.083	60	
$15R_1$	$cchch$	3.073	37.70	20	$E_g = 2.986$ eV
$15R_2$	$chhhh$	3.073	37.70	80	
21R	$ccchcch$	3.073	52.78	28	

Conventionally, each layer in a sequence is classified according to the arrangement of the layers above and below. If the layers above and below are the same, it is an *h* layer. If the layers above and below are not the same, it is a *c* layer. Thus the ABCABC sequence can be written simply as '*c*' and the ABABAB sequence can be written simply as '*h*'. The advantage of this nomenclature is that more complex sequences are easily described. For example, elemental La, Pr, Nd, and Pm all crystallize in a structure that has an *hc* stacking sequence which corresponds to ABCBABCB. The $D0_{24}$ structure, described in Section F. (v), also had *hc* stacking.

This notation is useful for specifying the structure of related polytypes (polytypes are materials with the same composition and structures that differ only in the stacking sequence). SiC, which finds uses as an abrasive material, a structural material, and as a semiconducting material, is also well known for its wide variety of polytypes. All of the SiC polytypes can be described as close packed Si with C in half of the tetrahedral interstices (analogous to sphalerite and wurtzite). However, more than 70 different polytypes have been differentiated on the basis of their different stacking sequences. The 15 polytypes with the shortest repeat distance along [001] are listed in Table 4.33.

The names for the SiC polytypes are selected based on the crystal class (for example, H for hexagonal, R for rhombohedral, etc.) and the number of layers it takes to form the repeat unit along [001]. In each case, you can always choose a hexagonal cell approximately 3.07 Å along the basal edge and $N \times 2.51$ Å high, where N is the number of layers per repeat unit. The lattice parameters for a number of the polytypes are listed in Table 4.33. Some polytypes have very long repeat distances. For example, the c dimension for the 174R polytype is 436.7 Å and for the 393R polytype it is 989.6 Å. The parameter η, listed in Table 4.33, is the so-called *hexagonality*. This quantity is defined as the number of h layers divided by the total number of distinct layers. The band gap of SiC is observed to increase with the hexagonality. Finally, we note that similar polytypism has been observed in ZnS; 3C, 2H, 4H, 6H, 8H, 10H, and 15R polytypes are all known.

(J) Extensions of the close packing description to more complex structures

In the previous examples, we have mentioned only crystal structures that have all of the close packed positions filled. However, there are many other structures where only a subset of the eutactic positions are filled and these can also be described in terms of eutactic arrangements. A good example of this the ReO_3 (DO_9) structure (a description of this structure can be found in Table 3B.15). In this case, the anions (the O) occupy only three quarters of the ccp eutactic positions and the Re(VI) occupy one quarter of the octahedral interstices. There is also an anti-ReO_3 structure whose prototype is Cu_3N.

There are, of course, many more exotic crystal structures that can be thought of as eutactic arrangements. One example is Na-β-alumina, which has the approximate composition $NaAl_{11}O_{17}$. This structure can be thought of as a ccp arrangement of oxygen with three quarters of the O in every fourth layer missing. The Al cations occupy octahedral and tetrahedral sites in the densely packed regions and the Na occupy sites in the layer that contains the vacancies. The high vacancy concentration in the layer that contains the Na allows rapid ionic diffusion. Thus, this material is used as a solid electrolyte and these layers are known as conduction planes.

i. *Structures derived from ReO$_3$*
The ReO_3 structure can be thought of as a three dimensional network of corner-sharing octahedral units, as shown in Fig. 4.31. Using this defective ccp framework as a starting point, it is possible to derive many other metal oxide structures; a few examples are given below.

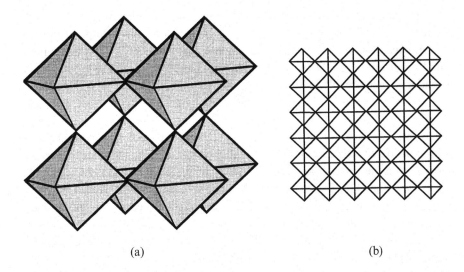

(a)   (b)

Figure 4.31. The structure of ReO_3. (a) An oblique view of the connected polyhedral network. A 12-coordinate interstitial site exists in the central cavity formed by the eight connected polyhedra. (b) A projection down the cube axis.

As mentioned in the last section, the oxygen anions in the ReO_3 structure fill only three quarters of the ccp positions. This leaves rather large 12-fold interstices that can be clearly seen in the polyhedral representation. These structural vacancies allow compounds derived from the ReO_3 structure, such as WO_3, MoO_3 (Table 3B.5), and V_2O_5 (Table 3B.8), to react with a variety of electropositive elements including H, Na, Li, and K to form a range of nonstoichiometric compounds collectively known as *oxide bronzes*. The name dates back to their discovery in the nineteenth century and comes from the fact that these compounds are highly reflective, conductive, and look like metals. The first composition discovered ($Na_{0.8}WO_3$) closely resembles the better known Cu–Sn alloy in its appearance. The metallic species (usually an alkali or alkaline earth metal) fills the interstitial site in the structure. For example, a series of tungsten bronzes can be formed with the general formula A_xWO_3 where A is an electropositive element and $0 < x < 1$. Note that when x goes to one the formula is AWO_3; this compound has the same stoichiometry and structure as a perovskite. Thus, an ReO_3-type compound can be transformed to a perovskite-type compound simply by filling the empty site on the ccp lattice with a metallic species. Closely related to the ReO_3 structure are the hexagonal tungsten bronze structure (Table 3B.16) and the tetragonal tungsten bronze structure (Table 3B.17).

Because the intercalation or ion insertion reaction that transforms a transparent compound such as WO_3 to a bronze can occur at room temperature and can be driven by either a chemical or electrical potential, these materials have found applications in electrochromic devices (so-called 'smart' windows) and secondary (rechargeable) batteries.

ii. *Structures derived by crystallographic shear*

A variety of structures that can be thought of as connected octahedra are generally considered to be closely related to the ReO_3 structure. These structures are derived from the three-dimensional corner-sharing network by crystallographic shear operations [25, 26]. From the most basic point of view, this amounts to changing some of the polyhedra from corner-sharing to edge-sharing arrangements. A more systematic description is that all of the atoms on one side of a plane are translated by a vector smaller than the repeat distance. Several examples are depicted in Fig. 4.32. For simplicity, only one layer of each structure is shown. In some cases, the layers are stacked and connected by corner-sharing, but in others they are connected to the next layer by edge-sharing or even weak van der Waals interactions.

The change in coordination that accompanies crystallographic shear must, of course, result in a change in the stoichiometry. By choosing different planes for the shear operation and different spacings for the planes, you can see that it is possible to create an infinite number of structures. Closely related phases with shear along the same plane, but different shear plane spacings, have different stoichiometries and are usually referred to collectively as shear phases or Magnéli phases. For example, there are a series of Magnéli phases derived from WO_3 by shear on the $\{\bar{1}30\}$ plane that have formulae W_nO_{3n-2}, where n is 20, 24, 25, or 40. The crystallographic shear concept is not limited to the ReO_3 type structure. The same concept can be applied to the rutile structure to produce phases with compositions Ti_nO_{2n-1}, where $4<n<10$. In this case, the shear plane is $\{\bar{1}20\}$. When rutile is 'saturated' with shear planes, it becomes Ti_2O_3 and it has the corundum structure (Table 3B.19). Thus, we say that crystallographic shear plane operations create a smooth transition between the rutile and corundum structures.

K) Van der Waals solids

Even though they are held together only by weak forces, it is not uncommon for molecules to crystallize in structures that have closest packed arrangements, particularly when the molecule has spherical symmetry or rotates freely and, thus, acts as a sphere. An interesting example of close packing in a molecular solid is

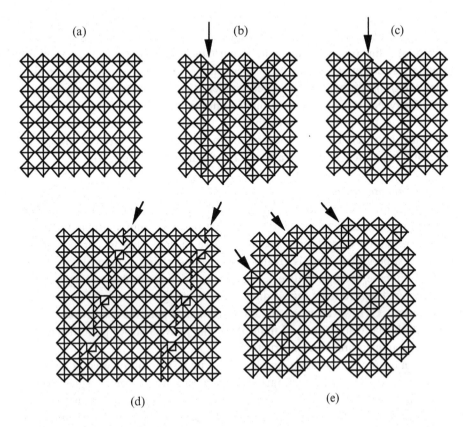

Figure 4.32. Structures derived from ReO_3 by crystallographic shear. In each case, the line along which the shear plane intersects the depicted plane of the structure is indicated with an arrow. (a) The ideal ReO_3 structure. (b) The V_2O_5 structure. (c) The closely related V_2MoO_8 structure [27]. (d) Projection along (100) of the $Mo_{18}O_{52}$ structure [28]. (e) Projection on (010) of the Mo_8O_{23} structure [29].

the fullerite, C_{60}, and the fulleride compounds that form when C_{60} is reacted with alkali or alkaline earth metals. Pure C_{60} forms a ccp structure (each molecule decorates one site on the cubic F lattice) and when up to three alkali metal atoms react with it to form a fulleride, these atoms simply fill the interstitial voids. A more distorted A_4C_{60} has bct structure and A_6C_{60} has a bcc structure [30].

(L) Noncrystalline solid structures

We conclude this chapter by taking note of the fact that no real solid materials are, according to the most strict definition, ideal crystals. For example, no random alloy nor solid solution specimen possesses true long range translational symmetry because of the chemical disorder. In fact, the point (vacancies and interstitials), line (dislocations), and planar (stacking faults, anti-phase domain boundaries) defects that invariably occur in even the best single crystals also break the translational symmetry requirement that is a necessary condition for crystallinity. Nevertheless, despite these common imperfections, most solids are a close enough approximation to the ideal crystal that we treat them as such. There are, however, a range of less common solids that are different enough to be treated separately and these are very briefly discussed below.

i. *Incommensurate structures*

Incommensurate phases occur in systems with competing periodicities [31]. The simplest way to define incommensurate structures is to compare them to super-lattice structures. Superlattice structures have, usually due to compositional ordering, a periodicity that is greater than the underlying structural configuration. In this case, the superlattice periodicity is said to be commensurate with the underlying structure if its repeat distance is an integer multiple of one of the lattice vectors. However, when the structure has a periodic distortion with a wavelength that is an irrational multiple of the Bravais lattice periodicity, it is said to be an incommensurate structure.

As an example, consider a simple layered material built of weakly interacting A and B type layers. The atoms in the A layer have an average spacing of a and the atoms in the B layer have an average spacing of b. A commensurate structure (Fig. 4.33a) is formed if b/a is a simple rational fraction. For example, if $3a = 2b$, then the repeat unit in the lateral direction is $2b$. On the other hand, if b/a is an irrational fraction, then a lateral repeat unit can not be defined and the structure is incommensurate. Such an arrangement is illustrated in Fig. 4.33b. Incommensurate structures can be found in materials with compositional incommensurability, such as intercalated graphite. Displacive incommensurability occurs when an incommensurate, periodic distortion is superimposed on an otherwise periodic lattice, as is the case for TaS_2 and K_2SeO_4. In other cases, it is the ordering of defects that creates an incommensurate superstructure.

ii. *Quasicrystalline structures*

Quasicrystals have many of the properties of crystals, including long range translational order and long range orientational order [32–34]. Furthermore, they produce sharp diffraction patterns. However, the translational order is non-periodic, the structures have no rotational point symmetry, and the diffraction patterns have five-fold rotational symmetry. For these reasons, they are called quasicrystals rather then crystals. Solids exhibiting quasicrystalline order were

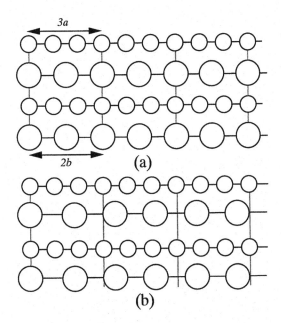

Figure 4.33. (a) A commensurate layered structure. (b) An incommensurate layered structure.

first discovered by Schectman in 1984 in rapidly cooled Al–M alloys (where M is Mn, Fe, or Cr). Quasicrystals are also found in Al–Cu–Li, Al–Cu–Fe, Al–Cu–Ru, Ga–Mg–Zn, Al–Pd–Mn, Ni–Ti–V, Pd–U–Si and many other alloys. A number of these materials appear to be thermodynamically stable in the quasicrystalline state and relatively large faceted crystals can be grown.

In conventional crystals, we fill space using identical copies of a single repeat unit. This results in long range rotational and translational order. In contrast, quasicrystalline order results from packing identical copies of more than one type of repeat unit. As an example in two dimensions, consider the Penrose tiling shown in Fig. 4.34. The tiling is built from two rhombuses that completely fill two-dimensional space. The first type of rhombus has complementary angles of 72° and 108° and the second has complements of 144° and 36°. You should notice that these angles are $(N \cdot 360)/10$, where N is 2, 3, 4, and 1 respectively. The packing of these tiles produces decagons, some of which are shaded in Fig. 4.34a. Because the decagons all have the same orientation, this tiling has long range orientational order. Furthermore, if we shade a subset of the rhombuses that have parallel edges, we see a kind of translational order. Figure 4.33b shows one subset of the rhombuses that forms a set of nearly parallel and approximately evenly spaced irregular lines. If one examines the other subsets that have parallel edges in common, four additional sets of lines are defined. These five lines are the same as would be generated by a five-fold rotation axis. This long

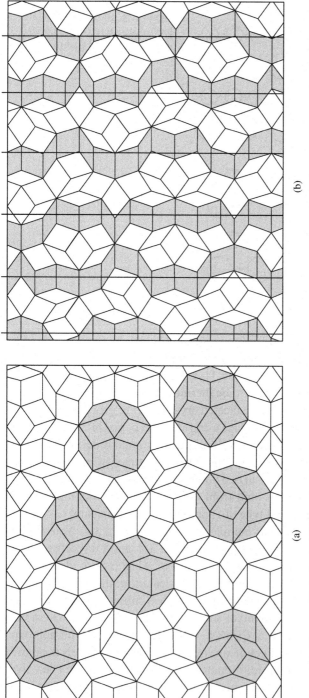

(a)

(b)

Figure 4.34. Penrose tilings that exhibit quasicrystalline order. (a) The decagons (shaded) have orientational order without translational order. (b) When all rhombuses with common parallel edges are shaded, we see that they form zig-zag chains that have approximately the same orientation and spacing [34].

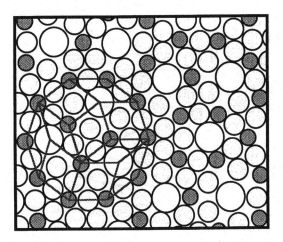

Figure 4.35. A ternary atomic structure built by decorating the rhombuses that make up a Penrose tiling.

range orientational order gives rise to constructive interference and diffraction patterns with five-fold symmetry.

The quasicrystal must, of course, be built of atoms, not tiles. To construct a crystal from patterns such as the one shown in Fig. 4.34, we have to add a basis to the tiles. In the conventional crystal, we added an identical basis to each lattice site. To build the quasicrystal, we use identical repeats of two different basis groups. For example, a ternary quasicrystal can be built by decorating each of the two rhombuses with a group of atoms, as illustrated in Fig. 4.35. In part of the structure, lines have been added to emphasize that while the atomic arrangement appears random, it is actually built according to rules imposed by the Penrose tiling. Three-dimensional generalizations of this arrangement are closely related to the structures of the quasicrystals observed in nature.

iii. *Liquid crystalline structures*
As the name implies, these materials have structures that are intermediate between liquids and crystals. This type of order has been found in rod-shaped molecular materials that are held together by relatively weak intermolecular bonds. Liquid crystals have orientational order, but no translational order. For example, in the nematic liquid crystalline states, rod-like molecules line up so that their axes are in the same direction, but otherwise the spacing of the molecules is random. Several types of liquid crystalline order are illustrated schematically in Fig. 4.36. Liquid crystals are commonly used in liquid crystal diode (LCD) displays. By applying an electric field, it is possible to produce alternate structural phases that have different (bright or dark) appearances.

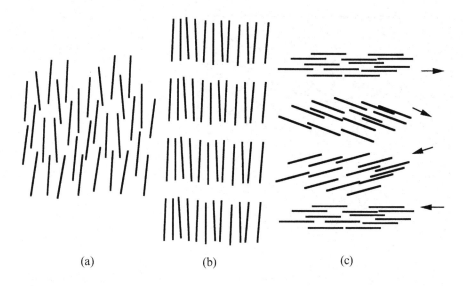

Figure 4.36. Schematic illustrations of different liquid crystalline states. Each line represents a rod-like molecule. (a) nematic (b) smectic (c) chiral nematic (the four groups shown are meant to lie in four parallel planes).

iv. Amorphous materials

The components of an amorphous or noncrystalline material lack both long range translational and orientational order. In the past, such materials have been modeled as being composed of randomly dispersed atoms or molecules. However, in the last few decades a considerable body of data has been amassed to support the idea that a previously unrecognized degree of short range order exists. For example, the local coordination number of Si and O in fused silica glass is the same as it is in quartz; only the bond angles are altered. Changes in this local order can be used to modify glass properties. Figure 4.37a shows a two-dimensional slice of amorphous silica. Because the stoichiometry of the glass is SiO_2, each Si makes four bonds to O and each O makes two bonds to Si and a well connected network is formed. If an alkali or alkaline earth oxide is added, with stoichiometry M_2O or MO, the network is disrupted and some of the O have fewer bonds to Si (see Fig. 4.37b). The relatively lower fraction of strong Si–O bonds weakens the network and makes the material melt at a much lower temperature. Thus, soda and lime are added to silica to make glasses easier to process. The network can also be modified without disruption by adding an oxide with the same stoichiometry, such as PbO_2 (see Fig. 4.37c).

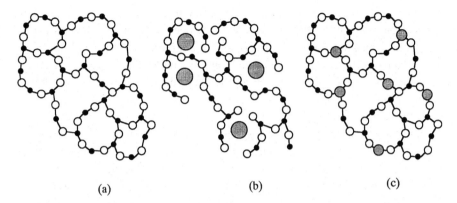

Figure 4.37. (a) Silica glass where the small black circles are Si and the larger white ones are O. (b) Addition of a larger alkali metal oxide changes the stoichiometry and disrupts the network. (c) Addition of an oxide of the same stoichiometry modifies the properties of the network without disruption.

Glasses are generally formed by cooling rapidly enough from the liquid state to prevent the crystalline phase from nucleating. While almost any material can form an amorphous solid, the cooling rates necessary to prevent crystallization vary greatly. Glass forming materials such as borates, silicates, phosphates, and germanates will solidify as amorphous structures under almost any conditions. Most metallic materials, on the other hand, must be splat cooled at rates of $\approx 10^{6\circ}$/sec to form a glass.

(M) Problems

(1) The tetragonal close packed (tcp) arrangement is described in Table 3B.18. In this arrangement, each atom has 11 nearest neighbors at identical distances.

    (i) What is the atomic density of this arrangement and how does it compare to the ccp and bcc arrangements?

    (ii) The tcp arrangement is similar to the hcp arrangement. If a structure with the tcp arrangement were to transform to the hcp structure, how do you think that the lattice vectors of the original cell would be related to the transformed cell?

    (iii) Describe the shifts that would be required to bring the spheres in the tcp arrangement to the hcp arrangement.

(2) The radius of the largest sphere that will fit into the six-fold interstitial site of a ccp eutactic structure depends on both the size of the ion in the eutactic site, r_e, and the cubic lattice constant, a_o. Determine an expression for this relationship.

(3) Test the formula that you derived in problem 2 on several rock salt compounds from Table 4.17. To do this, use the lattice constant and the anion size for a given compound to predict the size of the largest cation which should be allowed to fit. Is the cation that is there larger or smaller than what you predict? Make sure that you consider several monovalent, divalent, trivalent, and tetravalent compounds in your comparison.

(4) A number of ABO_3 compounds crystallize in the perovskite structure. Assuming the ideal packing of hard spheres (in other words, cations and anions are in contact, but anions are not necessarily in contact), determine the ratio $(r_a+r_o)/(r_b+r_o)$ where r_a is the radius of cation A, r_b is the radius of cation B, and r_o is the anion radius. Next, use this ratio to determine the lattice parameter, a_o. Compare your computed lattice parameters and radius ratios with some of the known values for compounds listed in Table 4.28 and discuss reasons for any differences.

(5) The O atoms in the olivine structure occupy hcp sites. What are the indices and locations of the planes of O atoms in the orthorhombic cell that correspond to the closest packed hexagonal layers? You will need to use the data in Table 4.24.

(6) Perovskite is only one of the structures that compounds with a metal-to-oxygen ratio of two to three (M_2X_3) might assume. Data for two other structures, corundum and ilmenite, are given in Tables 3B.19 and 3B.20 [36].

    (i) Describe the ilmenite and corundum structures based on their packing.

    (ii) For any compound, M_2O_3 or $MM'O_3$, can you predict which of the three structures will form, based on the cation radii? (Hint: consider the relative radii of the cations in the compounds listed in Tables 4.28, 3B.19 and 3B.20.)

(7) High pressure experiments have shown that oxides with the stoichiometry of ABO_3, in the high pressure limit, transform to the perovskite structure. Explain this by comparing calculated values for the packing fractions of the ilmenite (Table 2B.20), corundum (Table 2B.19), and perovskite structures (Table 4.28).

(8) According to the close packing description, NiAs and NaCl are very similar structures. However, the fact that the NaCl structure is mostly adopted by ionic compounds and the NiAs structure is mostly adopted by intermetallic compounds suggests that the 'chemistry' of these structures is very different. Explain this difference. (It may help to consider polyhedral representations of these structures.)

(9) K_3C_{60} can be thought of as a eutactic arrangement of C_{60} complex anions with K ions stuffed into all of the interstices. (fcc lattice constant $= 11.385$ Å.)

    (i) Why are no similar structures observed for oxides?

(ii) Do you think this structure could be formed by combining other alkali metals with C_{60}?

(iii) What are the similarities and differences between this structure and the $D0_3$ structure (structure of Fe_3Si) ?

(10) Compare the metal oxides that form the fluorite structure with those that form the rutile structure. Can the cations be separated into two groups based on size? Test this idea on the fluorides that form these structures. Explain this trend.

(11) The C16 crystal structure is specified in Table 3B.21. Sketch and describe this structure. How well does it compare to the three prototype packing sequences? What are the distances to the Cu atom's nearest Cu neighbors? What are the distances to the Cu atom's nearest Al neighbors?

(12) Table 3B.22 specifies the structure of the compound Ti_2CS. This is the $AlCCr_2$ structure type. Describe the coordination environment around each of the three atoms and describe the packing in this compound. In other words, describe it with respect to one of the common close packed structures. Compare this structure to one of the prototypes described earlier in this chapter.

(13) Based on the close packing description, describe a likely mechanism for a phase transformation from a sphalerite structure to the rock salt structure. Find an example of a polymorphic compound that can take either structure. Describe the conditions under which each of the polymorphs is stable.

(14) Oxides with the garnet structure are important for a number of optical and magnetic applications. The garnet structure is specified in Table 3B.23. Describe the garnet structure as a close packed arrangement. In other words, if close packed sites are occupied, specify which atoms occupy them and what the packing arrangement is (bcc, ccp, or hcp). If interstitial sites are occupied, specify the type of sites (O or T), the fraction that are occupied, and which atoms occupy them.

(15) Diaspore has the molecular formula AlO(OH). If we ignore the H, the compound has the stoichiometry AlO_2. The structural details of diaspore are given in Table 3B.24. Describe this structure in terms of its packing and compare it to the other AB_2 structures in this chapter (rutile, fluorite, CdI_2, $CdCl_2$).

(16) Boehmite is a polmorph of diaspore which also has the molecular formula AlO(OH). The structural details of boehmite are given in Table 3B.25. Describe this structure in terms of its packing and compare it to diaspore (see problem 15).

(17) The chromium boride crystal structure is described in Table 3B.26. Can this be described as a eutactic structure? Is it similar to any other structure that you know of?

(18) Use sketches to explain how the rutile structure (MO_2) can be transformed to the corundum structure (M_2O_3) by reduction (oxygen loss) and crystallographic shear. (Hint: consider O loss and shear along $(10\bar{1})_r$.)

(19) The crystal structure of $LiFeO_2$ is summarized in Table 3B.27.

(i) Describe this compound as a eutactic arrangement of atoms with filled interstices (specify ccp or hcp and which ions occupy which fraction of the sites).

(ii) Describe this structure in terms of one of the common prototypes discussed in this chapter.

(20) The fluorite structure (Table 4.27) has the stoichiometry AO_2, the pyrochlore structure (Table 3B.28) has the stoichiometry $A_2B_2O_7$, and the bixbyite structure (Table 3B.29) has the stoichiometry M_2O_3. Describe the packing in each of these three structures and the manner in which they are related.

(21) The high pressure form of TiO_2 (TiO_2 II) has a structure closely related to α-PbO_2, as described in Table 3B.30. Describe the packing in this structure and compare it to a common prototype. In other words, if close packed sites are occupied, specify which atoms occupy them and what the packing arrangement is (bcc, ccp, or hcp). If interstitial sites are occupied, specify the type of sites (O or T), the fraction that are occupied, and which atoms occupy them.

(22) The structure of γ-LaOF is described in Table 3B.31.

(i) What is the Bravais lattice of this structure?

(ii) How many general positions are there in this group?

(iii) Do any of the atoms in this structure occupy the general positions?

(iv) Sketch a projection of this structure down the [001] axis.

(v) Describe the packing in this structure. In other words, if close packed sites are occupied, specify which atoms occupy them and what the packing arrangement is (bcc, ccp, or hcp). If interstitial sites are occupied, specify the type of sites (O or T), the fraction that are occupied, and which atoms occupy them.

(vi) Compare this structure to a common binary prototype structure.

(23) The chalcopyrite structure is described in Table 3B.32.

(i) What is the Bravais lattice of this structure?

(ii) Draw a sketch showing the symmetry elements and general positions in the point group from which $I\bar{4}2d$ is derived.

(iii) How many general positions are there in the group $I\bar{4}2d$?

(iv) Explain the meaning of the symbol d in this space group. (Name the symmetry element, its position in the cell, and its action.)

(v) Sketch a projection of this structure down the [001] axis.

(vi) Describe the packing in this structure. In other words, if close packed sites are occupied, specify which atoms occupy them and what the packing arrangement is (bcc, ccp, or hcp). If interstitial sites are occupied, specify the type of sites (O or T), the fraction that are occupied, and which atoms occupy them.

(vii) Compare this structure to a common binary prototype structure.

(24) The structure of Cr_2Al is described in Table 3B.33.

(i) What is the Bravais lattice associated with this structure?

(ii) From which of the 32 crystallographic point groups was this space group derived?

(iii) Specify the pattern produced by this point group. Your answer should include a sketch using the conventional symbols and a list of coordinates (x, y, z) for the equivalent positions in this pattern.

(iv) How many general equivalent positions are there in the space group I4/mmm? (You don't have to list the coordinates, simply specify how many there are.)

(v) Do any of the atoms in this structure occupy special positions? (If so, which ones?)

(vi) Draw the structure, projected along [100] and along [001].

(vii) Describe the coordination of the Al atoms. Specify how many neighbors each Al atom has, the type of atom, and the bond distances.

(viii) Describe the Cr_2Al structure in terms of its packing. In other words, if close packed sites are occupied, specify the packing arrangement (bcc, ccp, or hcp), which atoms occupy them, and if there is ordering. If interstitial sites are occupied, specify the type of sites (O or T), the fraction that are occupied, and which atoms occupy them. Finally, compare this structure to one of the common binary prototype structures that we have discussed.

(25) The structure of Ga_2Zr is described in Table 3B.34.

(i) List the general equivalent positions of space group Cmmm.

(ii) Sketch a projection of this structure down the [001] axis.

(iii) Describe the packing in this structure. In other words, if close packed sites are occupied, specify which atoms occupy them and what the packing arrangement is (bcc, ccp, or hcp). If interstitial sites are occupied, specify the type of sites (O or T), the fraction that are occupied, and which atoms occupy them.

(iv) Compare this structure to a common binary prototype structure.

(26) The structure of β-Cu_2HgI_4 is described in Table 3B.36.

(i) What is the Bravais lattice associated with this structure?

(ii) This cell is non-primitive. Choose a set of primitive lattice vectors and use a calculation to prove that they define a primitive cell.

(iii) From which of the 32 crystallographic point groups was this space group derived?

(iv) Draw a picture showing the symmetry elements and general equivalent positions in the point group $\bar{4}2m$.

(v) How many general equivalent positions are there in the space group I$\bar{4}$2m?

(vi) Do any of the atoms in β-Cu$_2$HgI$_4$ occupy general equivalent positions? (If so, which ones?)

(vii) Draw the structure, projected along [100].

(viii) Describe the coordination of the Hg atoms. Specify the nearest neighbors (atom types), and the bond distances.

(ix) Describe the β-Cu$_2$HgI$_4$ structure in terms of its packing. In other words, if close packed sites are occupied, specify the packing arrangement (bcc, ccp, or hcp), which atoms occupy them, and if there is ordering. If interstitial sites are occupied, specify the type of sites (O or T), the fraction that are occupied, and which atoms occupy them. Finally, compare this structure to one of the common binary prototype structures that we have discussed.

(27) The C14 crystal structure is specified in Table 4.31. Sketch and describe this structure. How well does it compare to the three prototype packing sequences?

(N) References and sources for further study

[1] A.F. Wells, *Structural Inorganic Chemistry*, 5th edition (Clarendon Press, Oxford, 1984), Chapter 4, pp. 141–156 and Chapter 29, pp. 1274–1326. Fig. 4.1 is drawn after Fig. 4.1 on p. 142. This source provides a description of close packed arrangements.

[2] W.H. Baur, *Materials Research Bulletin* **16** (1981) 339. The tetragonal close packed arrangement.

[3] T. Hales, University of Michigan, 1998. Hales' claim was made on the internet and is widely accepted. However, at the time of this writing, the proof has yet to be substantiated by peer review.

[4] A.R. West, *Solid State Chemistry and its Applications* (J. Wiley & Sons, Chichester, 1984), Chapter 7, pp. 212–18.

[5] N.W. Ashcroft and N.D. Mermin, *Solid State Physics* (Holt Rinehart and Winston, New York, 1976).

[6] The data in this table was compiled from Table 7.1 on p. 218 of Ref. [4] and Tables 4.2, 4.3, and 4.4 on pp. 70–7 of Ref. [5].

[7] A.R. West, *Solid State Chemistry and its Applications* (J. Wiley & Sons, Chichester, 1984), Chapter 7, pp. 219–21. Fig. 4.7 is drawn after Fig. 7.8 on p. 221.

[8] W.B. Pearson, *A Handbook of Lattice Spacings and Structures of Metals and Alloys*, vol. 2 (Pergamon, Oxford, 1967). A source of crystal structure data.

[9] R.W.G. Wyckoff, *Crystal Structures*, Volumes 1, 2, and 3 (John Wiley & Sons, New York, 1964). A source of crystal structure data.

[10] M. O'Keeffe, *Acta Cryst.* **A33** (1977) 924.

[11] A.R. West, *Solid State Chemistry and its Applications* (J. Wiley & Sons, Chichester, 1984). Table 4.7 is based on Table 7.2, p. 221.

[12] A.R. West, *Solid State Chemistry and its Applications* (J. Wiley & Sons, Chichester, 1984). Fig. 4.18 is drawn after Fig. 7.21 on p. 253. Fig. 4.19 is drawn after Fig. 7.22 on p. 254. Fig. 4.20 is drawn after Fig. 7.23 on p. 256. AB_2 type structures.

[13] N.N. Greenwood, *Ion Crystals, Lattice Defects, and Nonstoichiometry* (Butterworths, London, 1968), pp. 92–102. The spinel structure.

[14] A.F. Wells, *Structural Inorganic Chemistry*, 5th edition (Clarendon Press, Oxford, 1984), pp. 334–51, pp. 1319–23. The data in Table 4.29 is from Table 29.13 on p. 1321.

[15] W. Hume-Rothery and G. Raynor, *The Structure of Metals and Alloys* (Institute of Metals, London, 1962), pp. 217–28. Interstitial compounds.

[16] C.S. Barrett and T.B. Massalski, *The Structure of Metals* (McGraw Hill, 1966), pp. 259–61. Interstitial compounds.

[17] W. Hume-Rothery and G. Raynor, *The Structure of Metals and Alloys* (Institute of Metals, London, 1962), Chapter V. The data in Table 4.32 comes from Table XXVI on p. 229. Laves phases.

[18] C.S. Barrett and T.B. Massalski, *The Structure of Metals* (McGraw Hill, 1966), pp. 256–9. Laves phases.

[19] A.F. Wells, *Structural Inorganic Chemistry*, 5th edition (Clarendon Press, Oxford, 1984), pp. 1304–10. Laves phases.

[20] A. Zunger and S. Mahajan, *Atomic Ordering and Phase Separation in Epitaxial III-V Alloys*, vol. 3 (Elsevier, Amsterdam, 1993). Fig. 4.28 is drawn after a fig. in this ref. Ternary structures based on corner-sharing tetrahedra.

[21] A.R. West, *Solid State Chemistry and its Applications* (J. Wiley & Sons, Chichester, 1984). Fig. 4.29 is drawn after Fig. 7.8 on p. 222. Ternary strutures based on corner-sharing tetrahedra.

[22] L.S. Ramsdell, *Am. Mineral.* **32** (1947) 64. SiC polytypes.

[23] R.W.G. Wyckoff, *Crystal Structures* (John Wiley & Sons, New York, 1964). SiC polytypes.

[24] M. O'Keeffe, *Chem. Mater.* **3** (1991) 332. SiC polytypes.

[25] A.F. Wells, *Structural Inorganic Chemistry*, 5th edition (Clarendon Press, Oxford, 1984), pp. 608–21, Chapter 13. Structures derived by crystallographic shear.

[26] L.A. Bursill and B.G. Hyde, in: *Progress in Solid State Chemistry*, Vol. 7. Eds. H. Reiss and J.O. McCaldin (Pergamon, New York, 1972), p. 177. Structures derived by crystallographic shear.

[27] R.C.T. Slade, A. Ramanan, B.C. West and E. Prince, *J. Solid State Chem.* **82** (1989) 65–9.

[28] L. Kihlborg, *Arkiv Kemi* **21** (1963) 443–60.

[29] L. Kihlborg, *Arkiv Kemi* **21** (1963) 461–9.

[30] J.E. Fischer, P. A. Heiney, and A.B. Smith III, Solid State Chemistry of Fullerene-Based Materials, *Acc. Chem. Res.* **25** (1992) 112–18. Fulleride structures.

[31] P. Bak, Commensurate Phases, Incommensurate Phases, and the Devil's Staircase, *Rep. Prog. Phys.* **45** (1982) 587. Incommensurate structures.

[32] D. Levine and P.J. Steinhardt, Quasicrystals. I. Definitions and Structure, *Phys. Rev. B* **34** (1986) 596. Quasicrystalline structures.

[33] J.E. Socolar and P.J. Steinhardt, Quasicrystals. II. Unit Cell Configurations, *Phys. Rev. B* **34** (1986) 617. Quasicrystalline structures.

[34] David R. Nelson, Quasicrystals, *Scientific American*, August 1986, p. 42. Fig. 4.34 is drawn after a fig. in this ref. Quasicrystalline structures.

[35] C.N.R. Rao and J. Gopalakrishnan, *New Directions in Solid State Chemistry* (Cambridge University Press, Cambridge, 1989), pp. 195–199, Chapter 4. Liquid crystals.

[36] D. Giaquinta and H.-C. zur Loye, Structural Predictions in the ABO_3 phase diagram, *Chem. Mater.* **6** (1994) 365–72. ABO_3 compounds.

Chapter 5
Diffraction

(A) Introduction

Diffraction is the principal means of determining the structure of crystalline matter. There are a number of experimental methods that employ a variety of radiation sources (X-ray, electron, neutron), but a single interference theory underlies them all. Rather than attempt a comprehensive survey of all the experimental techniques, the goal of this chapter is to describe the theoretical underpinnings of diffraction so that we can understand the relationship between diffraction data and crystal structures. In cases where there is a need to be more specific, emphasis is placed on powder X-ray diffraction using a scanning diffractometer; this is one of the most common applications of diffraction. Descriptions of other diffraction methods are limited to brief comments in Section F on the comparative advantages and disadvantages of selected techniques.

(B) Bragg's formulation of the diffraction condition

In 1913, Bragg devised a theory to explain the patterns that were observed when X-rays were scattered from crystalline materials [1,2]. At this time, it was already widely believed, based on other evidence, that atoms were arranged periodically in crystals. Although X-rays scatter from individual atoms and the most accurate model treats each atom in the crystal as a source of scattered radiation (this more complete model is developed in the next section), Bragg's observations can be explained using the simplified model illustrated in Fig. 5.1. Specifically, we will assume that X-rays are scattered by parallel planes of atoms. Each plane of atoms acts as a half-silvered mirror, reflecting 10^{-5} to 10^{-3} of the incident radiation, depending on its scattering power. In this model, each plane of atoms is a source of scattered radiation and diffraction effects are observed when the distance between the sources is comparable to the wavelength of the radiation. In contrast to reflected radiation, whose intensity is nearly constant with variations in the incident angle, diffracted radiation is produced by constructive and destructive interference and its intensity has a strong angular dependence. Specifically, when the phases of the waves scattered from atoms in parallel planes differ by an integer multiple number of wavelengths, these waves interfere constructively and a peak

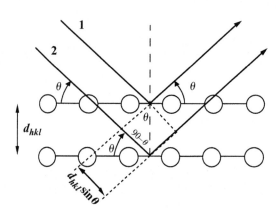

Figure 5.1. Geometry for the formulation of Bragg's law.

in the diffraction pattern is created. Otherwise, destructive interference leads to a low intensity.

In Fig. 5.1, the phase difference between wave 1 and wave 2 is determined by the path length difference, which can be found geometrically to be $2d_{hkl}\sin\theta$ (where d_{hkl}, the 'd-spacing', is the distance between identical planes specified by the index hkl). Thus, the condition for constructive interference (a large diffracted beam intensity) is that

$$\lambda = 2d_{hkl}\sin\theta. \tag{5.1}$$

This is known as Bragg's Law and its value is obvious: if you have monochromatic radiation with a single, known wavelength (λ), and you measure the angles at which peaks in the scattered radiation occur (θ), you can determine the interplanar spacings, d_{hkl}. This information, together with the equations in Chapter 2, Section G(ii), that relate d-spacings to a, b, c, α, β, and γ, can be used to determine the lattice parameters.

Unfortunately, this simple theory is a bit limited because it specifies only the interplanar spacings that characterize the lattice. In order to determine the crystal structure, we also need to specify the chemistry and geometry of the basis. This information shows up in the intensities of the diffraction peaks, so we need a more sophisticated theory that relates the geometric configuration of the atoms in the basis to the diffracted beam intensity.

Ⓒ The scattering of X-rays from a periodic electron density

Because of their small size, atomic nuclei have a negligible X-ray scattering cross section. It is the electrons within the solid that account for nearly all of the X-ray scattering. Thus, the scattering power of an atom increases with the atomic

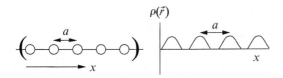

$\rho(\vec{r})$

Figure 5.2. The atomic positions and electron density for a one-dimensional crystal.

number (Z) and the scattering power of a solid increases with its (mass and electron) density. For this reason, Pb ($Z=82$) is a common material for X-ray shielding and Be ($Z=4$) and lithium borate glasses are used as X-ray transparent windows.

i. *The periodic electron density*

Because it is the electrons that do the scattering, it is the electron density, $\rho(\vec{r})$, that determines how the X-rays diffract from the crystal. Since the electrons are directly associated with the atoms, the electron density (in most cases) has the same periodicity as the atomic structure and we can define it in terms of the crystal lattice.

We begin by defining a position vector, $\vec{r}$, that ranges over all possible positions within the unit cell and has its origin at $x=y=z=0$.

$$\vec{r}=x\vec{a}+y\vec{b}+z\vec{c}. \tag{5.2}$$

As defined in Chapter 2, $\vec{a}$, $\vec{b}$, and $\vec{c}$ are the lattice vectors along the three crystallographic axes and x, y, and z are the fractional unit cell coordinates that are greater than or equal to zero and less than one. Next, we recall the definition of the lattice translation vector that identifies each unit cell in the crystal.

$$\vec{R}=u\vec{a}+v\vec{b}+w\vec{c}. \tag{5.3}$$

With these two vectors we can specify any arbitrary position in the crystal and write the electron density of the crystal as a periodic function:

$$\rho(\vec{r})=\rho(\vec{r}+\vec{R}). \tag{5.4}$$

For reasons that will become clear later in this section, we find it useful to describe the electron density in terms of a Fourier series. If you are unfamiliar with the Fourier series, you may want to review Appendix 5A or one of the cited references. The Fourier series representation of the electron density of the one dimensional crystal in Fig. 5.2 is:

$$\rho(\vec{x})=\sum_{h=-\infty}^{\infty} S_h e^{i(2\pi hx)} \tag{5.5}$$

where $0 \leq x < 1$ and the coefficients, S_h, are

$$S_h = \frac{1}{a}\int_0^a \rho(\vec{x})e^{i(2\pi hx)}dx.$$ (5.6)

The electron density in Eqn. 5.5 is easily generalized to three dimensions. In terms of fractional coordinates and the integers, h, k, and l,

$$\rho(\vec{r}) = \rho(x, y, z) = \sum_{h,k,l=-\infty}^{\infty} S_{hkl}e^{i2\pi(hx+ky+lz)}.$$ (5.7)

At this point, we make use of the reciprocal lattice concept established in Chapter 2, Section F. Specifically, recalling that:

$$\vec{G}_{hkl} = h\vec{a}^* + k\vec{b}^* + l\vec{c}^*,$$ (5.8)

and that

$$\vec{a}^* \cdot \vec{a} = 2\pi \qquad \vec{b}^* \cdot \vec{a} = 0 \qquad \vec{c}^* \cdot \vec{a} = 0$$
$$\vec{a}^* \cdot \vec{b} = 0 \qquad \vec{b}^* \cdot \vec{b} = 2\pi \qquad \vec{c}^* \cdot \vec{b} = 0$$
$$\vec{a}^* \cdot \vec{c} = 0 \qquad \vec{b}^* \cdot \vec{c} = 0 \qquad \vec{c}^* \cdot \vec{c} = 2\pi,$$ (5.9)

we can use Eqn. 5.2 to write:

$$\vec{G}_{hkl} \cdot \vec{r} = 2\pi(hk + ky + lz).$$ (5.10)

In this case, the integers h, k, and l are Miller indices that specify points in the reciprocal lattice and, therefore, refer to planes in the direct lattice. Substituting this into Eqn. 5.7, we have an expression for the electron density in terms of the reciprocal lattice vectors. This is an essential component of the diffraction theory described below.

$$\rho(\vec{r}) = \sum_{h,k,l=-\infty}^{\infty} S_{hkl}e^{i\vec{G}_{hkl}\cdot\vec{r}}$$ (5.11)

where

$$S_{hkl} = \frac{1}{V}\int_0^{\vec{r}} \rho(\vec{r})e^{i\vec{G}_{hkl}\cdot\vec{r}}d\vec{r}.$$ (5.12)

In the next section, we will see that it is possible to measure the Fourier coefficients, S_{hkl}, using the diffraction experiment and to sum these quantities according to Eqn. 5.11 to construct the electron density.

ii. *Elastic scattering*

By definition, elastic scattering is a collision where energy is conserved, but the direction of the scattered wave is allowed to change. So, for an elastically scattered plane wave, the orientation of the wave vector $\vec{k}$, can change, but its magnitude, $|k| = 2\pi/\lambda$, must remain constant. In our discussion of diffraction theory, we will assume that X-rays are plane waves and that they scatter elastically from the electrons in the crystal. We begin by considering the superposition (interference) of plane waves.

Consider two coherent plane waves (waves with the same $\vec{k}$), Ψ_1 and Ψ_2. In the general case, Ψ_1 and Ψ_2 can have different amplitudes $(A_1 \neq A_2)$ and Ψ_2 can be phase shifted with respect to Ψ_1 by an amount ϕ.

$$\Psi_1 = A_1 e^{i\vec{k}x} \tag{5.13}$$

$$\Psi_2 = A_2 e^{i(\vec{k}x+\phi)}. \tag{5.14}$$

The superposition of the two waves is written as the sum, Ψ_t.

$$\Psi_1 + \Psi_2 = A_1 e^{i\vec{k}x} + A_2 e^{i(\vec{k}x+\phi)} \tag{5.15}$$

$$\Psi_t = e^{i\vec{k}x}(A_1 + A_2 e^{i\phi}). \tag{5.16}$$

Because the phase difference, ϕ, is a constant, we can define a new constant amplitude, $A = A_1 + A_2 e^{i\phi}$, and rewrite Ψ_t:

$$\Psi_t = A e^{i\vec{k}x}. \tag{5.17}$$

This result says that the superposition of two coherent waves is simply another coherent wave with an amplitude that depends on the phase shift, ϕ. The phase shift depends only on the relative positions of the sources (the path length difference). Considering the allowed values for $e^{i\phi}$, the maximum value of the new amplitude is $A_1 + A_2$ and the minimum value is $A_1 - A_2$. For the case of $A_1 = A_2$, the maximum is $2A_1$ and the minimum is zero. The maximum is realized when the sources are separated by an integer number of wavelengths $(n\lambda)$ so that the phase shift (ϕ) is 0 or 2π $(e^{i2\pi} = e^0 = +1)$; this is known as the condition for constructive interference. On the other hand, when the separation of the sources is $n\lambda/2$ and the phase shift is π, the amplitude is minimized $(e^{i\pi} = -1)$; this is the condition for destructive interference. We can extend this result

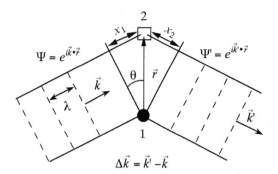

Figure 5.3. Geometry for determining the relative phases of waves scattered from two volume elements in the unit cell. The volume element labeled 1 is at the origin [3].

to the superposition of N coherent waves, originating at N sources, each with a phase shift ϕ_N, to form Ψ_t.

$$\Psi_t = A e^{i\vec{k}x}, \quad \text{where } A = \sum_N A_N e^{i\phi_N}. \tag{5.18}$$

In summary, we have found that the critical data needed to determine the amplitude that results from the superposition of many coherent waves (as happens in a scattering or diffraction experiment) are the relative phase differences between the waves. Based on what we found in Section B, we can say that the phase differences between coherent waves are determined by the relative positions of the sources (the path length difference). Thus, the resultant amplitude (the diffracted beam intensity) is determined by the configuration of the scattering sources (the atomic positions).

We are interested in the problem of coherent waves scattering from a crystal. In this case, the scattering sources are the electrons in the crystal and their configuration is described by the electron density, $\rho(\vec{r})$. The most general solution to this problem has already been expressed in Eqn. 5.18. We need only to determine the relative phases (ϕ_N) and amplitudes (A_N) of the waves scattered from the different possible sources to determine the new amplitude, A. We determine each of the relative phases from the path length difference using the construction in Fig. 5.3.

Consider all the space in a single unit cell to be partitioned into volume elements, ΔV. Each of these volume elements represents a scattering center. The location of each element, with respect to the unit cell origin, is specified by the position vector, $\vec{r}$, as illustrated in Fig. 5.3. If the volume elements are sufficiently small, the electron density in each element is given by a constant value, $\rho(\vec{r}_i) = \rho_i$, where the index, i, specifies the particular volume element. Furthermore, we will assume that the amplitude of the wave scattered from a particular volume element is proportional to the number of electrons within that element, $\rho_i \Delta V$.

The phase difference between the wave scattered from a reference volume element at the origin and any other volume element in the unit cell labeled by $\vec{r}_i$ can be determined with reference to Fig. 5.3.

Figure 5.3 shows an incoming plane wave, Ψ, characterized by the wave vector, $\vec{k}$, and the scattered plane wave with a new wave vector, $\vec{k}'$. Remember that because we consider only elastic scattering, the magnitudes of $\vec{k}$ and $\vec{k}'$ are equal, only the direction differs. The phase difference between a wave scattered from source 1 and source 2 is ϕ and is determined from the path length difference, $L = x_1 + x_2$, simply by multiplying by the degrees of phase per unit length, $2\pi/\lambda$:

$$\phi = (x_1 + x_2)\frac{2\pi}{\lambda}. \tag{5.19}$$

From Fig. 5.3, we see that x_1 and x_2 are equal to $r\sin\theta$ so that:

$$\phi = \frac{2\pi}{\lambda}r\sin\theta + \frac{2\pi}{\lambda}r\sin\theta. \tag{5.20}$$

From Fig. 5.3, we can see that:

$$\vec{k}\bullet\vec{r} = (2\pi/\lambda)r\cos(90 - \theta) \text{ and } \vec{k}'\bullet\vec{r} = (2\pi/\lambda)r\cos(90 + \theta). \tag{5.21}$$

Since $\cos(90 - \theta) = \sin\theta$, and $\cos(90 + \theta) = -\sin\theta$, we can rewrite Eqn. 5.20 as the difference between scalar products of the position vector and the wave vectors.

$$\begin{aligned}\phi &= \vec{k}\bullet\vec{r} - \vec{k}'\bullet\vec{r}\\ &= (\vec{k} - \vec{k}')\bullet\vec{r}\\ &= -\Delta\vec{k}\bullet\vec{r}.\end{aligned} \tag{5.22}$$

At this point, we can specify the amplitude and phase of waves scattered from each volume element in the unit cell, identified by the index i. The superposition of all of these waves gives the total scattered amplitude which is calculated according to Eqn. 5.18. The sum is written in the following way:

$$A = \sum_i \rho(\vec{r}_i)\Delta V e^{-i\Delta\vec{k}\bullet\vec{r}_i}. \tag{5.23}$$

Finally, we assume infinitesimal volume elements and rewrite this sum as an integral:

$$A = \int \rho(\vec{r})e^{-i\Delta\vec{k}\bullet\vec{r}}dV. \tag{5.24}$$

Next, we want to use our definition of the electron density, as expressed in Eqn. 5.11, to rewrite the amplitude

$$A = \int \sum_{hkl} S_{hkl} e^{i\vec{G}_{hkl}\cdot\vec{r}} e^{-i\Delta\vec{k}\cdot\vec{r}} dV \qquad (5.25)$$

$$= \sum_{hkl} S_{hkl} \int e^{i(\vec{G}_{hkl}-\Delta\vec{k})\cdot\vec{r}} dV.$$

Without further analysis, inspection of Eqn. 5.25 allows us to identify an important condition on the amplitude of the scattered wave. For any given value of *hkl*, which specifies the reciprocal lattice vector, $\vec{G}_{hkl}$, we can see that if the scattering vector, $\Delta\vec{k}$, is equal to the reciprocal lattice vector, then this term in the sum is simply proportional to S_{hkl}, a Fourier coefficient of the electron density. More importantly, when $\Delta\vec{k} \neq \vec{G}_{hkl}$, the amplitude goes to zero (this is demonstrated in the next section). So, the amplitude of the scattered wave is finite only if $\Delta\vec{k} = \vec{G}_{hkl}$ and in this condition (the Bragg condition), the amplitude of the scattered wave is proportional to a Fourier coefficient of the electron density. The significance of this can not be overstated. By measuring the intensity (proportional to $|A_{hkl}|^2$) of a diffracted Bragg peak, we determine a Fourier coefficient of the electron density. By measuring enough of these coefficients (diffracted peak intensities), we can construct the electron density function and determine the atomic structure of the crystal's basis.

iii. *Scattering from many centers*

In this section, we will examine how the scattered wave amplitude changes with the scattering vector and the number of scattering centers. For simplicity, the argument is made in one dimension with the understanding that the result applies to three-dimensional systems as well. We begin with the assumption that our linear crystal has identical scattering centers localized exactly at the lattice points. In other words, the electron density has a constant value (ρ) at each lattice site and is zero in between. Thus, instead of integrating $\rho(\vec{r})dV$ over all space, we can perform a discrete summation over the lattice points. If the points in the one-dimensional crystal are specified by the vector $\vec{r}_n = n\vec{a}$, then the amplitude of the wave created by the superposition of the waves scattered from each source is:

$$A_0 = \rho \sum_{n=1}^{N} e^{-i\vec{r}_n\cdot\Delta\vec{k}}. \qquad (5.26)$$

Normalizing by the constant factor ρ and substituting for $\vec{r}_n$, we have

$$A = \sum_{n=1}^{N} e^{-in\vec{a}\cdot\Delta\vec{k}}. \qquad (5.27)$$

Computing the sum over the N lattice points is simplified by using the following relationship:

$$\sum_{n=0}^{N-1} x^n = \frac{1-x^N}{1-x}.$$
(5.28)

Using Eqn. 5.28, the expression for the amplitude is:

$$A = \frac{1 - e^{-iN\vec{a}\cdot\Delta\vec{k}}}{1 - e^{-i\vec{a}\cdot\Delta\vec{k}}}.$$
(5.29)

The experimentally observed quantity is actually the intensity of the scattered wave, $I = A^*A$. Computing A^*A, using the identity in Eqn. 5.14 and the fact that $\sin^2\theta = 1/2(1 - \cos 2\theta)$, we find that the intensity is

$$I = \frac{\sin^2 \frac{1}{2}N(\vec{a}\cdot\Delta\vec{k})}{\sin^2 \frac{1}{2}(\vec{a}\cdot\Delta\vec{k})}.$$
(5.30)

The intensity is plotted as a function of $\Delta\vec{k}$ for three values of N in Fig. 5.4. The increasing values of N correspond to a crystal growing in size. In the calculation, the lattice parameter (a) was taken to be 3.0 Å. Thus, the lengths of the reciprocal lattice vectors are equal to $n2\pi/a$ or $2.1n$. Note that the function peaks when $\Delta\vec{k} = 2.1n$. Most importantly, note that as N increases to the size of a realistic crystal (300 Å), the peaks become narrower and more intense. From this analysis, we can say that when $\Delta\vec{k} = \vec{G}_{hkl}$, the scattered radiation is very intense and when $\Delta\vec{k} \neq \vec{G}_{hkl}$, the intensity goes to zero. This intense scattered radiation is, of course, a diffraction peak and our diffraction condition is that $\Delta\vec{k} = \vec{G}_{hkl}$. In Section (iv), we will show that this condition is equivalent to the Bragg condition.

As a final note, it should be pointed out that the dependence of Bragg peak width on particle size is an easily observable phenomenon. As the average particle size in a powder decreases below 2000 Å, the diffracted peaks clearly increase in breadth and diminish in intensity. This line broadening effect is illustrated by the comparison shown in Fig. 5.5. Finely divided MgO, with particle sizes of about 200 Å, was formed by the thermal decomposition of a hydrotalcite. The diffraction peaks from this material are broader and less intense than those from a coarser grained specimen.

It follows that the width of a diffraction peak can be used as a measure of particle dimensions. The quantitative relationship is given by the Scherrer formula [4]:

$$t = \frac{0.9\lambda}{B\cos\theta_B},$$
(5.31)

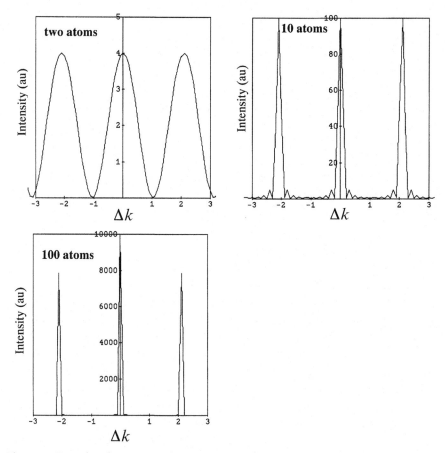

Figure 5.4. Intensity of a superposition of plane waves scattered from a one-dimensional periodic array of atoms. Note the diffraction peaks intensify and sharpen as more atoms or scattering centers are added to the crystal.

where t is the thickness of the particle in the direction perpendicular to the planes from which the diffraction peak originates, B is the width of the diffraction peak at one half of the maximum intensity, measured in radians, λ is the wavelength of the X-rays, and θ_B is the angular position of the peak's maximum. Equation 5.31 can be used to estimate the sizes of powder particles, under the assumption that all particles have about the same size. Furthermore, it is possible to determine the anisotropy of the particle shape by examining the breadth of peaks originating form different sets of planes.

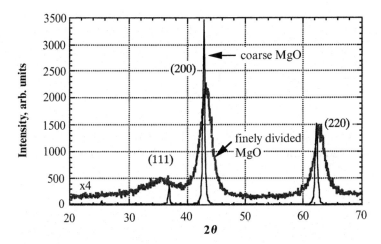

Figure 5.5. Powder X-ray diffraction patterns for coarse grained and finely divided MgO. The peaks produced by the larger crystallites are narrower and much more intense (the pattern from the finely divided material has been magnified by a factor of four for clarity). The finely divided material contains stabilizing aluminum impurities, which accounts for the small angular shifts in the peak positions.

iv. *Equivalence of the Bragg condition and the elastic scattering condition*

Equating the diffraction condition derived from elastic scattering theory with Bragg's diffraction condition is a matter of definitions. In Section (ii), we found that the diffraction condition is $\Delta \vec{k} = \vec{G}$, which can also be written as:

$$\vec{k} + \vec{G} = \vec{k}'. \tag{5.32}$$

Because it is an elastic process, we know that $|\vec{k}| = |\vec{k}'|$ and that $k^2 = k'^2$. So, from Eqn. 5.32,

$$(\vec{k} + \vec{G})^2 = k'^2 = k^2$$
$$k^2 + 2\vec{k}\cdot\vec{G} + G^2 = k^2$$
$$2\vec{k}\cdot\vec{G} + G^2 = 0. \tag{5.33}$$

Because $\vec{G}$ is a reciprocal lattice vector, there must be another reciprocal lattice vector of the same magnitude, $-\vec{G}$, for which:

$$2\vec{k}\cdot\vec{G} = G^2. \tag{5.34}$$

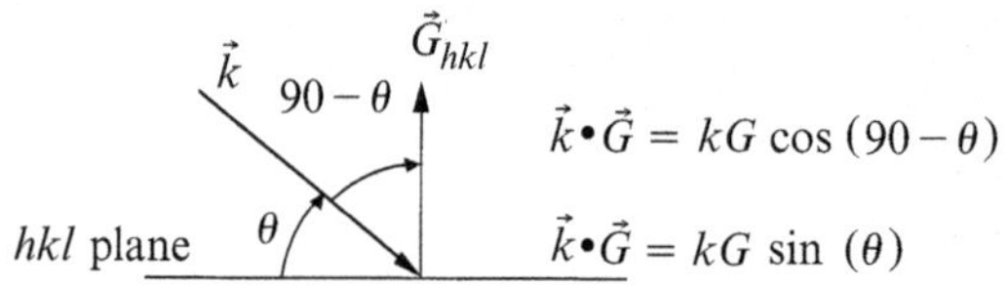

$$\vec{k} \cdot \vec{G} = kG \cos(90 - \theta)$$

$$\vec{k} \cdot \vec{G} = kG \sin(\theta)$$

Figure 5.6. Relation of the elastic scattering geometry to the Bragg formulation.

As demonstrated in Fig. 5.6, $2\vec{k} \cdot \vec{G}$ can be written as $2kG\sin\theta$, where θ is the angle of incidence as defined by Bragg's law. Furthermore, based on our definition of $\vec{G}$,

$$d_{hkl} = 2\pi/|G|. \tag{5.35}$$

Substituting into Eqn. 5.34,

$$2kG\sin\theta = G^2$$
$$2(2\pi/\lambda)\sin\theta = 2\pi/d_{hkl}$$
$$2d_{hkl}\sin\theta = \lambda. \tag{5.36}$$

Thus, the conditions $\lambda = 2d\sin\theta$ and $\Delta\vec{k} = \vec{G}$ are equivalent.

v. The Ewald construction

Just as the Bragg condition can be understood in terms of a simple geometric construction, the condition that $\Delta\vec{k} = \vec{G}$ can be understood in terms of the Ewald construction. Consider a hypothetical monoclinic crystal with $a = 2$ Å, $b = 3$ Å, $c = 4$ Å, and $\beta = 45°$. A superposition of the real and reciprocal lattices of this crystal is illustrated in Fig. 5.7, projected along the [010] axis. Note that in this illustration, the points are labeled only by their h and l indices and the reciprocal lattice is scaled by a multiplicative factor of 6 for clarity.

Consider the diffraction condition for the $(\bar{1}0\bar{1})$ plane. We know that $\vec{G}_{\bar{1}0\bar{1}}$ is the vector pointing from the origin to the $(\bar{1}0\bar{1})$ point in the reciprocal lattice (labeled $\bar{1}\bar{1}$). For an incoming wave, $\vec{k}$, to create a diffraction peak, $\Delta\vec{k}$ or $\vec{k}' - \vec{k}$ must equal $\vec{G}_{\bar{1}0\bar{1}}$. Geometrically, therefore, $\vec{G}_{\bar{1}0\bar{1}}$, $\vec{k}$, and $\vec{k}'$ must form the edges of a triangle, as shown in Fig. 5.7. Furthermore, since the length of $\vec{k}$ and $\vec{k}'$ must be equal, then their common origin defines the center of a sphere, called the Ewald sphere, and their endpoints lie on the surface of this sphere. This construction is appealing because for a fixed X-ray beam ($\vec{k}$), the size of the Ewald sphere is constant. Thus, to diffract from any set of planes, (hkl), we need only rotate the crystal until the reciprocal lattice point specified by the vector $\vec{G}_{hkl}$ is on the surface of the Ewald sphere. Figure 5.7 also shows that this is the same as the

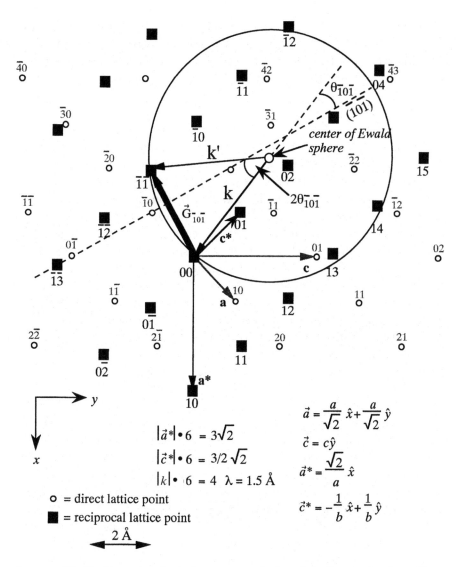

$$\vec{a} = \frac{a}{\sqrt{2}}\,\hat{x} + \frac{a}{\sqrt{2}}\,\hat{y}$$

$$\vec{c} = c\hat{y}$$

$$|\vec{a}*|\cdot 6 = 3\sqrt{2}$$

$$|\vec{c}*|\cdot 6 = 3/2\,\sqrt{2}$$

$$|k|\cdot 6 = 4 \quad \lambda = 1.5\,\text{Å}$$

$$\vec{a}* = \frac{\sqrt{2}}{a}\,\hat{x}$$

○ = direct lattice point

■ = reciprocal lattice point

$$\vec{c}* = -\frac{1}{b}\,\hat{x} + \frac{1}{b}\,\hat{y}$$

2 Å

Figure 5.7. The Ewald construction is shown on a projection of the direct and reciprocal lattices. A complete description of the figure can be found in the text.

Bragg condition. The $(\bar{1}0\bar{1})$ plane in the direct lattice is highlighted by the dashed line. The angle that $\vec{k}$ makes with this plane satisfies the Bragg condition (23°). The angle between the incoming beam $(\vec{k})$ and the diffracted beam $(\vec{k}')$ is always $2\theta_{hkl}$.

vi. *Concluding remarks*

The last few sections of this chapter have described the essential elements of diffraction theory. It is important to emphasize how these ideas relate to practical experiments. Recall that knowledge of the electron density, as expressed in Eqn. 5.11, is equivalent to knowledge of the crystal structure. We divide the structure into a lattice and a basis; if the lattice is known, then all of the reciprocal lattice vectors are also known. Specifying the basis amounts to determining a sufficient list of the Fourier coefficients of the electron density, S_{hkl}. From our analysis of the diffraction problem, we found that when the scattering vector is *not* equal to a reciprocal lattice vector, the scattered amplitude is zero, and when the scattering vector *is* equal to a reciprocal lattice vector (i.e. the Bragg condition is satisfied), the scattered amplitude is proportional to the Fourier coefficient of the electron density, S_{hkl}. In principle, therefore, the Fourier coefficients can be measured and the electron density function can be constructed. In practice, we determine these coefficients from the relative intensities, I_{hkl}, of the diffracted peaks. Unfortunately, the intensity and the amplitude are not equal:

$$I_{hkl}\alpha(A^*A = |S_{hkl}|^2). \tag{5.37}$$

Thus, relative intensity measurements give no information about the sign and phase of S_{hkl}. Methods have been developed to work around this problem, but will not be described here [5]. In conclusion, understanding the relationship between the Fourier coefficients of the electron density and the diffracted beam intensity is essential for the determination of crystal structures.

We summarize the preceding sections with the following statements. A crystal structure should be thought of as a lattice plus a basis. The measurement of diffraction peak angular *positions* provides the information necessary to specify the lattice. The measurement of diffraction peak *intensities* allows the positions of the atoms in the basis to be determined.

(D) The relationship between diffracted peak intensities and atomic positions

i *The structure factor*

In practice, the analysis of diffraction data is simplified by separating the diffracted peak intensity into two contributions, that which comes from each of the

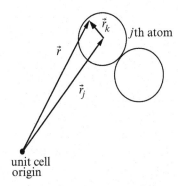

Figure 5.8. Geometry for the position vectors describing the atomic configuration.

atoms in the basis (called the *atomic form factor*), and that which comes from the arrangement of the atoms in the basis. The diffraction peak amplitude is thus expressed as a function of the atomic coordinates of the atoms in the basis, and in this form it is known as the *structure factor* [6–8]. We begin by recalling Eqn. 5.24 for the scattered amplitude. Inserting the diffraction condition, we have:

$$A_{hkl} = \int \rho(\vec{r}) e^{-i\vec{G}_{hkl}\cdot\vec{r}} dV. \tag{5.38}$$

In Eqn. 5.38, the position vector, $\vec{r}$, ranges over all positions within a single unit cell and each unit cell is identical. Next, we rewrite the electron density as a super-position of the electron densities of each atom in the basis, $\rho_j(\vec{r}_k)$, where $\vec{r}_k$ is a vector whose origin is at the center of the jth atom. To express this quantity in terms of our usual position vector, $\vec{r}$, we also define vectors, $\vec{r}_j$, that point from the cell origin to the center of each of the j atoms in the basis (see Fig. 5.8). So, the electron density due to the jth atom, $\rho_j(\vec{r}_k)$, is $\rho_j(\vec{r}-\vec{r}_j)$. Note that when $\vec{r}-\vec{r}_j$ is larger than the atomic radius, ρ_j goes to zero. Within this framework, we assume that the total electron density is the sum of all of the atomic electron densities:

$$\rho(\vec{r}) = \sum_{j=1}^{J} \rho_j(\vec{r}-\vec{r}_j). \tag{5.39}$$

We can now rewrite the total scattered amplitude as a sum of integrals over each atom in the basis.

$$A_{hkl} = \sum_{j=1}^{J} e^{-i\vec{G}\cdot\vec{r}_j} \int \rho_j(\vec{r}-\vec{r}_j) e^{-i\vec{G}\cdot(\vec{r}-\vec{r}_j)} dV. \tag{5.40}$$

Note that each of the integrals, although performed over the entire range of $\vec{r}$ (the entire unit cell), rapidly goes to zero far from the jth atom. Each of these integrals represents the electron density due to the jth atom. Thus, the atomic form factor is defined as:

$$f_j = \int \rho_j(\vec{r} - \vec{r}_j) e^{-i\vec{G} \cdot (\vec{r} - \vec{r}_j)} dV. \tag{5.41}$$

We assume that the form factor is an atomic quantity that is invariant from crystal to crystal, and a function only of the scattering vector. This assumption is supported by the fact that when atoms form bonds in crystals, only the valence electrons are redistributed. The majority of the electrons (the core electrons) remain unchanged and, thus, make the same contribution to the electron density, irrespective of the crystal in which the atom is situated. So, from this point onward, we will take the values of f_j to be atomic constants, like size or mass, roughly proportional to atomic number (number of electrons) and with some angular dependence. Values of f_j as a function of scattering angle are tabulated in the International Tables and an approximate method for computing them is described in Section E(iii) of this chapter.

The amplitude, which we now call the structure factor, is written in the following way:

$$A_{hkl} = S_{hkl} = \sum_{j=1}^{J} f_j e^{-i\vec{G}hkl \cdot \vec{r}_j}. \tag{5.42}$$

Note that the structure factors are identical to the Fourier coefficient of the electron density. From Eqn. 5.42, we can see that the intensity of each Bragg peak, specified by h, k, and l, is a function of the chemical identity of the atoms in the basis, f_j, and the positions of each atom, specified by $\vec{r}_j$ (the fractional coordinates). In other words, within this framework, each atom is simply a point with a fixed scattering power (f_j) and the diffracted beam intensity is a function of how these points are arranged in the unit cell. In practice, it is convenient to make the following substitution:

$$\vec{G}_{hkl} \cdot \vec{r}_j = (h\vec{a}^* + k\vec{b}^* + l\vec{c}^*) \cdot (x_j\vec{a} + y_j\vec{b} + z_j\vec{c})$$
$$= 2\pi(hx_j + ky_j + lz_j), \tag{5.43}$$

so that Eqn. 5.42 is rewritten:

$$S_{hkl} = \sum_{j=1}^{J} f_j e^{-i2\pi(hx_j + ky_j + lz_j)}. \tag{5.44}$$

Note that while the intensity of a diffraction peak is always real:

$$I_{hkl} \alpha S_{hkl}^* S_{hkl}, \tag{5.45}$$

S_{hkl} need not be real. In conclusion, the importance of Eqn. 5.44 is that we can now express the diffracted peak intensity, a readily measurable quantity, in terms of the positions of the atoms in the unit cell.

Example 5.1

The rutile structure, described in Table 4.19, has only three variable parameters, a, c, and x. The values for a and c are determined from the positions of the Bragg peaks. How can x be measured?

1. To measure x, we need to know how the structure factor and, therefore, the diffracted peak intensities vary with the value of x. In practice, only relative intensities are measured so, for a single measurement of x, we need to identify one peak intensity that is independent of x and another that is dependent on x.

2. Inspection of the data in Table 4.19 indicates that the structure factor of the $(00l)$ peaks is independent of the value of x. Following eqn. 5.44:

$$S_{00l} = \sum_{j=1}^{6} f_j e^{-i2\pi(lz_j)} = f_{Ti}(e^0 + e^{-i\pi l}) + 2f_0(e^0 + e^{-i\pi l})$$

$$S_{00l} = f_{Ti}(1 + (-1)^l) + 2f_0(1 + (-1)^l).$$

When l is odd, S_{00l} vanishes. However, when l is even, S_{00l} is finite and independent of x. Using the index e to represent any even integer, we have:

$$S_{00e} = 2f_{Ti} + 4f_0.$$

3. The intensity of the $h00$ peaks does depend on the value of x:

$$S_{h00} = \sum_{j=1}^{6} f_j e^{-i2\pi(hx_j)} = f_{Ti}(e^0 + e^{-i\pi h})$$

$$+ f_0\left(e^{i2\pi(hx)} + e^{-i2\pi(hx)} + e^{-i2\pi h(1/2-x)} + e^{-i2\pi h(1/2+x)}\right)$$

$$S_{h00} = f_{Ti}(1 + (-1)^h) + f_0\left(e^{i2\pi hx} + e^{-i2\pi hx} + e^{-i\pi h}e^{i2\pi hx} + e^{-i\pi h}e^{-i2\pi hx}\right)$$

$$S_{h00} = f_{Ti}(1 + (-1)^h) + f_0\left((e^{i2\pi hx} + e^{-i2\pi hx}) + (e^{i2\pi hx} + e^{-i2\pi hx})e^{-i\pi h}\right)$$

$$S_{h00} = f_{Ti}(1 + (-1)^h) + f_0\left((1 + e^{-i\pi h})(e^{i2\pi hx} + e^{-i2\pi hx})\right)$$

$$S_{h00} = f_{Ti}(1 + (-1)^h) + f_0\left((1 + (-1)^h)2\cos(2\pi hx)\right).$$

4. We can see that S_{h00} vanishes for all odd h and that for even h:

$$S_{e00} = 2f_{Ti} + 4f_0\cos(2\pi hx).$$

5. So, the ratio of the measured intensities is:

$$\frac{I_{e00}}{I_{00e}} \propto \frac{S_{e00}{}^{*}S_{e00}}{S_{00e}{}^{*}S_{00e}} = \frac{[2f_{\mathrm{Ti}} + 4f_{\mathrm{o}}\cos(2\pi hx)]^{2}}{[2f_{\mathrm{Ti}} + 4f_{\mathrm{o}}]^{2}}.$$

The left-hand side of this equation can be measured experimentally for various values of h and l, as long as they are even. The values of f_{Ti} and f_{o} are tabulated constants, so x is the only unknown variable on the right-hand side of the equation and it can, therefore, be computed. In practice, it would be best to measure as many peak intensities as possible and find an average or best fit value of x. Finally, we note that the atomic scattering factors, f_{Ti} and f_{o}, depend on the angular position of the peaks and will not be the same on the top and bottom of the equation. Furthermore, for simplicity, we have ignored a number of additional factors that affect the intensity. These factors are discussed in detail in Section E.

ii. *Systematic absences due to lattice centering*

The calculation of structure factors for different crystal systems leads us to the conclusion that some diffraction peaks are systematically absent (the phenomenon was already encountered in Example 5.1) [9,10]. A few useful relationships to remember when computing structure factors are:

$$e^{in\pi} = \begin{cases} -1 \text{ if } n \text{ is odd} \\ 1 \text{ if } n \text{ is even} \end{cases}$$
$$e^{0} = 1. \tag{5.46}$$

We first consider the P (primitive) lattice. There is one atom per cell and the coordinates of this atom (the $j=1$ atom) are $(0,0,0)$. Thus, following from Eqn. 5.44,

$$S_{hkl} = \sum_{j=1}^{1} f_{j}e^{0} = f_{1}e^{0} = f_{1} \quad \text{(P lattice).} \tag{5.47}$$

This is, of course, the trivial case. The results are that there are no absences, that there should be a Bragg peak for every possible value of h, k, and l, and that the intensity should scale with the angular variation of f_{j}.

Next, consider an I lattice. There are two atoms in this unit cell that have the fractional coordinates $(0,0,0)$ and $(1/2,1/2,1/2)$. Labeling the atoms $j=1$ and 2, respectively, the structure factor is:

$$S_{hkl} = f_{1}e^{-i2\pi(0)} + f_{2}e^{-i\pi(h+k+l)} \quad \text{(I lattice).} \tag{5.48}$$

Table 5.1. *Systematic absences due to lattice centering.*

lattice type	conditions on reflections
P	all hkl
I	$h+k+l=2n$
F	h, k, and l all odd or all even
A	$k+l=2n$
B	$h+l=2n$
C	$h+k=2n$
R	$-h+k+l=3n$

For an I lattice, $f_1=f_2=f$ and $h+k+l=n$, then

$$S_{hkl}=f\left[1+(-1)^n\right]$$
$$=2f \quad \text{when } h+k+l \text{ is even}$$
$$=0 \quad \text{when } h+k+l \text{ is odd.} \tag{5.49}$$

This result implies that half of the diffraction peaks have an intensity that scales with the angular variation of f [for example, (110), (200), or (211)] and the other half are 'missing' or have zero intensity [including (100), (111), and (210)]. In this case, we say that the $n=$ odd reflections are systematically absent and that the condition for a plane to diffract is $h+k+l=2n$. Similar conditions exist for other lattice centering types and these are summarized in Table 5.1. These conditions are also summarized graphically in Figs. 5.9 and 5.10.

iii. *Systematic absences due to glide operators*
Consider the case of an a type axial glide plane perpendicular to the b-axis. If there is an atom at (x,y,z), then there is also an atom at $(x+1/2,\bar{y},z)$ and the structure factor is:

$$S_{hkl}=f\left\{e^{-i2\pi(hx+ky+lz)}+e^{-i2\pi(h/2+hx-ky+lz)}\right\}. \tag{5.50}$$

Now we consider the set of reflections for which $k=0$ (the set of reflections that are observed for a diffraction pattern recorded when the incident beam is parallel to [010]).

$$S_{h0l}=f\left\{e^{-i2\pi(hx+lz)}\cdot(1+e^{-i\pi h})\right\}. \tag{5.51}$$

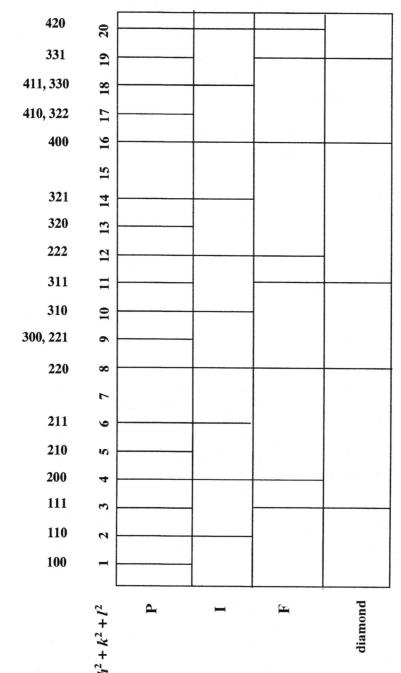

Figure 5-9. Systematic absences in the diffraction patterns of four cubic structures.

From Eqn. 5.51, we can see that when h is odd, S_{h0l} is 0. So, for the $h0l$ reflections, the condition for the existence of a reflection is that $h = 2n$.

Similar conditions can be derived for other types of glide. For example, for diagonal glide (n) with a mirror perpendicular to $\vec{b}$, the translation $\vec{R} = a/2 + c/2$ makes the condition for $h0l$, $h + l = 2n$. For diamond glide (d) in an orthorhombic system, the condition on $h0l$ is that $h + l = 4n$. In a cubic or tetragonal system, the condition for hhl is that $2h + l = 4n$.

iv. *Systematic absences due to the presence of screw axes*
Consider a crystal structure with a 2_1 screw diad, parallel to $\vec{c}$. If there is an atom at (x,y,z), then there is also one at $(\bar{x},\bar{y},z + 1/2)$. Therefore, the structure factor is:

$$S_{hkl} = f\left\{e^{-i2\pi(hx+ky+lz)} + e^{-i2\pi(-hx-ky+l/2+lz)}\right\}. \tag{5.52}$$

From this, we find that for the $00l$ reflections, all reflections with l odd are absent. In other words, the condition on the $00l$ reflections is that $l = 2n$.

$$\begin{aligned}S_{00l} &= f\left\{e^{-i2\pi lz} + e^{-2\pi(l/2+lz)}\right\} \\ &= f\left\{e^{-i2\pi lz} \cdot (1 + e^{-i\pi l})\right\}.\end{aligned} \tag{5.53}$$

Example 5.2

Table 3B.3 in Appendix 3B describes the scheelite structure, which has the space group $I4_1/a$. Determine the systematic absences that occur in the diffraction pattern of this phase and list the indices of the planes that have the six largest d-spacings and create diffracted beams.

1. The symmetry elements in this group that lead to systematic absences are the lattice centering operation, I, the screw tetrad, 4_1, and the axial glide plane, a. It is simplest to individually consider the general equivalent positions created by each of these operators; regardless of the details of the S_{hkl}, these terms can always be separated as multiplicative factors from the complete summation.
2. The centering operation takes a point at (x, y, z) and moves it to $(x + 1/2, y + 1/2, z + 1/2)$. Therefore, the following term can be factored from the structure factor:

$$\begin{aligned}S_{hkl} &= f\left\{e^{-i2\pi(hx+ky+lz)} + e^{-i2\pi(hx+h/2+ky+k/2+lz+l/2)}\right\} \\ S_{hkl} &= f\left\{e^{-i2\pi(hx+ky+lz)}[1 + e^{-i\pi(h+k+l)}]\right\} \\ S_{hkl} &= f\left\{e^{-i2\pi(hx+ky+lz)}[1 + (-1)^{(h+k+l)}]\right\}\end{aligned}$$

which gives the condition that $h+k+l$ must be an even number. This is expressed as $h+k+l=2n$.

3. From the symbol 'a' we know that the translation of the axial glide plane is $a/2$ and that the mirror lies normal to the c axis. So, a point at (x,y,z) is moved to $(x+1/2, y, \bar{z})$. Therefore, the structure factor is:

$$S_{hkl}=f\left\{e^{-i2\pi(hx+ky+lz)}+e^{-i2\pi(hx+h/2+ky-lz)}\right\}.$$

This will lead to condition on $hk0$ peaks:

$$S_{hk0}=f\left\{e^{-i2\pi(hx+ky)}\bullet[1+e^{-i\pi h}]\right\}$$

$$S_{hk0}=f\left\{e^{-i2\pi(hx+ky)}\bullet[1+(-1)^h]\right\}.$$

This gives the condition that h must be an even number, $h=2n$. Since this is a tetragonal group, we know that the a and b axes are the same and that there is also a b axial glide leading similarly to the condition that $k=2n$. The two conditions can be summarized as $h,k=2n$.

4. The 4_1 screw tetrad creates points at (x, y, z), $(\bar{x}, \bar{y}, z+1/2)$, $(\bar{y}, x, z+1/4)$, and $(y, \bar{x}, z+3/4)$. We can see that the translations along [001] will create conditions on the $00l$ diffraction peaks. The structure factor can be written in the following way:

$$S_{00l}=f\left\{e^{-i2\pi lz}[1+e^{-i\pi l}+e^{-i\pi l/2}+e^{-i\pi 3l/2}]\right\}.$$

From this we can see that cancellation occurs for all but $l=4n$.

5. The interplanar (d) spacing can be calculated according to Eqn. 2.21. For a tetragonal crystal, 2.21 simplifies to:

$$\frac{1}{d_{hkl}^2}=\frac{h^2+k^2}{a^2}+\frac{l^2}{c^2}.$$

It is clear that the largest d-spacings occur for planes with the lowest indices, hkl. To find the largest six, we need to consider which of the possible hkl will appear (apply the conditions found above), calculate d_{hkl} for each one, and order them. To do this, we can systematically increment the indices and eliminate the absent ones.

6. Since $c\approx2a$, we'll consider increments of h and k from zero to two and increments of l from zero to four. This process is carried out in the table opposite:

trial index	observed	d-spacing, Å	trial index	observed	d-spacing, Å
100	X		211	211	2.30
001	X		220	220	1.85
110	X		202	202	2.38
101	101	4.76	221	X	
111	X		122	X	
200	200	2.62	222	222	1.76
002	X		300	X	
201	X		003	X	
210	X		103	103	3.07
102	X		004	004	2.84
112	112	3.10	104	X	

So, the largest six d-spacings, in decreasing order, are:

101	4.76 Å
112	3.10 Å
103	3.07 Å
004	2.84 Å
200	2.62 Å
202	2.38 Å

In conclusion, we can say that any crystal structure whose symmetry includes a lattice centering operation (I, F, R, A, B, or C), a screw operator, or a glide operator, has some systematic absences in the diffraction pattern. Screw operators lead to systematic absences of reflections emanating from planes perpendicular to the rotation axis, and glide planes lead to absences of reflections emanating from planes parallel to the mirror.

v. *Systematic absences and the reciprocal lattice*

Recall that there is one point in the reciprocal lattice for every plane in the direct lattice. Since the condition for diffraction is that the scattering vector be equal to a reciprocal lattice vector, there is a diffraction peak associated with every point in the reciprocal lattice. Furthermore, because all of the scattering vectors that lead to diffraction intensities are equal to reciprocal lattice vectors, the diffraction pattern, or distribution of scattered intensities (diffraction peaks), must reproduce the structure of the reciprocal lattice.

It is important to remember that the symmetry operations that lead to systematic absences in the diffracted peaks also lead to systematic absences in the reciprocal lattice. At first, the idea of systematic absences in the reciprocal lattice

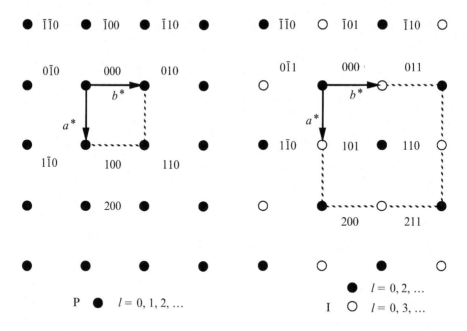

Figure 5.10 (a & b). Schematic representations of the reciprocal lattice of the primitive and body centered orthorhombic cells [9].

seems to contradict the rule that each plane in the direct lattice creates a point in the reciprocal lattice. However, in the case of the lattice centering operations, the so-called absence is really just an artifact based on the choice of a non-primitive conventional cell; if a unit cell is reindexed in terms of a primitive cell, there will be no missing points in the reciprocal lattice and no systematically absent diffracted peaks. An obvious point that is worth remembering is that the diffraction pattern, or the reciprocal lattice of a given arrangement of atoms, is always the same, no matter what unit cell you choose. The only thing that changes are the indices or labels on the diffracted peaks (reciprocal lattice points).

It is possible to make direct experimental measurements of the reciprocal lattice using precession X-ray photographs or electron diffraction patterns. Selected examples of projected reciprocal lattices are presented in Fig. 5.10. Fig. 5.10a shows the $l=0$ layer [this is the $(hk0)$ plane] of the reciprocal lattice of an orthorhombic P direct lattice. All lattice points are present and each layer, at ± 1 along the z axis, looks the same. Fig. 5.10b shows the reciprocal of an I lattice with the $l=$ even and $l=$ odd layers superimposed. You should be able to see that this is an F arrangement and that the reciprocal lattice of the F cell (Fig. 5.10c) has an I arrangement. In general, the density of the reciprocal lattice points is

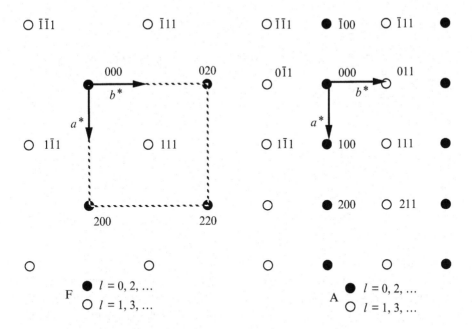

Figure 5.10 (c & d). Schematic representations of the reciprocal lattice of the face-centered and end-centered orthorhombic cells [9].

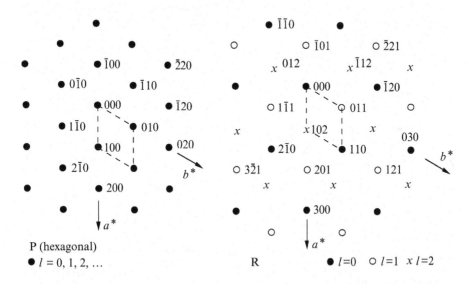

Figure 5.10 (e & f). Schematic representations of the reciprocal lattice of the primitive hexagonal and rhombohedral cells [9].

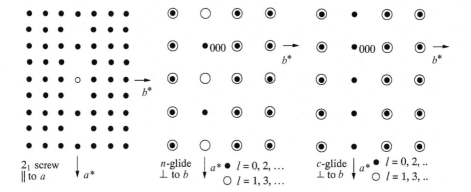

Figure 5.10 (g–i). Schematic representations of the reciprocal lattice of structures with selected screw and glide symmetry [9].

reduced as we go from P to F. The missing row structure of the A centered reciprocal lattice is shown in Fig. 5.10d and the relationship between the hexagonal P and the R lattices are shown in Fig. 5.10e and Fig. 5.10f. The reciprocal lattice absences due to selected screw and glide operators are shown in Figs. 5.10 g, h, and i.

Experimental diffraction patterns recorded using TEM or a precession camera look similar to the patterns in Fig. 5.10 and can lead to deduction of the Bravais lattice vectors and the space group symmetry, an essential step in the determination of a crystal structure.

Example 5.3

Consider a binary alloy with the composition AB. The atoms in this structure occupy the sites of a cubic F lattice. At low temperature, the A atoms occupy the positions at $(0,0,0)$ and $(1/2,1/2,0)$ and the B atoms occupy the positions at $(1/2,0,1/2)$ and $(0,1/2,1/2)$. At high temperature, the A and B atoms occupy all four of the sites with equal probability. At intermediate temperatures, the structure is partially ordered. How can the intensities of diffracted beams be used to measure the degree of order in this alloy?

1. We begin by assigning the probability of an A atom being at an A site as p and the probability of a B atom being at an A site as $(1-p)$. So, the scattering factor for the A site (which we will call f_1) is an average of f_A and f_B, weighted by the occupancy, p. Specifically, $f_1 = pf_A + (1-p)f_B$. It follows that the average scattering factor for the second site is $f_2 = (1-p)f_A + pf_B$.

2. Using the definitions given above, we can write the structure factor in terms of the occupation probability. For example, for an S_{hk0} peak, the structure factor reduces to:

$$S_{hk0} = f_1 + f_1(-1)^{h+k} + f_2(-1)^h + f_2(-1)^k$$
where $f_1 = pf_A + (1-p)f_B$ and $f_2 = (1-p)f_A + pf_B$

3. Note that when h and k are even, as for (200), S_{hk0} is independent of the occupancy, p.

$$S_{ee0} = 2f_1 + 2f_2$$
$$S_{ee0} = 2[pf_A + (1-p)f_B] + 2[(1-p)f_A + pf_B]$$
$$S_{ee0} = 2pf_A + 2f_B - 2pf_B + 2f_A - 2pf_A + 2pf_B$$
$$S_{ee0} = 2f_A + 2f_B.$$

4. However, when h and k are odd, as for (110), S_{hk0} depends on the occupancy, p.

$$S_{oo0} = 2f_1 - 2f_2$$
$$S_{oo0} = 2[pf_A + (1-p)f_B] - 2[(1-p)f_A + pf_B]$$
$$S_{oo0} = 2pf_A + 2f_B - 2pf_B - 2f_A + 2pf_A - 2pf_B$$
$$S_{oo0} = 4p(f_A - f_B) - 2(f_A - f_B)$$
$$S_{oo0} = (4p-2)(f_A - f_B).$$

When A and B are distributed randomly, $p = 0.5$ and S_{oo0} vanishes. This makes sense, since a completely random distribution makes all of the sites equivalent, and the structure is then identical to cubic F and should have the absences associated with cubic F. On the other hand, as p increases towards 1, the structure becomes cubic P and the oo0-type reflections become increasingly intense.

5. Finally, remembering that the intensity is proportional to the complex conjugate of the structure factor, we can write an expression that relates the occupancy to the intensity ratio of two diffracted beams.

$$\frac{I_{oo0}}{I_{ee0}} = \frac{(4p-2)^2(f_A - f_B)^2}{4(f_A + f_B)^2}.$$

Since the left-hand side of this equation can be measured, and f_A and f_B are known, the occupancy factor, p, can be computed. Because the intensities of the ee0-type diffracted beams are independent of the amount of order, they are usually called the fundamental reflections. Peaks that result from ordering are called superlattice reflections. In this case, the intensity of the oo0 peaks increases with p (order). As in Example 5.1, we have ignored several additional factors that affect the intensity. These factors are described below.

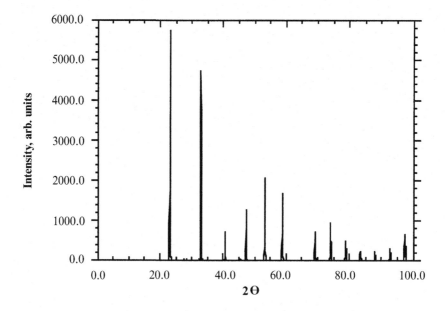

Figure 5.11. A powder X-ray diffraction pattern of Na_xWO_3, a cubic material with the perovskite structure. The X-rays were Cu K_α radiation with a wavelength of 1.54056 Å.

(E) Factors affecting the intensity of diffracted peaks

While the structure factor is the most important influence on diffracted beam intensity, there are a number of other factors that must be considered if we wish to determine the arrangement of atoms in the unit cell through a detailed comparison of calculated and observed diffracted beam intensities [11,12]. These factors will be described with reference to the commonly used powder diffraction experiment. After describing relevant aspects of the powder diffraction experiment, the factors affecting peak intensity will be described and a sample calculation will be carried out.

i.   Measuring diffracted beam intensities

X-ray powder diffraction experiments use a monochromatic source. While true single wavelength sources are not attainable, carefully filtered and monchromatized radiation serves as an excellent approximation. Descriptions of the practical aspects of generating and conditioning X-ray beams can be found in other sources and will not be discussed further [13]. Powder patterns (such as the one shown in Fig. 5.11) typically show the diffracted beam intensity as a function of

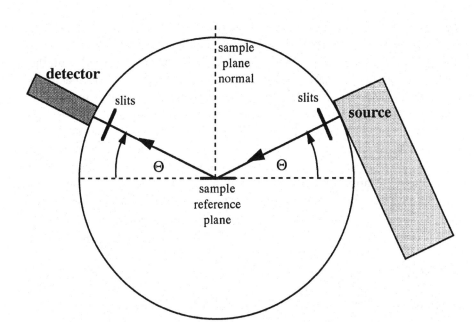

Figure 5.12. The geometry for the diffractometer. The source and the detector move synchronously so that both have the same angular relationship to the reference plane.

the angle between the diffracted beam and a reference plane that contains the surface of a polycrystalline sample (this is illustrated in Fig. 5.12). The polycrystalline nature of the specimen is important to the experiment. In a sufficiently fine powder without texture, crystallites with all possible orientations are present. Thus, for every angle that satisfies Bragg's law, some fraction of the sample (a subset of the particles) will have the correct lattice planes oriented parallel to the reference plane. For example, the first intense peak in the pattern in Fig. 5.11, at $2\theta - 23°$, was caused by a small subset of crystallites in the powder that had their {100} planes oriented exactly parallel to the reference plane of the diffractometer. A different subset that had their {111} planes oriented parallel to the reference plane produced the third peak at $2\theta = 40.5°$, and so on. Because of the random orientation of the crystallites, all members of a family of planes contribute to the same peak.

The diffractometer geometry is shown schematically in Fig. 5.12. At the source, X-rays are produced in the water-cooled vacuum tube. Before reaching the sample, the beam must exit the tube through a beryllium window, and then pass through a set of slits which transmits only the beams traveling parallel to the plane of the drawing. If the angle between the sample reference plane and the incident beam is θ, then the angle between the diffracted beam and the sample reference plane must also be θ. Thus, to measure the intensity of the diffracted beam, the detector is placed at this position. During the experiment, the angle, θ, is varied systematically. The source and detector move together, each taking the angle θ, so that the detector is always positioned correctly to measure the intensity of any diffracted beams that are produced. Because the absolute intensity is a function of many experimental parameters (slits, scan rate, detector characteristics, etc.), intensities are usually compared in a normalized format.

The data shown in Fig. 5.11 are for a compound with the cubic perovskite structure (see Chapter 4, Section F(xiii)). Based on the information provided in the following sections, we will compute the intensities of the peaks in this powder pattern. It will be noted when the discussion is specific to the powder diffraction geometry. In other cases, the discussion applies more generally.

ii. *Example of a structure factor calculation*
The first step in computing the intensities of diffracted beams is to specify the structure factor. For $NaWO_3$, this process is illustrated as an example, below.

Example 5.4

$NaWO_3$ has the perovskite structure and the atomic positions are given below. Specify the structure factor.

Na	$(0, 0, 0)$	O	$(1/2, 1/2, 0)$
W	$(1/2, 1/2, 1/2)$	O	$(1/2, 0, 1/2)$
		O	$(0, 1/2, 1/2)$

1. Using the atomic positions listed above, and Eqn. 5.44, reproduced below,

$$S_{hkl} = \sum_{j=1}^{J} f_j e^{-i2\pi(hx_j + ky_j + lz_j)},$$

we compute the structure factor by summing over the five-atom basis:

$$S_{hkl} = f_{Na} + f_W e^{-i\pi(h+k+l)} + f_O\left[e^{-i\pi(h+k)} + e^{-\pi(h+l)} + e^{-i\pi(k+l)} \right]. \tag{5.54}$$

2. Using Eqn. 5.54, we can compute the structure factor for any reflection specified by the indices, (hkl). Because the structure is cubic P, there will be no systematic absences due to cell centering. A partial structure factor list is generated in Table 5.2.

3. Note that to compute actual values for S_{hkl}, we need to know the atomic scattering factors, f_{Na}, f_W, f_O. These parameters are described below.

Table 5.2. *Structure factors for a cubic perovskite.*

$h\,k\,l$	S_{hkl}	$\sin\theta/\lambda$	S_{hkl}	$|S_{hkl}|^2$
1 0 0	$f_{Na} - f_{W} - f_{O}$	0.13	$9.25 - 66.2 - 6.6$	4038
1 1 0	$f_{Na} + f_{W} - f_{O}$	0.184	$8.5 + 62.1 - 5.6$	4225
1 1 1	$f_{Na} - f_{W} + 3f_{O}$	0.225	$7.8 - 58 + 14.7$	1260
2 0 0	$f_{Na} + f_{W} + 3f_{O}$	0.26	$7.3 + 55.8 + 13.4$	5852
2 1 0	$f_{Na} - f_{W} - f_{O}$	0.29	$6.85 - 53.6 - 4.0$	2576
2 1 1	$f_{Na} + f_{W} - f_{O}$	0.318	$6.4 + 51.5 - 3.7$	2938
2 2 0	$f_{Na} + f_{W} + 3f_{O}$	0.367	$5.65 + 48.1 + 9.6$	4013

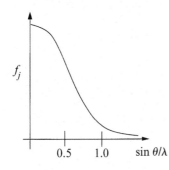

Figure 5.13. The approximate form of the angular dependence of the atomic scattering factor.

iii.   *The atomic scattering factor*
To complete the structure factor calculation we need to know the values of f_{Na}, f_{W}, and f_{O}. As mentioned earlier, these scattering factors are atomic quantities that are tabulated in the International Tables. When actually looking up the values in the tables, one must remember that f_j depends on the scattering vector. In other words, the values of f_j depend on the angle of the diffracted beam and they are different for each peak in the pattern in Fig. 5.11. To make the tables independent of the λ, the scattering factors are listed as a function of $\sin\theta/\lambda$. The typical functional dependence of f_j is illustrated schematically in Fig. 5.13. Reduced scattering in the normal direction is one of the reasons that higher angle diffraction peaks usually have low intensities, as we see in the data presented in Fig. 5.11. It is useful to remember that at any given angle, the scattering factor is proportional to the atomic number, or the number of electrons that are associated with the atom. The scattering factors needed for our intensity calculation are provided in Table 5.2, where they are used to complete the structure factor calculation.

The atomic scattering factors can also be approximated [14] by

$$f(s) = Z - 41.78214 \cdot s^2 \cdot \sum_{i=1}^{4} a_i e^{-b_i s^2}. \tag{5.55}$$

In Eqn. 5.55, Z is the atomic number, $s = \sin\theta/\lambda$, and the coefficients (a_i and b_i) can be found in Table 5B.1 in Appendix 5B.

iv. *The temperature factor*

The oscillatory vibration of atoms in a crystal leads to an angle dependent effect on the diffracted peak intensity that is known as the temperature factor. As the temperature or the diffraction angle increases, the intensity of the diffracted peak is diminished. The magnitude of the intensity reduction depends on the amplitude and shape of the atom's thermal oscillations. The simplest way to understand this is that as an atom spends less time in its ideal position, the probability that it will scatter an X-ray in a constructive way is diminished. The correction factor,

$$I_{hkl} \propto e^{-B_j \left(\frac{\sin\theta}{\lambda}\right)^2}, \tag{5.56}$$

can be derived using classical approximations [15]. The coefficient B_j is the 'isotropic temperature factor' and is proportional to the atom's mean squared displacement about its equilibrium position. Values of B_j are not expected to be atomic quantities independent of the crystal structure, but instead should be dependent on the bonding geometry and elasticity of the solid. In the simplest approximation, this factor is ignored altogether. An improved approximation is to assume that B_j is the same for all atoms in the structure (a value between 0.5 and 1.0 is usually appropriate). In the most sophisticated structure determinations, best fit values of B are determined from the data during the refinement process, where differences between a model structure and the experimentally determined intensities are minimized using a least squares technique. Refined values that are much larger than 1.0 at room temperature suggest positional disorder, a bifurcated site, or simply a flawed structural model. For sites with unusual coordination geometries, anisotropic thermal models reflecting the point symmetry of the site are used. Values of intensity v. temperature for some Al reflections are shown in Fig. 5.14.

v. *The Lorentz-polarization factor*

The Lorentz-polarization factor is a combination of two geometric factors described more completely elsewhere [11]. The Lorentz factor depends on experimental geometry and the polarization factor reflects the change in scattering

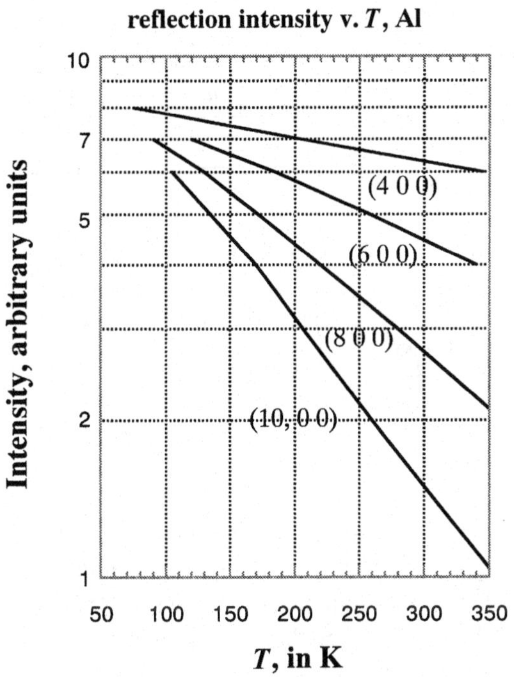

reflection intensity v. T, Al

Figure 5.14. Temperature dependence of the intensity of the $(h00)$ reflections of Al [15].

efficiencies of electrons with the incident beam angle. Basically, electrons are more likely to re-radiate the scattered photons in the forward direction (the direction of the incident beam) or the reverse direction. This has the effect of increasing the intensity of low and high angle diffraction peaks with respect to peaks at intermediate angles. The angular dependence of the scattered intensity (I_p) is given by the Thomson equation.

$$I_p \propto 1/2(1 + \cos^2 2\theta). \tag{5.57}$$

The so-called Lp factors can be found tabulated in the International Tables or they can be approximated by the following simple equation:

$$Lp = \frac{1 + \cos^2 2\theta}{\sin^2 \theta \cos \theta}. \tag{5.58}$$

vi. *Multiplicity*
The multiplicity factor, M_{hkl}, accounts for the fact that in the powder diffraction experiment, each diffracted beam is produced by a different subset of crystals and these subsets do not have the same number of members. The random

237

Table 5.3. *The multiplicity of reflections in the cubic system.*

$n\,0\,0$	$n_1\,n_1\,n_1$	$n_1\,n_1\,0$	$n_1\,n_2\,0$	$n_1\,n_1\,n_2$	$n_1\,n_2\,n_3$
6	8	12	24	24	48

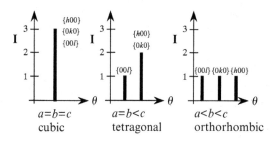

Figure 5.15. The effect of changes in symmetry on the multiplicity and peak intensity of powder diffraction peaks.

orientation of crystallites implies that equivalent crystallographic orientations are not discriminated. So, while it is assumed that there are the same number of particles in each possible orientation, the total number of particles contributing to each Bragg peak (*hkl*) is potentially increased by the number of orientations that are equivalent. Thus, more equivalent orientations means that more crystallites contribute to the Bragg peak, resulting in a peak of greater intensity. For example, all planes in the {220} family of a cubic crystal contribute to the peak that would be indexed as (220). In this case, there are twelve equivalent orientations, so the multiplicity is 12:

$$
\begin{array}{cccc}
(220) & (\bar{2}20) & (\bar{2}02) & (0\bar{2}2) \\
(202) & (2\bar{2}0) & (20\bar{2}) & (02\bar{2}) \\
(022) & (\bar{2}\bar{2}0) & (\bar{2}0\bar{2}) & (0\bar{2}\bar{2})
\end{array}
$$

The maximum multiplicity is 48 for a reflection in the cubic system where $h \neq k \neq l \neq 0$. The minimum multiplicity would be two, in the triclinic system. Thus, the maximum variation of intensities due to this effect is a factor of eight, which can occur in cubic systems. The multiplicities for crystals in the cubic system are given in Table 5.3.

Multiplicity changes account for intensity reductions when systems transform to a lower symmetry. For example, consider an *h*00 reflection in the cubic system with a multiplicity of six. If the cubic crystal transforms to a tetragonal crystal, there will now be two reflections, an *h*00 with a multiplicity of four and a 00*l* with a multiplicity of two. If the crystal continues to reduce its symmetry and becomes orthorhombic, there will be three separate peaks, *h*00, 0*k*0, and 00*l*, each with a multiplicity of two. These changes are illustrated schematically in Fig. 5.15.

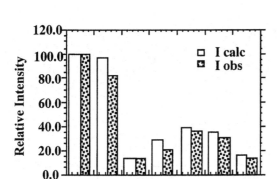

Figure 5.16. A comparison of the calculated and observed intensity data.

vii. Computing the intensities

To compute intensities for the diffracted beams in the powder experiment, we simply combine the factors described in the last six sections according to the equation:

$$I_{hkl} = |S_{hkl}|^2 \cdot M_{hkl} \cdot Lp(\theta) \cdot e^{-B\left(\frac{\sin\theta}{\lambda}\right)^2}. \tag{5.59}$$

Note that I_{hkl} represents the *integrated* intensity, not the maximum intensity. The integrated intensity is the area under the trace of the diffraction peak. Because the absolute value of the measured intensity is an arbitrary function of the experimental conditions, intensities are reported in a normalized format. Conventionally, peak intensities are expressed as a percentage of the most intense peak in the pattern. Thus, the intensity of the most intense peak is assigned a value of 100 % and all the rest are lower. The favorable comparison of measured and computed intensities shown in Fig. 5.16 (see Example 5.5) demonstrates the validity of the approximations. Intensity calculations are easily carried out using computer programs designed for this purpose or within 'spread sheet' type environments.

Example 5.5

Compute the intensities of the first seven diffracted beams that you expect to observe in a powder pattern of $NaWO_3$ and compare them to the observations in Fig. 5.11.

1. Part of the information that we need is found in Table 5.2. The temperature factor is computed from Eqn. 5.56 by assuming that the isotropic temperature factor, B, is 1. The Lorentz-polarization factor is computed using Eqn. 5.58, and the multiplicity

values are determined from Table 5.3. All that remains is to compute the results using Eqn. 5.59.

2. The computed data are tabulated in Table 5.4 and compared to the observed intensity in Fig. 5.16.

Table 5.4. *Comparison of observed and calculated intensities for $Na_{0.8}WO_3$.*

$h\,k\,l$	$|S_{hkl}|^2$	$\sin\theta/\lambda$	Lp	Temp	M_{hkl}	I_o	I_{calc}	$I_{observed}$
1 0 0	4038	0.13	46.6	0.983	6	1110105	100	100
1 1 0	4225	0.184	22.0	0.967	12	1078268	97	82
1 1 1	1260	0.225	14.0	0.951	8	134154	12	13
2 0 0	5852	0.26	9.86	0.935	6	323574	29	21
2 1 0	2576	0.29	7.64	0.919	24	434236	39	36
2 1 1	2938	0.318	6.09	0.904	24	388117	35	30
2 2 0	4013	0.367	4.27	0.874	12	179715	16	12

viii. *Preferred orientation*

The calculation described in the preceding section was based on the assumption that particles in the powder sample are distributed randomly and of approximately the same size. In practice, however, this is not always the case and preferred orientation or texture can affect the distribution of intensities. The reason for this is that each peak in the pattern is caused by diffraction from a different subset of the particles in the powder. If the particles are distributed in a truly random fashion, then the number of particles in each orientation should be identical. However, if the particles have a shape anisotropy, this might not be true. Assume, for example, that the particles are hexagonal platelets. In a packed powder, plate-like particles are most likely to lie with their basal plane, (0001), parallel to the reference plane. Thus, the (0001) diffraction peak will originate from a greater number of particles than other peaks, such as the (1000). As a result of this preferred orientation, the (000l)-type reflections will be intensified relative to (hki0)-type reflections. In general, anything that changes the assumed random distribution of particles will affect the distribution of relative intensities. We should also note that a few large particles, in an otherwise fine powder, can seriously affect the distribution of intensities. See Fig. 5.17. In monolithic specimens produced by sintering, casting, or deformation, the processing frequently imparts some texture to the specimen.

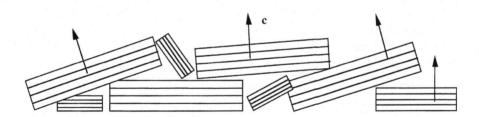

Figure 5.17. Preferred orientation in a powder diffraction sample. Highly anisotropic particles are likely to have similar alignments.

ix.   *The absorption correction*

While the relative intensities that derive from a powder experiment are not influenced by absorption, it is important to correct transmission data. During such an experiment, some portion of the diffracted beam intensity is absorbed in the crystal by inelastic processes before it leaves the solid. Because the attenuation due to absorption depends on the path length through the crystal and each beam takes a different path depending on the scattering vector, each beam is attenuated by a different amount and a correction must be made to the intensities. The correction must be made knowing the shape of the crystal, the absorption coefficient of the sample, and the path of the diffracted beam. X-ray absorption coefficients for each element can be found in the International Tables. From these atomic parameters, an absorption coefficient per length can be computed for any crystal and used for the correction. The accuracy of the absorption correction is typically limited by the accuracy with which the shape of the crystal is known. To limit this source of error, the most careful intensity measurements can be performed on crystals that have been ground to a spherical shape.

x.   *The extinction correction*

Extremely perfect crystals show reduced intensities due to 'extinction'. Basically, a crystal's degree of perfection influences peak intensity. During very long measurements, the crystal can accumulate considerable radiation damage and become 'less perfect' as the measurement proceeds. Thus, intensities measured at the end of the experiment may be affected less by extinction than those at the beginning and an extinction correction is needed. In practice, the need for an extinction correction is determined by periodically measuring a small subset of intensities during the measurement and seeing if their intensities remain constant.

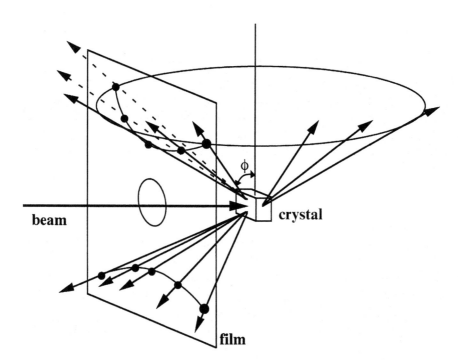

Figure 5.18. The geometry of the Laue back-reflection experiment. The diffracted beams from a single zone intersect the film along a hyperbola [13].

(F) Selected diffraction techniques and their uses

i. *Laue back-reflection*

Laue back-reflection is used to determine the orientation of a single crystal. The ability to determine the orientation of exposed surfaces is important not only for systematic studies of the anisotropy of physical properties, but also for the production of materials that are used in single crystal form. Such materials include substrates used for thin film growth, semiconductors for integrated circuits, and even the high temperature alloys used in turbine blades.

Perhaps the most important thing to remember about the Laue technique is that it uses 'white' (polychromatic) X-ray radiation. In other words, the beam contains a wide spectrum of X-ray wavelengths, λ. So, although the angular relationship between the beam and the crystal is fixed, all planes can diffract in a single orientation. Each set of planes produces its own diffraction spot from a different wavelength in the beam.

The experimental arrangement is shown schematically in Fig. 5.18. The

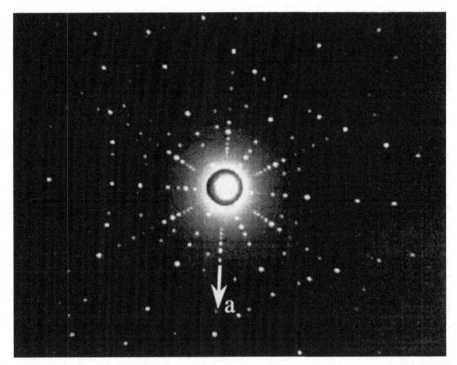

Figure 5.19. A Laue back-reflection photograph from α-Al_2O_3, with the c axis normal to the paper.

beam of X-rays passes through a photographic film and strikes the sample crystal. Diffracted beams emerge in all directions, and some of them expose the film to create a white diffraction spot. The planes in a single zone produce diffracted beams that lie on the surface of a cone whose axis is the zone axis. The angle between the cone axis and the cone surface, ϕ, is the angle at which the zone axis is inclined to the transmitted beam. Because the groups of diffracted beams from a single zone are arranged in this conical geometry, the spots on the film lie either along hyperbolas or straight lines. The orientation of the crystal can then be established based on the indices of the observed reflections, their positions on the film, and the experimental geometry. An example of a Laue pattern is shown in Figure 5.19.

ii.   *Powder diffraction*
Powder diffraction patterns can be obtained using either a Debye–Scherrer camera or an automatic scanning diffractometer [17]. The following discussion focuses on the common diffractometer method. However, it is worth noting that

the Debye–Scherrer technique, in which data is recorded on a film rather than by an electronic detector, is still used for the analysis of very small samples. Good powder diffraction data can be obtained on a scanning diffractometer in about two hours, much faster if a position sensitive detector is used. Powder diffraction is most commonly used for phase identification and lattice parameter determination. However, it can also be used for particle size and shape analysis and, in certain cases, it can be used for the solution of unknown structures.

Because every crystalline material has a unique diffraction pattern, in the same way that every individual has a unique fingerprint, these patterns can be used for identification. The angular positions and relative intensities of the powder diffraction lines of many thousands of materials are cataloged in a generally available powder diffraction data base, and the identification of an unknown specimen amounts to comparing the observed diffraction pattern with those recorded in the data base. Although hard copies of the data (consisting of a card catalogue and several indices) are still available, it is now common to use the CD-ROM version. Searching the data base is simplified by the fact that, in most cases, a given specimen is not completely unknown. For example, you may have a reasonable idea of the elements in your sample. The visual appearance of the sample (faceting of crystallites, habit, color) provides additional information and the initial stages of this comparison can, of course, be carried out by computer.

Comparisons are made in terms of the relative intensities of the peaks and the d-spacings, quantities that should be the same for all experiments, regardless of the details of the source or detector. As an example, the d-spacings and relative intensities of the peaks in the experimental pattern in Fig. 5.11 are shown in Table 5.5. Comparison with the information from the powder diffraction data base, shown in Table 5.6, confirms the identity of the specimen and allows Miller indices to be assigned to the peaks. Knowledge of the Miller index of each reflection is necessary to determine lattice parameters.

The sample shown in Fig. 5.11 is a single phase specimen. In many applications, multiphased materials are studied. Patterns from mixed phase materials show a superposition and it is often possible to identify more than one phase in a sample. As a rule of thumb, the specimen must contain at least 1 to 5 weight percent (w/o) of a phase to allow detection by this method. As mentioned earlier, the diffraction peaks broaden when individual grains become very small (20Å to 2000Å). Quantitative measurements of the peak width can then be used with the Scherrer formula (Eqn. 5.31) to estimate the average grain size.

Although it is most common to determine the structure of unknown crystals using single crystal data, it is also possible to develop accurate structural models using powder data. Algorithms have been developed to deduce the

Table 5.5 *Tabulated data from the pattern in Fig. 5.11.*

2Θ	*d*-spacing	I/I_o	2Θ	*d*-spacing	I/I_o
23.12	3.84	100	73.91	1.28	16
32.94	2.72	82	78.67	1.21	8
40.58	2.22	13	83.33	1.15	4
47.24	1.92	21	88.00	1.11	4
53.23	1.72	36	92.65	1.06	5
58.82	1.57	30	97.26	1.03	11
69.12	1.36	12			

Table 5.6 *Selected information from card 28–1156 in the JCPDS-ICDD file.*

	d Å	Int.	*h k l*
Na_xWO_3	3.82	100	1 0 0
Sodium tungsten oxide	2.703	80	1 1 0
	2.207	20	1 1 1
Rad: CuKα1 Lambda: 1.5405 Filter: Mono *d*-sp: Guinier	1.912	50	2 0 0
Cutoff: Int: Visual *I/I*cor:	1.710	30	2 1 0
Ref. Salje, Institute of Mineralogy, University of Hanover,	1.561	28	2 1 1
Hanover, Germany, private communication, (1974)	1.352	10	2 2 0
	1.274	15	2 2 1
Sys: cubic S. G. P*m*3*m* (221)	1.209	15	3 1 0
a: 3.8232 b: c: A: C:	1.153	3	3 1 1
A: B: C: Z: 1 mp:	1.104	4	2 2 2
Ref: Salje, Hafami, *Z. Anorg. Allg. Chem.*, **396** (1974) 267	1.060	6	3 2 0
Dx: 7.228 Dm: SS/FOM: F18=46(.022, 18)	1.022	10	3 2 1

Color: violet to red

Bravais lattice, the indexing sequence, probable space group, and lattice parameters from powder diffraction data. [18–21] Furthermore, atomic positions and thermal factors can be determined from powder data using a fitting procedure known as the Rietveld method [22]. Powder studies are usually used in cases where the relatively larger specimens necessary for single crystal studies are not available.

iii. Single crystal methods

Most unknown structures are determined using single crystal X-ray diffraction. Intensity data from single crystal diffraction experiments are typically collected on an automated four circle diffractometer. The collection and analysis of hundreds of peaks are carried out with the aid of a computer. Film methods, including precession, rotating-crystal, and Wiessenberg, are still used to determine cell size, shape, and space group [23]. These are all transmission methods and require only small crystals of the order of 0.1 mm in diameter.

Well established algorithms (not described here) can be applied to the diffraction data to determine the Bravais lattice, the cell parameters, and probable space group for a structure. The intensity data is used to construct a Fourier series representing the electron density (for example, see Fig. 9.1). The positions of the basis atoms in the cell are determined by finding peaks in the electron density function. The most significant challenge in the process is the so-called 'phase problem'. As mentioned earlier, I_{hkl} is proportional to $|S_{hkl}|^2$ and, therefore, contains no information about the sign and phase of the coefficient S:

$$S_{hkl} = A_{hkl} + iB_{hkl}$$
$$|S_{hkl}|^2 = (A_{hkl} + iB_{hkl})(A_{hkl} - iB_{hkl}) = A_{hkl}^2 + B_{hkl}^2. \tag{5.60}$$

So, intensity measurements can not be used to specify uniquely the values of A_{hkl} and B_{hkl}.

The difficulty is probably best demonstrated by considering Friedel's law, which says that all diffraction patterns are centrosymmetric, regardless of whether or not the structure has a center of symmetry. This can be demonstrated by considering the structure factor of a crystal without a center of symmetry:

$$S_{hkl} = \cos 2\pi(hx + ky + lz) + i\sin 2\pi(hx + ky + lz)$$
$$= A_{hkl} + iB_{hkl}$$
$$S_{\bar{h}\bar{k}\bar{l}} = \cos 2\pi(-hx - ky - lz) + i\sin 2\pi(-hx - ky - lz). \tag{5.61}$$

Because $\cos(\theta) = \cos(-\theta)$ and $\sin(-\theta) = -\sin(\theta)$,

$$S_{\bar{h}\bar{k}\bar{l}} = A_{hkl} - iB_{hkl}$$
$$|S_{hkl}|^2 = (A_{hkl} + iB_{hkl})(A_{hkl} - iB_{hkl}) = A_{hkl}^2 + B_{hkl}^2$$
$$|S_{\bar{h}\bar{k}\bar{l}}|^2 = (A_{hkl} - iB_{hkl})(A_{hkl} + iB_{hkl}) = A_{hkl}^2 + B_{hkl}^2. \tag{5.62}$$

So, we see that $|S_{hkl}|^2 = |S_{\bar{h}\bar{k}\bar{l}}|^2$. It follows that $I_{hkl}^2 = I_{\bar{h}\bar{k}\bar{l}}^2$.

There are many ways of getting around the phase problem that are all based on the same principle: the structure measurement is usually highly over-determined. Consider that all the details of a crystal structure are specified by the unit cell parameters (there are a maximum of six), the atomic coordinates (the maximum number is three per atom) and the thermal parameters (one per atom for the isotropic condition). So, for a triclinic structure with 10 atoms per cell, there are 46 parameters. However, in a good single crystal X-ray diffraction study, it is not unusual to measure 1000 to 2000 reflection intensities. The fact that there are far more experimental observations than there are structural variables allows some simple mathematical relationships and rules of self consistency to be applied.

Once a working model of the crystal structure has been obtained, a refinement process can be used for optimization and evaluation of the structure's variable parameters [24]. The refinement is a least squares fitting algorithm that can be used to minimize the difference between the proposed model and the observed data by adjusting the set of structural parameters (a, b, c, α, β, γ, x_j, y_j, z_j, B_j). The result of a successful structural refinement is an accurate set of parameters with an estimate of the standard deviation. The degree to which the refined model agrees with the observations is measured in terms of an 'R' factor:

$$R = \frac{\overset{\text{all reflections}}{\sum} \left| S_{hkl}^{\text{obs}} - S_{hkl}^{\text{calc}} \right|}{\sum S_{hkl}^{\text{obs}}}. \tag{5.63}$$

Although an R-factor of 0.2 indicates that a structure is basically correct, the final arbiter should be whether or not the bond lengths and bond angles are chemically reasonable. Models that have an R-factor less than 0.1 are usually very reliable; even for carefully made measurements, $R < 0.03$ is rare.

iv.   Electron diffraction

Materials scientists and engineers frequently work with polycrystalline specimens. To determine the orientation or structure of a single grain (with micronscale dimensions) in a polycrystal, it is necessary to have a focused and directed source of radiation; in general, X-rays do not suffice when high spatial resolution is required. Electrons, on the other hand, can be focused and directed with nanometer-scale precision. This is normally accomplished in a scanning or transmission electron microscope.

In a scanning electron microscope, the beam can be used to create an electron backscattered diffraction pattern (also known as a backscatter Kikuchi diffraction pattern) [25]. When the electron beam penetrates the sample, it is elastically scattered in all directions. For each lattice plane, some of the scattered

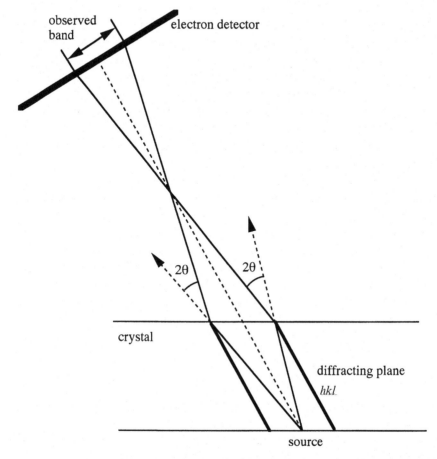

Figure 5.20. Schematic illustration of the formation of backscattered diffraction bands. Each diffracted beam that defines a band edge is a section through a cone of diffracted electrons with a very large apex angle. For presentation purposes, the Bragg angle, θ, is greatly exaggerated. Typically, this angle is less than 1°.

electrons will satisfy the Bragg condition and diffract. From each plane, the diffraction of electrons creates two cones of radiation; a schematic two-dimensional illustration of this process is shown in Fig. 5.20. The diffracted cones have very large apex angles so that a band is formed where they intercept the detector. A typical backscattered diffraction pattern is shown in Fig. 5.21. Because of this geometry, the band center is a trace of the plane in real space and the width of the band is inversely proportional to the interplanar spacing.

Electron backscattered diffraction is now widely used for both orientation

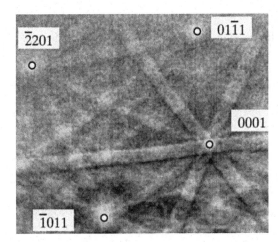

Fig. 5.21. A backscattered electron diffraction pattern from α-Al_2O_3, obtained using an SEM. The low index poles have been labeled.

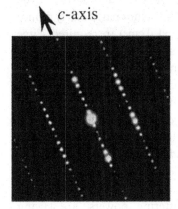

Figure 5.22. A selected area diffraction pattern of 6H-SiC, courtesy of M. DeGraef, CMU.

determination and phase identification. Spatially resolved backscattered patterns can be recorded and indexed automatically by computer to form an orientation map of the surface of the specimen. Orientation mapping has become known as orientation imaging microscopy [26].

When the size of the crystallites is smaller than one micron, transmission electron microscopy (TEM) is used. TEM is ideal for very small crystals in two phase microstructures (typically in ceramic and metallurgical specimens) or when the sample is in short supply (for example, a thin film). When the area contributing to the diffraction is determined by the beam size, transmission Kikuchi patterns (similar to the backscattered Kikuchi patterns described above) can be produced. This technique is usually called microdiffraction. When the area contributing to the beam size is determined by an aperture, a spot pattern is formed (see Fig. 5.22). This technique is usually called selected area diffraction. The spot patterns produced by selected area diffraction are much like precession photos and give a view of the reciprocal lattice. Thus, this method can be useful for the determination of lattice type, cell size (approximate), and space group. The one significant disadvantage is that secondary diffraction (the diffraction of an already diffracted beam) makes the interpretation of intensity data unreliable. Although the powder method is the best way to identify phases, electron diffraction is useful for specimens with very small volumes, for thin film specimens, or for the detection of secondary phases that make up only a small fraction of the sample volume.

v. Neutron diffraction

Neutron sources intense enough for diffraction experiments can only be generated using nuclear reactors. Such reactors are found at large multi-user facilities such as Brookhaven, Oak Ridge, and Argonne National Laboratories. Even with a large reactor, the neutron beam is rather weak and not particularly monochromatic. For these reasons, very large specimens (~ 1.0 cm^3) are required. Time of flight measurements are usually used to increase efficiency. In this case, the entire 'white' beam is used and energies are separated based on differences in velocity.

There are two distinct advantages of the neutron diffraction experiment. First, in this experiment, the neutrons scatter from the nuclei of the atoms in the crystal, not from the electrons. The scattering efficiency (f_j) is thus proportional to the neutron–nucleus interaction and is not proportional to the atomic number. In fact, the distribution of scattering powers is almost a random function of atomic number and, fortunately, light nuclei such as H and Li, which have extremely low X-ray scattering powers, are strong neutron scatterers. Thus, neutron diffraction is the method of choice for locating the positions of light atoms in crystals. Another advantage of neutron scattering behavior is that two atoms with nearly the same atomic number (say Co and Ni) have nearly the same X-ray scattering power and are thus difficult to discriminate in X-ray diffraction measurements, but they usually have very different neutron scattering powers. This can be very important for determining the structure of intermetallic compounds and ternary compounds.

Another advantage of neutron diffraction is that neutrons have a magnetic dipole moment and can thus interact with unpaired electrons. This allows the study of magnetic order in compounds that exhibit antiferromagnetic (NiO) or ferrimagnetic behavior (for example, in spinels and garnets). So, while X-rays can probe only chemical order, neutrons can probe chemical and magnetic order.

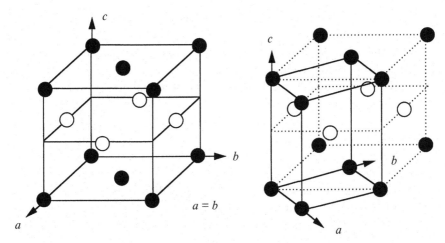

Figure 5.23. The $L1_0$ structure. Dark circles are Cu, open circles are Au.

G Problems

(1) Write a general expression for the structure factor of a rock salt compound with the composition AB. Assuming that $f_A = 2f_B$ and that the cubic lattice constant is 4.28 Å, draw a schematic of the powder pattern that would be observed using radiation with a wavelength of 1.54 Å for $0° < \theta < 30°$.

(2) $BaTiO_3$ has a structure that is completely analogous to $SrTiO_3$, except that it is tetragonal. Assume that $SrTiO_3$ and $BaTiO_3$ form solid solutions, $(Sr, Ba)TiO_3$, but at an undetermined concentration of Ba, the solid solution takes on a tetragonal structure. Outline a powder diffraction experiment that could be used to determine the concentration of Ba at which the material transforms from cubic to tetragonal. Be sure to explain the signature that the transformation would have on the diffraction experiment.

(3) Assume that you are asked to analyze a powdered sample of $SrTiO_3$. This material has the perovskite structure, described in Table 4.28 in the previous chapter. Chemical analysis tells you that in addition to Sr and Ti, the specimen contains 5 to 10 weight percent Ba. This Ba might be dissolved in the $SrTiO_3$, or it might occur as a separate oxide phase ($BaTiO_3$, for example). Explain how a powder X-ray diffraction experiment would discriminate between the two possibilities.

(4) Consider the intermetallic compound, CuAu. In its 'disordered' state, the Au and Cu atoms randomly occupy sites in the fcc structure with equal probability. In its 'ordered' state, the sites are occupied in the manner shown in Fig 5.23a. This is the so-called CuAu or $L1_0$ structure.

(i) Referring to Fig. 5.23a, specify the lattice and basis of both the ordered and disordered forms.

(ii) Draw the $(hk0)$ plane of the reciprocal lattice for both forms (with all of the points labeled) including all points with h or $k \leq 2$.

(iii) Based on the reciprocal lattices, suggest a method for evaluating the degree of order in CuAu.

(iv) Next, consider the alternative unit cell shown in Fig. 5.23b. Draw a new picture of the $(hk0)$ plane of the reciprocal lattice of this structure and index the points with h or $k \leq 2$. Show the reciprocal lattice vectors on the diagram.

(5) A crystal structure has a glide plane perpendicular to $\vec{b}$ (its normal is parallel to $\vec{b}$) and has a translation component $a/2 + c/2$. Does this create any special conditions (systematic absences) on reflections? If so, what are they?

(6) Use a structure factor calculation to demonstrate that for a crystal in space group $P4_2/mnm$, the following conditions exist on diffraction peaks.

for $0kl$: $k + l = 2n$

for $00l$: $l = 2n$

for $h00$: $h = 2n$.

(7) Use a structure factor calculation to determine the indices of the first five reflections observed in the diffraction pattern of diamond.

(8) Consider a structure consisting of stacked planes of atoms. The planes are spaced at regular intervals, 3 Å apart. However, the atoms within the planes are not periodically arranged. What would the powder diffraction pattern of such a material look like?

(9) The ReO_3 and $CaTiO_3$ structures are described in Tables 3B.15 and 4.28, respectively.

(i) Sketch a c-axis projection of the $D0_9$ structure, which is described in Table 3B.15.

(ii) What is the coordination number of each atom in this structure?

(iii) Describe this structure as a polyhedral network.

(iv) The $D0_9$ structure is closely related to the well known perovskite structure, described in Table 4.28. The BO_3 ($D0_9$) structure transforms into the ABO_3 ($E2_1$) structure simply by placing an A-type atom at the center of every cubic cell. Between these two limiting cases exist a series of random solid solutions with the general formula A_xBO_3, where $0 \leq x \leq 1$. Assuming that the cubic lattice constant does not change as a function of x, explain how you could use the diffraction pattern of A_xBO_3 to determine x.

(10) Consider the reaction of Al powder and Fe powder to form FeAl. The structural data for these materials are given in the table below.

Material	space group	lattice param.	structure type
Fe	$Im\bar{3}m$	$a_0 = 2.87$ Å	bcc
Al	$Fm\bar{3}m$	$a_0 = 4.08$ Å	fcc
FeAl	$Pm\bar{3}m$	$a_0 = 3.01$ Å	B2 or CsCl

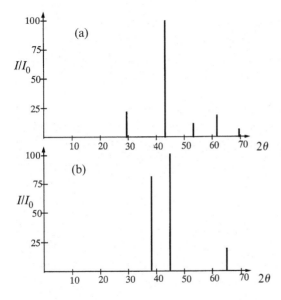

Figure 5.24. Schematic diffraction patterns for Problem 10.

One of the schematic diffraction patterns shown in Fig. 5.24 is of the combined, unreacted powders of Al and Fe, and the other is of the intermetallic phase, FeAl. Distinguish between the two. All computations and reasoning on which you base your decision must be included with your answer. Assume that the X-ray wavelength is 1.54 Å and the experimental resolution is limited to $\pm 0.5°$ in 2θ.

(11) Al and Cu form many intermetallic phases. Assume that during microscopic analysis of an Al–Cu alloy specimen in the TEM, you have noticed three distinct phases. By diffracting from each of these phases independently, you observe four d-spacings for each phase. The observed data are tabulated below. You should assume that these lists are incomplete (in other words, not all of the allowed reflections were observed in the experiment).

Observed d-spacings for phases in Cu–Al specimen, in Å

Cu–Al phase 1	Cu–Al phase 2	Cu–Al phase 3
2.78	4.304	3.89
2.556	3.037	3.56
2.087	2.374	2.90
1.98	2.146	2.51

(i) Assuming that one of the three phases is $CuAl_2$ (the structure of this phase is described in Table 3B.21), that another is $AlCu_4$ (a cubic P phase with $a = 6.26$ Å), and that another is Al_4Cu_9 (a cubic P phase with $a = 8.704$ Å), determine the identity of phases 1, 2, and 3 (show the work used to reach your conclusions).

(ii) Index (assign an hkl) each of the observed d-spacings.

(12) The powder diffraction patterns of three cubic crystals are shown in Fig. 5.25. Determine which has a P cell, which has an I cell, and which has an F cell ($\lambda = 1.54$ Å).

(13) Many III–V type compounds crystallize in the sphalerite structure (see Table 4.25). The space group symmetry of this structure is $F\bar{4}3m$ and the atomic arrangement is shown in Fig. 4.22.

(i) What are the indices of the first five Bragg peaks in the X-ray powder pattern of a compound with this structure?

(ii) If the atomic scattering factors of the group III and group V element are identical, how will the pattern be different?

(14) Consider a binary metallic alloy with the composition AB. The atoms in this structure occupy the sites of cubic I lattice. However, at low temperature, the A atoms occupy the positions at (0,0,0) and the B atoms occupy the positions at (1/2,1/2,1/2). At high temperature, the A and B atoms occupy both of the sites with equal probability. At intermediate temperatures, the structure has intermediate order. How can the intensities of diffracted beams be used to measure the degree of order in this alloy? Your answer should be as specific as possible and include the relevant equations. You can assume that the cubic lattice constant does not vary with the degree of order.

(15) $LiGaO_2$ crystallizes in two polymorphic forms. One structure is described in Table 3B.14, the other has a hexagonal cell such that $a = 2.911$Å and $c = 14.17$Å. Schematic powder diffraction patterns for both phases are shown in Fig. 5.26 (each pattern includes only the first five observed peaks and was recorded with $\lambda = 1.54060$ Å, see caption for details).

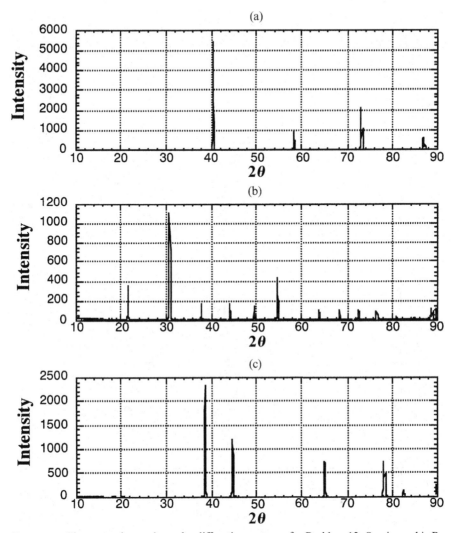

Figure 5.25. Three experimental powder diffraction patterns for Problem 12. One is a cubic P, another is a cubic I, and yet another is a cubic F.

(i) Define the orientation and action of the '2_1' operation and determine the systematic absences that will created in the diffraction pattern by this operator.

(ii) Define the orientation and action of the 'a' operation and determine the systematic absences that will created in the diffraction pattern by this operator.

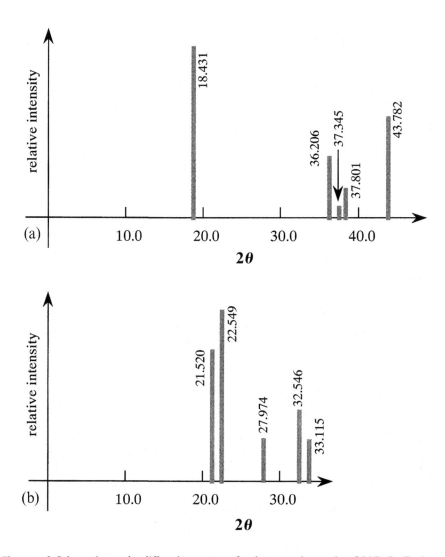

Figure 5.26. Schematic powder diffraction patterns for the two polymorphs of LiGaO$_2$. Each pattern shows the first five observed peaks. The numbers beside each peak are the observed values of 2θ. Each of the patterns was recorded with CuK$_\alpha$ radiation that had a wavelength of 1.54060 Å.

(iii) Define the orientation and action of the 'n' operation and determine the systematic absences that will be created in the diffraction pattern by this operator.

(iv) Determine which of these patterns is associated with each polymorph and index (specify *hkl* values) the first five peaks of each pattern. Be sure to describe the process you use for identification and selection.

(16) Consider a two-dimensional oblique unit cell, with lattice parameters $\{a = 3\text{ Å}, b = 4\text{ Å}, \gamma = 60°\}$.

(i) Make an accurate scale-drawing of the lattice (from $-2a$ to $+2a$ and $-2b$ to $+2b$) and label each point.

(ii) On the same drawing, superimpose the reciprocal lattice. (Scale the reciprocal lattice by a factor of six so that its size is comparable to the direct lattice.) Label the lattice points.

(iii) Perform the Ewald construction for the reciprocal lattice points $(2,1)$ and $(-1, -1)$, assuming that the X-ray wavelength is 1.5 Å. In other words, draw the Ewald circle for this wavelength such that $(2,1)$ is in Bragg orientation. Also draw the incident and diffracted beams; then repeat for $(-1, -1)$. Remember to scale the radius of the Ewald sphere as well.

(iv) Finally, draw the (21) and $(-1\ -1)$ planes on the unit cell and show the Bragg condition in the crystal lattice.

(17) The Cr_2Al structure is described in Table 3B.33. Based on this information, answer the following questions:

(i) In this structure, the I operator creates a systematic absence in the diffraction pattern. Use the structure factor to determine the pattern of absences.

(ii) Although not in the space group symbol, this group has a 2_1 operator parallel to (but not coincident with) the tetrad axis. Use the structure factor to determine if this operator leads to a pattern of absences, and specify the pattern.

(iii) At what x,y coordinate does the 2_1 operator intersect the (001) plane?

(iv) Assume that you measure the powder X-ray diffraction pattern of this material. What are the d-spacings of the five lowest angle peaks that you observe?

(v) How could you determine z (the position of the Cr atom) using X-ray diffraction data?

(vi) Assume that as the temperature increases, the Cr and Al atoms disorder. In other words, the same sites are occupied, but each site can be occupied by Cr or Al with equal probability. Describe the changes that you expect to see in the X-ray diffraction pattern.

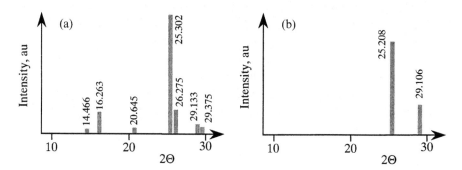

Figure 5.27. Schematic diffraction patterns of two polymorphs of Cu_2HgI_4. See Problem 21.

(18) The γ-LaOF structure is described in Table 3B.31. In addition to the symmetry elements listed in the name of the group, this structure has a screw diad along$\langle 100 \rangle$. What systematic absences will occur in an X-ray diffraction pattern of this crystal?

(19) The chalcopyrite structure is described in Table 3B.32. Based on this information, answer the following questions:

(i) In addition to the symmetry elements listed in the name of the group, this structure has a screw diad along$\langle 001 \rangle$. Specify the position of these elements in the unit cell and the systematic absences that they cause in the diffraction pattern.

(ii) When this material is annealed at high temperature, the Cu and Fe atoms can each occupy the $4a$ and $4b$ sites with equal probability. If the material is then quenched, this disorder on the metal sublattice can be 'frozen in'. Explain how the diffraction pattern of this quenched sample would differ from that of a slowly cooled and well ordered sample.

(20) The structure of graphite is described in Table 3B.35. Determine the three largest interplanar spacings that would be observed in an X-ray diffraction pattern of graphite.

(21) The structure of β-copper mercury tetraiodide is described in Table 3B.36. When β-Cu_2HgI_4 is heated to above 70 °C, it transforms to a cubic structure, called α-Cu_2HgI_4. One of the two schematic powder diffraction patterns shown below (see Fig. 5.27) represents the room temperature β-phase and the other represents the pattern from the α-phase observed at higher temperatures. The patterns were recorded using X-rays with a wavelength of 1.540598 Å.

(i) Determine which pattern represents each phase.

(ii) Which of the cubic Bravais lattices does the cubic phase have?

(iii) Estimate a lattice parameter for the cubic phase.

(iv) Based on the answers above, describe a plausible structure for the cubic phase.

(v) Index all of the peaks in each pattern.

(H) Review problems

(1) The parameter λ describes the degree of inversion in the spinel structure [see Chapter 4, Section F(x)]. Explain how you would use X-ray diffraction to measure this parameter in any spinel with the general formula AB_2O_4.

(2) SiC crystallizes in a number of different polymorphic forms. Two of these polymorphs are very common structure types, sphalerite (B3) and wurtzite (B4), and are described in Tables 4.25 and 4.26. Answer the following questions.

(i) Sketch a projection of each of these structures along [001] and [0001], respectively.

(ii) What is the coordination number of each of the atoms in each structure?

(iii) How are the atoms packed? In other words, describe these structures in comparison to common eutactic arrangements.

(iv) Compare the d-spacing of the first observed diffraction peak in the sphalerite form of SiC with the d-spacing of the first $(000l)$-type peak from the wurtzite polymorph.

(v) Assuming that you were using $Cu_{k\alpha}$ radiation with $\lambda = 1.54$ Å, at what angular position would you expect to observe these two peaks?

(vi) During a vapor deposition experiment, the sphalerite polymorph is found to grow on the (0001) plane of the wurtzite polymorph. Predict the orientation of the sphalerite layer.

(3) The commercial production of α-Al_2O_3, and ultimately Al metal, involves the thermal decomposition of hydrous precursors (Al hydroxides and oxy-hydroxides) extracted from Al bearing ore. One of the compounds, AlO(OH), occurs in two polymorphs which decompose along different paths. The structure of the diaspore polymorph is specified in Table 3B.24 and the structure of the boehmite polymorph is specified in Table 3B.25. (Note: in each case, we have ignored the H. For the purposes of this exercise, you can assume that the two materials are AlO_2.)

(i) What are the Bravais lattices of these two crystals?

(ii) Which of the two polymorphs has a greater density?

(iii) Do the atoms in these structures occupy general or special positions?

(iv) How is Al coordinated in each of these structures?

(v) Describe the packing of O in each of these structures.

(vi) In both diaspore and boehmite, the smallest angle Bragg peak (the one with the longest *d*-spacing) is (020). Using the structure factor, explain why the (010) peak is absent from each pattern.

(vii) The diffraction pattern from only one of these two polymorphs contains a (110) peak. Which polymorph is it? (Explain your reasoning.)

(viii) Determining the positions of the H atoms in these structures is significantly more challenging than determining the Al and O positions. Explain the theoretical concepts or experimental methods that you would employ to determine the H-atom positions.

(ix) When either of these materials is heated, they evolve H_2O and ultimately yield α-Al_2O_3. (The α-Al_2O_3 can be described as hcp eutactic O with Al in 2/3 of the octahedral interstices.) However, one of the polymorphs undergoes this transformation directly at 700 °C, without any transition phases. The other polymorph forms three different transition phases before finally forming α-Al_2O_3 at 1100 °C. Which of the polymorphs transforms directly? (Explain your reasoning.)

(4) The most common form of TiO_2 is rutile. Titania also occurs naturally in two other polymorphs, anatase and brookite, that have completely different structures. The anatase structure is specified in Table 3B.43.

(i) What is the Bravais lattice of anatase? What is the Bravais lattice of rutile (space group = $P4_2/mnm$)? What is the Bravais lattice of brookite (space group = P*cab*)?

(ii) From which of the 32 crystallographic point groups is the space group of anatase derived?

(iii) Draw a picture that specifies the general positions in this point group. Do any of the atoms in the anatase structure occupy the general positions?

(iv) Does this point group have an inversion center?

(v) Explain the meaning of the 'a' in the space group symbol.

(vi) Many of the diffracted beams from crystals of this structure are systematically absent. Show that the Bravais lattice of this structure causes systematic absences.

(vii) The basis of this structure also creates systematic absences in the diffraction pattern. Demonstrate that there is a condition on the (00*l*) reflections.

(viii) Sketch a picture of this structure, projected down the [100] axis (note that this is not the usual orientation of our drawings).

(ix) With respect to the atoms in the basis, specify the location of the 4_1 axis.

(x) Describe the coordination environment of the Ti atoms (you can assume that Ti and O atoms are bonded if they are 1.8 to 2.2 Å apart).

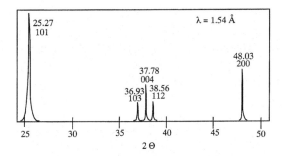

Figure 5.28. A portion of the anatase powder diffraction pattern, recorded with $Cu_{k\alpha}$ radiation, with a wavelength of 1.54 Å. The vertical scale is the relative intensity, in arbitrary units. The numbers above each peak specify the angular position and the index.

(xi) Sketch a polyhedral representation of the structure and describe the manner in which the polyhedra are connected.

(xii) Describe this as a eutactic structure. Can the structure be described in comparison to a common binary prototype structure? [Hint: consider the changes that occur as z goes to 1/4.]

(xiii) Compounds with the anatase structure are completely determined by three experimental parameters: a, c, and z. Explain, using the appropriate equations, how you would use the information in the powder diffraction pattern to determine a and c.

(xiv) Explain how you would use the information in the powder diffraction pattern to determine z. (While you do not actually have to determine z from the data, your explanation should include all of the necessary equations.)

(xv) On heating, anatase transforms to the rutile structure. Describe (in general terms) how the diffraction pattern in Fig. 5.28 would change if a small portion of the specimen (≥ 5 wt %) transformed to the rutile structure.

References and sources for further study

[1] C. Kittel, *Introduction to Solid State Physics*, 5th edition (J. Wiley & Sons, New York, 1976) p. 39. Description of Bragg diffraction.

[2] B.D. Cullity, *Elements of X-ray Diffraction*, 2nd edition (Addison Wesley, Reading, Mass, 1978) p. 86. Description of Bragg diffraction.

[3] C. Kittel, *Introduction to Solid State Physics*, 5th edition (J. Wiley & Sons, New York, 1976) p. 45–49. Fig. 5.3 is drawn after Fig. 13, p. 49. Scattering from a periodic electron density.

[4] B.D. Cullity, *Elements of X-ray Diffraction*, 2nd edition (Addison Wesley, Reading, Mass, 1978) pp. 101–2. The Scherrer formula.

[5] M.J. Burger, *Contemporary Crystallography* (McGraw-Hill, New York, 1970), Chapter 12. A discussion of the phase problem.

[6] B.D. Cullity, *Elements of X-ray Diffraction*, 2nd edition (Addison Wesley, Reading, Mass, 1978) Section 4.4. A discussion of the structure factor.

[7] A.R. West, *Solid State Chemistry and its Applications* (J. Wiley & Sons, Chichester, 1984), Section 5.5.3. A discussion of the structure factor.

[8] C. Kittel, *Introduction to Solid State Physics*, 5th edition (J. Wiley & Sons, New York, 1976) p. 58–65. A discussion of the structure factor.

[9] A.R. West, *Solid State Chemistry and its Applications* (J. Wiley & Sons, Chichester, 1984), Section A7.2. Systematic absences. Fig. 5.10 is drawn after Fig. A7.3, pp. 700–1.

[10] D.E. Sands, *Introduction to Crystallography* (W.A. Benjamin, New York, 1969) pp. 100–11. Systematic absences.

[11] B.D. Cullity, *Elements of X-ray Diffraction*, 2nd edition (Addison Wesley, Reading, Mass, 1978) pp. 126–43. Diffracted beam intensities.

[12] A.R. West, *Solid State Chemistry and its Applications* (J. Wiley & Sons, Chichester, 1984), Section 5.5. Diffracted beam intensities.

[13] D.K. Bowen and B.K. Tanner, *High Resolution X-ray Diffractometry and Topography* (Taylor and Francis, London, 1998), Chapter 2. X-ray sources and beam conditioning.

[14] M. DeGraef and M. McHenry, *Crystallography, Symmetry, and Diffraction*, to be published by Cambridge University Press, 2002. The atomic form factor.

[15] C. Kittel, *Introduction to Solid State Physics*, 5th edition (J. Wiley & Sons, New York, 1976) pp. 63–5. Thermal factors. Fig. 5.14 was adapted from Fig. 27 on p. 63. Kittel cites the original source as Nicklow and Young, *Phys. Rev.* **B 152** (1966) 591.

[16] B.D. Cullity, *Elements of X-ray Diffraction*, 2nd edition (Addison Wesley, Reading, Mass, 1978) Chapter 5. A discussion of Laue diffraction. Fig. 5.18 is drawn after Fig. 3.7 on p. 94.

[17] A.R. West, *Solid State Chemistry and its Applications*, (J. Wiley & Sons Chichester, 1984), Section 3.2.1 and Section 5.6. A discussion of powder diffraction.

[18] P.-E. Werner, L. Eriksson, and M. Westdahl, *J. Appl. Cryst.* **18** (1985) 367.

[19] A. Boultif and D. Louer, *J. Appl. Cryst.* **24** (1991) 987.

[20] P.-E. Werner, *Arkiv Kemi* **31** (1969) 513.

[21] J.W. Visser, *J. Appl. Cryst.* **2** (1969) 89.

[22] R.A. Young, ed., *The Rietveld Method* (Oxford University Press, Oxford, 1993).

[23] M.J. Burger, *Contemporary Crystallography* (McGraw-Hill, New York, 1970), Chapters 7–9. A discussion of several useful diffraction methods.

[24] M.J. Burger, *Contemporary Crystallography* (McGraw-Hill, New York, 1970), Chapter 14. A discussion of the refinement process.

[25] V. Randle, *Microtexture Determination and its Applications* (The Institute of Materials, London, 1992) pp. 11–15.

[26] B.L. Adams, S.I. Wright, K. Kunze, Orientation Imaging: The Emergence of a New Microscopy, *Met. Trans.*, **24A** (1993) 819–31.

Chapter 6
Secondary Bonding

(A) Introduction

This chapter and the three that follow describe the cohesive forces that stabilize crystals. Each chapter concentrates on one of four limiting cases. While the limiting cases have the advantage of being easy to describe, it is important to keep in mind that real chemical bonds rarely fit exactly into one of these categories.

Our discussion of cohesive forces begins in this chapter with a description of the van der Waals bond. A brief description of dipolar bonding and hydrogen bonding is found at the end of this chapter. All three of these cohesive forces are considered to be weak and are known as secondary bonds. In comparison, the stronger ionic, metallic, and covalent bonds are considered to be primary bonds. The key assumption in the models describing secondary bonding is that the electronic energy levels of the bonded atoms are insignificantly perturbed. In other words, bonded atoms are very nearly indistinguishable from free atoms.

Each of the four chapters on bonding has been developed with a similar structure. First, the subject will be described phenomenologically, so that an intuitive understanding is developed. Second, a physical model is introduced and used, when possible, to predict measurable quantities.

i. Substances held together by van der Waals bonds

Van der Waals bonding plays a significant role in the cohesion of three types of solids. The first are solids containing uncharged atoms or molecular species without polar bonds, including inert gases such as He, Ne, Ar, Xe, and Kr. Also, while the bonding between the atoms in elemental diatomic molecules such as O_2, N_2, and F_2 is covalent, it is the van der Waals force that binds the molecules together when they crystallize at low temperature.

The second type are solids that contain uncharged molecular species with polar bonds, but where the bonded ligands have a roughly symmetric arrangement so that the molecule has no net dipole moment. This category includes small symmetric molecules such as CCl_4, alkanes [$CH_3(CH_2)_xCH_3$], and some polymers, such as polyethylene.

Layered compounds such as MoO_3, V_2O_5 (shown in Fig. 6.1), MoS_2, TiS_2, and graphite are the third type of solid where van der Waals bonding plays an

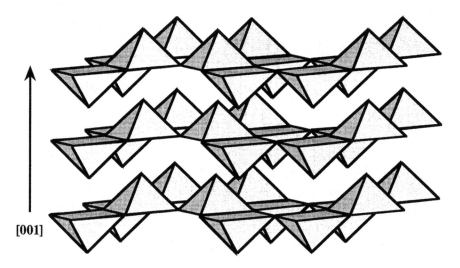

[001]

Figure 6.1. Vanadium pentoxide is a layered compound formed by corner- and edge-sharing VO_5 square pyramids. The layers are held together only by weak van der Waals bonds.

important role in cohesion. While atoms within the layers of these compounds are held together by strong, primary bonds, the layers are held together by weaker van der Waals forces. Although these materials can have very high melting points, they generally have very low mechanical stability and cleave easily at the interlayer bonds (like mica). Because of this 'flaky' property, graphite and MoS_2 are used as solid lubricants.

Because van der Waals bonds are weak, solids held together by these forces typically melt or decompose at low temperatures. As a lower limit for stability, you can consider He, which boils at 4.2 K. As an upper limit for stability, you can consider long chain alkanes which melt in the neighborhood of 100 °C. Two factors that affect the strength of the bonding or the cohesive energy of the solid are the molecular weight and the molecular shape.

ii. *The effect of molecular weight*
The cohesive energy of a crystal can be defined as the energy required to dismantle the components and separate them until there is no longer any interaction. It can be computed by subtracting the total energy of the crystal from the total energy of the unbonded atoms. Because the unbonded state is frequently taken as zero energy, the cohesive energies of stable solids are negative. We can think of the cohesive energy as a crystal's bond strength. The relationship between the cohesive energy and the melting point is intuitive. Generally, as the cohesive energy increases, more thermal energy is required to melt the solid, so

Table 6.1. *Inert gas data.*

gas	cohesive energy per atom, eV	T_m, K	atomic number
Ne	0.020	24	10
Ar	0.080	84	18
Kr	0.116	117	36
Xe	0.170	161	54

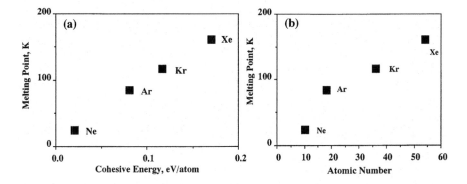

Figure 6.2. The melting point and cohesive energy of the inert gases increase with the atomic mass and number of electrons. These solids are held together by van der Waals bonds.

the melting point is higher. Figure 6.2(a) and Table 6.1 show the relationship between the melting point (T_m) and the cohesive energy per atom for the inert gases. Figure 6.2(b) shows the relationship between the melting point and the atomic number, or number of electrons. We find that both T_m and the cohesive energy increase with the number of electrons. Figure 6.3 and Table 6.2 show the same trend for the alkanes, where molecular weight is proportional to the number of electrons.

iii. *The effect of molecular shape*
The effect of steric factors on the strength of the van der Waals interaction is illustrated by the trend in the boiling points of the isomers of pentane, shown in Fig. 6.4. Each of these molecules has exactly the same molecular weight and number of electrons. Therefore, the change in the cohesive energy (as measured by the boiling point) must be due to the difference in the molecular shape or configuration. Using these molecules as examples, we see that the linear n-pentane

Table 6.2. *Alkane data.*

carbons	mol. wt.	T_m, °C	carbons	mol. wt.	T_m, °C	carbons	mol. wt.	T_m, °C
1	16	−182.0	9	128	−51.0	20	283	36.8
2	30	−183.3	10	142	−29.0	22	311	44.4
3	44	−189.7	12	170	−9.6	23	325	47.6
4	58	−138.4	13	184	−5.5	24	339	54.0
5	72	−130.0	14	198	5.9	26	367	56.4
6	86	−95.0	16	226	18.2	27	381	59.5
7	100	−90.6	17	240	22	29	409	63.7
8	114	−56.8	19	269	32.1	30	423	65.8

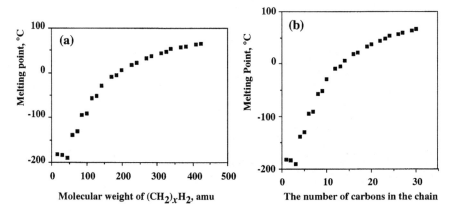

Figure 6.3. The melting points of the alkanes vary with their mass. These solids are held together by van der Waals bonds and the strength of this force increases with the molecular mass.

is more strongly bound than the nearly spherical neopentane. Examination of additional molecules leads one to the conclusion that configurations which have large aspect ratios form stronger intermolecular bonds. The reason for this is that the linear configuration (in comparison to a more compact geometry) allows more of the electrons in each molecule to be in a close proximity with electrons in the adjacent molecules; they can, therefore, interact more strongly.

To conclude this introduction, we can state the following generalizations. Van der Waals bonds increase in strength as the molecular weight and number of electrons increase. Also, molecules with larger aspect ratios are bound with greater strength. To understand why these generalizations hold, we need to consider the physical origin of the van der Waals bond.

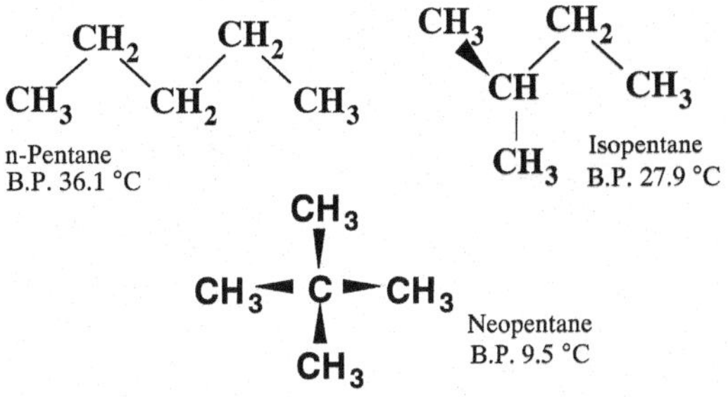

Figure 6.4. The strength of the van der Waals bonds in the isomers of pentane, as indicated by the boiling points, varies with the molecular shape. Drawn after [1].

(B) A physical model for the van der Waals bond

In all bonds, the attractive force that holds atoms together is electrostatic. Since van der Waals bonds form between uncharged, closed-shell species, the electrostatic attraction must be dipolar in origin.

i. The origin of the molecular attraction and repulsion

Random fluctuations of the electron density around an otherwise spherically symmetric atom or molecule can create transient or temporary dipoles. The strength of the electric field around a dipole is proportional to the dipole moment (the product of the nuclear charge and its separation from the center of the electron density) and inversely proportional to the cube of the distance, r, from the center of the dipole [2]. The dipolar field created by one temporary dipole is thus able to 'induce' a dipole in a neighboring atom by polarizing its electron density. The magnitude of the induced dipole moment is proportional to the temporary field and, therefore, decreases with r^{-3}. The interaction energy between the two dipoles is the product of the two dipole moments divided by r^3, so it varies as r^{-6}. This attractive energy between uncharged species is what causes inert gases to condense. Because of the r^{-6} dependence, this attraction is weak and acts only at relatively short ranges. As a result, solids held together by van der Waals bonds usually have low melting points.

The attractive van der Waals force brings atoms together until the electron distributions on adjacent atoms begin to overlap. This creates a repulsive force. When the repulsion becomes strong enough to compensate the attraction, an equilibrium separation is established. The origin of this repulsive force is in both the Coulombic repulsion of the like-charged electrons that surround each atom and the Pauli exclusion principle. The exclusion principle says that two electrons with the same energy can not occupy the same space. Thus, as the electron densities from the two atoms begin to overlap, electrons associated with the first atom have a tendency to occupy states on the second atom that are already

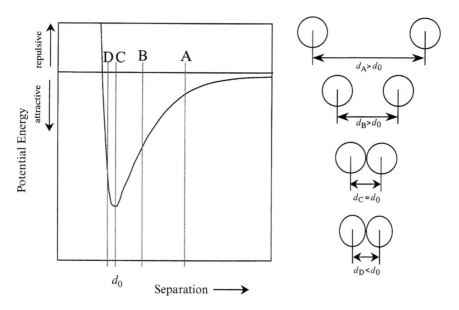

Figure 6.5. See text for a description.

occupied and vice versa. To prevent multiple occupancy, electrons must be pro-moted to higher energy states at a significant cost of energy. This interaction is strongly repulsive at short separations, but decreases rapidly at larger separa-tions.

ii. *The Lennard-Jones potential energy model*
The attractive dipolar and repulsive short-range force have been incorporated into a parameterized model known as the Lennard-Jones model or the '6–12' potential model. The potential energy between two atoms, V_0, as a function of separation, r, is

$$V_0(r) = \frac{-A}{r^6} + \frac{B}{r^{12}}$$
(6.1)

where A and B are positive constants. Lennard-Jones originally described this model in the mid-1920s [3]. This potential satisfies our requirements, since it shows an attractive r^{-6} character and a strong, short-range repulsion, r^{-12}. Interestingly, the exponent 12 was chosen for computational simplicity. Generally, any exponent greater than eight gives about the same result.

The qualitative effect of the function described in Eqn. 6.1 is shown sche-matically in Fig. 6.5. When atoms are separated by distances much larger than an atomic diameter (d_A), there is only a very weak attractive interaction. This increases as they get closer together (d_B) and at the point where they contact (d_C), the attraction is maximized. Finally, if the atoms are forced closer together (d_D), the attraction diminishes and eventually they repel each other.

The more common (and most convenient) form of the 6–12 potential is:

$$V_0(r) = 4\varepsilon \left[\left(\frac{\sigma}{r}\right)^{12} - \left(\frac{\sigma}{r}\right)^{6} \right].$$
$$(6.2)$$

In this case, σ and ε are constants that depend on the two bonding atoms. However, the new constants are defined in a more physically significant way than those in Eqn. 6.1. σ is the value of r at which $V_0 = 0$. Thus, this can be taken as the 'size' of the repulsive core or the diameter of the hard sphere atom. The minimum value of the function or the depth of the energy well is given by ε. Therefore, ε can be taken as a measure of the bond strength. Four 6–12 potentials for the inert gases are shown on the graph in Fig. 6.6, where σ and ε are marked.

The parameters for this model, σ and ε, can be obtained from two sources. One set of parameters comes from experimental measurements of the properties of the gaseous phase of Ne, Ar, Kr, and Xe [4]. The second comes from the calculations of Gordon and Kim [5], mentioned earlier in this chapter. Both sets of parameters are listed in Table 6.3.

Knowing that the dipole interaction is the origin of the attractive van der Waals force, it is easy to see why the bond strength increases with the atomic number. Atoms with a larger atomic number have more electrons, separated by a greater distance from the center of positive charge. This allows for larger dipole moments and, therefore, greater interaction energies.

iii. *Calculating the lattice constant, cohesive energy, and bulk modulus*
To find the equilibrium separation between two atoms, simply differentiate to find the minimum of $V_0(r)$ (Eqn. 6.2):

$$\frac{dV_0}{dr} = 4\varepsilon\sigma^{12} \cdot -12\left(\frac{1}{r}\right)^{13} - 4\varepsilon\sigma^{6} \cdot -6\left(\frac{1}{r}\right)^{7} = 0.$$
$$(6.3)$$

Solving for r in Eqn. 6.3 allows the equilibrium separation, r_0, to be determined:

$$r_0 = 2^{1/6}\,\sigma = 1.12\sigma.$$
$$(6.4)$$

A graphical solution to this problem is illustrated in Fig. 6.7. Note that the value of r_0 depends only on the parameter σ. Thus, it is the size parameter that determines the interatomic separation. Table 6.4 compares calculations of r_0 for inert gases, using the two sets of constants in Table 6.3, with the known values.

Table 6.3. *Parameters for the 6–12 potential [6].*

inert gas atom	experiment		theory	
	σ in Å	ε in 10^{-3} eV	σ in Å	ε in 10^{-3} eV
He			2.20	3.9
Ne	2.74	3.1	2.71	3.5
Ar	3.40	10.4	3.28	10.9
Kr	3.65	14.0	3.48	15.5
Xe	3.98	20.0		

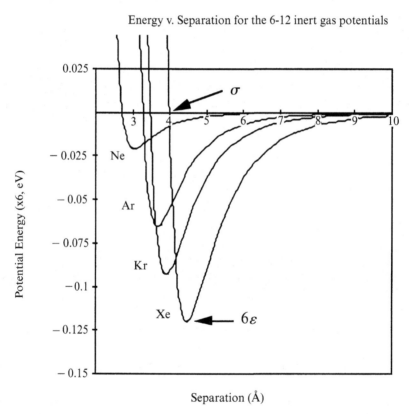

Figure 6.6. Energy v. separation curves for the inert gases based on Eqn. 6.2 (multiplied by a factor of six). The energy is the sum of a long range attractive contribution (−) and a short range repulsion (+). The minimum of the curve is a measure of the bond strength. At separations less than σ, the energy is positive, so this defines the effective diameter of the atom.

Table 6.4. *Experimental and calculated values of r_0.*

gas	r_0 (exp) (Å)	$1.12\sigma_{exp}$ (Å)	$1.12\sigma_{th}$ (Å)
Ne	3.13	3.08	3.04
Ar	3.76	3.82	3.67
Kr	4.01	4.10	3.90
Xe	4.35	4.47	4.46

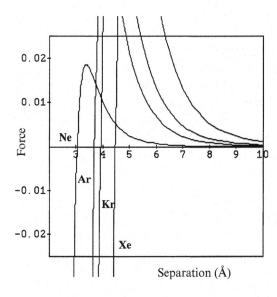

Figure 6.7. The first derivative of the interatomic potential, with respect to distance, gives the force displacement curve. The positive forces are attractive, the negative forces are repulsive, and the zero force point is the equilibrium separation.

The cohesive energy is computed by summing the potential energy between each atom and all of the other atoms in a crystal. The interaction energy, U_j, of the j-th atom with all of the other N-1 atoms in the crystal is:

$$U_j = \sum_{i=1, i\neq j}^{N} V(d_i),$$

(6.5)

where d_i is the distance to the i-th atom. There is a similar term for every atom in the crystal, so the complete sum, U, is:

$$U = \frac{1}{2}\sum_{j=1}^{N} U_j.$$

(6.6)

The factor of 1/2 is included to avoid counting each interaction twice. Assuming every atom is indistinguishable (this is true when they are situated at Bravais lattice positions), we can write:

$$U = 1/2\,NU_j. \tag{6.7}$$

Finally, the cohesive energy per atom, U', is simply the total energy, as in Eqn. 6.7, divided by N. (Note that primed quantities will be taken to be molar quantities in the remainder of the text.)

$$U' = 1/2\,U_j. \tag{6.8}$$

How do we actually go about computing the sum in Eqn. 6.5? There are several ways to do this. We will use the simplest approximation first, and see how it works out. Specifically, we will assume that because the van der Waals forces are very weak, only the nearest neighbors make a significant contribution to the cohesive energy. In other words, we only need to carry out the sum over the nearest neighbors. All inert gases have the fcc structure, so the cohesive energy per atom is:

$$U' = 1/2 \sum_{i=1}^{12} V(d_i). \tag{6.9}$$

For all 12 neighbors, $V(d_i) = V(d_0) = -\varepsilon$. When we substitute the result in Eqn. 6.4, we have:

$$U' = -6\varepsilon. \tag{6.10}$$

The calculated and measured values of the cohesive energy are compared in Table 6.5. In general, we have underestimated U'. This is reasonable, since we have ignored all the additional interactions with atoms outside of the nearest neighbor shell. Including these interactions improves the estimate.

Next, we compute the bulk modulus B (compressibility $= B^{-1}$). By definition:

$$B = -V \left(\frac{\partial P}{\partial V} \right)_T \tag{6.11}$$

where V is volume and P is pressure. Since:

$$P = \frac{-\partial U}{\partial V} \text{ at } T = 0,$$

$$B = V \frac{\partial}{\partial V} \left(\frac{\partial U}{\partial V} \right). \tag{6.12}$$

Table 6.5. *Calculated and measured cohesive energies of the inert gases.*

gas	$-U'$ (exp), eV/atom	$6\varepsilon_{exp}$, eV/atom	$6\varepsilon_{th}$, eV/atom
Ne	0.02	0.019	0.021
Ar	0.08	0.062	0.065
Kr	0.116	0.084	0.093
Xe	0.17	0.120	0.120

To work in terms of molar quantities and the interatomic separation, we remember that the inert gases have the fcc structure and write the volume per atom in terms of d:

$$a^3/4 = V' \text{ and } a = \sqrt{2}\, d,$$

$$V' = d^3/\sqrt{2}.$$

Substituting into Eqn. 6.12, we find,

$$B = \frac{\sqrt{2}}{9} d \left[\frac{1}{d^2} \frac{\partial^2 U'}{\partial d^2} + \frac{\partial U'}{\partial d} \frac{(-2)}{d^3} \right]. \tag{6.13}$$

Because we are interested in B near equilibrium, we consider B at $d = d_0$:

$$\frac{\partial U'(d_0)}{\partial d} = 0$$

and

$$B = \frac{\sqrt{2}}{9} \frac{1}{d} \frac{\partial^2 U'(d_0)}{\partial d^2}. \tag{6.14}$$

Evaluation of 6.14 leads to:

$$B = \frac{48\varepsilon}{\sigma^3}. \tag{6.15}$$

For Kr, we compute $B = 2.21 \times 10^{10}$ dyne cm^{-3}. This is not very close to the accepted value (3.5×10^{10}) and suggests the need to account for the long range interactions.

iv. *Accounting for long range interactions*
To correct errors stemming from our neglect of long range interactions, we begin by expressing all of the interatomic distances, d_i, as multiples of the shortest one, d_0.

$$U' = 1/2 \sum_{i=1}^{N} V(\rho_i d_0). \tag{6.16}$$

Therefore, determining the cohesive energy per atom,

$$U' = 2\varepsilon \left[\sum_{i=1}^{N} \left(\frac{\sigma}{\rho_i d} \right)^{12} - \sum_{i=1}^{N} \left(\frac{\sigma}{\rho_i d} \right)^{6} \right], \tag{6.17}$$

amounts to evaluating the sums:

$$A_{12} = \sum_{i=1}^{N} (1/\rho_i)^{12}; \qquad A_6 = \sum_{i=1}^{N} (1/\rho_i)^6. \tag{6.18}$$

It is interesting to note that the sums in Eqn. 6.18 depend only on the crystal structure and are thus the same for all isostructural crystals. These constants have been evaluated for common structures [7]. Minimization of Eqn. 6.17 to find the equilibrium separation leads to:

$$d_0 = (2A_{12}/A_6)^{1/6} \sigma. \tag{6.19}$$

For the fcc structure, $A_{12} = 12.13$ and $A_6 = 14.45$ and:

$$d_0 = 1.09\sigma. \tag{6.20}$$

This result differs from the nearest neighbor estimate (Eqn. 6.4) by only 3%. When the corrected value for the cohesive energy is computed using our new value for d_0 and Eqn. 6.17, we find:

$$U' = -1/2\varepsilon \frac{A_6^2}{A_{12}} = -8.6\varepsilon. \tag{6.21}$$

So, if we compare the new value with the nearest neighbor result (Eqn. 6.10), we see that the 12 nearest neighbors contribute only 70% of the total cohesive energy. A quick calculation can be used to show that the next nearest neighbors contribute 8% to the total and the final 22% is supplied by the rest of the crystal.

We should also correct our estimate of the bulk modulus. Following the procedure outlined in the previous section, but using our new equation for the cohesive energy, we find:

$$B = \frac{75\varepsilon}{\sigma^3}. \tag{6.22}$$

Substituting appropriate values for Kr, we compute $B_{Kr} = 3.46 \times 10^{-10}$ dyne-cm^{-2}, which is identical to the measured value. The differences between the values computed using the nearest neighbor model and the long range interaction model are

Table 6.6. *Comparison of the nearest neighbor and the long range model.*

parameter	nearest neighbor	long range	difference
d_0	1.12σ	1.09σ	3%
U	-6ε	-8.6ε	30%
B	$48\varepsilon/\sigma^3$	$75\varepsilon/\sigma^3$	37%

Table 6.7. *Measured and calculated* physical parameters for the inert gases.*

	d_0 (observed) Å	$d_0 = 1.09\sigma$ Å
Ne	3.13	2.99
Ar	3.76	3.71
Kr	4.01	3.98
Xe	4.35	4.34
	$-U'$ (observed) eV/atom	$-U' = 8.6\varepsilon$ eV/atom
Ne	0.02	0.027
Ar	0.08	0.089
Kr	0.116	0.120
Xe	0.17	0.172
	B (observed) $\times 10^{-10}$ D/cm^3	$B = 75\varepsilon/\sigma^3 \times 10^{-10}$ D/cm^3
Ne	1.1	1.81
Ar	2.7	3.18
Kr	3.5	3.46
Xe	3.6	3.81

Note:
*values of σ and ε are the 'experimental' values in Table 6.3.

given in Table 6.6. A review of the predicted and observed values is given in Table 6.7.

Finally, we note that the model works best as mass increases and quantum effects (primarily zero-point oscillations) become less and less important. Also, similar 'pair potential' models have been successfully used to describe the bonding in many other systems.

v.   *Connecting the physical models with the phenomenological trends*
As noted earlier, the van der Waals bond energy increases as the atomic mass and molecular weight increase. This is because the total number of electrons and size

increase with both of these parameters; larger species with more electrons can generate a greater dipole moment and stronger van der Waals bond because it is possible to separate more charge over a greater distance.

All other things being equal, molecules of the same molecular weight have greater bond strengths if they have more anisotropic (less compact) shapes. The origin of this trend is that the bonding force is weak and very short range (d^{-6}) so that electrons on adjacent molecules must get very close before the dipolar attractive force provides substantial cohesion. If a molecule has a compact shape, the electrons on atoms near the center can not contribute to the cohesive force.

Finally, we should note that while cohesive dipolar interactions are present in all solids, they are usually so weak in comparison to primary bonds that they constitute no more than a few percent of the total cohesive energy. Thus, we consider the van der Waals interaction to be significant only for crystals of uncharged atoms or molecules that are not held together by primary bonds.

Example 6.1 Using the 6–12 potential to compare structural stability

Using the 6–12 potential, compare the cohesive energy per atom of a five atom cluster and a seven atom cluster, assuming that the five atom cluster consists of four atoms tetrahedrally coordinating a fifth and the seven atom cluster consists of six atoms octahedrally coordinating a seventh.

1. First, we consider the geometries of these clusters and label the atoms for clarity.

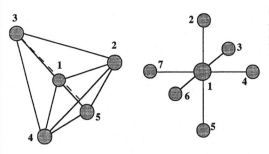

Figure 6.8. The geometry of the five and seven atom clusters for Example 6.1.

2. Next we note that the interaction potential between all of these atoms is given by Eqn. 6.2, shown below,

$$V_0(r) = 4\varepsilon \left[\left(\frac{\sigma}{r} \right)^{12} - \left(\frac{\sigma}{r} \right)^{6} \right],$$

and the total cohesive energy is given by the sum of the mutual interaction energies:

$$U = \frac{1}{2} \sum_{j=1}^{N} \sum_{\substack{i=1 \\ i \neq j}}^{N} V(d_{ij}).$$

3. Computing this energy amounts to specifying the separations, d_{ij}. In the tetrahedral cluster, every atom interacts with four other atoms. However, there are only two possible separations. The separation between the central atom and a vertex atom ($d_1 = d_{ij}$ where i or $j = 1$) and the separation between any two vertex atoms ($d_2 = d_{ij}$ where i and $j > 1$).

4. Based on the geometry of a cube with edge length a, we can write the ratio of these two lengths:

$$\frac{d_1}{d_2} = \frac{1/2\sqrt{3}a}{\sqrt{2}a} = \frac{\sqrt{3}}{2\sqrt{2}} = \sqrt{\frac{3}{8}}.$$

5. Now, the sum for the cohesive energy only has two types of term:

$$U = \frac{1}{2} \cdot 4\varepsilon \left[8\left(\frac{\sigma}{d_1}\right)^{12} - 8\left(\frac{\sigma}{d_1}\right)^{6} + 4 \cdot 3\left(\sqrt{\frac{3}{8}} \cdot \frac{\sigma}{d_1}\right)^{12} - 4 \cdot 3\left(\sqrt{\frac{3}{8}} \cdot \frac{\sigma}{d_1}\right)^{6}\right].$$

If $\eta = \sigma/d_1$,

$$U = 8\varepsilon \left[2\eta^{12} - 2\eta^{6} + \frac{3 \cdot 3^{6}}{8^{6}}\eta^{12} - \frac{3 \cdot 3^{3}}{8^{3}}\eta^{6}\right].$$

6. To find d_1, we minimize U with respect to η:

$$\frac{\partial U}{\partial \eta} = 0 = 24\eta^{11} - 12\eta^{5} + \frac{12 \cdot 3^{7}}{8^{6}}\eta^{11} - \frac{6 \cdot 3^{4}}{8^{3}}\eta^{5}.$$

7. Solving for the distance, we find:

$$d_1 = \sigma \left(\frac{1 + \dfrac{81}{1024}}{2 + \dfrac{3^{7}}{8^{6}}} \right)^{-1/6} = 1.1091\sigma.$$

Substituting this back into our equation for the total cohesive energy, we find that:

$$U_5 = -4.6383\varepsilon.$$

8. We can apply exactly the same logic to determine the cohesion in the octahedron. Here, there are three different distances:

$d_1 = d_{ij}$ where i or $j = 1$,

$d_2 = d_{ij}$ where $i \neq j \neq 1$,

$d_3 = d_{ij}$ where $i \neq j \neq 1$ and $|i - j| = 3$.

9. From Fig. 6.7, we can see that:

$$d_2 = \sqrt{2}d_1; \qquad d_3 = 2d_1.$$

10. Substituting these three distances (all in terms of d_1) into the cohesive energy as in step five, minimizing with respect to d_1, and solving, we find:

$$d_1 = 1.0859\sigma,$$

and

$$U_7 = -9.2038\varepsilon.$$

11. Finally, in order to make an appropriate comparison between the stability of the tetrahedral and octahedral cluster, we need to normalize these energies by the number of atoms in the cluster:

for the tetrahedral cluster, $U'_5 = -0.928\varepsilon$/atom,

for the octahedral cluster, $U'_7 = -1.315\varepsilon$/atom.

12. So, the octahedral configuration is more stable for a group of atoms which all have the same size. This is sensible both in terms of radius ratio rules and because the cluster of seven atoms has more bonds per atom.

(C) Dipolar and hydrogen bonding

Molecules with permanent dipole moments, such as those shown in Fig. 6.9, are attracted to each other by dipolar bonding forces. Consider, for example, the data in Table 6.8 that show the effect of dipolar bonding. Because all of these molecules are isoelectric and of similar lateral dimension, the increase in T_m can not be explained by the van der Waals interaction. The relative strengths of the dipolar bonding force can be gauged by the electronegativity difference between the covalently bonded atoms and the molecular geometry. We assume that the electropositive atom will be the center of positive charge and the electronegative

Table 6.8. *Effect of dipolar bonding on the melting point of molecular materials [1].*

compound	dipole moment	T_m °C
SiH_4	0	−185.0
PH_3	0.55	−132.5
H_2S	0.94	−82.9

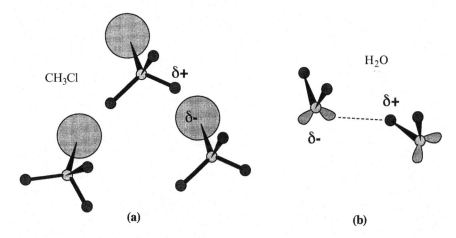

Figure 6.9. Dipolar bonding. Some molecules, as shown in (a), have a fixed dipole moment. Dipolar bonding is the attractive interaction between the partially negative side of one molecule (at the most electronegative element) and the positive side of another (the location of the least electronegative element). (b) Hydrogen bonding is a special case (for H–O, H–N, and H–F containing molecules) that is much stronger than normal dipolar bonding. The interaction between two water molecules is shown in (b).

atom will be the center of negative charge. Table 6.8 clearly shows that as the electronegativity difference and molecular geometry act to increase the dipole moment, the bonding force also increases.

One special type of dipolar interaction, which is given a separate name due to its greater strength, is the hydrogen bond. This interaction is much stronger than the other secondary bonding types. When H bonds with an electronegative element, its single electron is stripped away and a bare, unshielded proton is left. This center of positive charge now interacts with electronegative atoms on other molecules. The strength of this interaction is on the order of 0.3 eV. The effect of hydrogen bonding only needs to be considered when H is bonded to F, O, or

Table 6.9. *Evidence for H-bonding in the melting points of molecular solids [1].*

	molecular weight	T_m °C
H_2O	18	0
H_2S	34	−83
H_2Se	81	−66
H_2Te	130	−48
HF	20	−83
HCl	36.5	−111
HBr	81	−86
HI	128	−51
NH_3	17	−78
PH_3	34	−132
AsH_3	78	−114
SbH_3	125	−88

Table 6.10. *Effect of H-bonding on the melting points of molecular materials [1].*

name	structure	molecular weight	T_m °C
n-hexane	$CH_3(CH_2)_4CH_3$	86	−95.3
1-pentanol	$CH_3(CH_2)_3CH_2OH$	88	−78.9
glycerol	CH_2 CH CH_2 | | | OH OH OH	88	18.2

N. Thus, amines (RNH_2), alcohols (ROH), and carboxylic acids (RCOOH) are all important hydrogen bonded systems. All other things being equal, the strength of a hydrogen bond decreases with the electronegativity difference between H and the electronegative ligand. Therefore, the H-bond strength is HF > HO > HN. Evidence for H-bonding is shown in Tables 6.9 and 6.10.

(D) The use of pair potentials in empirical models

While the Lennard-Jones pair potential (Eqn. 6.2) is intended as a physical model for van der Waals bonds, it is often extended empirically to model other types of crystals by adjusting the constants σ and ε. The assumption here is that

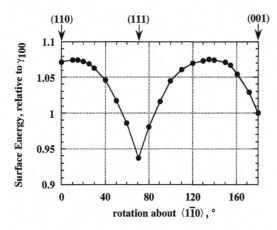

Figure 6.10. The relative surface energies of an fcc material, computed using a Lennard-Jones potential, for surfaces perpendicular to $\langle 1\bar{1}0 \rangle$ [9].

regardless of the physical mechanism governing the interactions among atoms, the total energy of the system can be treated as if it were the sum of attractive and repulsive pairwise contributions. Because of its computational simplicity, the Lennard-Jones potential is frequently used as a model for pairwise interactions, with the constants, σ and ε, chosen such that the pairwise potentials reproduce known properties with reasonable accuracy. Empirical models constructed in this way can then be used to compute physical properties that are difficult, tedious, or impossible to measure by experiment. While quantitative accuracy can not be expected, the relative energies computed from such models are often qualitatively meaningful and instructive.

For example, a Lennard-Jones model for Cu was used to compute the surface energy per unit area (γ) as a function of the surface normal [9]. The surface energy can be defined as the work required for the creation of a unit area of surface. Creating new surface requires that one break bonds, and this is why surfaces have positive excess energies. For a surface with a unit cell that contains N atoms in an area A, the surface energy is:

$$\gamma = \frac{\left[NU' \sum_i \sum_{j>i} V_0(d_{ij}) \right]}{A}.$$ (6.23)

In Eqn. 6.23, U' is the bulk potential energy per atom, V_0 is the Lennard-Jones potential defined in Eqn. 6.2, and d_{ij} is the distance between the i-th and j-th atom. Creating a surface breaks the symmetry of the Bravais lattice so that all positions that were lattice sites in the bulk structure no longer have identical environments and we must compute the sum over atom pairs. Results obtained for the fcc structure are summarized in Fig. 6.10.

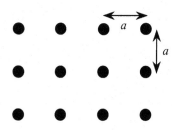

Figure 6.11. See Problem 1.

Within the simple interpretation presented here, the relative energies in Fig. 6.10 scale with the relative density of broken bonds per unit area; the (111), (100) and (110) surfaces have 6.93, 8, and 8.48 broken bonds per a^2. Using Wolf's [9] Lennard-Jones parameters for Cu, we find that γ_{100} is 0.896 J/m². The energies given in Fig. 6.10 are computed assuming that all of the atoms are fixed in bulk-like positions. In fact, we expect that the change in the coordination environment at and near the surface should lead to changes in atomic positions and inter-atomic distances. These energy lowering atomic relaxations can be evaluated by iterative calculations wherein the energy is minimized as a function of atomic positions. Lennard-Jones potentials predict that the outer atomic layer of the crystal relaxes away from the bulk; i.e. an increase in interlayer separation is pre-dicted. This result is counter-intuitive and not consistent with experiment, illus-trating the hazards of using pairwise Lennard-Jones potentials to model strong primary bonds in systems where nonlocal many-body effects are important. Nevertheless, qualitative trends in the anisotropy of the surface energy of Cu do reproduce experimental observations [10].

(E) Problems

(1) Assume that Ar crystallizes in the two-dimensional structure shown in Fig. 6.11.

(i) Including only nearest neighbor effects, compute the equilibrium separation of the atoms.

(ii) Including only nearest neighbor effects, compute the cohesive energy.

(iii) Compute the contribution to the cohesive energy from the next nearest neighbors.

(iv) Based on your knowledge of inert gas bonding, propose a more likely two-dimensional structure and estimate its cohesive energy.

(2) Two possible configurations for the hypothetical Ar_3 molecule are shown in Fig. 6.12.

(i) Compute the nearest neighbor separation.

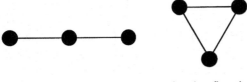

Figure 6.12. Two possible configurations for the hypothetical Ar$_3$ molecule.

linear configuration   trigonal configuration

(ii) Compute the cohesive energy. You can leave your answer in terms of σ and ε.

(iii) Which configuration is more stable?

(3) Using methods similar to those illustrated in Example 6.1, calculate the interatomic distances from an atom with three nearest neighbors in trigonal planar coordination, five nearest neighbors in trigonal bipyramidal coordination, seven nearest neighbors in pentagonal bipyramidal coordination, and eight nearest neighbors in cubohedral coordination.

(4) Assuming only nearest neighbor interactions, the cohesive energy of a fcc structured inert gas crystal is -6ε, (Eqn. 6.10) where ε is the Lennard-Jones parameter in eV. Determine the fraction of the cohesive energy contributed by the second nearest neighbors, the third nearest neighbors, and all neighbors greater than third.

(5) Compare the stability of a five atom cluster (one central atom with four surrounding it) in tetrahedral and square planar coordination.

(6) Compare the stability of a six atom cluster (one central atom with five surrounding it) in trigonal bipyramidal and square pyramidal coordination.

(7) Compare the stability of a seven atom cluster (one central atom with six surrounding it) in octahedral and trigonal prismatic coordination.

(8) Consider an atom bonded in a ccp structure and its 12 nearest neighbors to be an independent cluster of 13 atoms. Using the 6–12 potential that we have used to describe inert gas bonding, compare the total cohesive energy of this cluster (per atom) to the cohesive energy of a 13 atom cluster representing the nearest neighbor coordination shell in the hcp structure.

(9) Explain the observed trend in the melting points of the halogens:

X$_2$	T_m °C	X$_2$	T_m °C
F$_2$	−218 °C	Br$_2$	−7 °C
Cl$_2$	−101 °C	I$_2$	114 °C

(10) What would be the most likely structure for the hypothetical compound, NeXe? What is the cohesive energy of this compound? (Assume that the potential

energy parameters, σ and ε, for the Ne–Xe interaction are $\sigma_{Ne-Xe} = 1/2(\sigma_{Ne} + \sigma_{Xe})$ and $\varepsilon_{Ne-Xe} = (\varepsilon_{Ne}\varepsilon_{Xe})^{1/2}$.)

(11) Using an isotropic Lennard-Jones potential which includes only nearest neighbor interactions, show that tetragonal distortions of a simple cubic arrangement of identical atoms are always energetically unfavorable.

(12) The structure of graphite is described in Table 3B.35.

(i) Compute the distances to the nearest neighbor carbon in the a–b plane and along the c-axis.

(ii) Which of the C–C separations correspond to van der Waals bonds and which correspond to covalent bonds?

(iii) Consider a one-dimensional chain of C atoms, linked by van der Waals bonds, that interact according to the usual 6–12 potential. Determine a value of σ that can be used to reproduce the van der Waals bond length in graphite. (Note: include the long range interactions in your model.)

(iv) What is the cohesive energy per atom (in terms of ε) of this one-dimensional C chain?

(13) Use the Lennard-Jones potential (Eqn. 6.2) for the following calculations:

(i) Determine the cohesive energy, per atom, of a two-dimensional square lattice extending infinitely in the x and y directions. Make sure that your answer is accurate to three significant figures.

(ii) Next, repeat this calculation for a three-dimensional structure consisting of five layers. In other words, the system extends infinitely in the x and y directions, but is only five layers thick along the z direction. Furthermore, the atoms from adjacent layers form lines parallel to the z direction.

(iii) Repeat this calculation for the three-dimensional, infinite, cubic P lattice.

(F) References and sources for further study

[1] C.H. Yoder, F.H. Suydam, and F.H. Snavely, *Chemistry*, 2nd edition (Harcourt Brace Jovanovich, New York, 1980) pp. 240–3. Phenomenological aspects of secondary bonding.

[2] P. Lorrain and D.R. Corson, *Electromagnetic Fields and Waves* (W.H. Freeman and Company, San Francisco, 1970) pp. 61–4. The dipolar interaction.

[3] J.E. Lennard-Jones, *Proc. Roy. Soc.* **A106** (1924) 441 & **A109** (1925) 584.

[4] N. Bernardes, *Phys. Rev.* **122** (1958) 1534. Experimental values for the 6–12 parameters.

[5] R.G. Gordon and Y.S. Kim, *J. Chem. Phys.* **56** (1972) 3122. The origin of the 6–12 parameters.

[6] W.A. Harrison, *Electronic Structure and the Properties of Solids: The Physics of the Chemical Bond* (Dover Publications, Inc., New York, 1989) Chapter 12. Table 6.3 is constructed from data in Table 12.2. Harrison cites references [4] and [5] as the original sources.

[7] N.W. Ashcroft and N.D. Mermin, *Solid State Physics* (Holt Rinehart and Winston, New York, 1976) pp. 398–402.

[8] C. Kittel, *Introduction to Solid State Physics*, 5th edition (J. Wiley & Sons, New York, 1976) pp. 73–85.

[9] D. Wolf, *Surface Science* **226** (1990) 389–406. The results in Fig. 6.10 were adapted from this source.

[10] M. McLean and B. Gale, *Phil. Mag.* **20** (1969) 1033.

Chapter 7
Ionic Bonding

(A) Introduction

The ionic bonding model describes the limiting case in which an electron is transferred from the outer orbital of an electropositive atom to an empty outer orbital of an electronegative atom. The results is two charged species, an anion ($-$) and a cation ($+$). Again, we see that the cohesive force is electrostatic in origin. Ideally, there is no further interaction between the electrons on each of the ions. In other words, only the occupation of the electronic states associated with each atom changes, not the energy levels of the states. As with the other bonding models, this describes only the limiting case and does not accurately reflect all that is known about ionic crystals. Nevertheless, we will see that a quantitative model based on these assumptions works surprisingly well. We should also note that while secondary, metallic, and covalent bonds can be formed between identical atoms, ionic bonds only form between different types of atoms.

i. Solids that are held together by ionic bonds

Any compound that forms between a halide (group VIIA) or chalcogenide (group VIA) and an alkali metal (group IA), an alkaline earth metal (group IIA), or a transition metal (B groups) is usually said to be ionic. The criterion defined in Chapter 1, Section D, is that the Pauling electronegativity difference must be greater than 1.7. Ionic bonds are primarily found in extended crystalline structures, but they occasionally also occur in simple molecules. Our interests lie with the crystalline solids.

ii. The effect of charge and separation on bond strength

It is obvious that the bonding force in ionic systems is electrostatic; the bond strength increases as the bonds become shorter and the charge on the ions increases [2–4]. These trends are illustrated in Table 7.1, which shows the lattice energies of selected ionic materials. We use the lattice energy (the energy required to separate all of the ions in a crystal to infinity) as a measure of the crystal's bond strength.

Note in Table 7.1 that LiCl ($d_0 = 2.57$Å) and SrO ($d_0 = 2.58$Å) have approximately the same interatomic spacing and the same crystal structure (rock salt),

Table 7.1. *Lattice energies and nearest neighbor distances in selected ionic compounds [1].*

MX	lattice E, eV	d_0	MX	lattice E, eV	d_0
LiF	10.73	2.014	BeO	44.49	1.656
LiCl	8.84	2.570	MgO	39.33	2.106
LiBr	8.36	2.751	CaO	35.38	2.405
LiI	7.84	3.000	SrO	33.34	2.580
			BaO	31.13	2.770
NaF	9.56	2.320			
NaCl	8.14	2.820	CaF_2	27.04	2.365
NaBr	7.74	2.989	SrF_2	25.66	2.511
NaI	7.30	3.236	CeO_2	99.76	2.343
KF	8.51	2.674			
KCl	7.41	3.146			
KBr	7.07	3.298			
KI	6.72	3.533			
RbF	8.13	2.826			
RbCl	7.14	3.290			
RbBr	6.84	3.444			
RbI	6.53	3.671			
CsF	7.67	3.007			
CsCl	6.83	3.571			
CsBr	6.54	3.712			
CsI	6.26	3.955			

but different lattice energies. In fact, the lattice energy of SrO is roughly four times that of LiCl. This is the same ratio as the ratio of the products of the ionic charges, (in LiCl, $q_1q_2 = 1$ and in SrO, $q_1q_2 = 4$), a demonstration that the bond strength (lattice energy) is proportional to the strength of the electrostatic attraction between ions with opposite charges.

iii. *The effect of size on the structure*

Steric (size) factors influence the structure of ionic compounds. These constraints are reflected in the radius ratio rules that were described in detail in Chapter 1, Section E(ii). Basically, the rules say that a given geometric configuration becomes unstable with respect to a lower coordination number configuration

Table 7.2. *Critical radius ratios.*

radius ratio, $\rho = r_+/r_-$	cation coordination number
$1.00 \geq \rho \geq 0.732$	8
$0.732 \geq \rho \geq 0.414$	6 or 4 square planar
$0.414 \geq \rho \geq 0.225$	4 tetrahedral
$0.225 \geq \rho \geq 0.155$	3
$0.155 \geq \rho \geq 0$	2

Table 7.3. *Cation size influences the crystal structure of fluorides.*

rutile		fluorite	
compound	cation size, Å	compound	cation size, Å
CoF_2	0.65	CaF_2	1.0
FeF_2	0.61	SrF_2	1.18
MgF_2	0.72	BaF_2	1.35
MnF_2	0.67	CdF_2	0.95
NiF_2	0.69	HgF_2	1.02
PdF_2	0.86	EuF_2	1.17
ZnF_2	0.74	PbF_2	1.19

when the like-charged ligands contact each other and the central ion is no longer in contact with all of the coordinating ligands. The critical radius ratios are summarized in Table 7.2.

Although the rules are not followed strictly, the data for the MX_2 oxides and fluorides that crystallize either in the rutile structure (with six-coordinate cations) and fluorite structure (eight-coordinate cations) are compelling. When ρ is below a critical value, the rutile structure is always adopted. These data are summarized in Tables 7.3 and 7.4. For the metal fluorides, the critical cation radius lies between 0.86 Å (r_{Pd}) and 0.95 Å (r_{Cd}). For the metal oxides, the boundary lies at 0.775 Å, the radius of the tetravalent Pb cation. PbO_2 is polymorphic and can adopt either the rutile or the fluorite structure. Cations smaller than Pb form the rutile structure and cations larger than Pb form the fluorite structure. Note that in order to make a meaningful comparison, only octahedral cation radii are considered.

Table 7.4. *Cation size influences the crystal structure of oxides.*

rutile		fluorite	
compound	cation size, Å	compound	cation size, Å
TiO_2	0.605	PbO_2	0.775
CrO_2	0.55	CeO_2	0.87
IrO_2	0.625	PrO_2	0.85
MnO_2	0.53	PaO_2	0.90
MoO_2	0.65	ThO_2	0.94
NbO_2	0.68	UO_2	0.89
OsO_2	0.63	NpO_2	0.87
PbO_2	0.775	PuO_2	0.86
RuO_2	0.62	AmO_2	0.85
SnO_2	0.69	CmO_2	0.85
TaO_2	0.68		
WO_2	0.66		

(B) A physical model for the ionic bond

In this section, a physical model is developed that can be used to rationalize the generally observed properties of ionically bonded crystals and to compute physical properties with reasonable accuracy.

i. *The formation of an ionic bond*

We begin by assuming that ionic bond formation can be described by a series of fundamental steps to which independent energies can be assigned [5–7]. For example, consider the formation of an ionically bonded crystal, MX, from solid metallic M and gaseous diatomic X_2. The first step is to disassemble the reactants into individual atoms. The energy required to create gaseous M atoms is S, the sublimation energy per atom for the solid. The energy to form an isolated X atom is $1/2D$, where D is the dissociation energy for an X_2 molecule. From the free atoms, we create free ions by removing an electron from M to form a cation (M^+) and adding an electron to X to create an anion (X^-). The energy required to remove an electron from the M atom is the ionization energy, I_M, and the energy to add an electron to the X atom is the negative of the electron affinity, E_X. (The electron affinity is the energy required to remove an electron from an anion.) The free ions then condense under the influence of the attractive electrostatic force

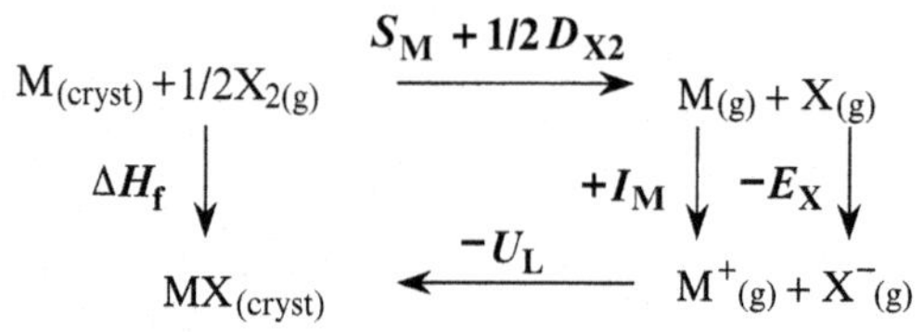

$$M_{(cryst)} + 1/2X_{2(g)} \xrightarrow{\;\;S_M + 1/2\,D_{X2}\;\;} M_{(g)} + X_{(g)}$$

$$\Delta H_f \downarrow \qquad\qquad +I_M \downarrow\; -E_X \downarrow$$

$$MX_{(cryst)} \xleftarrow{\;\;-U_L\;\;} M^+_{(g)} + X^-_{(g)}$$

Figure 7.1. Schematic illustration of a Born–Haber cycle. See Eqn. 7.1.

between ions with opposite charge. Finally, a fixed separation is established by the short range repulsive interactions, as described in Chapter 6. Since the lattice energy (U_L) is defined as the energy required to separate all of the ions in the crystal to a noninteracting state, the energy for the condensation step is $-U_L$. The sum of the energies for each step of the process must equal the enthalpy of formation of MX, ΔH_f.

$$\Delta H_f = S + \frac{1}{2}D + I_M - E_X - U_L. \tag{7.1}$$

The idea of equating the energy from two different mechanistic paths is called a thermochemical cycle; this particular cycle is known as a Born–Haber cycle. The cycle is illustrated in Fig. 7.1 and can be used to compute the lattice energy, U_L, from known elemental (I_M, E_X), molecular (D), and thermodynamic (ΔH_f) quantities. Before applying this method, we take a moment to describe the ionization energy and electron affinity in greater detail.

ii. *The ionization energy and electron affinity*

The ionization energy and the electron affinity are both electron removal energies. The term ionization energy is used when an electron is removed from a neutral atom or from a cation. The term electron affinity is used when the electron is removed from an anion. The relevant processes are illustrated schematically in Fig. 7.2.

The ionization energy, I_M, is always a positive quantity because energy is required to separate an electron from the attractive field created by the positive charge on the nucleus. Selected ionization energies are listed in Table 7.5; a more complete listing can be found in [8]. The ionization energies exhibit obvious periodic trends, similar to those discussed in Chapter 1. As you move down a column within each main group, I_M decreases as Z (atomic number) increases. This is because of the change in size. As the atom gets larger, the influence of the nuclear potential on the outermost valence electrons (those that will be removed) is diminished and the valence electrons are, therefore, more easily removed. Also,

Table 7.5. *Ionization energies, eV/atom [8].*

$M \rightarrow M^+ + e^-$				$M \rightarrow M^{2+} + 2e^-$				$M \rightarrow M^{3+} + 3e^-$			
Li	5.39			Be	27.53			B	71.38		
Na	5.14			Mg	22.68			Al	53.26		
K	4.34	Cu	7.73	Ca	17.98	Zn	27.36	Sc	44.10	Ga	57.22
Rb	4.18	Ag	7.58	Sr	16.72	Cd	25.90	Y	39.14	In	52.68
Cs	3.89	Au	9.22	Ba	15.22	Hg	29.19	La	35.81	Tl	56.37

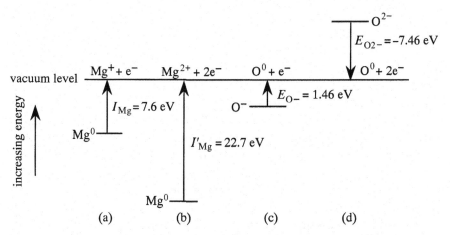

Figure 7.2. Schematic illustration of the definitions of the ionization energy and electron affinity for Mg and O, respectively. The vacuum level refers to a state of separation where the separated species no longer affect each other. (a) The first ionization energy of Mg to form a monovalent ion. (b) The second ionization energy to form the stable divalent ion. (c) The electron affinity of a monovalent O ion. Note that the monovalent anion is more stable than the neutral atom. (d) The electron affinity of the doubly charged oxide ion is negative, indicating that the neutral atom is more stable than the doubly charged anion.

the ionization energy increases rapidly with increasing ionic charge. In this case, it is the increased electrostatic attraction that makes it increasingly difficult to remove electrons.

As we traverse a single row in the periodic chart from left to right, the nuclear charge increases, but the distance between the valence electrons and the center of positive charge remains nearly the same. Thus, the valence electrons are more strongly bound and have higher ionization energies. For example, consider the formation of a monovalent Cu cation and a monovalent K cation. In each case, it is the 4s electron that is removed from the neutral atom to form the cation. The Cu

Table 7.6. *Electron affinities of non-metals (eV/atom) [10].*

F^-	3.45 ± 0.09	O^-	1.46 ± 0.04	O^{-2}	-7.46 ± 0.2
Cl^-	3.71 ± 0.06	S^-	2.15 ± 0.13	S^{-2}	-4.34 ± 0.09
Br^-	3.49 ± 0.04	Se^-	2.20 ± 0.13	Se^{-2}	-5.07 ± 0.09
I^-	3.19 ± 0.04	Te^-	$2.30 \pm ?$	Te^{-2}	$-4.21 \pm ?$
At^-	$2.64 \pm ?$				

atom binds the 4s electron more strongly because it has ten additional positive charges in the nucleus.

Finally, note that the ionization energy varies in approximately the same way as the electronegativity. Also, atoms that are easily ionized (those that have low values of I_M or electronegativity) are likely to form ionic or metallic compounds, while those that are difficult to ionize form covalent bonds.

It is important to remember that the electron affinity, E_X, is the ionization energy for the anion. When considering the ionic bond, it is actually the negative of this value (the energy required to form the anion) that we are interested in. Electron affinities for typical anions are listed in Table 7.6.

It is interesting to note that the electron affinities of the monovalent halides and chalcogenides are positive quantities. This illustrates the fact that the charged anionic state is sometimes more stable than the neutral atomic state. The preference for the charged state occurs when the energies deriving from the electron–nuclear attraction and the change in the orbital angular momentum exceed the energy increase associated with the added electron–electron repulsions. Just as more highly charged cations are less stable (have higher ionization energies), the divalent chalcogenide anions are less stable than either the monovalent or neutral species. In this case, the repulsive electron–electron interactions exceed the attractive electron–nuclear interaction.

iii. Estimating lattice energies and compound stability using the Born–Haber cycle
As mentioned earlier, the Born–Haber cycle can be used to estimate lattice energies. For example, consider the formation of fluorite (CaF_2) from solid Ca and gaseous F_2. Using the cycle and quantities listed in Fig. 7.3, we can say that the lattice energy is:

$$U_L = S_{Ca} + D_{F_2} + I_{Ca+2} - 2E_F - \Delta H_f. \tag{7.2}$$

This sums to 27.3 eV per formula unit.

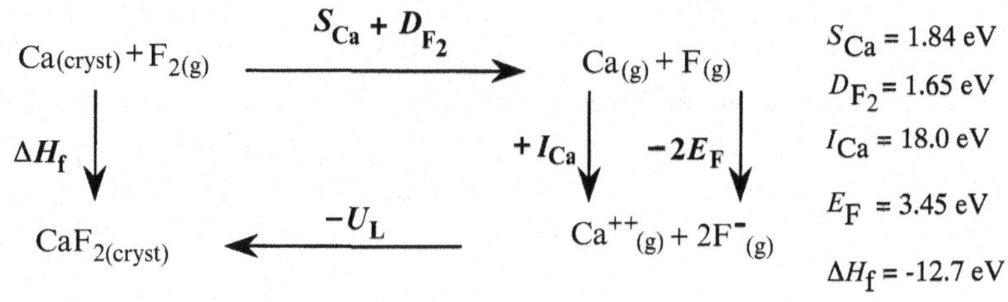

Figure 7.3. A Born–Haber cycle for CaF_2. See text for a description.

In the next section, we shall see that it is also possible to compute lattice energies by an independent physical model. Using these computed values for U_L, it is then possible to use the Born–Haber cycle to estimate thermodynamic quantities such as the heat of formation. This approach can be used to determine whether a hypothetical compound is stable. In fact, calculations of this sort were used to predict the stability of the first ionic crystals formed from 'inert' gas cations. The observation that O_2PtF_6 was ionically bound as $(O_2)^+(PtF_6)^-$, and that O_2 and Xe have approximately the same ionization energy, led Bartlett to predict the stability of $XePtF_6$ in advance of its synthesis [13].

Example 7.1

Use the thermochemical data presented in Table 7.7 to explain why NaCl forms a stable compound with the rock salt structure, but NeCl does not.

Table 7.7 *Thermochemical data for the formation of NeCl and NaCl (eV)* [6].

	S	$1/2D_{Cl2}$	I_M	E_{Cl}	U_L	$\Delta H^0_{f\,(calc)}$
NaCl	1.04	1.26	5.14	3.71	8.14	−4.41
NeCl	0	1.26	21.6	3.71	8.50	10.65

1. Based only on the data presented in Table 7.7 and Eqn. 7.1, we can compute the formation enthalpy in the right-hand column of Table 7.7.
2. We see from the summation that NaCl has a large negative heat of formation, but that NeCl's heat of formation is positive. Considering the magnitude of the positive

ΔH term for NeCl and that $\Delta G = \Delta H - T\Delta S$, there is no reasonable entropic term that could possibly stabilize this compound.

3. Comparison of the various energies reveals that it is the large ionization energy of Ne (the energy needed to broach the inert gas core) that makes this compound unstable. This is the same reason that compounds with alkali metals as divalent cations, such as $NaCl_2$, are never observed. It is generally not energetically practical to penetrate the inert gas core.

It should be pointed out that a negative formation enthalpy does not automatically imply the existence of a hypothetical compound. For example, the calculated ΔH for MgCl is -1.3 eV per formula unit. However, ΔH for $MgCl_2$ is -6.6 eV per formula unit. Thus, MgCl will always be unstable with respect to disproportionation to $1/2\ MgCl_2 + 1/2\ Mg$ [6].

iv. *The Coulombic bonding force and the ionic pair potential*

When we consider the magnitudes of the energies involved in preparing the free ions, we see that the lattice energy is essential for the stabilization of the crystal. Consider, for example, that the energy to form the isolated ions that will make up an ionic crystal is always $I_M - E_X$. If we examine the data in Tables 7.5 and 7.6, it is clear that there is no combination of ionization energies and electron affinities for which the ion formation process is favored (in other words, the sum is always positive and, therefore, always costs energy). The 'most favorable' case is for CsCl, where $I_{Cs} - E_{Cl} = 0.2$ eV. In other words, the tendency for atoms to form closed shell or inert gas electron configurations can never by itself drive compound formation. Only when ions with closed shell configurations act in concert with electrostatic cohesive forces between ions with opposite charges is it possible to stabilize ionic solids.

A simple way to write a pair potential for the ionic bond is to add the attractive Coulomb potential to the Lennard-Jones potential used in Chapter 6 [17, 18]. The Coulombic potential between two ions is

$$V_c(d) = \frac{k\, Z_1\, Z_2\, e^2}{d}, \tag{7.3}$$

where Z_1 and Z_2 are the integer values of the anion and cation charges, e is the unit charge of an electron, and d is the interatomic separation. The proportionality constant, k, is $1/(4\pi\varepsilon_0)$, where ε_0, the permittivity of free space, is 8.854×10^{-12} C/V-m. Thus, $k = 9 \times 10^9$ N-m^2/C^2. In units more relevant to the atomic-scale, we find that the constant $ke^2 = 14.4$ eV-Å. When the electrostatic contribution is added to the Lennard-Jones contribution, we have:

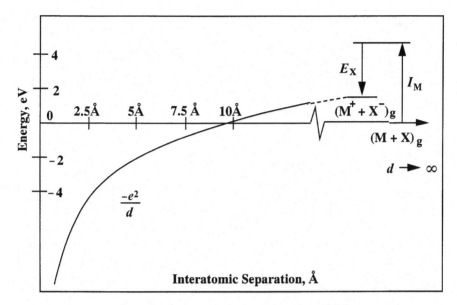

Figure 7.4. Energy/configuration diagram for the formation of NaCl [16].

$$V(d) = \frac{k\,Z_1\,Z_2\,e^2}{d} + 4\varepsilon\left[\left(\frac{\sigma}{d}\right)^{12} - \left(\frac{\sigma}{d}\right)^{6}\right]. \tag{7.4}$$

The total energy to form, for example, NaCl is:

$$I_{Na} - E_{Cl} + V(d). \tag{7.5}$$

Considering Eqn. 7.5, it is interesting to see how close the two ions must approach each other before the electrostatic stabilization energy overcomes the energy required for the ionization steps. To estimate this critical separation, we will ignore the short-range forces in the Lennard-Jones part of the potential because, at large separations, their contribution will be very small. Using $I_{Na} = 5.14$ eV, $E_{Cl} = 3.71$ eV, $Z_1 = 1$, and $Z_2 = -1$:

$$1.43\ eV = \frac{ke^2}{d}$$

$$\text{so, } d = \frac{ke^2}{1.43eV} = 0.7ke^2 = 10\text{Å}.$$

Therefore, the ionic species are stabilized by the electrostatic energy only at molecular-scale separations, as illustrated in Fig. 7.4.

295

5	6	7	8	9	10	11	electrons
		1	2	3	4	5	
		H	He	Li	Be	B	
7	8	9	10	11	12	13	
N	O	F	Ne	Na	Mg	Al	
15	16	17	18	19	20	21	
P	S	Cl	Ar	K	Ca	Sc	
33	34	35	36	37	38	39	
As	Se	Br	Kr	Rb	Sr	Y	
51	52	53	54	55	56	57	
Sb	Te	I	Xe	Cs	Ba	La	
83	84	85	86	87	88	89	
Bi	Po	At	Rn	Fr	Ra	Ac	
−3	−2	−1	0	+1	+2	+3	ion charge

Figure 7.5. Chart of the isoelectronic ions [17].

To accurately determine the minimum energy of the system or the equilibrium separation, one needs to use the full form of the potential, including the repulsive interaction which we are taking here to have the Lennard-Jones form. Therefore, before we can actually compute anything, we need a set of Lennard-Jones parameters (σ and ε) that describe the interaction between any two atoms. Although it is possible to develop parameterized pair-wise potentials for any combination of atoms, a satisfactory choice (and the one we will employ here) is to use the appropriate parameters for the inert gas atom with the same electronic configuration as the ion in question [17]. For example, for NaF we use the parameters for Ne and for KCl we use the parameters for Ar. While this choice of parameters will lead to marginally inferior quantitative predictions, it has the advantage of allowing us to easily complete a wide range of calculations without having to develop new potentials for every compound of interest. A version of the periodic chart that illustrates this principle by emphasizing the relationships between isoelectronic ions is shown in Fig. 7.5. For compounds made up of atoms with different inert gas cores, we have to combine the parameters of each. This is done by taking the average of the values for σ and the geometric mean for ε. Some examples are shown in Table 7.8.

The total energy of two monovalent ions with opposite charges with reference to the neutral bodies at infinite separation is:

$$E = I_M - E_X - \frac{ke^2}{d} + 4\varepsilon\left[\left(\frac{\sigma}{d}\right)^{12} - \left(\frac{\sigma}{d}\right)^{6}\right]. \tag{7.6}$$

Table 7.8. *L-J parameters for some inert gas combinations.*

atoms	$1/2\,(\sigma_1 + \sigma_2)$, Å	$\sqrt{\varepsilon_1\varepsilon_2}$, eV
HeAr	2.74	0.0065
NeAr	3.00	0.0062
NeKr	3.10	0.0074
ArKr	3.38	0.0130

To find the equilibrium atomic separation in a NaCl molecule, we differentiate the energy with respect to d, set the differentiated energy equal to zero, and solve for d. Because the attractive van der Waals component is small compared to the other energies involved, we will neglect it (it can be shown that this approximation is valid when $\sigma/d > 1$)

$$d_0 = 1.116\ \sigma(\sigma\varepsilon)^{1/11}$$
$$d_0 = 2.33\ \text{Å}.$$

This compares reasonably well with the experimental value of 2.51 Å ($\Delta = -7\%$).

The energy of the diatomic pair at the equilibrium separation is found by substituting d_0 into Eqn. 7.6. Doing this, we find that $E = -4.1$ eV, which compares well with the experimentally determined value of -4.3 eV. Repeating this calculation for other atomic pairs does not necessarily produce satisfactory results. The main reason for this is that complete charge separation is not a realistic condition for small molecules. As a general rule, we expect that as the coordination number decreases, the bonding becomes more directional and the fractional covalency increases.

v. Ionic bonding in the crystal

We now turn to the more interesting case, the ionically bonded crystal. To compute the total electrostatic contribution to the energy, we must sum both the attractive and repulsive interactions between all of the ions. We begin by assuming that the basis of our crystal has a total of n atoms, that there are n_i atoms of the i-th type, and that the charge on each atom is Z_i. The composition of the basis must be consistent with the rule of charge balance:

$$0 = \sum_{i=1}^{n} n_i Z_i. \tag{7.7}$$

Furthermore, if there are N_L lattice sites in the crystal and N_f formula units in each basis, then the total number of formula units in the crystal is $N_L N_f$ and the total number of atoms in the crystal is $n N_L$. We will use Bravais lattice vectors labeled $\vec{R}_k$ to specify each lattice site in the crystal and position vectors labeled $\vec{r}_{ij}$ to specify separations between two atoms (i and j) in the same basis. So that our result can be applied to all compounds with the same structure type, we normalize all possible interatomic spacings in the crystal by the shortest spacing, d_0. All of the interatomic separations can be written in terms of a dimensionless parameter, ρ, as we did when computing long range forces in the last chapter (see Chapter 6, Section B(iv)).

$$|\vec{R}_k + \vec{r}_{ij}| = \rho(\vec{R}_k + \vec{r}_{ij}) d_0.$$

With these definitions, we can sum all of the electrostatic interactions in the crystal.

$$E_c = N_L \frac{ke^2}{d_0} \left\{ \frac{1}{2} \sum_{i=1}^{n} \left[\sum_{\substack{j=1 \\ j \neq i}}^{n} \frac{Z_i Z_j}{\rho(\vec{r}_{ij})} + \sum_{k=1}^{N_L - 1} \left[\sum_{j=1}^{n} \frac{Z_i Z_j}{\rho(\vec{R}_k + \vec{r}_{ij})} \right] \right] \right\}. \tag{7.8}$$

The factor of N_L is introduced in the prefactor since there are N_L identical sums at each Bravais lattice point. The factor of 1/2 is included to account for the fact that each pairwise interaction is counted twice. The first term in the sum accounts for the interactions among atoms in the same basis. The second term accounts for the interactions between atoms that are not in the same basis. The terms in brackets depend only on the structure type and the charges on the constituents and, therefore, sum to a constant value for each crystal structure. This constant is usually referred to as the Madelung constant, A (the evaluation of this constant is described in detail in Appendix 7A) [20, 21]. If we normalize the total energy by the total number of formula units ($N_L N_f$) and take N_f as one, the energy per formula unit is:

$$E = A \frac{ke^2}{d_0}. \tag{7.9}$$

vi. *The reduced Madelung constant*
For most common structures, the Madelung constant has already been evaluated; a sample of such data is reproduced in Table 7.9. Although many forms of the Madelung constant are used in the literature, they are not interchangeable, and before attempting a lattice energy calculation, it is important to make sure

Table 7.9. *The Madelung constants for selected structures [19].*

compound	space group	a (Å)	b (Å)	c (Å)	$\beta°$	d_0 (Å)	A	α
$AlCl_3$	$C2/m$	5.93	10.24	6.17	108	2.2953	8.303	1.40
α-Al_2O_3	$R\bar{3}c$	4.76		13.01		1.8478	24.242	1.68
$BeCl_2$	$Ibam$	9.86	5.36	5.26		2.0170	4.086	1.36
BeO	$P6_3mc$	2.698		4.38		1.5987	6.368	1.64
BiSCl	$Pnma$	7.70	4.00	9.87		2.7226	10.388	
$CaCl_2$	$Pnnm$	6.25	6.44	4.21		2.7083	4.730	1.60
CaF_2	$Fm\bar{3}m$	534626				2.3604	5.03879	1.68
$CaTiO_3$	$Pm\bar{3}m$	3.84				1.9200	24.755	
$CdCl_2$	$R\bar{3}m$	3.86		17.50		2.6633	4.489	1.50
CdI_2	$P\bar{3}m1$	4.24		6.855		2.9882	4.3819	1.46
CsCl	$Pm\bar{3}m$					3.5706	1.76268	1.76
Cu_2O						1.8411	4.44249	1.48
$LaCl_3$	$P6_3/m$	7.483		4.375		2.9503	9.129	1.54
LaF_3	$P6_3/mmc$	4.148		7.354		2.3533	9.119	1.54
La_2O_3	$P\bar{3}m1$	3.937		6.1299		2.3711	24.179	1.63
LaOCl	$P4/nmm$	4.119		6.883		2.3964	10.923	
β-LaOF	$R\bar{3}m$	4.0507		20.213		2.4194	11.471	
γ-LaOF	$P4/nmm$	4.091		5.837		2.4214	11.3914	
$MgAl_2O_4$	$Fd\bar{3}m$	8.080				1.9173	31.475	
MgF_2	$P4_2/mnm$	4.623		3.052		1.9677	4.762	1.60
NaCl	$Fm\bar{3}m$					2.8138	1.74756	1.75
β-SiO_2	$P6_222$	5.02		5.48		1.6191	17.609	1.47
SiF_4	$I\bar{4}3m$	5.41				1.5461	12.489	1.25
$SrBr_2$	$Pnma$	11.44	4.31	9.22		3.1605	4.624	1.59
$TiCl_2$	$P\bar{3}m1$	3.561		5.875		2.5267	4.347	1.45
TiO_2 rutile	$P4_2/mnm$	4.5929		2.9591		1.9451	19.0803	1.60
TiO_2 anatase	$I4_1/amd$	3.785		9.514		1.9374	19.0691	1.60
TiO_2 brookite	$Pbca$	9.184	5.447	5.145		1.8424	18.066	1.60
UD_3	$Pm\bar{3}n$	6.64				2.0584	8.728	1.64
V_2O_5	$Pmmn$	11.519	3.564	4.373		1.5437	44.32	1.49
YCl_3	$C2/m$	6.92	11.94	6.44	111	2.5845	8.313	1.41
YF_3	$Pnma$	6.353	6.85	4.393		2.163	8.899	1.59
Y_2O_3	$Ia3$	10.604				2.2532	24.844	1.67
YOCl	$P4/nmm$	3.903		6.597		2.2844	10.916	
ZnO	$P6_3mc$	3.2495		5.2069		1.7964	5.99413	1.65
ZnS sphalerite	$F\bar{4}3m$					2.3409	6.55222	1.64
ZnS wurtzite	$P6_3mc$	3.819		6.246		2.3390	6.56292	1.64

that the constant is properly defined. For example, the values of A in Table 7.9 are determined based on the shortest interatomic spacing in the crystal. Other authors have used average spacings (in the case that the first coordination shell is not isotropic) and some authors normalize the constant against a lattice parameter rather than an interatomic distance. For binary structures, it is common to use a reduced Madelung constant, α:

$$\alpha = \frac{2A}{(n_1 + n_2)|Z_1 Z_2|}. \tag{7.10}$$

The reduced Madelung constant leads to a convenient expression for the total electrostatic energy which separates the chemical parameters such as charge (Z_i), stoichiometry (n_i), and ionic sizes (d_0) from the structural information in the Madelung constant:

$$E = \frac{k\, e^2 \alpha (n_1 + n_2) Z_1 Z_2}{2d_0}. \tag{7.11}$$

However, it is important to remember that Eqn. 7.11 applies only to binary compounds.

It is interesting to note that the reduced Madelung constants in Table 7.9 vary within a very small range, from $1.25 \leq \alpha \leq 1.76$. All other factors being equal, we would say that the structure with the highest α is the most stable. Two interesting trends are noteworthy. First, α generally increases with the coordination number of the structure. For example, $\alpha_{CsCl} > \alpha_{NaCl} > \alpha_{ZnS}$. Second, compounds with layered structures (more directional bonding) such as CdI_2, $CdCl_2$, and V_2O_5 have lower Madelung constants. This implies that the purely electrostatic contribution to the bonding is diminished while the covalent contribution is increased.

vii. *Calculating the equilibrium separation, the lattice energy, and the compressibility*
The total energy is written by modifying Eqn. 7.4 to include the ion formation energies and the electrostatic energy for all the atoms in the crystal. As an approximation for the Lennard-Jones portion of the energy, only the nearest neighbor contributions are included. Long range contributions can be added, but in most cases are insignificant. Our new expression for the energy as a function of separation for binary crystals is:

$$V(d) = I_M - E_X + \frac{\alpha Z_1 Z_2 (n_1 + n_2) k e^2}{2d} + 4NN\varepsilon \left[\left(\frac{\sigma}{d} \right)^{12} - \left(\frac{\sigma}{d} \right)^6 \right]. \tag{7.12}$$

In this formula, the number of nearest neighbors (NN), the reduced Madelung constant (α), and the interatomic separation (d) are the crystal structure sensitive parameters. Similarly, we can consider I_M, E_X, Z_1, Z_2, n_1, n_2, ε, and σ to be the parameters that reflect the chemistry of the compound. Note that because the Madelung constant varies over a relatively small range, variations in the chemistry of the compound affect the total energy much more than variations in the structural configuration.

Example 7.2

Use Eqn. 7.12, above, to compute the equilibrium spacing of the Na and Cl atoms in NaCl and the lattice energy of the crystal. Compare the computed values with the known values.

1. To find the equilibrium separation, we minimize Eqn. 7.12 with respect to d. Before beginning, we will make the following substitutions to simplify the calculation:

$$\eta = \frac{\sigma}{d} \quad \text{and} \quad C = \frac{\alpha Z_1 Z_2 (n_1 + n_2) k e^2}{2\sigma}.$$

2. We can simplify our numerical calculations by using only eV and Å as units. Under these conditions, the term ke^2 is equal to 14.4 eV-Å. This allows Eqn. 7.12 to be rewritten in the following way:

$$V(\eta) = I_M - E_X - C\eta + 4NN\varepsilon[\eta^{12} - \eta^6] \tag{7.13}$$

3. In our minimization of Eqn. 7.12, we equate $dV/d\eta$ with zero, use the approximation that when $\eta > 1$, $\eta^{11} \gg \eta^5$, and solve for d. (You should note that this same approximation is not valid for van der Waals bonding, where η is typically < 1. Here, the extra binding power of the Madelung attraction makes d smaller than σ and $\eta > 1$.)

$$\frac{dV}{d\eta}\bigg|_{\eta_0} = -C + 24NN\varepsilon[2\eta_0^{11} - \eta_0^5] = 0$$

$$\eta_o = \frac{\sigma}{d_0} = \left[\frac{C}{48NN\varepsilon}\right]^{1/11}$$

$$d_0 = \sigma\left[\frac{2\sigma 48NN\varepsilon}{\alpha Z_1 Z_2(n_1 + n_2)ke^2}\right]^{1/11}. \tag{7.14}$$

4. For a rock salt structure alkali halide, $n_1 = n_2 = Z_1 = Z_2 = 1$, $\alpha = 1.75$, and $NN = 6$. Therefore, the interatomic separation, d_0^{RS}, is:

$$d_0^{RS} = \sigma(11.4\sigma\varepsilon)^{1/11}. \tag{7.15}$$

5. Using this rule, we find that the equilibrium separation in NaCl ($\sigma = 3.0$, $\varepsilon = 0.0062$) is 2.61 Å. This is significantly larger than the separation calculated in the molecule and

---

deviates by -7.6% from the known value for the crystal, 2.81 Å. Computing for KCl ($\sigma = 3.28$, $\varepsilon = 0.0109$), we find that $d_0 = 3.02$ Å. This deviates by -4% from the known value of 3.15 Å.

6. Using these values of d_0, we can also calculate the lattice energy, U_L, or the cohesive energy, E_{NaCl}. The cohesive energy is the energy difference between the atoms at infinite separation and the ions arranged in the crystal.

$$
\begin{aligned}
E_{NaCl} &= I_M - E_X - A_0\eta + 4NN\varepsilon\eta^{12} \\
&= 5.14 - 3.71 - 9.67 + 0.796 \\
&= -7.44 \text{ eV} \\
U_L &= -(-9.67 + 0.796) = 8.87 \text{ eV.}
\end{aligned}
$$

7. The accepted lattice energy is 8.15 eV. It is reasonable that we have overestimated the lattice energy since we underestimated the interatomic spacing. If, instead, we use the accepted value for the interatomic separation, closer agreement is achieved.

We can also use the results from the method presented above to determine the lattice energy for MgO. Using the actual interatomic spacing (2.106 Å), $\sigma = 2.74$, and $\varepsilon = 0.0031$, we find that $U_L = -(-47.88 + 1.72) = 46$ eV. This is close enough to the accepted value of 40 eV to validate the ionic bonding description for crystalline MgO.

The bulk modulus is computed in a manner similar to that used for the inert gas crystals:

$$
B = -\Omega_p \frac{d^2 V(d)}{d\Omega_p^2}\bigg|_{d_0}, \tag{7.16}
$$

where Ω_p is the volume of a formula unit. For the rock salt structure, $a = 2d$ and $\Omega_p = (2d)^{3/4} = 2d^3$, so

$$
B = -\frac{1}{18d} \frac{d^2 V(d)}{dd^2}\bigg|_{d_0}. \tag{7.17}
$$

All that remains for the calculation is determining the differential of $V(d)$.

(C) Other factors that influence cohesion in ionic systems

i. Refinements to the model

There are a number of second order effects that can be included in our model [22]. First among these is the van der Waals attraction that we have been ignoring in our calculation of the interatomic separation. The term is:

$$4\varepsilon \sum_{i=1}^{N} \left(\frac{\sigma}{d_i}\right)^6.$$

(7.18)

Because this is an attractive force, it has the effect of decreasing the total energy. It is also possible to add a term that describes the (weak) attractions due to quadrapole interactions. This is an attractive term that varies as $1/d^8$. These factors account for no more than 5% of the total energy and, in most instances, significantly less than 5%.

The so-called 'zero-point' energy can also be included in refined models. This is the residual vibrational energy of the solid at absolute zero temperature. At 0 K, it is not possible for all of the vibrational modes of the crystal to occupy the lowest energy level. Instead, they pack into the lowest sequence of levels. Thus, when the energy per atom in a crystal is compared to the energy of isolated atoms at the same temperature, the zero-point energy is a positive contribution to the total energy. For an ideal Debye solid, it is approximately $2.25h\nu_{max}$, where ν_{max} is the frequency of the highest occupied vibrational mode of the crystal.

Because the van der Waals term is negative and the zero point term is positive, the two small effects tend to cancel each other out when they are included in the lattice energy calculation. Even when cancellation is incomplete, they represent only small portions of the total energy. For example, in alkali halides, the van der Waals components vary from 0.5% of the total energy in LiF to 5.0% in CsI. The zero-point energy, on the other hand, varies from 1.5% of the total energy in LiF to 0.5% in CsI. The contributions are even smaller when considering oxides and sulfides, where van der Waals components are less than 0.5% of the total energy and the zero-point energy is between 0.2 and 0.5% of the total energy.

ii. *The crystal field stabilization energy*

The crystal field stabilization energy [23, 24] can represent a much larger fraction of the total cohesive energy of the solid than the van der Waals and zero-point energy contributions. However, this effect is only important for crystals containing ions with partially filled d-orbitals.

Partially filled s- and p-orbitals interact so strongly with the neighboring atoms in a solid that they participate in ionic or covalent bonding. In contrast, partially filled f-orbitals are much closer to the nucleus. Because of this, they are well shielded from surrounding ligands and do not interact with neighboring atoms. The d-orbitals represent an intermediate situation. When a transition metal atom with d-orbitals is surrounded by a neighbor with only s and p valence electrons, the d-orbitals interact in a weak, but sometimes influential manner. The electrostatic interaction is treated as a perturbation to the simple ionic bond.

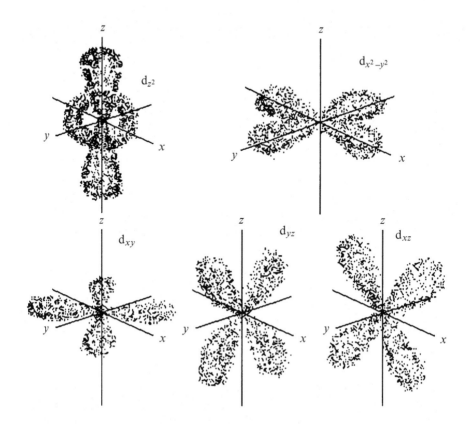

Figure 7.6. The distribution of electrons in the five d-orbitals. The d_{xy} orbitals lie in the x–y plane, the d_{yz} orbitals lie in the y–z plane, and the d_{xz} orbitals lie in the x–z plane.

The contribution to the total energy is known as the crystal-field stabilization energy (CFSE).

The interaction can be understood geometrically by referring to the spatial distribution of electrons in the d-orbitals (see Fig. 7.6). Consider, for example, octahedrally coordinated metal cations which are said to be situated in an octahedral 'crystal field'. This happens whenever the metal cations occupy the octahedral interstices of a eutactic arrangement of anions. Two of the d-orbitals, the d_{z^2} and the $d_{x^2-y^2}$, point directly at the negatively charged ligands, while the other three point between the anions, at the next nearest neighbor cations. Thus, valence electron–nearest neighbor anion repulsions will destabilize the $d_{x^2-y^2}$ and the d_{z^2} orbitals with respect to the d_{xy}, d_{yz}, and d_{xz} orbitals. The d_{xy}, d_{yz}, and d_{xz} oribitals will, in turn, be stabilized by the attractive valence electron–next nearest neighbor cation interaction. This not only breaks the degeneracy of the d-levels

Table 7.10. *Relative crystal field stabilization energies in an octahedral field [23].*

electrons	0	1	2	3	4	5	6	7	8	9	10
high spin CFSE	0	$4\Delta_o$	$8\Delta_o$	$12\Delta_o$	$6\Delta_o$	0	$4\Delta_o$	$8\Delta_o$	$12\Delta_o$	$6\Delta_o$	0
low spin CFSE	0	$4\Delta_o$	$8\Delta_o$	$12\Delta_o$	$16\Delta_o$	$20\Delta_o$	$24\Delta_o$	$18\Delta_o$	$12\Delta_o$	$6\Delta_o$	0
difference	0	0	0	0	$10\Delta_o$	$20\Delta_o$	$20\Delta_o$	$10\Delta_o$	0	0	0

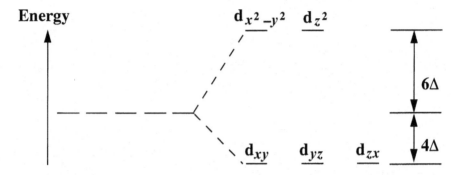

Figure 7.7. The influence of an octahedral field on the energy levels of electrons in d-orbitals.

(see Fig. 7.7), but it may also lower the total energy, depending on the manner in which the orbitals are occupied.

The energy difference between the separated d-levels is taken to be 10Δ, where Δ is an energy that scales with the strength of the crystal field. In general, we assume that increasing the charge on the ions or decreasing the interatomic spacing leads to a relatively higher value of Δ. The relative CFSE of 10Δ can be as high as 1 eV for oxides and hydrates of first row transition elements and as high as 2 eV in compounds with trivalent ions. There are two possible distributions for the d-electrons in the d-orbitals that lead to differences in the stabilization energy, as illustrated in Table 7.10. The high spin case occurs when electron–electron repulsions are great enough to force all five d-levels to be singly occupied before electrons pair in the same orbitals. This occurs when 10Δ is small compared to the electron–electron repulsions. In the low spin case, electrons doubly occupy the lowest levels before filling higher levels. This situation occurs when 10Δ is larger than the electron–electron repulsions.

The CFSE also depends on the coordination environment and can, therefore, influence the crystal structure assumed by a transition metal compound. For example, the tetrahedral and cubic crystal fields lower the energy of the d_{z^2}

Table 7.11. *Comparison of high spin CFSE for octahedral and tetrahedral sites [24].*

number of electrons	CFSE	
	octahedral	tetrahedral
0	0	0
1	$4\Delta_o$	$6\Delta_t$
2	$8\Delta_o$	$12\Delta_t$
3	$12\Delta_o$	$8\Delta_t$
4	$6\Delta_o$	$4\Delta_t$
5	0	0
6	$4\Delta_o$	$6\Delta_t$
7	$8\Delta_o$	$12\Delta_t$
8	$12\Delta_o$	$8\Delta_t$
9	$6\Delta_o$	$4\Delta_t$
10	0	0

Note: $\Delta_t \leq 1/2\ \Delta_o$ for a given ion with the same ligands.

and $d_{x^2-y^2}$ orbitals with respect to the d_{xy}, d_{yz}, and d_{xz} orbitals. So, an ion with three d electrons in the high spin configuration could be better stabilized in an octahedral site than a tetrahedral site. For example, the CFSE is known to influence the distribution of the A and B cations in the spinel structure. A comparison of the relative crystal field stabilization energies that can be established at the different sites is provided in Table 7.11.

The CFSE contributes to the lattice energy and adds additional stability to the crystal that is not expected on the basis of the simple ionic model outlined in this chapter. The data in Fig. 7.8 clearly illustrate that the 'excess' lattice energy depends on the number of d-electrons. Although it is difficult to predict with quantitative accuracy, the CFSE can make up as much as 8% of the lattice energy. Note that the stabilization increases with anion size.

iii. *Polarization*
The ionic bonding model applies specifically to hard sphere ions with integer charges. This limiting condition is probably never actually realized. In every case, it is reasonable to expect that the charge density around an ion is distorted to some degree. These distortions are called polarization [25]. Polarization can be

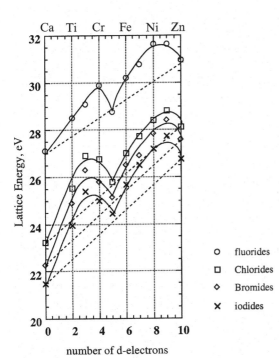

Figure 7.8. The lattice energies for the dihalides of the first row transition metals. For each series, the lattice energy expected on the basis of the ionic model is shown by the dashed line; the data points show the observed lattice energy. The excess lattice energy that occurs for cases other than zero, five, or ten d-electrons is a manifestation of the CFSE [23].

thought of as the link that bridges the ionic and covalent models. In the high polarization limit, valence electrons from one atom's 'sphere of influence' are actually shared between neighboring atoms; in this case, we would say that a covalent bond is formed. In the ionic model, the valence electron density should be spherically distributed around the cation. In most cases, it would probably be most accurate to visualize the charge density as being between these two limits. Here we are reminded of the continuous range of behavior represented on Ketelaar's triangle, as described in Chapter 1.

Although these polarization effects are not easily quantified, they are generally associated with increased covalency in bonding. Polarization increases as coordination number decreases and as the difference between the anion and cation polarizabilities increases. A proposed polarizability scale is shown in Table 7.12 [25]. The periodic trends in polarizability can be rationalized using the same arguments to which we have appealed in the past to explain electronegativity, size, and metallicity. Basically, as the size and negative charge increase, so does the polarizability.

Increasing polarization or covalency always leads to a decrease in the interatomic distance and an increase in the lattice energy, making the compound more stable. However, for a crystal with a significant polarization component in

Table 7.12. *Relative polarizability of ions (au) [25].*

He			Li$^+$	0.08	Be^{+2}	0.03	B^{+3}	0.01		
Ne	O^{-2}	3.1	F$^-$	1.0	Na$^+$ 0.2		Mg^{+2} 0.1		Al^{+3} 0.07	Si^{+4} 0.04
Ar	S^{-2}	7.3	Cl$^-$	3.1	K$^+$ 0.9		Ca^{+2} 0.6		Sc^{+3} 0.4	Ti^{+4} 0.3
Kr	Se^{-2}	7.5	Br$^-$	4.2	Rb$^+$ 1.8		Sr^{+2} 1.4		Y^{+3} 1.0	
Xe	Te^{-2}	9.6	I$^-$	6.3	Cs$^+$ 2.8		Ba^{+2} 2.1		La^{+3} 1.6	Ce^{+4} 1.2

the bonding, lattice energies calculated on the basis of a purely ionic model usually agree well with measured values. This is due to the self-compensating nature of the errors in the ionic model. A bond that has significant polarization is shorter than a purely ionic bond, an effect that alone would add to the apparent stability. However, polarization also equalizes the opposing ionic charges by moving some electron density from the anion to the cation. This effectively reduces the ionic component of the lattice energy. However, effective ionic charges are neither generally known nor accounted for in calculations. Thus, calculations using integer charges overestimate the true ionic bond energy and unwittingly compensate for the (unaccounted for) excess energy due to covalency. In fact, one might say that the self-compensating errors in the ionic bonding model are more responsible for its success than its accurate depiction of nature is. Problem 8 at the conclusion of this chapter illustrates how accurate information can be calculated from a model, even when the physical description is obviously incorrect.

(D) Predicting the structures of ionic compounds

i. *The minimum energy structure*

Because our total energy function (Eqn. 7.12) includes parameters sensitive to the crystal structure, we should be able to use it to compute the potential energy of trial structures and select the lowest energy structure. For example, if we evaluate the energy of KCl in three trial structures and take the lowest energy structure to be $E=0$, we find the results summarized in Table 7.13.

Without further consideration, our results look great because they predict that the rock salt structure is the most stable and this is actually the structure of KCl. However, a similar calculation gives the same results for all alkali halides, including the ones that crystallize in the CsCl structure. Furthermore, the energy difference between the NaCl and CsCl structures for KCl is not really significant when you consider the expected accuracy of the calculation. This illustrates an important weakness of this approach; the energy differences between

Table 7.13. *Energy of KCl in three trial structures.*

structure	NN	α	d_0	$E_{(relative)}$ eV
ZnS	4	1.64	2.95	+0.31
NaCl	6	1.75	3.04	0.0
CsCl	8	1.76	3.11	+0.05

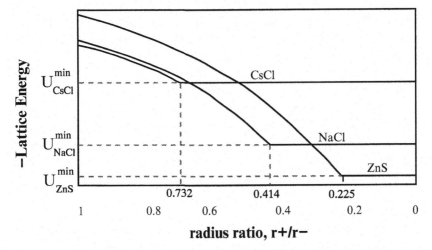

Figure 7.9. The lattice energy v. radius ratio for three trial structures [26].

different structures are often very small fractions of the total energy. Again, we find that the cohesive energy of a compound is relatively insensitive to the exact arrangement of the atoms. This observation implies that useful predictive calculations must be more accurate than these small differences. It must also be noted that the equilibrium structure is the one with the minimum *free energy*. Therefore, since the model does not include entropy, it is only expected to be correct at 0 K or when the energy difference is large enough to overwhelm entropic differences.

Another approach to determining the equilibrium energy (although still not entirely realistic) is to compute the energy of each structure as a function of radius ratio and assume that the actual radii are the same in all structures. These data are shown in Fig. 7.9.

First, note that the most favorable lattice energy for each specified structure is achieved at the radius ratios predicted on the basis of the simple rigid sphere model used in the first chapter. However, the CsCl configuration is only marginally preferable to NaCl and it becomes unpreferred if the radius expansion (for an ion in eight-fold coordination) is included. Second, note that as the radius ratio decreases below the critical value, the lattice energy does not decrease any further. It remains constant because the anions (treated as hard spheres) reach the point of contact, and when their positions become fixed, so do all of the interatomic separations and energies. This would not necessarily be the case if the cation were permitted to occupy an asymmetric position in the interstitial cavity between the anions. Third, and perhaps most interesting, we see the ill-defined nature of the critical values. For example, when the radius ratio falls below the critical value of 0.41 for the boundary between six- and four-coordinate structures, the stability of the ZnS structure does not surpass that of the NaCl structure until the radius ratio falls below 0.3.

The examples above were intended to illustrate the difficulties associated with making accurate predictions based on lattice energy models. Even with improved models, we face another problem. When testing trial configurations, we can only hope to pick the correct structure if it is already included in the set of structures to be tested. While it is difficult to be sure that you have explored every possible configuration, statistical approaches such as the Monte Carlo method can be used to sample a wide range of configuration space.

ii. *Ionic radius and coordination*

The primary problem with the radius ratio rules is that ions are not rigid and, thus, fixed radii are not realistic. Using our simple model, it is possible to examine the change in ionic radius with coordination number. We begin by assuming that an MX compound can exist in both the rock salt and CsCl structures and that the anion size is constant. We need only calculate the interatomic separation. If:

$$U = -\frac{\alpha Z_1 Z_2 (n_1 + n_2) k e^2}{2d} + 4NN\varepsilon \left(\frac{\sigma}{d}\right)^{12},$$

then let $\sigma/d = \eta$ and $n_1 = n_2 = Z_1 = Z_2 = 1$ so that we can write:

$$U = -\frac{\alpha k e^2 \eta}{\sigma} + 4NN\varepsilon \eta^{12}.$$

Recalling that at d_0, the derivative of U with respect to d is equal to 0,

$$\frac{\alpha k e^2}{\sigma} = 48NN\varepsilon \eta^{11}.$$

$$d_0 = \sigma \left(\frac{48NN\varepsilon\sigma}{\alpha k e^2}\right)^{1/11}. \tag{7.19}$$

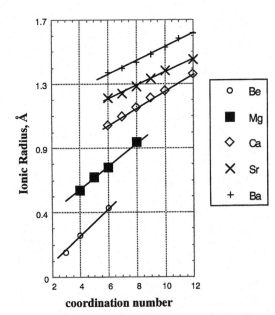

Figure 7.10. The influence of coordination number on the ionic radius of the divalent alkaline earth metal cations, as determined from X-ray diffraction measurements [27, 28].

We can write the ratio of d_0 in the eight-coordinate structure to d_0 in the six-coordinate structure as:

$$\frac{d_0^{8f}}{d_0^{6f}} = \left(\frac{NN_{CsCl}}{NN_{rs}} \cdot \frac{\alpha_{rs}}{\alpha_{CsCl}}\right)^{1/11} = \left(\frac{8}{6} \cdot \frac{1.75}{1.76}\right)^{1/11} = 1.026. \tag{7.20}$$

This numerical result has the physical implication that the interatomic separation 'expands' by 3% during the switch from a six- to an eight-coordinate configuration. We can use the same method to find that the ion contracts by 3% when it goes from a six- to a four-coordinate configuration.

   These observations justify Shannon's [27, 28] development of tables of ionic radii that depend on coordination number. These data can be found in Table 7B.1 in Appendix 7B. It should be emphasized that the data in Shannon's ionic radii table are based on experimental X-ray measurements, not the calculations described above. A small portion of Shannon's data, plotted in Fig. 7.10, emphasizes the idea that cations appear to shrink as the coordination number is reduced. It should be pointed out that anion–anion repulsions (next-nearest-neighbor interactions) probably have as much to do with these so-called expansions and contractions as any nearest neighbor interactions. In any case, the lesson from these observations is that the hard sphere model for atoms and ions has limited validity.

Table 7.14. *Periodic trends in the structures of oxides with group and period [29].*

I	II	III	IV	V	VI	VII	VIII
Li_2O	BeO	B_2O_3	CO_2	NO	O_2	F_2O_2	
Na_2O	MgO	Al_2O_3	SiO_2	P_2O_5	SO_3	ClO_2	
K_2O	CaO	Ga_2O_3	GeO_2	As_2O_5	SeO_3	BrO_2	
Rb_2O	SrO	In_2O_3	SnO_2	Sb_2O_5	TeO_3	I_2O_5	XeO_4
	ionic				**polymeric**		**molecular**
anti-fluorite	rock salt	corundum bixbyite	rutile	α-quartz	layered	chains	discrete

iii. *The influence of polarization on structure*

As noted earlier, increasing polarization of the charge around ions in a crystal implies increased covalency. The increase in covalency occurs as the fractional ionicity of the bonds decreases or the electronegativity difference diminishes. As bonding changes from ionic to covalent, structures change from isotropic close-packed to directional polymeric, and eventually to discrete molecules. In this context, we define *polymeric structures* as those composed only of vertex-sharing polyhedra (and, therefore, not closely packed) which lack a three-dimensionally bonded network; in other words, structures where covalently bonded layers or chains are held together by relatively weak intermolecular forces. This periodic change in structure types is illustrated in Table 7.14, which presents oxide compounds arranged according to the secondary element's position on the periodic chart. Compounds that O forms with elements on the left-hand side of the chart crystallize in close-packed structures with high coordination numbers (anti-fluorite, rock salt, rutile, corundum, bixbyite). These compounds possess elements with large electronegativity differences and exhibit highly ionic bonding. As the electronegativity difference diminishes, electronic polarization effects increase and structures appear which are built from corner-sharing tetrahedra (α-quartz) and loosely bound layers (P_2O_5, TeO_2) and chains (SeO_2). Finally, when the electronegativity difference is very small, only discrete molecules are formed.

The atomic size affects both electronegativity and polarization. Thus, in addition to electronegativity difference, we should consider the steric radius ratio effects. In general, as radius ratios decrease, there is an increase in polarization and a decrease in the coordination number. Thus, structures become increasingly discrete. This trend is illustrated schematically in Fig. 7.11.

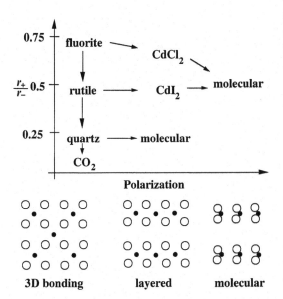

Figure 7.11. Polarization and radius ratios conspire to influence the crystal structure [30].

(E) Electronegativity scales

Clearly, electronegativity (or more exactly, electronegativity difference) has a considerable influence on the polarity of bonds in crystals. In this section, we revisit the concept of electronegativity, explaining its origin and some newer, alternative views. Pauling's original electronegativities were based on the observation that the energy of bonds between unlike atoms, E_{AB}, was usually greater than the average of the homopolar bond energies, E_{AA} and E_{BB}. He reasoned that this extra 'ionic resonance energy' came from the difference between the ability of the A and B atoms to attract electrons. The resonance energy, Δ, is calculated according to Eqn. 7.21. Pauling's electronegativity difference is then given by the empirical relationship in Eqn. 7.22.

$$\Delta = E_{AB} - \frac{E_{AA} + E_{BB}}{2} \tag{7.21}$$

$$|X_A - X_B| = 0.208\sqrt{\Delta}. \tag{7.22}$$

Due to the data available at the time, Pauling was able to determine electronegativities of a limited number of elements and with only limited accuracy. Over the years, as additional and more accurate thermochemical data became available, his tables were updated. Allred [34] presented an updated list in 1961 that is often used today. Most of these values are listed in the second column of Table 7.15.

Table 7.15 *Comparison of electronegativity scales [29].*

atom	Pauling	A&R (force)	spec (Allen)	Mulliken
H	2.20	2.20	2.300	3.059
Li	0.98	0.97	0.912	1.282
Be	1.57	1.47	1.576	1.987
B	2.04	2.01	2.051	1.828
C	2.55	2.5	2.544	2.671
N	3.14	3.07	3.066	3.083
O	3.44	3.50	3.610	3.215
F	3.98	4.10	4.193	4.438
Ne			4.787	4.597
Na	0.93	1.01	0.869	1.212
Mg	1.31	1.23	1.293	1.630
Al	1.61	1.47	1.613	1.373
Si	1.90	1.74	1.916	2.033
P	2.19	2.06	2.253	2.394
S	2.58	2.44	2.589	2.651
Cl	3.16	2.83	2.869	3.535
Ar			3.242	3.359
K	0.82	0.91	0.734	1.032
Ca	1.00	1.04	1.034	1.303
Ga	1.81	1.82	1.756	1.343
Ge	2.01	2.02	1.994	1.949
As	2.18	2.20	2.211	2.259
Se	2.55	2.48	2.424	2.509
Br	2.96	2.74	2.685	3.236
Kr			2.966	2.984
Rb	0.82	0.89	0.706	0.994
Sr	0.95	0.99	0.963	1.214
In	1.75	1.49	1.656	1.298
Sn	1.96	1.72	1.824	1.833
Sb	2.05	1.82	1.984	2.061
Te	2.10	2.01	2.158	2.341
I	2.66	2.21	2.359	2.880
Xe			2.582	2.586

A complete list of the updated values can be found in Appendix 7C (see Fig. 7C.1).

In 1934, Mulliken [32] proposed an alternative and more quantitative definition of the electronegativity. Specifically, he proposed that the electron attracting ability of an atom is the average of the ionization energy and the electron affinity, as given in Eqn. 7.23.

$$X_A = \frac{1}{2}(I_A + E_A).$$
(7.23)

Electronegativities based on this definition are listed in the last column of Table 7.15.

Another physically based definition of the electronegativity was proposed by Allred and Rochow in 1958 [33]. In this case, the electronegativity is scaled with the force of attraction between a nucleus and an electron on a bonded atom.

$$\text{Force} = \frac{e^2 Z_{\text{eff}}}{r^2}.$$
(7.24)

In this expression, Z_{eff} is the effective nuclear charge, or the charge seen by an electron outside the atom. This effective charge is much smaller than the actual nuclear charge because of the screening of the core electrons which lie between the bonding electron and the nucleus. The effective nuclear charge was determined using Slater's method [35]. Pauling's covalent radii were used for r, the distance from the nucleus to the bonding electron. In order to put this on the same scale as the Pauling electronegativities, constants of proportionality were selected so the electronegativity varies linearly with Z_{eff}/r^2:

$$X = 0.359 \frac{Z_{\text{eff}}}{r^2} + 0.744.$$
(7.25)

Using this definition, the electronegativity is not an energy.

The values of electronegativity computed in this way are given in Table 7.15. One interesting ramification of this definition is that it implies that the electronegativity will change with the size of the atom. Since the size of an atom changes with its oxidation state and coordination, this implies that electronegativity is not a fundamental atomic quantity.

More recently, Allen [29] proposed the spectroscopic electronegativity. In this case, as with Mulliken's definition, the electronegativity is an energy on a per electron basis.

$$X = \frac{m\varepsilon_p + n\varepsilon_s}{m+n}.$$
(7.26)

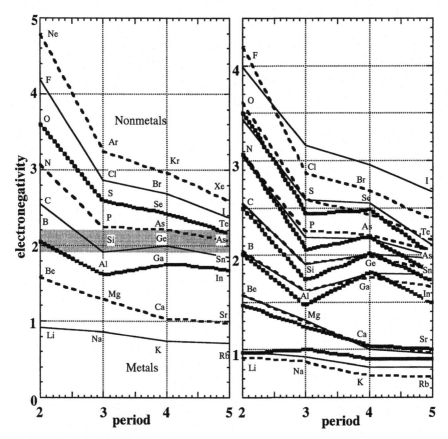

Figure 7.12. Electronegativity v. period. Left: Allen's spectroscopic electronegativities. The metalloid band is shaded. Right: comparison of the Pauling (thin line), Allred & Rochow (thickest line), and spectroscopic (medium thickness broken line) electronegativities [29].

In Eqn. 7.26, m is the number of p valence electrons, n is the number of s valence electrons, and ε_s and ε_p are the multiplet averaged, one electron, total energy differences between a ground state neutral and a singly ionized atom (these energies have been spectroscopically determined and are tabulated). The electronegativities computed in this way are listed in Table 7.15 and presented graphically in Fig. 7.12a. Note that a narrow band of electronegativity values includes all of the metalloid elements and separates the nonmetals (above) from the metals (below). In Fig. 7.12b, the spectroscopic electronegativities (medium-thick broken line) are compared to the Pauling (thinnest solid line) and Allred & Rochow (thickest line) values. In many cases, these values are similar. However, note that the spectroscopic and Allred & Rochow values produce oscillating electronegativities in the third, fourth, and fifth group; the Pauling values do this only in the fourth group. To conclude this discussion, it should be mentioned that it is the Pauling values that are used most frequently and should be considered suitable for most applications.

(F) The correlation of physical models with the phenomenological trends

In section A, we identified three phenomenological trends. First, ionic bonds form between atoms with large electronegativity differences. Second, the strength of a bond is determined by its length and the charges on the bonded atoms. Third, the geometric arrangement of atoms is determined by their relative ionic sizes.

To understand why ionic bonds require a large electronegativity difference, reconsider the Mulliken definition of electronegativity, X:

$$X = \frac{1}{2}(I + E). \tag{7.27}$$

By defining electronegativity in terms of ionization potential and electron affinity, we can see that a large electronegativity difference implies that one atom has a relatively small ionization energy and electron affinity (the electropositive atom), while the other has a relatively large ionization energy and electron affinity (the electronegative atom). Thus, a large electronegativity difference means that the energy needed to create ions $(I_M - E_X)$ is small and ionic bonds are energetically favorable.

The variation of bond strength with interatomic separation and ion charge follows from the fact that the largest component of the bond energy is the electrostatic energy:

$$V_c = -\frac{\alpha Z_1 Z_2 (n_1 + n_2) k e^2}{2d}.$$

This energy increases (becomes more negative) with increasing charge (Z_1 & Z_2) and diminishing separation, d.

Finally, we saw that the relative size of the ions affects the structures that are adopted. In general, the closest packed structures with the highest Madelung constant are most likely to form. However, repulsions from next nearest neighbor anions prevent a continuous increase in bond strength from a continued decrease in d (see Fig. 7.9). Eventually, lower coordination structures are adopted, even though they have lower Madelung constants, in order to reduce the repulsion from next nearest neighbor anions. We conclude this chapter by noting that the total energy of an ionic crystal is extraordinarily large compared to the energy differences between possible structures. Therefore, the ability to predict structures based on energy models relies on achieving an accurate description of electrostatic effects that are not well described by point charge approximations.

(G) Pair potential calculations of defect properties in ionic compounds

As described in Section D of Chapter 6, pair potential methods can be used to compute properties that are difficult to measure. Appropriate pair potentials for ionic materials must obviously contain an electrostatic component, a short-range repulsion, and an attractive van der Waals component. It is also possible to incorporate polarization effects in pair potential calculations by using the *shell model* [36]. Within the framework of this model, polarizable atoms are assumed to consist of a negatively charged shell and a positively charged core connected by a spring; the polarizability of the ion is specified by adjusting the spring constant. Regardless of the exact form of the potential, the parameters are chosen so that a model using these values reproduces known properties. It is optimistically assumed that the same model will then be capable of computing unknown properties. For example, potentials for alkaline earth oxides with the rock salt structure have been used to compute cohesive energies, phonon dispersion curves, and the energies of substitutional impurities [36].

There are many results available in the literature based on calculations of this type; one exemplary study of the defect properties of NiO is selected here to illustrate the results that can be obtained form these models [37]. Nickel oxide has the rock salt structure and Fig. 7.13 illustrates a $\Sigma5$ boundary, projected along the common [001] axis. The energy associated with this boundary plane is computed in a manner analogous to that used to compute free surface energies, described in Section D of Chapter 6. With the global geometry of the bicrystal and boundary plane fixed, the atomic positions are adjusted by an iterative computer algorithm that finds the minimum energy configuration. Taking the ideal crystal as the zero energy reference state, this particular boundary was found to have an energy of 1.88 J/m^2.

Computing the energies of aperiodic features, such as point defects, requires a special strategy such as the Mott–Littleton [38] approach. Basically, the crystal is divided into two segments. In the portion of the crystal nearest the defect, each atom is treated explicitly. The remainder of the crystal is treated as a polarizable dielectric continuum. For example, if we consider the dark atom near the boundary plane in Fig. 7.13, each atom within the sphere is treated individually and

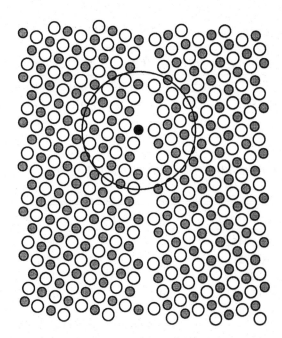

Figure 7.13. The symmetric Σ5 tilt boundary projected along the common [001] axis. this boundary has a 36.9° tilt around the [001] axis and the boundary is in the (310) plane. The dark spots correspond to cation positions and the light spots to anion positions [37].

interacts with the others within the sphere according to a pair potential. The segments of the model outside of the sphere are treated as a continuum.

Using a Mott–Littleton approach, Duffy and Tasker [37] computed the formation energies of O vacancies, Ni vacancies, and substitutional impurities both in the bulk and at several symmetric tilt grain boundaries. The defect–grain boundary interaction energy is defined as the difference between the defect energy in the bulk material and the defect energy in the grain boundary. The results, summarized in Table 7.16, illustrate that all of the grain boundaries considered attract the various point defects (the energies of the defects are lower at the boundary). Thus, the concentration of all defects is apparently enhanced in the grain boundary region.

(H) Problems

(1) In the physical description of the ionic bond that was developed in this chapter, we used a term from the Lennard-Jones model to account for short-range repulsions. This was chosen mainly for consistency. However, in practice, it is common to use a repulsive parameter of the form:

$$V_{\text{rep}}(r_{ij}) = A_{ij} e^{(-r_{ij}/\rho_{ij})}$$

Table 7.16. *Defect–grain boundary interaction energies (eV) in NiO [37].*

boundary	Ni vacancy	O vacancy	hole (Ni^{3+})	Co^{2+}	Al^{3+}	Ce^{4+}
Σ5 (310)/[001]	−0.36	−0.3	−0.25	−0.09	−0.18	−1.15
Σ25 (320)/[001]	−1.66	−1.25	−0.45	−0.22	−0.39	−1.73
Σ11 (211)/[001]	−0.95	−0.98	−0.23	−0.12	−0.24	−0.41

Table 7.17. *Born–Mayer–Huggins potential energy parameters for selected contacts.*

contact	A_{ij} (eV)	ρ_{ij} (eV)	C_{ij} (eV-Å^6)	source
Mg–O	1428.5	0.2945	0	[39]
Sr–O	959.1	0.3721	0	[39]
Ba–O	905.7	0.3976	0	[39]
Al–O	1460.3	0.29912	0	[40]
O–O	22764.0	0.149	27.88	[40]

where r_{ij} is the distance between the ith and jth atom and A_{ij} and ρ_{ij} are empirically derived parameters for specific atom pairs (see Table 7.17 above). The parameters are chosen so that a model using these values reproduces known properties. It is optimistically assumed that the same model will then be capable of computing unknown properties. When combined with an electrostatic term and a van der Waals term, this is the so-called 'Born–Mayer–Huggins' (BMH) form of the potential energy [39]. While the simple calculations described in this chapter can be carried out by hand, calculations involving more complex structures or defects are usually carried out by computer using the BMH potential.

(i) Combine the repulsive parameter above with the Madelung energy (ignoring the van der Waals part) and compute the equilibrium lattice constants and lattice energies for MgO, SrO, and BaO using only nearest neighbor repulsions. How do the values compare with the known values?

(ii) Repeat the first part, but include the O–O repulsions.

(iii) Next, include the attractive van der Waals component $(-C_{ij}/r^6)$.

(2) In 1956, Kapustinskii [41] noted that the lattice energy of a binary compound could be computed according to the following expression:

$$U = \frac{12.56(n_1 + n_2)Z_1Z_2}{r_c + r_a}\left(1 - \frac{0.345}{r_c + r_a}\right) \text{eV}$$

where the bond distance is assumed to be the sum of the cation and anion radii, $r_c + r_a$, in octahedral coordination. First, use Kapustinskii's equation to compute the lattice energies of MgO, SrO, and BaO. Repeat for ZnS and for CsCl. Compare this equation to that derived in problem 1 for the lattice energy and explain why it works, even though it does not depend on the coordination number or the Madelung constant.

(3) The normal spinel and inverse spinel structures were described in Chapter 4. Explain why $MgAl_2O_4$ is a normal spinel while $NiAl_2O_4$ is an inverse spinel. Why is magnetite (Fe_3O_4) an inverse spinel?

(4) In this exercise, we will apply the ionic bonding model to BaO.

(i) Begin by computing the equilibrium spacing between Ba and O and the lattice energy of BaO in the ZnS, NaCl, and CsCl structure.

(ii) On the basis of this data, which structure is most stable?

(iii) On the basis of the radius ratio rules, which structure should BaO take?

(iv) Assuming that the O radius is the same in all of the structures, how does the Ba ion radius change with structure?

(v) How do these numbers for the radii compare with those listed by Shannon?

(vi) Compute the lattice energy of BaO using a Born–Haber cycle (thermo-chemical cycle) and compare it to the values computed in (i).

(vii) In reality, BaO crystallizes in the NaCl structure and the equilibrium separation between Ba and O is 2.77 Å. Using this information, recalculate the lattice energy (substituting the experimental value of d_0 into your formula for the lattice energy) and compare it to that given by the thermo-chemical cycle.

(viii) Finally, comment on the performance of the ionic model and the likely sources of error.

(5) In this exercise, we will apply our model of the ionic bond to CaF_2.

(i) Begin by computing the equilibrium spacing between Ca and F, the equi-librium lattice constant, and the lattice energy, assuming that CaF_2 crystal-lizes in the fluorite structure. Compare the computed values to the known values and assess the validity of the model.

(ii) Compute the lattice energy of CaF_2 using a Born–Haber cycle (thermo-chemical cycle) and compare it to the value computed in (i).

(6) Consider an ionically bound compound, MX, that can crystallize in either the rock salt structure (M is octahedrally coordinated) or the zinc blende struc-ture (M is tetrahedrally coordinated). Using the following expression for the lattice energy:

$$-U(d) = \frac{-\alpha Z_1 Z_2 (n_1 + n_2) k e^2}{2d} + 4NN\varepsilon \left(\frac{\sigma}{d}\right)^{12}.$$

(i) Compute the ratio of d_0 in the rock salt structure to d_0 in the zinc blende structure where d_0 is the anion–cation bond distance.

(ii) Explain how the result of your calculation supports or refutes the hard sphere model (the assumption that the bond distance is the sum of the anion and cation radii).

(7) Madelung summations

(i) Examine the convergence of the Madelung constant for a rock salt crystal by computing the first ten nonidentical terms of the sum. To do this, note that the positive and negative atoms alternately occupy the nodes of a cubic P lattice. So, if we know the number of nth nearest neighbors, the distance to these neighbors, and the sign of the charge, we can compute the sum. (You might want to review Example 2.1 in Chapter 2). Plot a running sum of α versus the distance from the origin.

(ii) The sum in the Madelung constant converges more rapidly if it is taken over charge neutral volumes. In this case, the Madelung constant is computed by assuming that the neighbors form nested cubes around a single lattice point and only the charge within the largest cube is included. Calculate the Madelung constant using only the first 'coordination cube', then the first and second, and, finally, the first three cubes.

(iii) More recently, Wolf [*Phys. Rev. Lett.* **68** (1992) 3315] proposed an alternative scheme that simplifies this calculation. Explain how and why Wolf's method works and use it to compute the first six terms in the sum. Plot a running sum of α versus the distance from the origin.

(8) The metal, Li, crystallizes in the bcc structure, $a = 3.49$ Å. Assume that half of the Li atoms donate an electron to the other half, so that $Li+$ cations and Li$-$ anions are ordered on the bcc lattice, essentially forming a binary compound with the CsCl structure. Using the observed lattice constant for Li metal and the assumption that this new form of Li (the fictional 'dilithium') is ionically bound, compute its cohesive energy and compare it to the experimental value ($E_{Li} = 1.63$ eV/at). The electron affinity of Li is 0.62 eV.

(9) C_{60}, buckminster fullerene, is a spherical molecule of 60 C atoms that in many ways acts as a large, single atom. For the purpose of this exercise, ignore the molecular characteristics of C_{60} and treat it as a new atom with its own atomic properties such as radius (r_{60}), atomic scattering factor (f_{60}), electron affinity (E_{60}), etc.

(i) C_{60} crystallizes in the fcc structure (one C_{60} 'atom' at each of the fcc lattice sites). Assume that you are recording a powder diffraction pattern of pure fcc C_{60} (the cubic lattice parameter, $a_0 = 14.2$ Å) with $Cu_{k\alpha}$ radiation ($\lambda = 1.54$ Å). At what values of 2Θ would you expect to observe the first three Bragg reflections?

(ii) When C_{60} and Rb are reacted in a 1:1 ratio, the resulting compound still has an fcc lattice with $a_0 = 14.2$ Å, but the Rb atoms are in the interstitial positions. If they occupy the octahedral sites, then we would say that RbC_{60} has the rock salt (B1 or NaCl) structure. If, on the other hand, the Rb atoms occupy half of the tetrahedral sites in an ordered arrangement, we would say that RbC_{60} has the sphalerite (B3 or ZnS) structure. Explain how the first three peaks in the powder diffraction pattern could (or could not) be used to distinguish between these two possibilities.

(iii) Based on the fact that Rb was incorporated into the crystal without a change in the volume of the unit cell and that the ionic radius of Rb^+ is 1.7 Å, can you determine which interstitial site the Rb occupies?

(iv) Using the data below, and the assumption that RbC_{60} is an ionically bound compound with either the rock salt or sphalerite structure and a cubic lattice parameter of 14.2 Å, determine if these structures are stable and if so, which of the two is more stable.

Data:

Lennard-Jones parameter:

	σ(Å)	ε (eV)
C_{60}	9.21	0.105

Electron affinity of $C_{60} = 2.6$ eV.

(10) PbO_2 can crystallize in either the rutile (C4) or the fluorite (C1) structure. Using the model for ionic bonding, and assuming that the energy of the crystal depends on the bond length, compute the ratio of the Pb–O bond length in the fluorite structure (d_0^F) to the Pb–O bond length in the rutile structure (d_0^R): $d_0^F/(d_0^R)$. (Assume that in each structure, the nearest neighbor Pb–O bonds are all the same length, the Madelung constant for the rutile structure is 1.60, and the Madelung constant for the fluorite structure is 1.68.) Which structure of PbO_2 do you think is more stable at high pressure? Explain your reasoning carefully, using calculations where appropriate.

ⓘ References and sources for further study

[1] *CRC Handbook of Chemistry and Physics*, 61st edition, 1980–81, R.C. Weast and M.J. Astle, editors (CRC Press, Boca Raton, 1980). Tabulated lattice energies.

[2] A.R. West, *Solid State Chemistry and its Applications* (J. Wiley & Sons, Chichester, 1984) Section 8.1. Ionic bonding phenomenology.

[3] C.H. Yoder, F.H. Suydam, and F.H. Snavely, *Chemistry*, 2nd edition (Harcourt Brace Jovanovich, New York, 1980) pp. 81–96 (Chapter 6). Ionic bonding phenomenology.

[4] N.N. Greenwood, *Ion Crystals, Lattice Defects, and Nonstoichiometry* (Butterworths, London, 1968) Chapter 2, Sections 2.1 through 2.4. Ionic bonding phenomenology.

[5] J.A.A. Ketelaar, *Chemical Constitution* (Elsevier Publishing Co., Amsterdam, 1953) pp. 39–40. The Born–Haber cycle.

[6] N.N. Greenwood, *Ion Crystals, Lattice Defects, and Nonstoichiometry* (Butterworths, London, 1968) Chapter 2, Section 2.6. The Born–Haber cycle.

[7] A.R. West, *Solid State Chemistry and its Applications* (J. Wiley & Sons, Chichester, 1984) Section 8.2.7. The Born–Haber cycle.

[8] R.J. Borg and G.J. Dienes, *The Physical Chemistry of Solids* (Academic Press, San Diego, 1992) p. 131 (Sec. 5.5). Ionization energies. These authors cited ref. [9] as their source.

[9] T. Moeller, *Inorganic Chemistry* (Wiley, New York, 1982) p. 81.

[10] N.N. Greenwood, *Ion Crystals, Lattice Defects, and Nonstoichiometry* (Butterworths, London, 1968), Table 2.2, p. 11. The author cites references [11, 12] and references therein as the source of electron affinity data.

[11] M.F.C. Ladd and W.H. Lee, *Prog. Solid St. Chem.*, **1** (1963) 37.

[12] M.F.C. Ladd and W.H. Lee, *Prog. Solid St. Chem.*, **2** (1965) 378.

[13] N.N. Greenwood, *Ion Crystals, Lattice Defects, and Nonstoichiometry* (Butterworths, London, 1968), p. 24. The author cites references [14, 15] as the original source for the information on the stability of hypothetical compounds.

[14] N. Bartlett and D.H. Lohmann, *Proc. Chem. Soc.* (1962) 115.

[15] N. Bartlett, *Proc. Chem. Soc.* (1962) 218.

[16] J.A.A. Ketelaar, *Chemical Constitution* (Elsevier Publishing Co., Amsterdam, 1953). Figure 7.4 is drawn after this source.

[17] W.A. Harrison, *Electronic Structure and the Properties of Solids: The Physics of the Chemical Bond* (Dover Publications Inc., New York, 1989) Chapter 13. Model for the ionic bond. Figure 7.5 is drawn after Fig. 13.1, p. 300.

[18] N.N. Greenwood, *Ion Crystals, Lattice Defects, and Nonstoichiometry* (Butterworths, London, 1968) Chapter 2, Sections 2.1 through 2.4.

[19] Q.C. Johnson and D.H. Templeton, Madelung Constants for Several Structures, *J. Chem. Phys.* **34** (1961) 2004. The Madelung constant data are reproduced from this source.

[20] M.P. Tosi, *Solid State Physics* **16** (1964) 1. The Madelung constant.

[21] R.A. Jackson and C.R.A. Catlow, *Molecular Simulation*, **1** (1988) 207. The Madelung constant.

[22] N.N. Greenwood, *Ion Crystals, Lattice Defects, and Nonstoichiometry* (Butterworths, London, 1968) Chapter 2, Section 2.7, pp. 27–30. Second order effects on the energy.

[23] N.N. Greenwood, *Ion Crystals, Lattice Defects, and Nonstoichiometry* (Butterworths, London, 1968), pp. 30–32. Fig. 7.8 was drawn after Fig. 2.4.

[24] R.J. Borg and G.J. Dienes, *The Physical Chemistry of Solids* (Academic Press, San Diego, 1992), Chapter 6, Section 4, pp. 215–25. Table 7.11 is based on Table 6.5.

[25] N.N. Greenwood, *Ion Crystals, Lattice Defects, and Nonstoichiometry* (Butterworths, London, 1968) Chapter 2, Section 2.7, pp. 32–4, Chapter 3, Section 3.3, pp. 46–8. The data in Table 7.12 are from Table 2.10 on p. 33.

[26] J.A.A. Ketelaar, *Chemical Constitution* (Elsevier Publishing Co., Amsterdam, 1953). Fig. 7.9 is drawn after Fig. 4 on p. 32.

[27] R. D. Shannon and C. T. Prewitt, Effective Ionic Radii in Oxides and Fluorides, *Acta. Cryst.* **B25** (1969) 925.

[28] R. D. Shannon, Revised Effective Ionic Radii and Systematic Studies of Interatomic Distances in Halides and Chalcogenides, *Acta. Cryst.* **A32** (1976) 751.

[29] L.C. Allen, Electronegativity is the average one-electron energy of the valence shell electrons in ground-state free atoms, *J. Am. Chem. Soc.* **111** (1989) 9003–14.

[30] N.N. Greenwood, *Ion Crystals, Lattice Defects, and Nonstoichiometry* (Butterworths, London, 1968). Fig. 7.11 is drawn after Figs. 3.3 and 3.4 on p. 47.

[31] A.R. West, *Solid State Chemistry and its Applications* (J. Wiley & Sons, Chichester, 1984) Chapter 8.

[32] R.S.J. Mulliken, *J. Chem. Phys.* **2** (1934) 782.

[33] A.L. Allred and E.G. Rochow, A scale of electronegativity based on electrostatic force, *J. Inorg. Nucl. Chem.* **5** (1958) 264–8.

[34] A.L. Allred, Electronegativity values from thermochemical data, *J. Inorg. Nucl. Chem.* **17** (1961) 215–21.

[35] J.C. Slater, *Phys. Rev.* **36** (1930) 57.

[36] M.J.L. Sangster and A.M. Stoneham, Calculations of off-centre displacements of divalent substitutional ions in CaO, SrO, and BaO from model potentials, *Phil. Mag. B* **43** (1981) 597–608.

[37] D.M. Duffy and P.W. Tasker, Grain Boundaries in Ionic Crystals, *Physica* **131B** (1985) 46–52.

[38] N.F. Mott and N.J. Littleton, *Trans. Farad. Soc.* **34** (1938) 485.

[39] G.V. Lewis, Interatomic Potentials: Derivation of Parameters for Binary Oxides and Their Use in Ternary Oxides, *Physica* **131B** (1985) 114.

[40] J.R. Walker and C.R.A. Catlow, Structure and Transport in non-stoichiometric β-Al_2O_3, *J. Phys. C,* **15** (1982) 6151.

[41] A.F. Kapustinskii, *Quart. Rev.* **10** (1956) 283.

[42] Y. Wang, D. Tomanek, G. Bertsch, and R.S. Ruoff, Stability of C_{60} Fullerite Intercalation Compounds, *Phys. Rev. B* **47** (1993) 6711–20.

[43] R.S. Ruoff, Y. Wang, and D. Tomanek, Lanthanide- and Actinide-based fullerite compounds: Potential A_xC_{60} superconductors?, *Chem. Phys. Lett.* **203** (1993) 438.

Chapter 8
Metallic Bonding

(A) Introduction

When free atoms condense to form a solid, cohesion arises from a change in the occupation and/or distribution of electronic energy levels. In the ionic bonding model, valence electrons move from atomic orbitals on metallic atoms to atomic orbitals on relatively electronegative atoms. Electrostatic cohesion, therefore, results from a change in the occupation of previously existing electronic energy levels on the atoms. In our model for the metallic bond, valence electrons on metallic atoms will be removed from atomic energy levels and placed in crystal energy levels or *bands*. In this chapter, the band concept is introduced and these new crystal energy levels are described. We begin this chapter with a review of the types of materials that form metallic bonds and a summary of the trends in metallic bond strength.

i. *Materials that are held together by metallic bonds*
Figure 8.1 shows a periodic chart in which all of the metallic elements are shaded. The metal–nonmetal definition is the same as that proposed in Chapter 1. Based on this definition, we say that all metallic elements and combinations of metallic elements are bonded metallically.

ii. *Phenomenological trends in metallic bonding*
The strength of the metallic bond varies with the interatomic separation and the atomic valence. Specifically, the bond strength (as measured by the cohesive energy, E_c, and the melting temperature, T_m) increases as the interatomic separation decreases and as the number of valence electrons increases. These trends are illustrated by the data in Table 8.1 and reflect the fact that metallic solids are held together by the Coulombic attraction between delocalized valence electrons and positively charged, ionized cores. Thus, as the core charge and the density of the electron sea increases, the bond strength increases.

iii. *Qualitative free electron theory*
The most well known characteristic properties of metals are their reflectivity, their high electronic conductivity, and their high thermal conductivity. These are

Table 8.1. *The cohesive energy and melting points of the simple metals.*

	E_c (eV)	T_m °C	Z	d_0 (Å)	post-transition metals	E_c (eV)	T_m °C	Z	d_0 (Å)
Li	1.63	179	1	3.040	Zn	1.35	420	2	2.66
Na	1.11	98	1	3.716	Cd	1.16	321	2	2.98
K	0.934	64	1	4.544	Hg	0.67	−39	2	2.96
Rb	0.852	39	1	4.936					
Cs	0.804	28	1	5.265	In	2.52	157	3	2.88
Be	3.32	1283	2	2.286	Tl	1.88	303	3	3.36
Mg	1.52	650	2	3.209					
Ca	1.84	850	2	3.951	Sn	3.14	232	4	2.81
Sr	1.72	770	2	4.300	Pb	2.03	327	4	3.50
Ba	1.90	725	2	4.346					

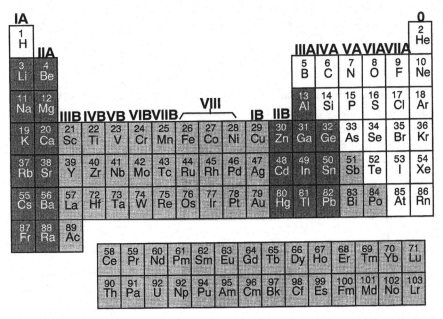

Figure 8.1. The periodic chart with metallic atoms shaded. The darker shading signifies those metals known as 'simple metals'.

the properties that led to the formulation of the free electron theory. The free electron model assumes that the metallic crystal is composed of positively charged ion cores in a 'sea' of delocalized electrons. Before formulating the physical model, we need to be a bit more specific about the features of the model. This model is best applied to the so-called simple metals that are labeled in Fig. 8.1 [1]. These include the alkali, alkaline earth, and some post-transition metals. The partially filled d-orbitals of the transition series present special challenges due to localization, charge transfer, and covalency, which will be dealt with in the next chapter on covalent bonding.

We will assume that the valence electrons in a metal are free to move throughout the volume of the solid without interruption from the ions. For example, consider Na, which has the electronic configuration: $1s^2 2s^2 2p^6 3s^1$. The electrons in the first and second principal shell are 'core' electrons, whereas the electron in the 3s orbital is the valence electron that is free to move about the crystal. One of the essential assumptions of the free electron theory is that while the core electrons remain tightly bound to the atom, the valence electrons are free to move about the crystal. In fact, we assume that the potential energy (the strength of the interaction) between the valence electrons and the ion cores is zero. Therefore, the electrons are not aware of the presence of the ions and the total energy of the solid is equal to the kinetic energy of the free electrons. This model will obviously have to be modified to explain cohesion in metallic crystals.

The assumption that the electrons are truly free represents a limiting case that is never actually realized. This is analogous to our descriptions of van der Waals and ionic bonds, which were also idealized limiting cases. However, we should note that there is reasonable experimental evidence for this assumption. For example, at low temperatures, electrons in pure metals can travel as far as 1 cm between collisions. In fact, if we take the size of the ionized Na core to be the ionic radius of Na^+ (0.98 Å) and note that for crystalline Na, $1/2d_0 = 1.38$ Å, we find that the ion cores fill only 15% of the volume of the crystal [1].

(B) A physical model for the metallic bond: free electron theory

The derivation of the free electron theory uses some basic quantum mechanical tools. While a detailed explanation of these tools lies outside the scope of this book, we will review the basic quantum mechanical assertions in Section (i). We will not need to use all of these assertions in this chapter; some will not be used until our discussion of the covalent bond in Chapter 9. However, it is best to present the assertions as a group.

Table 8.2. *Selected quantum mechanical operators.*

position	r
momentum	$p = \hbar/i\, \nabla$
potential energy	$V(\vec{r})$
kinetic energy	$\dfrac{p^2}{2m} = \dfrac{-\hbar^2}{2m}\nabla^2$
total energy or Hamiltonian	$H = \dfrac{-\hbar}{2m}\nabla^2 + V(\vec{r})$

i. *Wave functions, operators, and expectation values*

More detailed descriptions of the quantum mechanical assertions stated below can be found in any quantum mechanics textbook [2, 3]. The first assertion is that any electron can be completely described by a wave function, $\Psi(\vec{r})$, that can have both real and imaginary parts. This function contains all of the necessary information regarding an electron's position and energy and is the device we will use to describe and keep track of electrons in crystals. For example, the wave function for the 1s electron of an H atom is given by Eqn. 8.1:

$$\Psi_{100}(\vec{r}) = A_{1s}\, e^{-r/a_0}. \tag{8.1}$$

The three subscripts of the wave function are the three quantum numbers (nlm), a_0 is the Bohr radius:

$$a_0 = \frac{\hbar}{me^2} = 0.529\ \text{Å}, \tag{8.2}$$

and A_{1s} is a normalization constant,

$$A_{1s} = \frac{1}{\sqrt{\pi}}\left(\frac{1}{a_0}\right)^{\frac{3}{2}}. \tag{8.3}$$

Operators are used to obtain physical quantities such as position, momentum, and energy from a wave function. Five useful operators are listed in Table 8.2. Expectation, or average, values for observable physical quantities can be calculated by using wave functions and operators, as shown in Eqn. 8.4.

$$\langle o \rangle = \frac{\displaystyle\int \Psi^* O \Psi\, d\vec{r}}{\displaystyle\int \Psi^* \Psi\, d\vec{r}}. \tag{8.4}$$

In this expression, $\langle o \rangle$ is the expectation value and O is the operator. It is common practice to normalize the wave functions so that the denominator of Eqn. 8.4 is unity.

The best way to see how wave functions and operators are used to describe the physical properties of electrons is to perform a calculation. As an example, we can use the H 1s wave function given in Eqn. 8.1, the position operator given in Table 8.2, and the formula for the expectation value given in Eqn. 8.4 to compute the average position of the electron with respect to the nucleus. The wave function has already been normalized so that:

$$\int \Psi_{100}^* \, \Psi_{100} \, d\vec{r} = 1. \tag{8.5}$$

Therefore, determining the average radial position, $\langle r \rangle$, amounts to calculating the numerator of Eqn. 8.4 when Ψ is Ψ_{100} and the operator, O, is the position operator, r.

$$\langle r \rangle = A_{1s}^2 \int e^{-r/a_0} \, r \, e^{-r/a_0} \, d\vec{r}. \tag{8.6}$$

To evaluate Eqn. 8.6, we first convert it to spherical coordinates and substitute $\alpha = 2/a_0$:

$$\langle r \rangle = A_{1s}^2 \int_0^{2\pi} \int_0^\pi \int_0^\infty r^3 e^{-\alpha r} \, dr \, \sin\theta \, d\theta \, d\phi, \tag{8.7}$$

where $d\vec{r} = dr \, rd\theta \, r \, \sin\theta \, d\phi$. Solution of the inner integral gives $6a_0^4/2^4$. Completing the integration, we find that:

$$\langle r \rangle = 3/2 \, a_0 = 0.79 \, \text{Å}. \tag{8.8}$$

Therefore, our conclusion is that, on average, the electron can be found somewhere on a sphere of radius 0.79 Å from the nucleus.

At this point, we will introduce a shorthand notation for the integrals used to calculate expectation values. In this notation (called bra-ket notation), Eqn. 8.4 is:

$$\langle o \rangle = \frac{\langle \Psi | O | \Psi \rangle}{\langle \Psi | \Psi \rangle}, \tag{8.9}$$

and Eqn. 8.6 is

$$\langle r \rangle = \langle \Psi_{100} | r | \Psi_{100} \rangle. \tag{8.10}$$

Based on our definitions of the wave function, operator, and expectation value, it follows that there is a discrete value of energy (ε) associated with each electronic state or wave function. This one-to-one correspondence between a set of discrete energies, known as eigenvalues, and a set of electron wave functions, known as eigenstates, is one of the most important aspects of quantum mechanics. In other words, electrons are not able to assume a continuous range of energies. Our final assertion is that the eigenstates, $\Psi(\vec{r})$, are given by the solution to the following differential equation:

$$H\Psi(\vec{r}) = \varepsilon\Psi(\vec{r}), \tag{8.11}$$

which we call the time independent Schrödinger wave equation. We will soon see that the existence of boundary conditions, even those as simple as the requirement that a metal's electrons be contained within the crystal, lead to a discrete set of solutions. It is not necessary that an eigenstate actually be occupied by an electron.

ii. *Formulation of the free electron theory*
As described in Section A(iii), we imagine that the valence electrons in a metal are free to move about in a crystal, which we take to be a cube of length $L = V^{1/3}$. We assume that N electrons can occupy the N lowest energy eigenstates given by the time independent Schrödinger equation:

$$H\Psi_k(\vec{r}) = \varepsilon_k \Psi_k(\vec{r}), \tag{8.12}$$

where the index, k, labels a discrete energy level and the Hamiltonian operator, H, is the sum of the kinetic and potential energies, as in Table 8.2. However, in the free electron theory, we assume that the electrons do not interact with the ion cores or with each other [4–6]. Thus, there is no potential energy and the Hamiltonian is identical to the kinetic energy operator. We can, therefore, rewrite Eqn. 8.12 in the following way:

$$\frac{-\hbar^2}{2m}\nabla^2\Psi_k(\vec{r}) = \varepsilon_k \Psi_k(\vec{r}) \tag{8.13}$$

$$\frac{-\hbar^2}{2m}\left(\frac{\partial^2}{\partial x^2} + \frac{\partial^2}{\partial y^2} + \frac{\partial^2}{\partial z^2}\right)\Psi_k(\vec{r}) = \varepsilon_k \Psi_k(\vec{r}).$$

The important fact to emphasize is that each $\Psi_k(\vec{r})$ is an orbital that can be occupied in accordance with the Pauli exclusion principle and each ε_k is the energy eigenvalue of an electron in that orbital.

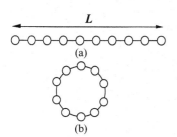

Figure 8.2. A linear array of 10 atoms (a) is converted to a periodic array of 10 atoms using cyclic boundary conditions. In this one-dimensional example, $\Psi(0) = \Psi(L)$ and $\Psi(x) = \Psi(x+L)$.

The boundary condition is that the electron must remain in the crystal, so $\Psi_k(\vec{r})$ must go to zero at values of $\vec{r}$ greater than L. In order to eliminate the difficulty of the surface, we use the traditional cyclic boundary conditions. This boundary condition matches one edge of the crystal with the other such that if you imagine an electron leaving one edge of the crystal, it immediately appears on the opposite face. In one dimension, this is equivalent to converting a line from $x = 0$ to $x = L$ to a ring of atoms with circumference L, as shown in Fig. 8.2. Although difficult to imagine in three dimensions, it is easily applied in the following way:

$$\Psi(x + L, y, z) = \Psi(x, y, z)$$
$$\Psi(x, y + L, z) = \Psi(x, y, z)$$
$$\Psi(x, y, z + L) = \Psi(x, y, z). \tag{8.14}$$

These are also known as the periodic or Born–von Karman boundary conditions.

Because we imagine the electron is free to travel throughout the crystal, we propose normalized plane waves as solutions to the wave equation labeled 8.13.

$$\Psi_K(\vec{r}) = \frac{1}{\sqrt{V}} e^{i\vec{k}\cdot\vec{r}}. \tag{8.15}$$

We can verify that this is an acceptable solution by differentiation:

$$\frac{-\hbar^2}{2m}\left(\frac{\partial^2}{\partial x^2} + \frac{\partial^2}{\partial y^2} + \frac{\partial^2}{\partial z^2}\right)\frac{1}{\sqrt{V}} e^{ik_x x} e^{ik_y y} e^{ik_z z} = \varepsilon_k \frac{1}{\sqrt{V}} e^{ik_x x} e^{ik_y y} e^{ik_z z}$$

$$\frac{-\hbar^2}{2m}(i\vec{k})^2\, e^{ik_x x} e^{ik_y y} e^{ik_z z} = \varepsilon_k\, e^{ik_x x} e^{ik_y y} e^{ik_z z}$$

$$\varepsilon_k = \frac{\hbar^2 \vec{k}^2}{2m}. \tag{8.16}$$

In Eqn. 8.16, we see that the energy of an electron in the k-th state is proportional to the square of the electron's wave vector, $\vec{k}$.

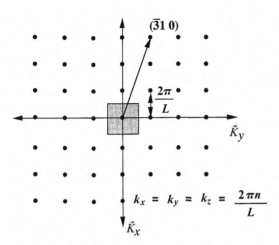

Figure 8.3. The allowed values of the free electron wave vector are limited to discrete values specified by integer multiples of three basis vectors. They can thus be used to define a lattice.

At this point, there is nothing to prevent ε and k from taking a continuous range of values. However, the application of the boundary conditions discretizes ε and k.

$e^{ik_x x} = e^{ik_x(x+L)}$ implies:
$e^{ik_x L} = 1.$
Likewise, $e^{ik_y L} = 1$ and $e^{ik_z L} = 1.$ $\qquad(8.17)$

Therefore,

$$k_x = \frac{2\pi n_x}{L}; k_y = \frac{2\pi n_y}{L}; k_z = \frac{2\pi n_z}{L}, \qquad(8.19)$$

and the total wave vector can be written:

$$\vec{k} = \frac{2\pi}{L}(n_x \hat{x} + n_y \hat{y} + n_z \hat{z}). \qquad(8.20)$$

Considering the fact that n_x, n_y, and n_z can be any integer, the components of the wave vector, $\vec{k}$, are limited to integer multiples of $2\pi/L$. Because $\vec{k}$ is limited to a discrete set of values, it is analogous to the direct and reciprocal lattice vectors defined in Chapter 2. Thus, it can be used to define a lattice in 'wave vector space', which we will call 'k-space'. Note that the dimension of this vector (inverse length) is the same as that of the reciprocal lattice vectors, $\vec{G}_{hkl}$. This lattice is illustrated in Fig. 8.3.

Note that the discretization of $\vec{k}$ leads to the discretization of the possible energy levels (see Eqn. 8.16). There are several interesting aspects to the arrangement of these crystal energy levels that are distinct from atomic energy levels. First, it is clear from Eqn. 8.20 that the energy difference between two incremental states is quite small. For example, if we consider a crystal with lateral dimensions of 1 cm, then the energy separation between the lowest state and the state with the next highest energy is $(2\pi^2\hbar^2)/mL^2$ or about 2×10^{-14} eV. This is much smaller than the thermal energy available at room temperature. Note also that the spacing between energy levels changes with the crystal size and that the spacing increases as the crystal gets smaller. It is also interesting to note that because the energy is determined by the magnitude of $\vec{k}$, there is significant degeneracy. For example, for unequal and nonzero values of n_x, n_y, and n_z, there are 48 states with the same energy that can hold 96 electrons.

The volume of k-space per allowed point, $(2\pi/L)^3$, is an important quantity (see Fig. 8.3). In a real system, N is very large and the volume of k-space occupied (proportional to the number of states needed to accommodate N electrons) is correspondingly large. It follows that the number of allowed levels in this volume is:

$$\frac{V}{8\pi^3}. \tag{8.21}$$

This is the density of levels or number of allowed values of $\vec{k}$ per unit volume.

Now that we have specified the density of $\vec{k}$ points and, subsequently, the corresponding density of energy levels, we can fill the levels up with the N electrons. According to the Pauli exclusion principle, two electrons of opposite spin can occupy each energy level. The lowest energy levels are filled first, beginning with $k=0$ ($n_x=0$, $n_y=0$, $n_z=0$). As the values of the indices are continuously incremented, higher and higher energy orbitals are occupied until all the electrons have been accommodated. The last electron fills the highest energy level, specified by the so-called 'Fermi wave vector', k_F. The filling of higher energy levels, specified by increasing values of k, is illustrated schematically in Fig. 8.4. Note that multiple vectors, $\vec{k}$, have the same magnitude and, therefore, there are many degenerate levels. For relatively large values of k (in macroscopic systems with many electrons), there are many levels spaced very closely together. Therefore, we can say that the occupied levels form a sphere in k-space with volume

$$V = \frac{4}{3}\pi k_F^3. \tag{8.22}$$

A projection of this sphere is depicted in Fig. 8.4.

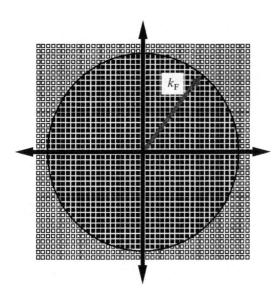

Figure 8.4. Because the distance between adjacent sites on the k-space lattice is extremely small in comparison to the length of the Fermi wave vector, k_F, the occupied sites (states) approximately form a sphere in k-space. k_F defines the boundary between occupied and unoccupied states. The occupied states are represented by the black squares and the unoccupied states by the white squares.

The total number of electrons in this volume of k-space, N, is equal to the number of electrons per orbital, multiplied by the total number of orbitals (the total volume divided by the volume per orbital).

$$N = 2 \cdot \frac{4}{3}\pi k_F^3 \cdot \frac{V}{8\pi^3} = \frac{V}{3\pi^2}k_F^3. \tag{8.23}$$

Now we can define the Fermi vector in terms of the valence electron density, n,

$$k_F = \left(\frac{3\pi^2 N}{V}\right)^{1/3} = (3\pi^2 n)^{1/3}, \tag{8.24}$$

and the Fermi energy, ε_F, the energy of the highest filled level, as:

$$\varepsilon_F = \frac{\hbar^2 k_F^2}{2m} = \frac{\hbar^2}{2m}(3\pi^2 n)^{2/3}. \tag{8.25}$$

Typical Fermi energies are between 1 and 15 eV. It is important to note that in each of these equations, n, the valence electron density, is the only materials parameter. The valence electron density depends on both the structure (the packing of the atoms) and the chemistry (the number of valence electrons) of the metal.

To find the total energy of N electrons in this system, we need to sum the energies of each electron:

$$E = 2 \sum_{k=0}^{k_F} \frac{\hbar^2 \vec{k}^2}{2m}. \tag{8.26}$$

The sum in Eqn. 8.26 is carried out over each eigenstate, labeled by a wave vector, and multiplied by two to account for the double occupancy of each state. The number of terms in this sum is equal to half of the number of valence electrons in the crystal. Because this is a very large number, and the energy separation between adjacent levels is very small when compared to ε_F, we transform the sum to an integral. However, we must first remember that each point in k-space has the following volume:

$$\Delta k = \frac{8\pi^3}{V}. \tag{8.27}$$

So, if we multiply E by $\Delta \vec{k}/\Delta k$, we have

$$E = \frac{V}{4\pi^3} \sum_{k=0}^{k_F} \frac{\hbar^2 \vec{k}^2}{2m} \Delta \vec{k}. \tag{8.28}$$

Finally, because the k-levels are so close together, we can assume that $E(k)$ is a continuous function and approximate Eqn. 8.26 as an integral:

$$E = \frac{V}{4\pi^3} \int_0^{\vec{k}_F} \frac{\hbar^2 \vec{k}^2}{2m} d\vec{k}. \tag{8.29}$$

After converting this from a volume integral to a line integral (assuming isotropic space), we have:

$$E = \frac{V}{4\pi^3} \frac{\hbar^2}{2m} \int_0^{k_F} k^2 \cdot 4\pi k^2 \, dk. \tag{8.30}$$

Integration of Eqn. 8.30 gives the kinetic energy per volume:

$$E/V = \frac{\hbar^2}{10m\pi^2} k_F^5. \tag{8.31}$$

It is important to remember that each electron has a different energy. The average energy per electron is:

$$\frac{E}{N} = \frac{E}{V} \cdot \frac{V}{N} = \frac{E}{V} \cdot \frac{1}{n} = \frac{\hbar^2}{10m\pi^2} k_F^5 \cdot \frac{3\pi^2}{k_F^3},$$

$$\frac{E}{N} = \frac{3}{5} \cdot \frac{1}{2} \frac{\hbar^2 k_F^2}{m} = \frac{3}{5} \varepsilon_F. \tag{8.32}$$

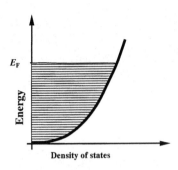

Figure 8.5. The energy density of free electron states. Electrons fill the energy levels of the crystal from lowest to highest, up to E_F. As the energy increases, there are more states with similar energies.

Note that the free electron model predicts that the energy per electron (and per atom) of a metallic crystal is positive. In other words, the energy of the crystal is greater than the energy of the free atoms and there is no cohesive force. We will address this problem in part *(iv)* of this section.

iii. *The energy density of states*
The final important element of the free electron theory is the distribution of energy levels or the energy density of states. The density of states is the number of states with energy between ε and $\varepsilon + \Delta\varepsilon$, where $\Delta\varepsilon$ is very small compared to ε_F. Recalling Eqn. 8.25, we can say that the total number of states with energy below ε is:

$$N = \frac{V}{3\pi^2}\left(\frac{2m\varepsilon}{\hbar^2}\right)^{3/2}.$$
(8.33)

Therefore, the energy density of orbitals is:

$$\frac{dN}{d\varepsilon} = D(\varepsilon) = \frac{V}{3\pi^2}\left(\frac{2m}{\hbar^2}\right)^{3/2} \cdot \frac{3}{2}\varepsilon^{1/2} = \frac{V}{2\pi^2}\left(\frac{2m}{\hbar^2}\right)^{3/2}\varepsilon^{1/2}.$$
(8.34)

The result is that the density of free electron states increases parabolically with energy. This result is illustrated graphically in Fig. 8.5, in the form of an energy level diagram. The diagram shows that as the energy increases, the number of states with similar energies increases.

The density of states at the Fermi level (the number of electrons energetically close to unoccupied states) is an important parameter in the free electron model, since only electrons in these states are able to participate in dynamic processes such as electronic conduction. Based on this idea, it is possible to explain a number of interesting physical properties including heat capacity, thermal and electrical conductivity, the Hall effect, and magnetic properties. Although these properties do not lie within the scope of this book, descriptions can be found in most text books on solid state physics [5, 6].

Example 8.1

We will soon see that there is a discontinuity in the distribution of energy levels when the wave vector extends beyond the first unit cell or Brillouin zone of reciprocal space. This has an important influence on structural stability. At what electron concentration does the Fermi sphere of a fcc metal touch the Brillouin zone boundary? Express the electron concentration in electrons per atom.

1. First, recall from Chapter 2 that the Brillouin zone is the primitive unit cell of the reciprocal lattice. It is the volume of space closest to each lattice point.

2. We saw earlier (Chapter 5) that the reciprocal lattice of the cubic F lattice has the arrangement of a cubic I lattice.

3. The boundary plane of the Brillouin zone is found by drawing a vector to the nearest neighbor points on the lattice and bisecting each vector with a plane. The nearest point in this case is at $(1,1,1)$ (the first allowed reflection for an fcc crystal). The zone's boundary plane must bisect this vector at $(1/2,1/2,1/2)$.

4. So, when the Fermi wave vector, k_F, is equal in magnitude to one half the length of the first reciprocal lattice vector (this distance in reciprocal space from $(0,0,0)$ to $(1/2,1/2,1/2)$), the Fermi sphere will just touch the zone boundary:

$$\text{if } |\vec{a}^*| = |\vec{b}^*| = |\vec{c}^*| = 2\pi/a, \text{ then}$$

$$k_F = \frac{2\pi}{a} \sqrt{\left(\frac{1}{2}\right)^2 + \left(\frac{1}{2}\right)^2 + \left(\frac{1}{2}\right)^2}$$

$$k_F = \frac{\sqrt{3}\,\pi}{a} = (3\pi^2 n)^{1/3}.$$

Assuming N atoms per cell and Z valence electrons per atom, we can substitute for the electron density, $n = NZ/a^3$:

$$\frac{\sqrt{3}\,\pi}{a} = \left(3\pi^2 \frac{NZ}{a^3}\right)^{1/3}.$$

$Z = 1.36$ electrons per atom.

5. Fractional electron concentrations can be realized in alloys. In Chapter 10, Section B(i), we shall see how this critical electron concentration affects structural stability.

---

iv. The free electron energy

Recalling Eqns. 8.25 and 8.32, we know that the total kinetic energy of the metallic crystal is proportional to the electron density raised to the power of two thirds.

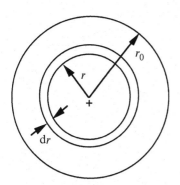

Figure 8.6. A schematic diagram for the integration of the electrostatic energy between a central positive charge and a constant electron density in a sphere with radius r_0.

$$E = \frac{3}{5}\varepsilon_F = \frac{3}{5}\frac{\hbar^2 k_F^2}{2m} = \frac{3}{5}\frac{\hbar^2}{2m}(3\pi^2)^{1/3}n^{2/3}. \tag{8.35}$$

Within the framework of the free electron theory, the kinetic energy is the only contribution to the total energy of the crystal. This suggests that as the electron density increases, the crystal becomes less stable. Considering the fact that there is no cohesive energy, this model does not provide an appropriate description of metallic bonding. We can assume that the missing component is the attractive force between the positive ion cores and the sea of valence electrons. This electrostatic energy can be determined in a way similar to the method that we used for ionic crystals. First, however, we'll consider an easy approximate solution to the problem and then we'll turn to the more exact result.

We begin by assuming that there are Z free electrons for every ion core and that these electrons are evenly distributed throughout a sphere of radius r_0, where r_0 is the radius of an atomic volume. According to Eqn. 8.24, the electron density, n, or number of electrons per volume is $k_F^3/(3\pi^2)$. We can also express the electron density as the number of valence electrons, Z, divided by the atomic volume, $4/3\,\pi r_0^3$. Equating the two expressions for n, we find that:

$$r_0 = \left(\frac{9\pi Z}{4}\right)^{1/3}\frac{1}{k_F}. \tag{8.36}$$

We also assume that the ion acts as a point charge in the center of the sphere, that the potential between the free electrons and the ion is Ze/r, and that there is no interaction between the atomic volumes. Using these assumptions, we can now apply basic electrostatics to compute the Coulombic energy (U_c) [7]. Using the fact that the electron density, $\rho(\vec{r})$, and the electric potential, $V(\vec{r})$, are spherically symmetric, and referring to the definitions in Fig. 8.6,

$$U_c = -\int_0^{\vec{r}_0} \rho(\vec{r}) V(\vec{r}) \, d\vec{r}$$

$$= -4\pi \int_0^{r_0} r^2 \, \rho(r) V(r) \, dr. \tag{8.37}$$

Next, we divide the electric potential into two components, the attractive part between the electrons and the core $[V_{ion}(r)]$ and the repulsive electron–electron part, $[V_e(r)]$. The term V_{ion} is simply Ze/r. The term V_e is the sum of the potential due to the charge within a sphere defined by r (a quantity we'll call V_e') and the potential due to the charge in the outer shell between r and r_0 (a quantity we'll call V_e'').

First, $V_e'(r)$, the potential due to the inner sphere of electron density, is

$$V_e'(r) = \frac{Q}{r} = \rho(r) \cdot \frac{4}{3} \pi r^3 \frac{1}{r}. \tag{8.38}$$

When $\rho(r)$ is constant $(\rho(r) = \rho_0)$, we assume that all the charge (Q) within a sphere of radius r, is at the center of the sphere.

Next, the potential due to the outer shell of charge between r and r_0, V_e'', is computed (for simplicity) at the center of the sphere. This is allowed because we know that the potential is the same at all points within the shell. Again, using the fact that ρ_0 is constant throughout the sphere:

$$V_e''(r) = \int_{\vec{r}}^{\vec{r}_0} \frac{\rho_0}{r} \, d\vec{r} = \int_r^{r_0} \frac{\rho_0}{r} 4\pi r^2 \, dr$$

$$= 4\pi\rho_0 \int_r^{r_0} r \, dr$$

$$= 2\pi\rho_0 (r_0^2 - r^2). \tag{8.39}$$

The total potential due to the electrons is:

$$V_e = \frac{1}{2}(V_e' + V_e''). \tag{8.40}$$

The factor of 1/2 accounts for the fact that we counted each interaction twice. We can rewrite Eqn. 8.40 by substituting Eqns. 8.38 and 8.39.

$$V_e = \frac{1}{2}\left[\frac{4}{3}\pi\rho_0 r^2 + 2\pi\rho_0(r_0^2 - r^2)\right]$$

$$V_e = \frac{1}{2}\left[2\pi\rho_0\left(r_0^2 - \frac{r^2}{3}\right)\right] \tag{8.41}$$

The total electrostatic potential energy is $V(r) = V_{ion} - V_e$, or,

$$V(r) = V_{ion} - V_e$$
$$= \frac{Ze}{r} - \frac{1}{2}\left[2\pi\rho_0\left(r_0^2 - \frac{r^2}{3}\right)\right]. \tag{8.42}$$

Finally, we can substitute the potential given in Eqn. 8.42 into the expression for the total electrostatic energy given by Eqn. 8.37:

$$U_c = -4\pi\int_0^{r_0} r^2\, dr\, \rho(r)\left\{\frac{Ze}{r} - \frac{1}{2}\left[2\pi\rho_0\left(r_0^2 - \frac{r^2}{3}\right)\right]\right\}. \tag{8.43}$$

Integrating Eqn. 8.43 leads to the following result:

$$U_c = -\frac{9}{10}\frac{(Ze)^2}{r_0}. \tag{8.44}$$

Note that in this expression, the physical parameters of the free electron theory (the electron density and the Fermi wave vector) are contained in r_0. Equation 8.44 can be rewritten in a way that is analogous to the form of the Madelung energy of the ionic solid.

$$U_c = -\frac{\alpha Z^2 e^2}{d_0} \tag{8.45}$$

where $d_0 = 2r_0$ and $\alpha = 9/5$.

It must be emphasized that this is an approximate result and neglects sphere-to-sphere interactions. Thus, we can say that this is the total electrostatic energy between the free valence electrons and the positively charged cores, assuming that the electrons interact only with the nearest ion core.

By combining this energy with the electron kinetic energy, we have significantly improved upon our earlier model; now we have some cohesive energy to hold the crystal together. Unfortunately, because the model does not include any sphere-to-sphere interactions, the binding energy is completely insensitive to the crystal structure. To get the longer range interactions, we must perform a calculation similar to the one used to determine Madelung constants in ionic crystals. As for the case of the ionic crystal, the result comes down to a single structure-sensitive parameter, α. The total electrostatic energy, including long range interactions, can be written as in Eqn 8.45 above, where α is a crystal structure-sensitive constant, analogous to the Madelung constant. For the fcc, hcp, and bcc structures, $\alpha = 1.79$. It is 1.67 for diamond and 1.76 for simple cubic. We should note that our simple approximate calculation gave $\alpha = 9/5 = 1.8$. This differs from the exact solution for close packed structures by less than 1% [10].

Considering the values for α given above, we note two important things about the relationship between a metal's structure and its stability. First, the most stable structure is the one that maximizes the total electrostatic cohesive force (the one with the highest α). In other words, the most stable structures are fcc, hcp, and bcc. This agrees with the phenomenological trends discussed at the beginning of this chapter. Second, the energy differences (on a per atom basis) between the different close packed structures are very small. Thus, within the framework of this model, the overall stability of a metallic crystal depends much more on the type of metal than on the exact structural configuration.

Now, we can combine the electrostatic (Eqn. 8.45) and kinetic (Eqn. 8.32) energies to get the total energy, U.

$$U = -\alpha\frac{Z^2e^2}{d_0} + \frac{3}{5}\varepsilon_F. \tag{8.46}$$

Before proceeding, we have to make a correction. Our calculation for the electrostatic interactions among the electrons overestimated the repulsions, an effect that artificially diminishes the cohesive energy predicted by the model. The inaccuracy stems from the assumption that the electron density is uniform and homogeneous. In fact, it should be thought of as a collection of rapidly moving point charges that tend to avoid each other. One reason that the electrons avoid each other is the exchange interaction, which prevents two electrons of the same spin from coming too close to each other. The electrons also avoid one another because of the Pauli exclusion principle. So, instead of moving independently and more or less at random, the motions of individual electrons are correlated. Since the electrons are trying to avoid one another, the distribution of charge is not uniform and we have, therefore, overestimated the size of the electron–electron repulsions. Thus, we correct the free electron energy with an exchange energy, E_{ex} [8].

$$U = -\frac{\alpha Z^2e^2}{2r_0} + \frac{3}{10}\frac{Z\hbar^2k_F^2}{m} - \frac{3Ze^2k_F}{4\pi}. \tag{8.47}$$

Considering the relationship between r_0 and k_F, given in Eqn. 8.36, we can rewrite the total energy in terms of k_F:

$$U = \frac{-\alpha Z^2e^2}{2}\left(\frac{9\pi Z}{4}\right)^{-1/3}k_F + \frac{3}{10}\frac{Z\hbar^2}{m}k_F^2 - \frac{3}{4}\frac{Ze^2}{\pi}k_F. \tag{8.48}$$

Because k_F, r_0, and the electron density, n, are related, either one can be used independently to determine the equilibrium lattice spacing. Thus, to determine the equilibrium spacing, d_0, we differentiate Eqn. 8.48 with respect to k_F (the Fermi vector) and find that:

Table 8.3. *Data for bcc alkali metals.*

atom	a, Å	$n = 2/a^3 \times 10^{-22}$	k_F, Å^{-1}	d_0, Å
Li	3.49	4.7	1.13	1.51
Na	4.225	2.65	0.91	1.83
K	5.225	1.4	0.73	2.26
Rb	5.585	1.15	0.69	2.42
Cs	6.045	0.91	0.63	2.62

$$k_F = \left[\frac{\alpha Z}{6} \left(\frac{9\pi Z}{4} \right)^{1/3} + \frac{1}{4\pi} \right] \cdot \frac{5}{a_0}. \tag{8.49}$$

This result tells us that the electron density, the Fermi wave vector, and the equilibrium spacing are the same for all metals with the same values of Z (periodic group) and α. Obviously, this model will not account for the phenomenological observations described at the start of the chapter. Using $Z=1$ and $\alpha=1.79$, as for an alkali metal, we find that $k_F = 2.2$ Å^{-1}. The interatomic separation corresponding to this value is unrealistically small.

Considering the actual data shown in Table 8.3, we see that there are two obvious problems with the theory. One is that it predicts exactly the same atomic volume and, therefore, the same equilibrium separation for all atoms with the same valence. The other is that the predicted value for k_F in the alkali metals is 2.2 Å^{-1}, much larger than the actual values that vary from 0.63 to 1.3 Å^{-1}. The inadequacy of this theory is really not too surprising considering the approximations that are used. Perhaps the greatest over-simplification is that the ion cores are point charges. In reality, the ion cores occupy a finite volume of space from which the valence electrons are excluded. Accounting for this excluded volume will not only modify the electron density and the electrostatic term, it will also provide the theory with some sensitivity to the ion core size, which is obviously a critical factor. The implementation of this correction is most easily understood by referring to Fig. 8.7.

Figure 8.7 compares the electrostatic potential used in our first calculation (a) with the empty core potential (b). Figure 8.7b shows the potential based on the same point charge at the center of the sphere, but incorporating the fact that the electrons must remain outside of this core region, the size of which is defined by r_c. In the allowed regions, the potential is identical to that shown in Fig. 8.7a. In other words:

$$w^0(r) = 0, \qquad r < r_c$$

$$w^0(r) = -\frac{Ze^2}{r}, \quad r > r_c. \tag{8.50}$$

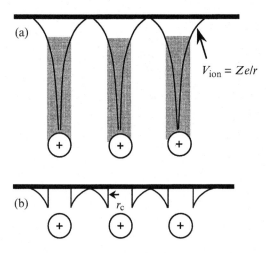

(a)

$V_{ion} = Ze/r$

(b) r_c

Figure 8.7. The electrostatic potential between electrons and ions assuming (a) that the ions are point charges and electrons can occupy any position. However, since the ions have a finite volume, it is not possible for electrons to occupy the most favorable positions (shaded regions). In (b), the potential is modified to reflect the fact that the ions have a finite size, r_c, and the valence electrons cannot enter the volume occupied by the cores. In our calculation, we subtract the electrostatic contributions from the shaded regions.

The variable w is used instead of V to indicate that it is not a real potential. This approximation, originally formulated by Ashcroft [9,10], is called the empty core pseudopotential model. We can correct our expression for the total energy (Eqn. 8.48) by correcting the electrostatic energy term. Our earlier approximation overestimated the attractive part of this energy because it assumed that electrons could very nearly approach the center of charge. Rather than recalculate the electrostatic term, we'll just determine the amount by which the energy was overestimated and deduct this energy from the first value. Using $n = N/V = \rho(r)$, the electrostatic energy of the empty core region is:

$$U_{ec} = \int_0^{\vec{r}_c} \rho(r) \frac{Ze^2}{r} \, d\vec{r}$$

$$= \int_0^{r_c} n4\pi r^2 \frac{Ze^2}{r} \, dr$$

$$= 2\pi n Z e^2 r_c^2. \tag{8.51}$$

Using this correction, the total electronic energy, in terms of k_F, is:

$$U = \frac{3}{10} \frac{Z\hbar^2 k_F^2}{m} - \frac{Z^2 e^2 \alpha k_F}{(18\pi Z)^{1/3}} + \frac{2Z}{3\pi} e^2 \, r_c^2 \, k_F^3 - \frac{3Ze^2 k_F}{4\pi}. \tag{8.52}$$

Note that the total electronic energy now depends on the identity of the atom in the crystal. By differentiating this equation with respect to k_F and setting it equal to zero, we can determine the equilibrium spacing. The result is a quadratic equation:

Table 8.4. *Empty core radii for the simple metals [10].*

atom	r_c, Å	atom	r_c, Å	atom	r_c, Å
Li	0.92	Be	0.58	Zn	0.59
Na	0.96	Mg	0.74	Cd	0.65
K	1.20	Ca	0.90	Hg	0.66
Rb	1.38	Sr	1.14		
Cs	1.55	Ba	1.60	In	0.63
				Tl	0.60
				Sn	0.59
				Pb	0.57

Table 8.5. *Comparison of calculated and computed values of d_0 for the alkali metals.*

atom	k_F (calc) Å^{-1}	$d_0 = 3.376/k_F$	d_0 (observed), Å
Li	0.89	3.79	3.040
Na	0.85	3.97	3.716
K	0.72	4.69	4.544
Rb	0.64	5.28	4.936
Cs	0.59	5.72	5.265

$$\left(\frac{2}{\pi}r_c^2\right)k_F^2 + 0.317k_F - 0.705 = 0, \tag{8.53}$$

whose roots give the equilibrium values of k_F. Values of r_c, which are similar to ionic radii, can be found in Table 8.4. To compute d_0 for a bcc metal, we note the following relationships that transform the equilibrium value of k_F to d_0.

$$\frac{k_F^3}{3\pi^2} = n = \frac{2}{a^3} = \frac{3\sqrt{3}}{4d_0^3} \tag{8.54}$$

$$d_0 = \left(\frac{9\sqrt{3}\pi^2}{4}\right)^{1/3}\frac{1}{k_F} = \frac{3.376}{k_F}. \tag{8.55}$$

The predicted and observed values are compared in Table 8.5.

Although the agreement is not as good as it was for solids held together by van der Waals or ionic bonds, it is fairly impressive considering the simplicity of the theory. We can also use this electronic energy function to estimate a bulk modulus.

Table 8.6. *Comparison of observed and computed binding energies (per atom) [11].*

metal	E_B (exp.) eV	E_c (exp.) eV	E_B (calculated) eV
Na	5.3	1.1	6.3
Mg	21.6	1.5	24.4
Al	52.2	3.3	56.3

$$B = \Omega^2 \frac{\partial^2 U}{\partial \Omega^2} = \frac{1}{9} k_F^2 \frac{\partial^2 U}{\partial k_F^2} \quad \text{per ion.} \tag{8.56}$$

By substituting in Eqn. 8.56, we find that B is:

$$B = (0.0275 + 0.1102\, k_F\, r_c^2)\, k_F^5 \times 10^{12} \frac{\text{erg}}{\text{cm}^3}. \tag{8.57}$$

While this expression is only valid at the minimum energy defined by k_F, reasonably accurate values for B are obtained by substituting observed values of k_F.

Finally, we note that it is also possible to use the electronic energy function to calculate the metal's binding energy, E_B, and cohesive energy, E_c. The binding energy is defined as the energy needed to separate the electrons and the ions to isolated states and the cohesive energy as that energy required to separate the crystal into neutral, noninteracting atoms. In this context, we take the crystal to be the reference state with zero energy. Note that the binding and cohesive energies differ by the ionization energy.

$$E_B - I_M = E_c. \tag{8.58}$$

Experimental and computed values for three elements are listed in Table 8.6.

Although there is reasonable agreement between the observed and calculated binding energies, the errors have the same order of magnitude as the total cohesive energies. Therefore, this model cannot be used to calculate the cohesive energy with any degree of reliability (although the trend in energies is accurately represented).

In summary, while the model described in this section does not have a satisfying quantitative accuracy, it does provide a sound physical basis for the explanation of the trends in the structures, bulk moduli, and binding energies of simple metals.

Example 8.2

By adding electrostatic terms to the electron kinetic energy, we provided a cohesive force to bind electrons in the crystal. In this example, we consider how well the electrons are bound to the crystal. Assume that the following simplified equation describes the energy of the electrons in a potassium crystal:

$$E(\text{eV}) = \frac{\hbar^2 \vec{k}^2}{2m} - 6.6$$

and the electrons with energy ≥ 0 are unbound. Draw an energy level diagram for a K crystal showing the lowest energy level, the highest (filled) level, and the unbound level. Next, compare the energy required to remove a single electron from a K atom to the energy required to remove a single electron from a K crystal. Determine the average energy required to remove an electron from a K crystal and the cohesive energy of K.

1. The lowest filled level is -6.6 eV, for $k=0$. The unbound level is at $E=0$. The highest filled level is the Fermi level, given by Eqn. 8.24. Knowing that K has the bcc structure, that the cubic lattice constant is 5.225 Å, and that it is an alkali metal with one valence electron per atom, we can use Eqn. 8.24 to determine that the Fermi level is 2.12 eV above the lowest level. When doing calculations of this sort, it is useful to remember that the value of the constant is:

$$\frac{\hbar^2}{m} = 7.63 \ \text{Å}^2\text{-eV.}$$

So, the highest filled level is at -4.48 eV. An energy level diagram representing these conclusions is shown in Fig. 8.8.

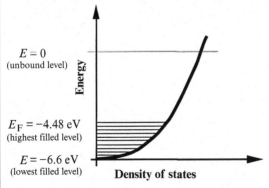

$E = 0$
(unbound level)

$E_F = -4.48$ eV
(highest filled level)

$E = -6.6$ eV
(lowest filled level)

Density of states

Figure 8.8. Energy level diagram for the electrons in K.

2. The minimum energy required to remove an electron from a K crystal is 4.48 eV (removing it from the highest filled state). The energy to remove an electron from an

isolated K atom is the ionization energy, 4.34 eV (see Table 7.5). So, within this model, even the highest energy electrons in the crystal are more stable than in the free atom. The minimum energy to remove an electron from a metallic crystal is usually called the work function.

3. The average energy to remove an electron is the total electronic energy divided by the number of electrons. The average energy per electron is $3/5\varepsilon_F$ (Eqn. 8.32). Correcting this for the electrostatic energy, we have:

$$6.6 - 3/5\varepsilon_F = 5.33 \text{ eV}.$$

As expected, this is larger than the minimum energy or work function.

4. The average energy to remove an electron is also known as the binding energy and the difference between the binding energy and the cohesive energy is the ionization energy. In other words, the cohesive energy is the energy required to separate all of the valence electrons and all of the ion cores, and then place the electrons back on the cores to form neutral atoms. Thus, the cohesive energy is $5.33 - 4.34 = 0.99$ eV.

(C) Failures of the free electron theory

The free electron theory is unable to account for the properties of transition metals. For the most part, this is because the theory does not adequately describe the electrons in d-orbitals. A more glaring failure is that it predicts that all elemental materials should be metallically bound. Metals, insulators, and semiconductors are not distinguished. Our first refinement of the free electron theory, described in Section D, accounts for the presence of the crystal lattice. The interaction between the electrons and the lattice creates discontinuities in the distribution of electrons which results in the distinction between metals and insulators.

(D) Electrons in a periodic lattice

i. *The relationship between reciprocal space and wave vector space*
The functional relationship between orbital energies and wave vectors is called a dispersion relationship. Free electrons have parabolic dispersion:

$$\varepsilon_k = \frac{\hbar^2 \vec{k}^2}{2m}.$$ (8.58)

The correction for the electrostatic interaction between the electrons and the cores lowers these energies, as demonstrated in Example 8.2. The wave vector takes a set of discrete values defined by the integers, n_x, n_y, and n_z.

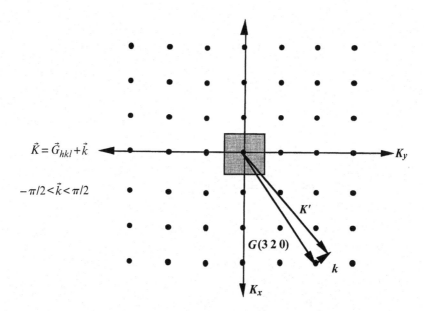

Figure 8.9. Reciprocal space is spanned by a sum over a reciprocal lattice vector, $\vec{G}_{hkl}$, and a wave vector, $\vec{k}$.

$$\vec{k} = \frac{2\pi}{L}(n_x\hat{x} + n_y\hat{y} + n_z\hat{z}). \tag{8.59}$$

By comparing Eqn. 8.59 and Eqn. 8.60 (below), it is easy to see the similarity between k-space and the reciprocal space described in terms of reciprocal lattice vectors, $\vec{G}_{hkl}$:

$$\vec{G}_{hkl} = \frac{2\pi}{a}(h\hat{a}* + k\hat{b}* + l\hat{c}*). \tag{8.60}$$

Note that both of these vectors have the units of inverse length. The main difference is the scale; because L is approximately 10^8 times the cell edge length (a), the k-vectors are very small compared to the G-vectors. So, in reciprocal (G-vector) space we can assume that $\vec{k}$ has an essentially continuous range of values.

Next, consider the section of reciprocal space plotted in Fig. 8.9. The set of $\vec{G}_{hkl}$ vectors can only point to discrete points specified by h, k, and l. However, $\vec{k}$ can point to any location. Thus, we can define a set of reciprocal space vectors, $\vec{K}$, that have a continuous range and are the sum of a $\vec{G}_{hkl}$ vector and a $\vec{k}$ vector.

$$\vec{K} = \vec{G}_{hkl} + \vec{k}. \tag{8.61}$$

Table 8.7. *Dispersion relations for the lowest energy zones, along [100].*

h k l	ε at $\vec{k}=0$	ε at $\vec{k}=(k_x 00)$	band labels
000	0	k_x^2	1
$\bar{1}00, 100$	$(2\pi/a)^2$	$(k_x \pm 2\pi/a)^2$	2, 3
$010, 0\bar{1}0, 001, 00\bar{1}$	$(2\pi/a)^2$	$k_x^2 + (2\pi/a)^2$	4, 5, 6, 7
$110, 1\bar{1}0, 101, 10\bar{1}$	$2(2\pi/a)^2$	$(k_x + 2\pi/a)^2 + (2\pi/a)^2$	8, 9, 10, 11
$\bar{1}\bar{1}0, \bar{1}10, \bar{1}0\bar{1}, \bar{1}01$	$2(2\pi/a)^2$	$(k_x - 2\pi/a)^2 + (2\pi/a)^2$	12, 13, 14, 15
$011, 01\bar{1}, 0\bar{1}1, 0\bar{1}\bar{1}$	$2(2\pi/a)^2$	$k_x^2 + 2(2\pi/a)^2$	16, 17, 18, 19

So, $\vec{K}$ is approximately continuous, $\vec{G}_{hkl}$ points to a specific lattice point, and $\vec{k}$ points to any place within a particular reciprocal space unit cell, specified by h, k, and l. Thus, any point in reciprocal space can be specified by Eqn. 8.61. Keep in mind that each discrete value of $\vec{k}$ and $\vec{K}$ represents an electronic orbital with a different energy; the $\vec{G}_{hkl}$ vectors simply provide an indexing mechanism.

ii. *Plotting dispersion in the reduced zone scheme*
Because the reciprocal lattice has translational periodicity, sections of the dispersion curve from parts of the reciprocal lattice specified by nonzero values of $\vec{G}_{hkl}$ can be translated to the first unit cell ($\vec{G}_{000}$) by the addition or subtraction of a G vector. The unit cell of the reciprocal lattice specified by $\vec{G}_{000}$ is known as the first Brillouin zone (see Chapter 2, Section F(iii)). The convention of plotting the entire energy dispersion curve in the first Brillouin zone is referred to as the reduced zone scheme. To depict the dispersion of the electronic energy in the reduced zone scheme, we use the wave vectors defined in Eqn. 8.61 with respect to the reciprocal lattice. For each unit cell in reciprocal space, specified by a reciprocal lattice vector, $\vec{G}_{hkl}$, we consider the dispersion along selected high symmetry directions using the following expressions for the parabolic dispersion (obtained by substituting Eqn. 8.58 into Eqn. 8.61).

$$\varepsilon_{\vec{k}} = \frac{\hbar^2}{2m}(\vec{k} + \vec{G}_{hkl})^2 \tag{8.62}$$

$$\varepsilon_{\vec{k}} = \frac{\hbar^2}{2m}\left[(\vec{k}_x + \vec{G}_x)^2 + (\vec{k}_y + \vec{G}_y)^2 + (\vec{k}_z + \vec{G}_z)^2\right]. \tag{8.63}$$

The dispersion relations in the different unit cells of the reciprocal lattice, specified by h, k, and l are given in Table 8.7. The segments of the dispersion

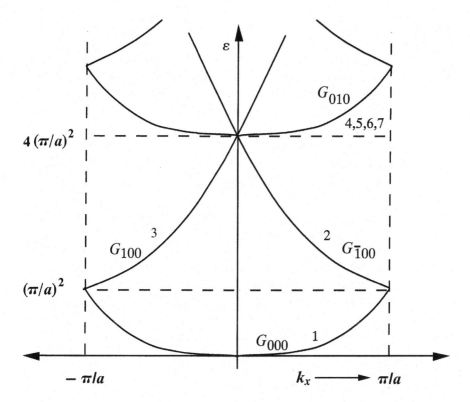

Figure 8.10. Dispersion of free electron energy levels, plotted in the reduced zone scheme. In the [100] direction, the limits of the first Brillouin zone are $\pm \pi/a$.

curves from each zone (unit cell of the reciprocal lattice) are referred to as bands. As an example, the data in Table 8.7 are plotted using this convention in Fig. 8.10. In Fig. 8.10, the band labeled 1 is the dispersion in the first zone. The band labeled 3, however, is the dispersion in the K range from π/a to $3\pi/a$ and has been translated by the reciprocal lattice vector $\vec{G}_{100}$. The band labeled 2 was brought to the first zone by a translation of $\vec{G}_{\bar{1}00}$.

Note that the dispersion changes with direction. While Table 8.7 gives the dispersion along $[k_x 00]$, the other high symmetry directions will be different. It is often of interest to plot the dispersion along other high symmetry directions, such as [110], or $[k_x k_x 0]$. On a dispersion or *band structure diagram*, such as the one shown in Fig. 8.10, the directions in the cell are usually specified by symbols defined with reference to the crystal structure. There are, however, some conventions. For example, the zone center is labeled Γ and the point at $\vec{k} = k_x = \pi/a$ is labeled X. Thus, the direction from Γ to X is [100]; the direction from Γ to K is usually [110]. Band structure diagrams showing multiple directions can be found in Chapter 9 (see, for example, Fig. 9.16).

As we will see in the remaining sections of this chapter and the next chapter, dispersion or band structure diagrams are an integral part of modern theories of chemical bonding and the properties of solids. It is useful to remember that *band structures* are nothing more than electron energy level diagrams for the solid. The bands shown in Fig. 8.10 are called free electron bands because they are plotted assuming the parabolic free electron dispersion. Implicit in the assumption of parabolic dispersion (Eqn. 8.58) is the assumption that the electrons still do not 'see' or int-eract with the lattice.

iii.   *The nearly free electron theory: the origin and magnitude of the energy gap*
The nearly free electron theory includes the effect of the lattice in a very simple way. Remember that X-rays were taken to be plane waves that scattered from the lattice. Within the framework of our model, the free electrons in a metal are also plane waves. Therefore, if we change our model and now assume that the electrons interact with the periodic array of ion cores, we expect them to scatter or diffract in much the same way as the X-rays. Recall the Bragg condition, $\Delta \vec{k} = \vec{G}_{hkl}$ (see Chapter 5, Section C(iv)). Here we consider the one-dimensional case, where $\vec{k}$ is parallel to $\vec{G}$, and determine the condition on $\vec{k}$ for diffraction:

$$(k + G)^2 = k^2$$
$$k^2 + 2kG + G^2 = k^2$$
$$k = 1/2\,G$$
$$k = \frac{\pm n\pi}{a}. \tag{8.64}$$

According to the Bragg condition, electrons diffract when $k = \pm \pi/a$. These are the points in K-space at the Brillouin zone boundaries (the edges of the reciprocal space unit cell). Since the electrons must be reflected at this point, we assume that equal amplitudes travel forward and backward, giving us a standing wave rather than a traveling wave. Thus, in the vicinity of the Brillouin zone boundary, the traveling plane wave description of the free electron wave function is no longer valid. We must instead consider the two possible standing waves:

$$\Psi = e^{ikx} \pm e^{-ikx}. \tag{8.65}$$

At the zone boundary, $k = \pm \pi/a$, so

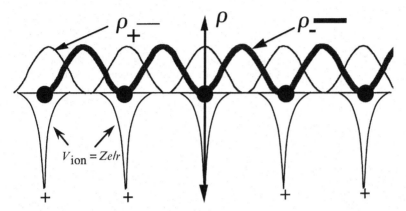

Figure 8.11. The two solutions at the zone boundary lead to different charge distributions. Based on electrostatics, the Ψ_+ solution has a lower energy. The electrostatic potential of the ion cores is shown in the lower part of the diagram.

$$\Psi = e^{i\pi\frac{x}{a}} \pm e^{-i\pi\frac{x}{a}},$$

$$\Psi_+ = 2\cos\left(\pi\frac{x}{a}\right), \tag{8.66a}$$

$$\Psi_- = 2i\,\sin\left(\pi\frac{x}{a}\right). \tag{8.66b}$$

The charge densities of the electrons in these states are:

$$\rho(+) = |\Psi_+|^2 \propto \cos^2 \pi\frac{x}{a} \tag{8.67a}$$

$$\rho(-) = |\Psi_-|^2 \propto \sin^2 \pi\frac{x}{a}. \tag{8.67b}$$

When this charge density is plotted (see Fig. 8.11), the Ψ_+ solution has the greatest charge density situated on the ion core positions, while Ψ_- puts the electrons between the cores. So, although both solutions are valid at $k = \pm\pi/a$, we have to assume that the energy of Ψ_+ is lower than that of Ψ_- and that away from the Brillouin zone boundary, the free electron dispersion still applies. The dispersion diagram in Fig. 8.12 reflects these conclusions.

By allowing the electrons to interact with the lattice, we have created a new and very important feature in the dispersion curve. There is now an energy gap at $k = \pm\pi/a$. Within this range of energies, there are no electronic states. Thus,

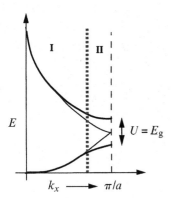

Figure 8.12. Comparison of the free electron dispersion (thin line) and the nearly free electron dispersion (the bold line). In region I, the dispersion is based on free electron, plane wave states. In region II, the dispersion is the result of standing wave states that occur because of diffraction from the lattice. Accounting for the interaction between the electrons and the lattice creates a gap at the zone boundary.

there is no longer a continuous range of energy levels that the electrons in a solid can occupy.

Our next task is to determine the size of this energy or band gap. We will assume that the crystal potential, or the strength of the interaction between the electrons and the ions, is given by the periodic potential:

$$U(x) = U \cos \frac{2\pi x}{a}. \tag{8.68}$$

This potential obviously has the same periodicity as the lattice. Based on the following normalized wave functions, we can compute ρ, the charge density.

$$\rho + = |\Psi_+|^2 = 4 \cos^2 \frac{\pi x}{a} \tag{8.69}$$

$$\rho - = |\Psi_-|^2 = 4 \sin^2 \frac{\pi x}{a}. \tag{8.70}$$

Given the charge density and the electric potential, we can simply compute the difference in the electrostatic energy between these two charge distributions. (It would also be possible to compute the difference in the energy expectation values for each wave function and get the same result.)

$$E_g = \int_0^a U(x)[\rho_+ - \rho_-]dx$$

$$E_g = 2U \int_0^a \cos \frac{2\pi x}{a} \left[\cos^2 \frac{\pi x}{a} - \sin^2 \frac{\pi x}{a} \right] dx$$

$$E_g = U. \tag{8.71}$$

So, the size of the energy gap, E_g, is equal to the amplitude of the crystal potential, U. Thus, in the weak interaction limit, U and E_g go to zero and we have the free electron situation. As the strength of the interaction between the valence electrons and the ions increases, the gap becomes larger and larger.

iv.   Metals and insulators

To determine how the electrons fill the allowed energy levels on our dispersion (energy level) diagrams, we must determine the number of possible values of k in each Brillouin zone. Because k is limited to integer multiples of $2\pi/L$, we can say that $k = n2\pi/L$. At the zone boundary, $k = \pi/a$, so $n = L/2a$. Also, since $L = Na$, $n = N/2$. Using this line of reasoning, the number of allowed k-states between 0 and π/a is $N/2$, and the number between $-\pi/a$ and $+\pi/a$ is N. Finally, because each k-state (orbital) can hold two electrons, the number of electrons in the first Brillouin zone is $2N$, where N is the number of primitive cells in the crystal. Each line or 'band' on the dispersion diagram represents dispersion in one particular Brillouin zone. For zones further away from $\vec{G}_{000}$, the energy increases. To distinguish metals, semiconductors, and insulators, we have to count the valence electrons and see how they fill the bands.

A necessary condition for a metal is that at least some fraction of the valence electrons are able to move easily under the influence of an externally applied field. Such motion is dependent upon the electrons' ability to change energy (and momentum) states in a nearly continuous manner. This means that some electrons must occupy states that are energetically very near unoccupied states so the activation barrier to move from one state to another is small compared to the available thermal energy. This is only true for electrons in the highest filled levels (i.e., at the Fermi level), since there are unoccupied states at very slightly higher energies.

Within the framework of the free electron model, there is a continuous range of states such that the electrons in the highest filled levels always have unoccupied states available just above the Fermi level. However, consider the energy levels in the nearly free electron model. In this model, certain levels are separated from the next available level by an energy gap. Thus, if electrons just fill all the states up to the gap (this requires an even number of electrons per atom), then there are no easily accessible states just above the highest level and the material should not conduct electricity or reflect light. The condition that energetically close empty states are available only occurs when a band is partially filled and, according to the electron counting argument posed above, this occurs only when there is an odd number of electrons.

As a general rule, elements with an odd number of valence electrons should be metals (group I and group III, for example) and elements with an even number of valence electrons should be insulators (group II and group IV, for example). Obviously, this is not well obeyed; group II elements are all metals. The reason for this is that the first and second bands overlap (there is no gap) so that there are unoccupied states just above the Fermi level (see Fig. 8.13). Obviously, our

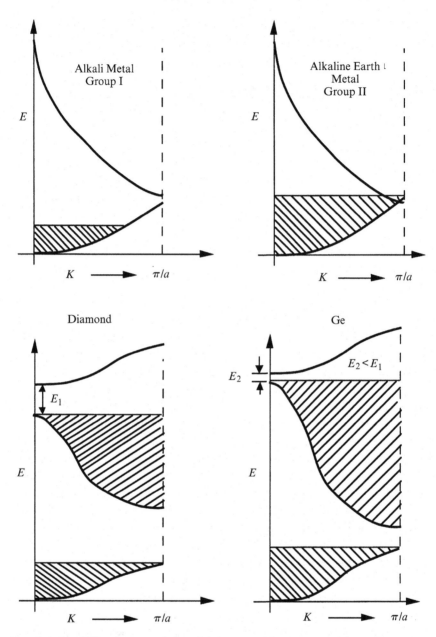

Figure 8.13. Band filling for group 1, 2, and 4 elements with one, two, and four valence electrons. The hatched areas show the filled levels. The gap between bands depends on the strength of the interaction between the electrons and the ion cores. Diamond is an insulator and Ge is a semiconductor.

simple nearly free electron model can not produce such effects. The width of the gap between the occupied and unoccupied states determines if a material will be insulating (large gap) or semiconducting (small gap). The limited electronic conduction in an intrinsic semiconductor is due to the thermal excitation of electrons across the gap. The filling of bands is illustrated in Fig. 8.13. Note also that the magnitude of the band gap should correlate with the strength of the crystal potential. Considering the bandgaps of the group IV elements (see Table 9.10), this implies that the strength of the electron–lattice interaction diminishes as one goes from smaller to larger atoms. This seems realistic when one considers the fact that in each case, four valence electrons are attracted to four nuclear charge and the separation is smallest in C and largest in Pb.

(E) Correlation of the physical models with the phenomenological trends

The reflectivity and conductivity of metals are explained by the fact that the valence electrons are not bound in localized states but occupy, instead, 'free' plane wave states. While they do have an electrostatic interaction with the ion cores, they are still free to move about the crystal in response to externally applied electric fields.

The trends in the melting point and cohesive energy have their origin in the electrostatic component of the bonding. As the electron density and the nuclear charge increase, the electrostatic attraction and the bond strength increase. As the atoms get larger, the separation between the centers of positive charge and the free electrons increases and the electrostatic attraction is diminished. Note that increasing the electron density also increases the kinetic energy and partially offsets the electrostatic component of the bonding. The electrostatic component of the bond energy also explains the fact that metals form dense, close packed structures. These structures minimize the total energy.

(F) Empirical potentials for calculating the properties of defects in metals

At the conclusions of Chapters 6 and 7, the use of pair potentials to compute the properties of defects in crystals was briefly discussed. It was previously noted that using an empirical Lennard-Jones model for Cu led to erroneous predictions. We can now see why such a model is not appropriate. The interactions among atoms in a metallic crystal are not pairwise, but are mediated by the local density of electrons surrounding the atoms. In the vicinity of heterogeneities such as a surface or grain boundary, the local atomic density and, therefore, the local density of electrons will not be uniform. Several empirical potentials have been developed to account for this problem; among the most well known are the Finnis–Sinclair (FS) [13] and embedded atom method (EAM) [14] potentials. These potentials have the following form:

$$U = -\sum_i f(\rho_i) + \sum_i \sum_{j>i} V_r(r_{ij}). \tag{8.72}$$

In Eqn. 8.72, $f(\rho_i)$ represents an empirical embedding function such as:

$$f(\rho_i) = f_e e^{-\beta[(r/r_e)-1]}, \tag{8.73}$$

which is the energy associated with placing the atom labeled i in the charge density arising from the surrounding atoms:

$$\rho_i = \sum_j \rho_{ij}(r_{ij}). \tag{8.74}$$

The second term in Eqn. 8.72 is the sum of the repulsive interatomic interactions and r_{ij} is the separation between atoms i and j. The repulsive interaction term is similar in spirit to the $(1/r)^{12}$ part of the Lennard-Jones potential. However, since the embedding energy is exponential, it is convenient to also assume an exponential function for the repulsive part:

$$V_r(r_{ij}) = V_e e^{-\gamma[(r/r_e)-1]}. \tag{8.75}$$

In Eqns. 8.73 and 8.75, r_e is the equilibrium interatomic separation and the remaining constants are adjustable parameters chosen so that a model based on Eqn. 8.72 reproduces known properties.

In Section D of Chapter 6, it was noted that based on Lennard-Jones potentials, the surface layer of Cu is incorrectly predicted to relax outward. When an EAM-type potential of the sort given by Eqn. 8.72 is used, the surface layers are predicted to relax inward, as observed experimentally. Potentials of this type have been used widely by materials scientists to compute unknown quantities. The results from one example of such a calculation are illustrated in Fig. 8.14, which shows the energies of symmetric tilt grain boundaries in gold, formed by rotations about $\langle 110 \rangle$ [15]. The relatively low energy cusps correspond to special CSL orientations (see Chapter 2, Section I(*iii*)).

Ⓖ Problems

(1) It is interesting to note that the Fermi energy in the free electron model depends only on the density of electrons, not on the number of electrons in the

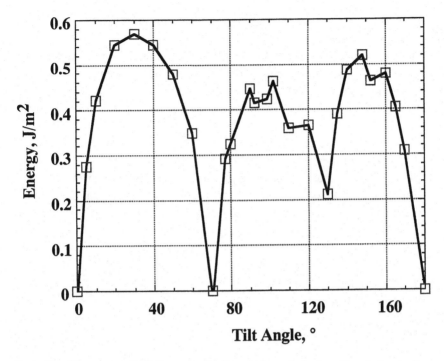

Figure 8.14. The energy of ⟨110⟩ symmetric tilt grain boundaries in gold, calculated using an EAM potential [15].

system. This implies that the density of energy levels must be a function of the volume of the system.

(i) Calculate the average spacing between free electron energy levels in a cube shaped crystal of Na that is 1 cm on each edge. How does this compare to the amount of available thermal energy at room temperature?

(ii) How does the situation change for a cube shaped crystal that is one micron on each side?

(iii) At what size will the separation between energy levels be larger than the thermal energy available at room temperature?

(iv) How will the properties of a crystal smaller than the critical size found in part (iii) differ from those of a macroscopic crystal?

(2) Using the model for cohesion in a metallic crystal, there are three components to the total energy of the simple metal. One is the free electron kinetic energy, the second is the electrostatic component (with the empty core correction), and the third is the exchange-correlation energy. For Na, Mg, and Al, determine the fractional contribution of each term to the total.

(3) Increasing the electron density in a metal increases the electrostatic bond strength. However, this stabilization is partially offset by an increase in the

electronic kinetic energy. Compare the relative stabilizing and destabilizing effects of a unit increase in the electron density.

(4) Develop an interatomic bonding potential for simple metals that includes only the electrostatic component of the energy (with the empty core correction) and the repulsive part of the 6–12 potential. Test this potential by trying to predict the lattice constant of several metals. Is this a realistic model?

(5) Consider a free electron gas in two dimensions, confined to a region of space of length a and width b.

(i) Use cyclic boundary conditions to determine the allowed k values for this situation. Indicate on a graph of k-space the allowed points and determine the density of points in k-space (that is, the number of points per area of k-space).

(ii) By analogy with the three-dimensional case, we can describe a 'Fermi-circle' such that within the circle, all the states are occupied and outside this circle, all of the states are empty. Indicate this circle on your graph.

(iii) If there are Z electrons per atom and the area per atom is Ω, what is the radius, k_F, of the Fermi-circle?

(iv) What is the Fermi energy?

(v) Determine the energy density of states.

(vi) Compute the average energy per electron in terms of the Fermi energy.

(6) In this exercise, we consider the bulk modulus of a simple metal.

(i) Assuming that the total energy of a simple metal is accurately described by the kinetic energy of the free electron gas, show that the bulk modulus of a simple metal is:

$$B = \{e^2 a_0 / 9\pi^2\} \, k_F^{\,5} \text{ (where } a_0 \text{ is the Bohr radius).}$$

(ii) Next, starting from the free-electron energy function corrected for the electrostatic energy, the exchange energy, and the empty core (Eqn. 8.52), show that the bulk modulus is:

$$B = (0.0275 + 0.1102 \, k_F \, r_c^2) \, k_F^{\,5} \times 10^{12} \, \frac{\text{erg}}{\text{cm}^3}.$$

(iii) Compare the accuracy of these two formulas by computing B for the alkali metals, Li, Na, K, Rb, and Cs.

(7) Consider the following observations regarding bonding in simple metallic materials:

(a) The cohesive energies of simple divalent metals are larger than those of simple monovalent metals.

(b) The cohesive energies of simple metals with the same number of valence electrons decrease as the size of the atom and the unit cell volume increases.

(c) Vacancy creation in simple metals is an endothermic process.

(d) Simple metals form dense, closely-packed crystal structures.

(i) Within the framework of the free electron theory, the total energy of a simple metal is given by Eqn. 8.32. Can this formula be used to explain the four observations cited at the beginning of the question? Describe the reasoning you used to reach your conclusion.

(ii) When we include an electrostatic bonding energy, the total energy is given by Eqn. 8.52. Does this new total energy function effectively explain the four observations cited at the beginning of the question? Again, describe the reasoning you used to reach your conclusion.

(8) At what electron concentration does the Fermi sphere of a bcc metal touch the Brillouin zone boundary? Express the electron concentration in electrons per atom.

(9) At what electron concentration does the Fermi sphere of a hcp metal touch the Brillouin zone boundary? Express the electron concentration in electrons per atom.

(10) The lowest occupied energy level in Na metal is -8.2 eV (8.2 eV below the unbound or vacuum level). What is the minimum amount of energy that can be used to remove an electron from solid Na? How does this compare to the energy needed to remove an electron from a single, free sodium atom? Next, what is the average energy for removing all of the electrons from a sodium crystal? How does this compare to the energy needed to remove an electron from a single, free sodium atom? Based on these calculations, what is the cohesive energy per atom for crystalline Na?

(11) Plot the dispersion, in the reduced zone scheme, along the [110] direction, for a cubic P material. Plot the three lowest energy bands.

(12) Plot the dispersion, in the reduced zone scheme, along the [111] direction, for a cubic P material. Plot the three lowest energy bands.

(H) References and sources for further study

[1] C. Kittel, *Introduction to Solid State Physics*, 5th edition (J. Wiley & Sons, New York, 1976) p. 155. Simple metals.

[2] W.A. Harrison, *Electronic Structure and the Properties of Solids: The Physics of the Chemical Bond*, (Dover Publications, Inc., New York, 1989) Chapter 1, section 1A. A discussion of the quantum mechanical assertions.

[3] R.L. Liboff, *Introductory Quantum Mechanics* (Holden-Day, Inc., San Francisco,

1980). Chapters 1, 2, and 3 provide a nice background for the quantum mechanical assertions.

[4] W.A. Harrison, *Electronic Structure and the Properties of Solids: The Physics of the Chemical Bond* (Dover Publications, Inc., New York, 1989) pp. 341–9. Discussion of free electron theory.

[5] C. Kittel, *Introduction to Solid State Physics*, 5th edition (J. Wiley & Sons, New York, 1976) pp. 155–63. Discussion of free electron theory.

[6] N.W. Ashcroft and N.D. Mermin, *Solid State Physics* (Holt Rinehart and Winston, New York, 1976) Chapter 2, pp. 30–42. Discussion of free electron theory.

[7] P. Lorrain and D. Corson, *Electromagnetic Fields and Waves*, 2nd edition (W.H. Freeman and Co., San Francisco, 1962) Chapter 2. Helpful electrostatics.

[8] W.A. Harrison, *Electronic Structure and the Properties of Solids: The Physics of the Chemical Bond* (Dover Publications, Inc., New York, 1989) pp. 539–42. A more detailed description of the exchange energy is included in this appendix to Harrison's book.

[9] N.W. Ashcroft and N.D. Mermin, *Solid State Physics* (Holt Rhinehart and Winston, New York, 1976) pp. 410–12, also see Problem 4 of Chapter 20. The electrostatic correction to the free electron theory.

[10] W.A. Harrison, *Electronic Structure and the Properties of Solids: The Physics of the Chemical Bond*, (Dover Publications, Inc., New York, 1989) pp. 349–57. The electrostatic correction to the free electron theory.

[11] N.W. Ashcroft and D.C. Langreth, *Phys. Rev.* **155** (1967) 682.

[12] C. Kittel, *Introduction to Solid State Physics*, 5th edition (J. Wiley & Sons, New York, 1976) pp. 185–90. A description of the properties of electrons in a periodic lattice.

[13] M.W. Finnis and J.E. Sinclair, *Philos. Mag. A* **50** (1984) 45.

[14] M.S. Daw and M.I. Baskes, *Phys. Rev. B* **29** (1984) 6443.

[15] D. Wolf, *J. Mater. Res.* **5** (1990) 1708. The results shown in Fig. 8.13 are based on Figure 7 of this paper.

Chapter 9
Covalent Bonding

(A) Introduction

The defining characteristic of a covalent bond is the existence of a local maximum in the valence electron density in the regions between the atomic cores. For example, the experimentally measured charge density in Si, illustrated in Fig. 9.1, shows peaks between the atomic positions [1–3]. From this phenomenon comes the simple idea that two atoms forming a covalent bond share their valence electrons. Concentrating the valence electrons in the spaces between the atomic cores is clearly distinct from the ionic bonding model, where the valence electrons are centered on the anion positions, and the metallic bonding model, where the valence electrons are uniformly distributed in the free electron sea. Therefore, we will have to adopt an alternative model for the description of the valence electrons in a covalently bonded crystal. In the ionic bonding model, it was assumed that valence electrons were transferred from atomic states on the cation to atomic states on the anion. In the metallic bonding model, it was assumed that valence electrons were transferred from atomic energy levels to free electron states. The objective of this chapter is to describe a model for the transfer of valence electrons from atomic energy levels to a new set of crystal energy levels which can be simply thought of as having properties that are intermediate between the atomic energy levels used in the ionic bonding model and the free electron energy levels used in the metallic bonding model. The new model is referred to as the linear combination of atomic orbitals or LCAO model and was originally proposed by Bloch [4].

The fact that the LCAO derived crystal energy levels have properties that are intermediate between atomic and free electron-like states allows them to be used, in the appropriate limits, to describe ionic and metallically bonded compounds as well as those that would normally be regarded as covalent. The versatility of the LCAO model is both valuable and appropriate when we consider the fact that the three types of primary chemical bonds (ionic, metallic, and covalent) are actually limiting cases of a single phenomenon. In each case, it is the electrostatic attraction between the valence electrons (occupying atomic states on the anion, free electron states, or crystal orbitals) and the partially or completely ionized atomic cores that provides the cohesion and stabilizes the crystal. It is also worth noting that the limiting cases for the three primary bond types are rarely realized

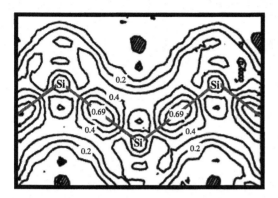

Figure 9.1. The experimentally determined valence electron density in the {110} plane of Si [1]. The contours are at 0.1 e/Å³. This is a difference electron density plot, where the electron density of the cores is subtracted from the total observed electron density. The total observed electron density is a finite Fourier series consisting of experimentally measured X-ray structure factors (see Eqns. 5.6 and 5.7). The straight gray lines connect the Si positions; note the peak in the electron density at the midpoint of the line connecting the Si atoms. This peak in the electron density between the two nuclei is the signature of the homopolar covalent bond. The shape of this peak in the electron density is theoretically predicted [2] and also found in many III–V semiconductors with the sphalerite structure [3].

in practice. For example, while the bonding between two identical nonmetallic atoms should be regarded as purely homopolar with an equal sharing of the valence electron density, we do not expect two different atoms to share electrons equally. In the so-called polar–covalent bond, the more electronegative atom gets a greater portion of the valence electron density than the electropositive atom. In the limiting case, where the electronegative atom completely strips the valence electrons from the more metallic element, the polar–covalent bond becomes ionic. Obviously, there are far more examples of the polar–covalent case than either the pure homopolar or the pure ionic. For this reason, a model for real chemical bonds must describe what happens between these two limits. The LCAO model described in this chapter gives us the ability to determine how the spacing and occupation of energy levels change when the free atoms form a crystal. Before beginning, we will review the materials that are best described as being covalently bonded and the trends in their properties.

i. *Which solids have covalent bonds*

Considering the discussion above, one might state that all solids have some degree of covalent bonding. In this section, however, we will concentrate on materials for which the ideal covalent bond approximation (shared, localized

Table 9.1. *Tetrahedral covalent radii, in Å [5, 6].*

	Be	B	C	N	O	F
Pauling	1.06	0.88	0.77	0.70	0.66	0.64
(Phillips)	(0.975)	(0.853)	(0.774)	(0.719)	(0.678)	(0.672)
	Mg	Al	Si	P	S	Cl
	1.40	1.26	1.27	1.10	1.04	0.99
	(1.301)	(1.230)	(1.173)	(1.128)	(1.127)	(1.127)
Cu	Zn	Ga	Ge	As	Se	Br
1.35	1.31	1.26	1.22	1.18	1.14	1.11
(1.225)	(1.225)	(1.225)	(1.225)	(1.225)	(1.225)	(1.225)
Ag	Cd	In	Sn	Sb	Te	I
1.52	1.48	1.44	1.40	1.36	1.32	1.28
(1.405)	(1.405)	(1.405)	(1.405)	(1.405)	(1.405)	(1.405)
	Hg					
	1.48					

valence electron density) is a better description than the ionic, metallic, or van der Waals approximations. First, we can say the chemical bonds within small molecules are almost always covalent. However, we are more interested in crystals. In this case, we say that covalent bonds occur in crystals of nonmetallic elements and in crystals of compounds containing both metallic and nonmetallic elements with an electronegativity difference ≤ 1.7. This is the same definition proposed in Chapter 1.

ii. *The crystal structures of covalent solids*
It is usually safe to assume that any structure that has most atoms in a coordination environment of four or fewer nearest neighbors is bonded covalently. For example, the most common structures for the group IV elements, group III–V compounds, and group II–VI compounds are the diamond, sphalerite, and wurtzite structures, respectively. Recall from Chapter 4 that all of the atoms in these structures are tetrahedrally coordinated. Silicate compounds are another example of a broad class of materials that have a significant degree of covalent bonding. In the silicates, Si is nearly always in tetrahedral coordination and oxygen has a lower coordination number.

Because the most common coordination environment for atoms in covalently bound crystals is tetrahedral, a special set of atomic radii were developed for these cases. The so-called tetrahedral covalent radii, listed in Table 9.1, can

Table 9.2. *Properties of group IV elements.*

Material	d_0(Å)	T_m(°C)	E_c(eV)	E_g(eV)
diamond	1.54	>3500	3.68	6.0
silicon	2.35	1410	2.32	1.1
germanium	2.44	937	1.94	0.7
tin (gray)	2.80	232	1.56	0.1

Table 9.3. *Effect of polarity on properties of an isoelectronic series.*

group	material	ionicity	T_m(°C)	E_c (eV)	E_g (eV)
IV	Ge	0 %	937	1.94	0.7 eV
III–V	GaAs	4 %	1238	1.63	1.4 eV
II–VI	ZnSe	15 %	1517	1.29	2.6 eV

be used to estimate size mismatch, strains, and lattice parameters in common semiconductors and semiconductor alloys.

iii. *Phenomenological trends in covalent bonding*
As with other bond types, it is possible to correlate interatomic distances and selected physical properties with the bond strength. For example, Table 9.2 shows that the bond strength, as reflected in the melting point, T_m, and cohesive energy, E_c, increases as the interatomic separation decreases. Note that the band gap (E_g) is inversely related to the interatomic separation. The cohesive energies of 30 polar covalent solid are tabulated in Appendix 9A.

The elements on either side of group IV (shown in Fig. 9.2) react to form the so-called III–V and II–VI compound semiconductors. The effects of bond polarity are illustrated in Table 9.3, where selected properties of isoelectric solids with nearly the same interatomic spacing are listed. As the polarity of the bond increases, the band gap increases. Therefore, the electronic and optical properties approach those of an insulating, transparent, ionic compound. The melting point also increases with polarity. Less energy is required to melt a crystal containing unpolarized species (Ge) than a crystal consisting of partially charged species (GaAs and ZnSe). However, the cohesive energy shows the opposite trend. In this case, the difference is rooted in the fact that these two quantities (T_m and E_c) measure the energy required to take the crystal to two different final states: a melt containing species that remain polarized versus a collection of neutral atoms. We will discuss this difference again in Section H of this chapter.

II	III	IV	V	VI
	5 **B** 2.0	6 **C** 2.5	7 **N** 3.0	8 **O** 3.5
	13 **Al** 1.5	14 **Si** 1.8	15 **P** 2.1	16 **S** 2.5
30 **Zn** 1.6	31 **Ga** 1.6	32 **Ge** 1.8	33 **As** 2.0	34 **Se** 2.4
48 **Cd** 1.7	49 **In** 1.7	50 **Sn** 1.8	51 **Sb** 1.9	52 **Te** 2.1
80 **Hg** 1.9	81 **Tl** 1.8	82 **Pb** 1.8	83 **Bi** 1.9	84 **Po** 2.0
s^2	s^2p^1	s^2p^2	s^2p^3	s^2p^4

Figure 9.2. The III–V and II–VI compounds are arranged symmetrically around group IV and usually crystallize in the sphalerite or wurtzite structures. The Pauling electronegativity is listed below the atomic number and symbol.

(B) A physical model for the covalent bond in a molecule

In the LCAO model, the electronic states in the crystal are formed by a superposition of atomic energy levels. This superposition is accomplished by constructing a linear combination of the atomic orbitals which hold the valence electrons on independent atoms. Since the electronic states on independent atoms are already well described (for a review of the electronic structure of the atoms, see Appendix 9B), specifying the new states for the electrons in the crystal amounts to determining a set of coefficients for the linear combination. The description of the LCAO model presented in this chapter largely follows Harrison's [7] treatment. The advantage of this particular model over more sophisticated approaches is that qualitatively realistic answers can be obtained without resorting to computer calculations. Further, the energies of the electronic states in the crystal are specified by three easily quantifiable parameters, the covalent energy, the polar energy, and the metallic energy. Before applying the LCAO model to a crystal, we will examine the principal features of the model by applying it to simple molecules. The results from this analysis will then be extended to periodic solids in Section C.

i.   The homopolar bond in a diatomic molecule

To describe the electronic energy levels in a diatomic molecule within the framework of the LCAO model, we begin by writing a molecular wave function as a linear combination of the two atomic wave functions.

$$\Psi(\vec{r}) = u_1 \, \Psi_1(\vec{r}) + u_2 \, \Psi_2(\vec{r})$$
$$|\Psi\rangle = u_1|1\rangle + u_2|2\rangle. \tag{9.1}$$

In Eqn. 9.1, $\Psi(\vec{r})$ is a wave function for the molecule, $\Psi_1(\vec{r})$ is the atomic orbital for the valence electron on the first atom, $\Psi_2(\vec{r})$ is the atomic orbital for the valence electron on the second atom, and u_1 and u_2 are the unknown coefficients. The unknown coefficients need not be real numbers. In each case we examine, the number of distinct electronic states in the bound system will match the number of orbitals used in the linear combination. In the present case, there will be two distinct sets of coefficients, u_1 and u_2.

It is common practice to choose a set of atomic orbitals that are mutually orthogonal and normalized, a condition called orthonormal. The condition for normality is:

$$\langle 1|1\rangle = \langle 2|2\rangle = \langle i|i\rangle = 1 \tag{9.2}$$

and the condition for orthogonality is:

$$\langle i|j\rangle = 0 \qquad \text{for } i \neq j. \tag{9.3}$$

In Eqns. 9.2 and 9.3, the states are assumed to be on the same atom. If i and j are on two different atoms (say, the H 1s wave function on two adjacent hydrogen atoms), we apply the 'zero overlap approximation' and say that $\langle s_1|s_2\rangle = 0$. While this is not strictly true, the errors introduced by this approximation are not considered significant [8].

Recalling the definition used to calculate expectation values (Eqn. 8.9), we can use the Hamiltonian operator to find the energy expectation value for an electron in this molecular orbital:

$$\varepsilon = \frac{\langle \Psi|H|\Psi\rangle}{\langle \Psi|\Psi\rangle}$$

$$\varepsilon = \frac{u_1^* u_1 \langle 1|H|1\rangle + u_1^* u_2 \langle 1|H|2\rangle + u_2^* u_1 \langle 2|H|1\rangle + u_2^* u_2 \langle 2|H|2\rangle}{u_1^* u_1 \langle 1|1\rangle + u_1^* u_2 \langle 1|2\rangle + u_2^* u_1 \langle 2|1\rangle + u_2^* u_2 \langle 2|2\rangle}. \tag{9.4}$$

Using the orthonormality conditions and simplifying the notation so that $\langle i|H|j\rangle = H_{ij}$, we can rewrite the energy expectation value in the following way:

$$\varepsilon(u_1^*, u_2^*) = \frac{u_1^* u_1 H_{11} + u_1^* u_2 H_{12} + u_2^* u_1 H_{21} + u_2^* u_2 H_{22}}{u_1^* u_1 + u_2^* u_2}. \tag{9.5}$$

According to this expression, we need to know u_1 and u_2 to determine ε. While we don't know these quantities, we can assume that minimizing Eqn. 9.5 with respect to u_1^* and u_2^* gives the lowest energy eigenvalue or lowest electronic ground state. This is accomplished by taking partial derivatives with respect to u_1^* and u_2^* and simultaneously equating them with zero.

$$\frac{\partial \varepsilon}{\partial u_1^*} = \frac{H_{11} u_1 + H_{12} u_2 - \varepsilon u_1}{u_1^* u_1 + u_2^* u_2} = 0$$

$$\frac{\partial \varepsilon}{\partial u_2^*} = \frac{H_{21} u_1 + H_{22} u_2 - \varepsilon u_2}{u_1^* u_1 + u_2^* u_2} = 0. \tag{9.6}$$

A nontrivial solution requires that:

$$(H_{11} - \varepsilon)u_1 + H_{12} u_2 = 0$$
$$H_{21} u_1 + (H_{22} - \varepsilon)u_2 = 0. \tag{9.7}$$

To solve for ε, we set the determinant of the coefficient matrix equal to zero.

$$\begin{vmatrix} H_{11} - \varepsilon & H_{12} \\ H_{21} & H_{22} - \varepsilon \end{vmatrix} = 0$$

$$(H_{11} - \varepsilon)(H_{22} - \varepsilon) - H_{12} H_{21} = 0$$
$$\varepsilon^2 - (H_{11} + H_{22})\varepsilon + H_{11} H_{22} - H_{12} H_{21} = 0. \tag{9.8}$$

The zeros of Eqn. 9.8, which define the energy levels of the system, are given by the solution to the quadratic equation, $\frac{1}{2a}\left[-b \pm \sqrt{b^2 - 4ac}\right]$:

$$\varepsilon_\pm = \frac{1}{2}(H_{11} + H_{22}) \pm \sqrt{\left[\frac{1}{2}(H_{11} - H_{22})\right]^2 + H_{12} H_{21}}. \tag{9.9}$$

A quick look at Eqn. 9.9 tells you that the ε_- solution has a lower energy than the ε_+ solution. Therefore, in its ground state, a single electron in this system would occupy the ε_- state. There are other interesting things you can learn by inspection of Eqn 9.9. For example, H_{11} and H_{22} represent the integrals that give the energy expectation values for electrons in the atomic orbitals labeled 1 and 2, when the atoms are at infinite separation. Thus, the values of H_{11} and H_{22} describe the energy levels of the atoms before a bond is formed (these energies are called atomic term values). The first quantity in Eqn. 9.9 is, therefore, the average energy of the two orbitals before the bond is formed; this describes the chemistry of the molecule rather than its structure. The second term in Eqn. 9.9 represents the change in energy that occurs when the orbitals are brought close enough together to interact. This interaction leads to the formation of one state that has an energy lower than the average of the two isolated atoms (this is ε_-, or the 'bonding' state) and another that has an energy that is higher than that of the two isolated atoms (this is ε_+, or the 'antibonding' state). Since H_{11} and H_{22} are constant for a given set of atoms, the energy difference between ε_+ and ε_- is determined by the size of H_{12} (which is equal to H_{21}). This quantity, as we shall see, depends on the separation between the atoms. In summary, the wave functions for electrons in a bound system are written as a linear combination of the atomic wave functions of the individual atoms, and the electronic energy levels are determined by minimizing the energy expectation value with respect to the unknown coefficients.

This model for the bond is intuitively sensible for the following reasons. If we consider a collection of atoms situated far enough apart that they are completely noninteracting, all the electronic states on each atom are easily enumerated and their energies can be specified. If we begin to bring the atoms closer together, at some distance the wave functions will begin to overlap and an electron on one atom will begin to be influenced by the nucleus of the other. At this point, we can be certain of two things. First, the total number of electrons and electronic states will remain constant. Second, the energies of the electrons will change. If this change reduces the total energy of the system, then chemical bonds will form. The LCAO approximation is a good way to simplify the complex problem of describing these changes in the electronic structure because it is necessary only to determine a set of unknown coefficients instead of unknown wave functions.

We can make this model a bit more tangible by considering a specific case. We will begin with the simplest possible situation, the H_2 molecule. In this case, $|1\rangle$ is the H 1s wave function on the first atom and $|2\rangle$ is the H 1s wave function on the second atom. Recall from Eqn. 9.9 that determining $\varepsilon \pm$ requires knowledge of only H_{11}, H_{22}, H_{12}, and H_{21}. These quantities are generally referred to as 'matrix elements' because of their position in the matrix that is ultimately used to solve the problem.

The value of H_{11} is simply the energy of an electron in the 1s orbital of an isolated H atom, with reference to the vacuum level. Because the two H atoms are identical,

$$H_{11} = H_{22} = \varepsilon_s. \tag{9.10}$$

Keep in mind that the atomic term values are well established characteristics of the free atoms experimentally determined by spectroscopic methods; selected term values are tabulated in Appendix 9B. For H, the term value ε_s is the same as the ionization energy of H (13.6 eV), which is simply the energy required to remove the electron from the atom.

The value of H_{12}, on the other hand, is a separation dependent 'overlap' integral. This value characterizes the interaction of the first and second wave function.

$$H_{12} = \langle 1|H|2 \rangle = \int \psi_1^* \, H \, \psi_2 \, d\vec{r}. \tag{9.11}$$

You can imagine that when the spatial separation between 1 and 2 is many times larger than the atomic dimension, H_{12} should tend to zero; in other words, there is very little overlap between the two wavefunctions. On the other hand, as the two atoms are brought into closer proximity, H_{12} increases. The exact value of H_{12} depends on the Hamiltonian operator (H) and, therefore, the potential energy. Since the appropriate potentials for this operator remain unknown, calculations of H_{12} are, at best, approximations. We shall avoid the calculation altogether and take H_{12} to be equal to $-V_2$, where the magnitude of V_2 depends on the interatomic separation (the discussion of how V_2 is parameterized for quantitative calculations is withheld until Section C). Assuming further that $H_{12} = H_{21}$, we can rewrite the determinant in Eqn. 9.8 in terms of ε_s and V_2.

$$\begin{vmatrix} (\varepsilon_s - \varepsilon) & -V_2 \\ -V_2 & (\varepsilon_s - \varepsilon) \end{vmatrix} = 0. \tag{9.12}$$

Solving for ε leads to the following simple result:

$$\varepsilon \pm = \varepsilon_s \pm V_2. \tag{9.13}$$

Considering that each of the two states described by Eqn. 9.13 is capable of holding two electrons and that the H_2 molecule has only two electrons, both will go to the low energy ε_- state. If we neglect some of the more subtle aspects of this problem, the total energy of the two atoms before bonding is $2\varepsilon_s$ and the energy after bond formation is $2\varepsilon_s - 2V_2$. The amount of energy gained by bond formation (this is also the energy that would be required to separate the two atoms) is $2V_2$. We can now see that this quantity, V_2, the interaction between states 1 and 2, is a measure of the bond strength. Following Harrison [7], we will call this quantity the *covalent energy*. Keep in mind that V_2 is sensitive to the interatomic separation and the exact form of the potential in the Hamiltonian.

If you substitute the solutions for ε into the original linear equations with u_1 and u_2, you find that the coefficients of the bonding wave function are $u_1 = u_2 = 1/\sqrt{2}$ and that the antibonding wave function coefficients are $u_1 = -u_2 = 1/\sqrt{2}$. The bonding process is illustrated schematically in Fig. 9.3. Note that the lower energy bonding orbital is symmetric around a point between the two atoms. On the other hand, the higher energy antibonding orbital passes through zero at a point midway between the atoms. Since the electron density at any point in space is proportional to $|\psi|^2$, the symmetric form of the bonding orbital places a concentration of negative charge between the two positively charged nuclei, as we expect for a covalent bond. However, the higher energy antibonding orbital has zero electron density at the midpoint on the internuclear axis. It is this difference in the charge distribution that causes the energy difference between the two states.

ii. *The polar covalent bond*

The model for the polar covalent bond is developed in a parallel fashion and we can consider LiH as an example. Li and H each have one valence electron in an s-orbital, the 2s and the 1s, respectively. Therefore, these two orbitals will serve as the basis with which we construct molecular orbitals. If we label H as 1 and Li as 2, we can begin by writing the determinant:

$$\begin{vmatrix} H_{11} - \varepsilon & H_{12} \\ H_{21} & H_{22} - \varepsilon \end{vmatrix} = 0. \tag{9.14}$$

We will again use the definition that $H_{21} = H_{12} = -V_2$. However, this time $H_{11} \neq H_{22}$. H_{11} is the energy level of the H 1s electron and H_{22} is the energy level of the Li 2s electron. These are, of course, the atomic term values for these orbitals.

$$H_{11} = \varepsilon_{1s} \text{ and } H_{22} = \varepsilon_{2s}. \tag{9.15}$$

For convenience, we define an energy equal to the average of these levels and one equal to half the difference:

$$\bar{\varepsilon}_s = \frac{1}{2}(\varepsilon_{1s} + \varepsilon_{2s}) \tag{9.16}$$

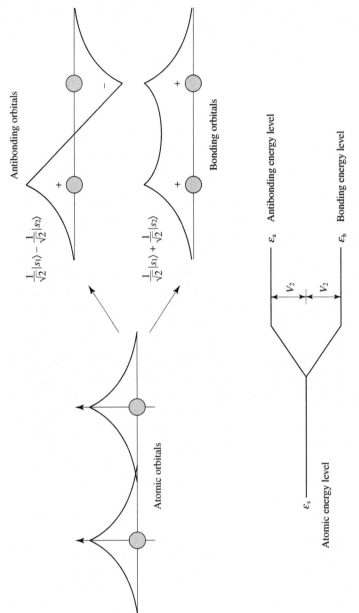

Antibonding orbitals

$\frac{1}{\sqrt{2}}|s_1\rangle - \frac{1}{\sqrt{2}}|s_2\rangle$

Bonding orbitals

$\frac{1}{\sqrt{2}}|s_1\rangle + \frac{1}{\sqrt{2}}|s_2\rangle$

Atomic orbitals

ε_a Antibonding energy level

V_2

V_2

ε_b Bonding energy level

ε_s

Atomic energy level

Figure 9.3. Energy level and charge distribution diagram for the formation of a homopolar diatomic molecule [9].

$$V_3 = \frac{1}{2}\left(\varepsilon_{2s} - \varepsilon_{1s}\right) \tag{9.17}$$

$$\varepsilon_{1s} = \bar{\varepsilon}_s - V_3$$

$$\varepsilon_{2s} = \bar{\varepsilon}_s + V_3. \tag{9.18}$$

Substitution into Eqn. 9.14 gives us the following determinant:

$$\begin{vmatrix} (\bar{\varepsilon}_s - V_3 - \varepsilon) & -V_2 \\ -V_2 & (\bar{\varepsilon}_s + V_3 - \varepsilon) \end{vmatrix} = 0 \tag{9.19}$$

which has the solutions:

$$\varepsilon_\pm = \bar{\varepsilon}_s \pm \sqrt{V_2^2 + V_3^2}. \tag{9.20}$$

This bonding process is depicted schematically in Fig. 9.4. Note that in the bonding (ε_-) state, the electron density on the anion is higher than on the cation, as one would expect for the lower energy state. The reverse is true for the higher energy antibonding (ε_+) state.

The quantity V_3 measures the energy difference between the valence electrons on the isolated, unbound, atoms. If $V_3 = 0$, the valence electrons have the same energy before bonding, there will be no electronic transfer, and the bond is purely covalent. As V_3 increases, more and more of the electron density is transferred from the higher energy orbital (on the cation) to the lower energy orbital (on the anion). Thus, V_3 is a measure of the bond polarity or ionicity and is called the 'polar energy'. In some ways, this term is analogous to the $I_M - E_X$ term in the model for the ionic bond (see Eqn. 7.6). For the bonding level, the energy is lowered by $\left(\sqrt{V_2^2 + V_3^2}\right)$. Therefore, both the ionic component to the bond (V_3) and the covalent component to the bond (V_2) act to lower the energy of the molecular orbital. Remember, at the beginning of this section we noted that the bond strength increases with polarity and this is reflected in the model. As the polarity increases, the splitting of the bonding and antibonding levels $\left(\sqrt{V_2^2 + V_3^2}\right)$ also increases. We will soon see that for the case of solids, the increasing polarity increases the splitting between the bonding and antibonding bands, or the bandgap.

V_2 and V_3 are clearly important quantitative parameters in the LCAO model and their relative values have a profound influence on the structure and properties of the bound system. For example, if V_3 is large with respect to V_2, we expect our description of the bonding to be near the ionic limit. On the other hand, if V_3 is negligible, then we are near the homopolar limit and expect structures and properties typical of covalent systems. As a quantitative measure of these relative values, we can consider the probability of finding the valence

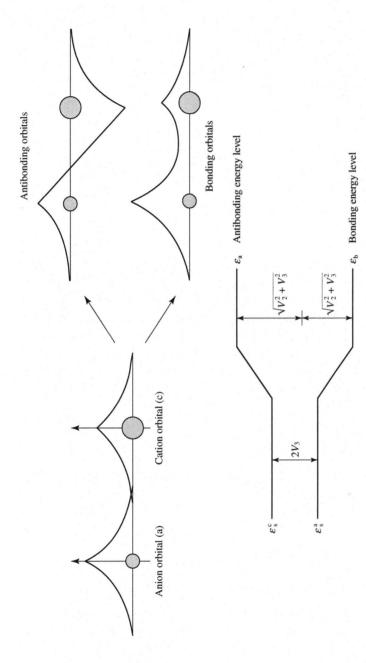

Figure 9.4: Energy level and charge distribution diagram for the formation of a polar bond [10].

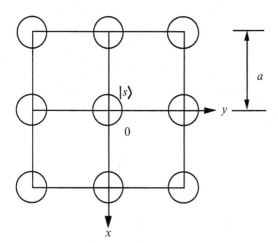

Figure 9.5. The geometry of s-orbitals on the square lattice. See text for discussion.

electron near either one of the two atoms. We begin by defining the probability of finding the electron in the vicinity of atom 1 (the anion) as $1/2(1 + \alpha_p)$ and the probability of finding it in the vicinity of atom 2 (the cation) as $1/2(1 - \alpha_p)$, where α_p is the polarity [7]. When α_p is 0, the electron is equally likely to be found on either atom. When α_p is 1, it is localized on the anion. Assuming no overlap between the atomic wave functions, Eqn. 8.4 can be used to show that the probability of finding the electron on the anion is $u_1^2/(u_1^2 + u_2^2)$. If u_1 and u_2 are determined by returning the value of ε_- to Eqn. 9.19, it can also be shown that the polarity is:

$$\alpha_p = \frac{V_3}{\sqrt{V_2^2 + V_3^2}}.$$
(9.21)

In summary, the most important parameters for our model of the polar covalent bond are V_2, which depends on the interatomic separation, and V_3, which reflects the relative electronegativity of the two atoms. The polarity, specified by Eqn. 9.21, gives us a way to quantify the relative importance of the two parameters.

(C) A physical model for the covalent bond in a homopolar crystal

To extend the LCAO model to crystals, we will begin with the simplified case of a two dimensional square lattice, as illustrated in Fig. 9.5. If we assume that each atom is identical and that the valence electrons are in s-orbitals, then the crystal wave function is formed by the linear combination of the atomic s-orbitals:

$$|\Psi\rangle = \sum_{i=1}^{N} u_i |s_i\rangle, \tag{9.22}$$

where the sum is carried out over the N atoms in the crystal. The energy expectation value is then:

$$\varepsilon = \frac{\sum_{i,j} u_i^* u_j H_{ij}}{\sum_{i,j} u_i^* u_j \langle i|j\rangle}. \tag{9.23}$$

Assuming the atomic basis functions are orthonormal, we take $\langle i|i\rangle = 1$ and $\langle i|j\rangle = 0$. Therefore, Eqn. 9.23 simplifies to:

$$\varepsilon = \frac{\sum_{i,j} u_i^* u_j H_{ij}}{\sum_{i} u_i^* u_i}. \tag{9.24}$$

While minimizing the energy expectation value for the diatomic molecule led to two equations, minimizing ε with respect to u_i^* leads to N independent linear equations of the form:

$$\sum_{j=1}^{N} [H_{ij} u_j - \varepsilon u_i] = 0. \tag{9.25}$$

Considering the fact that there are N such equations (for the different values of i) and that each equation has N terms, determining the N energy eigenvalues requires one to find the determinant of an $N \times N$ matrix. It is clear that this problem becomes intractable for any realistic value of N.

We can solve this problem, and all similar problems involving solids with periodic structures, by applying Bloch's theorem. Bloch's theorem states that for a periodic lattice potential, the solutions of the time independent Schrödinger equation (the eigenstates of H) obey the following relationship:

$$\Psi(\vec{r} + \vec{R}) = e^{i\vec{k}\cdot\vec{R}} \Psi(\vec{r}), \tag{9.26}$$

where $\vec{R}$ is a lattice translation vector and $\vec{k}$ is a wave vector in reciprocal space [11]. In other words, the wave function at a general position $\vec{r} + \vec{R}$ is the same as the wave function at the analogous location in the representative unit cell, multiplied by a factor of $e^{i\vec{k}\cdot\vec{R}}$. Thus, Bloch's theorem specifies the coefficients, u_j in Eqn. 9.25.

If the crystal has N_1 atoms in the x direction and N_2 in the y direction, then $N_1 N_2 = N$ is the total number of atoms. The linear combination of the atomic wave functions (Eqn. 9.22) can now be written in the following form:

$$|\Psi\rangle = \frac{1}{\sqrt{N}} \sum_{i}^{N} e^{i\vec{k}\cdot\vec{R}_i} |s_i\rangle. \tag{9.27}$$

In Eqn. 9.27, $|s_i\rangle$ represents the s-wave function of atom i at the location specified by $\vec{R}_i$ and the multiplicative factor of $(N)^{-1/2}$ is present for normalization (so that $\langle\Psi|\Psi\rangle = 1$).

We shall soon see that by inserting the wave function given in Eqn. 9.27 into Eqn. 9.25 leads to a tractable set of equations. However, note that the wave function in Eqn. 9.27 is a continuous function of k. The energy expectation value computed from such a wave function will be discrete only if k is discrete. As we found in Chapter 8, k is discretized by the application of periodic boundary conditions.

For a crystal with N_1 atoms in the x direction and length $N_1 a$, we can express the periodic boundary conditions in one dimension by the following relationship:

$$\Psi(\vec{r}) = \Psi(\vec{r} + N_1 a). \tag{9.28}$$

Writing the position vector as $\vec{r} = xa$, where a is the interatomic distance and x is a fractional coordinate, we can write:

$$e^{ik_x xa} |s_1\rangle = e^{ik_x(x+N_1)a} |s_{1+N_1}\rangle. \tag{9.29}$$

Since $|s_1\rangle = |s_{1+N_1}\rangle$,

$$e^{ik_x xa} = e^{ik_x(x+N_1)a}. \tag{9.30}$$

This implies that

$$e^{ik_x N_1 a} = 1 = e^{i2\pi} \quad \text{and} \quad k_x = \frac{2\pi}{N_1 a}. \tag{9.31}$$

So, as we found earlier in the free electron model in Chapter 8, $\vec{k}$ is quantized:

$$\vec{k} = \frac{2\pi}{a}\left(\frac{n_1}{N_1}\hat{x} + \frac{n_2}{N_2}\hat{y}\right), \tag{9.32}$$

where n_1 and n_2 are integers that must fall in the range:

$$\frac{-N_1}{2} \leq n_1 \leq \frac{N_1}{2}, \qquad \frac{-N_2}{2} \leq n_2 \leq \frac{N_2}{2}. \tag{9.33}$$

Using the lattice translation vector, $\vec{R}$,

$$\vec{R} = a(u\hat{x} + v\hat{y}), \tag{9.34}$$

the coefficients of Eqn. 9.25 can be written in the following way:

$$u_i(\vec{k}) = \frac{1}{\sqrt{N}}e^{i\vec{k}\cdot\vec{R}_i} = \frac{1}{\sqrt{N_1 N_2}}e^{i2\pi\left(\frac{n_1 u}{N_1} + \frac{n_2 v}{N_2}\right)}. \tag{9.35}$$

Remember that $\vec{R}_i$ specifies the location of the atom and there is a value of $\vec{k}$ (specified by n_1 and n_2) for each of the N atoms. Now that we have specified all of the coefficients in the linear combination, we can substitute them into the determinant in Eqn. 9.25:

$$\frac{1}{\sqrt{N}}\left[\sum_j H_{ij}e^{i\vec{k}\cdot\vec{R}_j} - \varepsilon e^{i\vec{k}\cdot\vec{R}_i}\right] = 0. \tag{9.36}$$

Because there are N such equations and each of them has N terms, we have to make some additional simplifying assumptions.

The assumptions we apply here are characteristic of those that we will use for all problems involving the LCAO method. Basically, we assume that only atoms in close proximity to each other interact. Interactions between atoms separated by greater distances are ignored. To see how this simplifies the problem, consider the first linear equation ($i=1$) in the set of equations specified by Eqn. 9.36. There are N terms in the sum which contain the matrix elements H_{11}, H_{12}, H_{13}, ... H_{1N}. The first term, H_{11}, is easy to specify. This is the atomic term value for the electron in the isolated atomic orbital, ε_s. The other $N-1$ terms are matrix elements for the interaction of the $i=1$ atom with every other atom in the crystal. Earlier, we assigned the covalent energy, V_2, to these H_{ij} ($i \neq j$) terms. Our critical assumption is that for the nearest neighbors of the $i=1$ atom, H_{ij} ($i \neq j$) $= -V_2$. Knowing that V_2 decreases with increasing interatomic separation, we make the assumption that for more distant atoms, $H_{ij} = 0$. This approach has validity when the electrons are localized in the vicinity of the ion cores and, therefore, we expect the interatomic potential energy to decrease rapidly towards zero as the interatomic separation increases. Applying these assumptions, the $i=1$ equation for the square lattice becomes:

$$H_{11}e^{i\vec{k}\cdot\vec{R}_1} + \sum_{j=2}^{5} H_{1j}e^{i\vec{k}\cdot\vec{R}_j} - \varepsilon e^{i\vec{k}\cdot\vec{R}_1} = 0, \tag{9.37}$$

where the indices in Eqn. 9.37 are defined in Fig. 9.6. Substituting in values for the matrix elements, Eqn. 9.37 becomes:

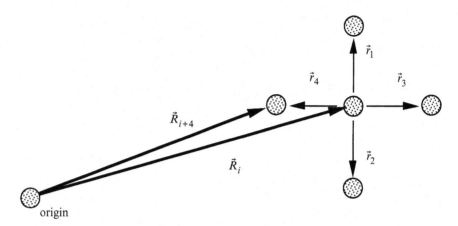

Figure 9.6. Definition of the position vectors used in Eqns. 9.39 and 9.40.

$$(\varepsilon_s - \varepsilon)e^{i\vec{k}\cdot\vec{R}_1} - V_2\left(e^{i\vec{k}\cdot\vec{R}_2} + e^{i\vec{k}\cdot\vec{R}_3} + e^{i\vec{k}\cdot\vec{R}_4} + e^{i\vec{k}\cdot\vec{R}_5}\right) = 0. \tag{9.38}$$

Dividing both sides by $e^{i\vec{k}\cdot\vec{R}_1}$, we obtain:

$$(\varepsilon_s - \varepsilon) - V_2\left(e^{i\vec{k}\cdot(\vec{R}_2-\vec{R}_1)} + e^{i\vec{k}\cdot(\vec{R}_3-\vec{R}_1)} + e^{i\vec{k}\cdot(\vec{R}_4-\vec{R}_1)} + e^{i\vec{k}\cdot(\vec{R}_5-\vec{R}_1)}\right) = 0. \tag{9.39}$$

Here, the vector differences, $(\vec{R}_j - \vec{R}_i)$, are simply the vectors that point from the central atom to the four nearest neighbor atoms, as shown in Fig. 9.6. We label these vectors $\vec{r}_i$ ($i = 1$ to 4) so that Eqn. 9.39 can be written:

$$\varepsilon_s - V_2 \sum_{i=1}^{4} e^{i\vec{k}\cdot\vec{r}_i} = \varepsilon. \tag{9.40}$$

The best part about this simplification is that all of the i equations specified by Eqn. 9.25 are identical to Eqn. 9.40, so that we only need to consider one.

Note that in Eqn. 9.40, the energy, ε, is a function of $\vec{k}$, so that this equation is a dispersion relation. We will, therefore, depict it in the same way that we depicted the free electron (Eqn. 8.16) and nearly free electron dispersion. We begin by considering how the energy of electron states changes as $\vec{k}$ increases from 0 to π/a along the [100] direction. If we write out the sum in Eqn. 9.40 and recognize from Fig. 9.6 that $\vec{r}_1 = -\vec{r}_2$ and $\vec{r}_3 = -\vec{r}_4$, we have:

$$\varepsilon = \varepsilon_s - V_2(e^{i\vec{k}\cdot\vec{r}_1} + e^{-i\vec{k}\cdot\vec{r}_1} + e^{i\vec{k}\cdot\vec{r}_3} + e^{-i\vec{k}\cdot\vec{r}_3})$$
$$\varepsilon = \varepsilon_s - 2V_2\left(\cos(\vec{k}\cdot\vec{r}_1) + \cos(\vec{k}\cdot\vec{r}_3)\right). \tag{9.41}$$

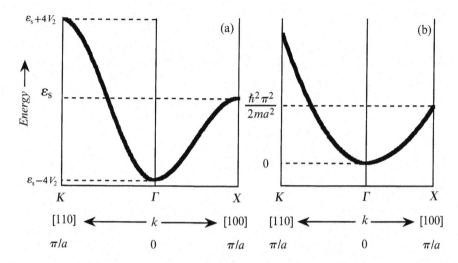

Figure 9.7. (a) The dispersion of electronic states for a square lattice of s orbitals derived from the LCAO model. (b) The free electron dispersion.

At $\vec{k} = 0$,

$$\varepsilon = \varepsilon_s - 4V_2, \tag{9.42}$$

and along $\vec{k} = [100]$,

$$\varepsilon = \varepsilon_s - 2V_2 - 2V_2\cos k_x a. \tag{9.43}$$

Based on Eqns. 9.42 and 9.43, the dispersion of the electron energy levels is plotted in Fig. 9.7.

When one compares the relative stability of the bonded system to the free atoms, the relevant reference energy is ε_s, the energy of electrons on the isolated atoms. Based on the dispersion plot in Fig. 9.7, we can see that along the [100] direction, the crystal states all have relatively lower energies than the free atom states (ε_s). However, along other directions, the same statement is not necessarily true. For example, the dispersion along the [110] direction is also shown in Fig. 9.7 and here we see that one half of the electronic states have energies that are larger than the energies of the valence electrons on the free atoms. The relative stability of the crystal state must be determined by examining the dispersion of the electronic states along all possible directions.

It is important to remember that the energy is a discrete function of $\vec{k}$. There are actually N closely spaced (yet distinct) orbitals with different energies. There is one orbital for every atom included in the linear combination and each has a slightly different energy level. For comparison, the free electron energy bands are also depicted in Fig. 9.7; while there are differences (especially at the zone boundaries) there is also an obvious similarity.

When we studied the diatomic molecule in the last section, we noted a distinct difference between bonding and antibonding states. The same difference can be said to occur in the crystal orbitals that make up the bands [12]. Consider, for example, the band of orbitals formed by the overlap of the s states.

$$\Psi(\vec{r}) = \sum_{i=1}^{N} e^{i\vec{k}\cdot\vec{R}_i}|s_i\rangle. \tag{9.44}$$

At the center of the Brillouin zone, where $\vec{k}=0$, all of the s orbitals will add together 'in phase' (the coefficient for each state in the sum will be $+1$). At the zone boundary, on the other hand, where we take $\vec{k}_x$ to be π/a and $\vec{R}_i$ to be an integer multiple of a, the crystal wave function is:

$$\Psi(\vec{r}) = \sum_{i=1}^{N} (-1)^i|s_i\rangle. \tag{9.45}$$

In this case, the s orbitals add together with alternating signs, just like the antibonding wave function of the diatomic model. This will produce zero electron density between the cores, a situation that is clearly less stable. So, we conclude that the lower energy states near the zone center are more 'bonding' in character while the higher energy states near the zone boundary are more 'antibonding' in character.

The important quantitative parameters which determine the electron energy levels are the values of ε_s (which depends only on the type of atom) and V_2. The value of ε_s sets an energy range and the size of V_2 determines how stable the electrons in the bonded state are with respect to the atomic state. The calculation of V_2 is nontrivial and depends on the exact form of the potential energy operator used in the Hamiltonian. In practice, it is common to use values derived through semi-empirical parameterizations. Such values allow the electronic structure to be determined in a qualitatively, if not quantitatively, correct manner. We use the parameterization suggested by Froyen and Harrison [13] to set numerical values for the covalent energy. This parameterization is based on the similarity of the LCAO bonds and the free electron bands and the assumption that in the limit of each model, the results should agree. Specifically, the band widths predicted by each model are equated. Since the LCAO band width along [100] from $\vec{k}=0$ to $\vec{k}=\pi/a$ is $4V_2$ and the free electron band width over

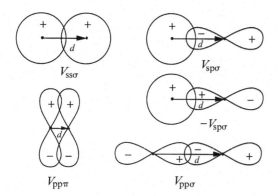

Figure 9.8. The fundamental types of orbital overlap for s and p orbitals [16]. For $V_{pp\sigma}$ and $V_{pp\pi}$, the overlap of opposite lobes changes the sign of the matrix element. The interatomic separation is d.

the some range (assuming the parabolic dispersion in Eqn. 8.16) is $\hbar^2\pi^2/2ma^2$, we find that:

$$V_2 = \frac{\hbar^2}{m}\frac{\pi^2}{8}\frac{1}{a^2}. \tag{9.46}$$

The ratio $\hbar^2 / m$ is equal to 7.62 eV-Å^2. The geometric factor ($\pi^2/8$) is specific to matrix elements for s–s type overlaps. The most important aspect of this parameterization is that it makes V_2 inversely proportional to the square of the interatomic spacing (in this case, a), which agrees with experimental observations [14, 15]. Since we expect the amplitude of wave functions to decrease exponentially, this approximation is probably not valid for more distant neighbors.

Equation 9.46 specifies a numerical value for V_2 only when the bands are formed by the overlap of s-orbitals. If the linear combination is formed from s and p orbitals, there are four fundamental types of overlaps that might be encountered. The overlaps are illustrated schematically in Fig. 9.8. The three subscripts specify the angular momentum of the two orbitals and the type of the overlap, respectively. The σ overlaps are centered on the internuclear axis and the π overlaps are parallel to the axis. Because the factor of $\hbar^2 / ma^2$ appears in all of the matrix elements, they differ only by a constant, which is labeled $\eta_{ll'm}$ for the matrix element $V_{ll'm} = \langle l|H|l'\rangle$. These constants are listed in Table 9.4 for several different geometric arrangements. Note also that the sign of the matrix element is negative for bonding types of overlap, where wave functions with the same sign meet, and positive for antibonding overlaps where the signs of the wave functions cancel. For example, recall that in Eqn. 9.12 we assumed that V_2 (in fact, this was $V_{ss\sigma}$) was a negative number. The values in Table 9.4 have the signs included. For example, $\eta_{ss\sigma}$ is $-\pi^2/8$ and this means that $V_{ss\sigma}$ (V_2) is a negative number, as previously assumed.

Table 9.4. *Dimensionless coefficients for approximate interatomic matrix elements for atoms in selected structures [13].*

Coefficient	simple cubic	diamond	fcc	bcc
$\eta_{ss\sigma}$	$-\pi^2/8 = -1.23$	-1.40	-0.62	-0.93
$\eta_{sp\sigma}$	$\dfrac{\pi}{2}\sqrt{\dfrac{\pi^2}{4}-1} = 1.90$	1.84	2.33	1.75
$\eta_{pp\sigma}$	$3\pi^2/8 = 3.70$	3.24	2.47	4.63
$\eta_{pp\pi}$	$-\pi^2/8 = -1.23$	-0.81	0	-0.93

Example 9.1

The dispersion relation we derived in the previous section considered only nearest neighbor interactions. Modify this equation to include second, third, and fourth nearest neighbors and compare the result with the nearest neighbor result.

1. We begin by determining the locations of the more distant near neighbors. If there are four neighbors at a distance of a, then there are four more at a distance of $\sqrt{2}a$, four more at $2a$, and eight at $\sqrt{5}a$. This arrangement is illustrated in Fig. 9.9.

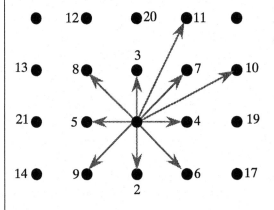

Figure 9.9. Indices of the neighbors on the two-dimensional square lattice.

2. The equations including the first and second nearest neighbors are:

$$\varepsilon(\vec{k}) = \varepsilon_s - \sum_{i=2}^{9} H_{1i}e^{i\vec{k}\cdot\vec{r}_i}, \tag{9.47}$$

where the meaning of the indices is shown in Fig. 9.9. If the matrix elements (H_{1i}) for the first four nearest neighbor terms are $-V_2$, it follows from Eqn. 9.46 that the matrix elements for the next four at $\sqrt{2}a$ are $-V_2/2$. Therefore, Eqn. 9.47 becomes:

$$\varepsilon(\vec{k}) = \varepsilon_s - V_2 \sum_{i=2}^{5} e^{i\vec{k}\cdot\vec{r}_i} - \frac{1}{2}V_2 \sum_{i=6}^{9} e^{i\vec{k}\cdot\vec{r}_i}. \tag{9.48}$$

The first part of Eqn. 9.48 was already given in Eqn. 9.41, and the second part is:

$$-\frac{1}{2}V_2\{2\cos(k_x a + k_y a) + 2\cos(-k_x a + k_y a)\}. \tag{9.49}$$

The complete dispersion relation is, therefore,

$$\varepsilon(\vec{k}) = \varepsilon_s - 2V_2\{\cos(k_x a) + \cos(k_y a)\}$$
$$- V_2\{\cos(k_x a + k_y a) + \cos(-k_x a + k_y a)\}. \tag{9.50}$$

3. Following similar reasoning, we find that the third and fourth near neighbors simply add correcting terms to the dispersion relation. The contributions from the third and fourth neighbors are in the third and fourth terms of Eqn. 9.51, respectively.

$$\varepsilon(\vec{k}) = \varepsilon_s - 2V_2\{\cos(k_x a) + \cos(k_y a)\}$$
$$- V_2\{\cos(k_x a + k_y a) + \cos(-k_x a + k_y a)\}$$
$$- \frac{1}{2}V_2\{\cos(2k_x a) + \cos(2k_y a)\}$$
$$- \frac{2}{5}V_2\{\cos(2k_x a + k_y a) + \cos(k_x a + 2k_y a)$$
$$+ \cos(-k_x a + 2k_y a) + \cos(-2k_x a + k_y a)\}. \tag{9.51}$$

4. Following this method, we see that the dispersion relation is a series of trigonometric functions and a new term is added for each new set of neighbors. Furthermore, the successive terms have diminishing amplitudes and increasing frequencies because of the increasing atomic distances. In this way, the solution resembles a Fourier series.

(D) A physical model for the covalent bond in a polar crystal

i.   *The centered square lattice with a two atom basis*
With the mechanics of the LCAO method established, we can now consider the case of a polar crystal. As an example, we use the centered square lattice with a two atom basis depicted in Fig. 9.10. We can proceed in a way entirely analogous to the diatomic polar molecule (Section B(*iii*)). For example, for the crystal MX, the linear combination of atomic orbitals is:

$$|\Psi\rangle = u_m \sum_{i=1}^{N} u_i |s_m\rangle + u_x \sum_{j=1}^{N} u_j |s_x\rangle. \tag{9.52}$$

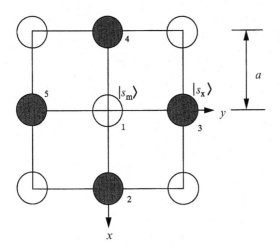

Figure 9.10 Atomic arrangement for the polar centered square lattice.

As before (Eqn. 9.27), the coefficients of any two orbitals of the same type (u_i, for example) have the form $e^{i\vec{k}\cdot\vec{R}_i}$. However, we must now find additional coefficients (u_m and u_x) between the orbital types. We do this by constructing Bloch sums for each of the orbital types [17]:

$$|\chi_m(k)\rangle = \frac{1}{\sqrt{N}} \sum_{i=1}^{N} e^{i\vec{k}\cdot\vec{R}_i} |s_m\rangle. \tag{9.53}$$

This allows us to write Eqn. 9.52 as:

$$|\Psi\rangle = u_m |\chi_m(k)\rangle + u_x |\chi_x(k)\rangle, \tag{9.54}$$

where m is identified with the metallic atom and x is identified with the nonmetal. Since there are now two unknown coefficients, we can proceed as we did for the polar molecule.

Replacing m and x with 1 and 2, respectively, and minimizing the energy expectation value for the wave function in Eqn. 9.54 leads to:

$$\begin{vmatrix} H_{11} - \varepsilon & H_{12} \\ H_{21} & H_{22} - \varepsilon \end{vmatrix} = 0, \tag{9.55}$$

where

$$H_{12} = \sum_{i}^{N} \sum_{j}^{N} e^{-i\vec{k}\cdot\vec{R}_i} e^{i\vec{k}\cdot\vec{R}_j} \langle s_i | H | s_j \rangle = \sum_{i}^{N} \sum_{j}^{N} e^{i\vec{k}\cdot(\vec{R}_j - \vec{R}_i)} \langle i | H | j \rangle. \tag{9.56}$$

As before, we make it possible to solve Eqn. 9.55 by limiting the number of terms that we include. In this case, we assume that the matrix elements are nonzero only for the on-site terms ($i=j$ in Eqn. 9.56) and for the nearest neighbor terms. The on-site terms in Eqn. 9.55 are found in H_{11} and H_{22}, which are simply the atomic term values for electrons in the s orbitals of the free atoms. For this general case, we assign these levels to be ε_{sm} for the more metallic atom and ε_{sx} for the more electronegative atom. Quantitative values for specific elements can be found in tabulated data [18]; a selection of atomic term values can be found in Table 9B.2 in Appendix 9B.

The evaluation of H_{12} is simplified by considering only the four nearest neighbors. This approximation leads to N identical equations of the form:

$$H_{12} = H_{21} = \sum_{i=1}^{4} e^{i\vec{k}\cdot\vec{r}_i}\langle s_m | H | s_x \rangle. \tag{9.57}$$

In Eqn. 9.57, the vectors $\vec{r}_i$ point from any atom to the four nearest neighbors. Taking $\langle s_m | H | s_x \rangle$ to be $V_{ss\sigma}$, Eqn. 9.57 can be simplified to:

$$H_{12} = 2V_{ss\sigma}\{\cos(\vec{k}\cdot\vec{r}_1) + \cos(\vec{k}\cdot\vec{r}_3)\}. \tag{9.58}$$

Before solving for the energies, it is convenient to define values for the average on-site energy, $\bar{\varepsilon}$, and the polar energy, V_3.

$$\bar{\varepsilon} = \frac{1}{2}(\varepsilon_{sx} + \varepsilon_{sm})$$

$$V_3 = \frac{1}{2}(\varepsilon_{sx} - \varepsilon_{sm}). \tag{9.59}$$

Thus, Eqn. 9.55 becomes:

$$\begin{vmatrix} \bar{\varepsilon} - V_3 - \varepsilon & H_{12} \\ H_{12} & \bar{\varepsilon} + V_3 - \varepsilon \end{vmatrix} = 0, \tag{9.60}$$

which leads to a dispersion relation for two bands:

$$\varepsilon(\vec{k}) = \bar{\varepsilon} \pm \sqrt{V_3^2 + [2V_{ss\sigma}\{\cos(\vec{k}\cdot\vec{r}_1) + \cos(\vec{k}\cdot\vec{r}_3)\}]^2}. \tag{9.61}$$

Based on Eqn. 9.61, Fig. 9.11 shows the dispersion of the two bands along the [100] direction. Assuming that each of the s-orbitals on the unbound atoms contained a single electron, both electrons will occupy the states in the lower energy band. Along this direction of the Brillouin zone, all of the occupied crystal energy levels are below the average of the two atomic energy levels ($\bar{\varepsilon}$). Furthermore, a gap in the energy level diagram opens up between the highest filled levels in the lower energy band and the lowest unoccupied levels in the

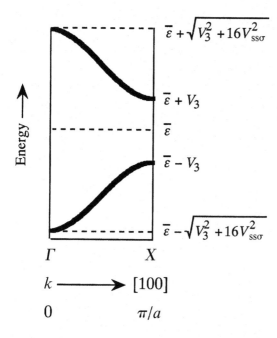

$$\bar{\varepsilon} + \sqrt{V_3^2 + 16V_{ss\sigma}^2}$$

$$\bar{\varepsilon} + V_3$$

$$\bar{\varepsilon}$$

$$\bar{\varepsilon} - V_3$$

$$\bar{\varepsilon} - \sqrt{V_3^2 + 16V_{ss\sigma}^2}$$

Figure 9.11. Dispersion along the [100] direction for the bands of a polar binary square lattice.

higher energy band (although, in general, this gap might not exist at all points along the Brillouin zone boundary). The magnitude of the gap depends on the size of V_3, which depends on the energy difference between the two atomic term values. It is instructive to see how the band structure changes with the polarity of the compound; this issue is illustrated by Example 9.2.

Example 9.2

Compare the dispersion for LiMg, LiAl, and LiCl. For the purpose of this exercise, assume that they are two-dimensional compounds which crystallize in the centered square lattice illustrated in Fig. 9.10. For each compound, determine the band gap and band width. Is there a systematic variation with polarity?

1. We begin by determining V_3 and $\bar{\varepsilon}$ for each compound. For this, all we need to do is take the appropriate term levels from Table 9A.2 and substitute them into Eqn. 9.59. The results are shown in Table 9.5.
2. Next, we need to determine values for $V_{ss\sigma}$. Considering Table 9.4 and Eqn. 9.46, we need only determine the interatomic distances. Taking $r_{Li} = 1.52$, $r_{Mg} = 1.60$, $r_{Al} = 1.43$, and $r_{Cl} = 1.10$ Å, we assume that each interatomic distance is the sum of the

Table 9.5. *Energy levels for three Li compounds, eV.*

	ε_{sm}	ε_{sx}	$\bar{\varepsilon}$	V_3	$V_{ss\sigma}$	ΔX
LiMg	-5.48	-6.86	-6.17	-0.69	-0.96	0.2
LiAl	-5.48	-10.11	-7.80	-2.32	-1.08	0.5
LiCl	-5.48	-24.63	-15.55	-9.58	-1.46	2.0

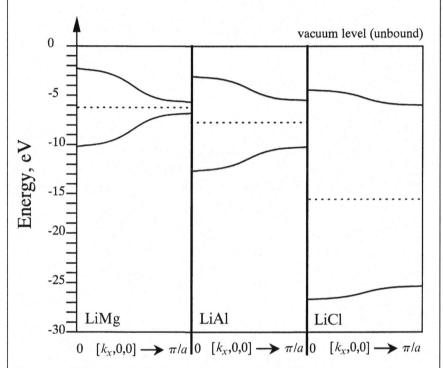

Figure 9.12. The relative band energies in three hypothetical Li compounds.

atomic radii and compute $V_{ss\sigma}$ for each compound. The final column gives the Pauling electronegativity difference for each combination.

3. To produce the dispersion plot, we graph Eqn. 9.61 using the data in Table 9.5. The result is shown in Fig. 9.12.

4. The band gaps and widths, also calculated using Eqn. 9.61, are shown in Table 9.6.

5. While the situation posed by this problem is purely hypothetical, the accepted trends are accurately reproduced by the model. Bands formed from states on atoms with nearly the same electronegativities are more homopolar or covalent, as exemplified by a shrinking band gap and increasing band width.

Table 9.6. *Band gaps and band widths for two-dimensional Li compounds.*

	Band gap @ $k = (\pi/a, 0, 0)$ (eV)	Band width (eV)
LiMg	1.38	3.2
LiAl	4.64	2.6
LiCl	19.16	1.6

On the other hand, as the bands become increasingly polar (ionic), the band gap increases and the bands become more narrow. Note that in the limit of zero band width, we have the pure ionic model in which electrons are simply transferred from one atomic state on the more metallic atom to another atomic state on the less metallic atom. Based on the change in the energies of the electrons from the free state to the bound state, we conclude that bonds get stronger as the polarity increases.

We should note that these generalizations do not include the effects associated with altering the interatomic distance. For example, while the covalent energy ($V_{ss\sigma}$) is highest for LiCl, it is certainly not the most covalent (in fact, the opposite is true). Of these three combinations, LiCl has the highest covalent energy because it has the smallest interatomic separation. Relative covalency or polarity can not be judged based on V_2 or V_3 alone. Computing the polarity according to Eqn. 9.21 shows that LiCl is the most polar, LiMg is the least, and LiAl is intermediate, as expected based on the electronegativity differences.

ii. The rock salt structure

We now have the tools to consider a slightly more complex LCAO description of a binary crystal. In this case, we examine the rock salt structure and use MgO as an example. We will assume that the crystal wave functions are formed by the mixing of the highest occupied valence orbitals, the Mg 3s and the O 2p. The geometric arrangement of these orbitals, projected along [001], is shown in Fig. 9.13.

If there are N formula units in the structure, the crystal wave functions are:

$$|\Psi\rangle = u_{px}|\chi_{px}(\vec{k})\rangle + u_{py}|\chi_{py}(\vec{k})\rangle + u_{pz}|\chi_{pz}(\vec{k})\rangle + u_s|\chi_s(\vec{k})\rangle, \qquad (9.62)$$

where the $|\chi_\alpha(\vec{k})\rangle$ are Bloch sums, as in Eqn. 9.53. Since there are four terms in Eqn. 9.65, partial derivatives of the energy expectation value lead to four equations and a 4×4 Hamiltonian matrix.

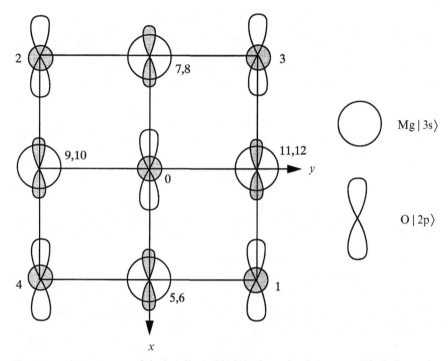

Figure 9.13. Arrangement of the bonding orbitals in the rock salt structure. The large open figures are at $z=0$. The smaller, shaded figures are at $z=\pm 1/2$. For clarity, only the p_x orbitals are shown. The numbers refer to the indices of the vectors in Eqn. 9.68.

	O p_x	O p_y	O p_z	Mg 3s
O p_x	$H_{11}-\varepsilon$	H_{12}	H_{13}	H_{14}
O p_y	H_{21}	$H_{22}-\varepsilon$	H_{23}	H_{24}
O p_z	H_{31}	H_{32}	$H_{33}-\varepsilon$	H_{34}
Mg 3s	H_{41}	H_{42}	H_{43}	$H_{44}-\varepsilon$

$$(9.63)$$

The matrix in Eqn. 9.63 will give four dispersion relations as solutions, each of which describes a band of electronic states. In each band, there is one state for each pair of atoms in the structure. At any particular value of $\vec{k}$, however, there are exactly four energy levels. In order to get to a set of dispersion relations or even a set of energy levels at a fixed $\vec{k}$, we need to specify all of the matrix elements.

To simplify the matrix elements, we assume that only certain orbitals interact. Specifically, we assume that $\langle i|H|j\rangle = 0$ for all but three special cases. First, when $i = j$, the matrix element between an orbital and itself is the atomic term value. For different types of orbitals on the same atom, the condition of orthogonality guarantees that these matrix elements are zero. For example, $\langle p_x i | H | p_y i \rangle$ = 0. Second, H_{ij} is nonzero when i and j label nearest neighbors. This accounts for the overlap between the Mg 3s and O 2p orbitals. Third, H_{ij} is nonzero when i and j label next nearest neighbors. This accounts for the Mg 3s–Mg 3s and the O 2p–O 2p overlaps.

Using these assumptions, all of the 16 sums in Eqn. 9.63 reduce to an expression with a manageable number of terms. For example, H_{11} reduces to N identical equations of the form:

$$H_{11} = \varepsilon_p + \sum_{i=1}^{12} e^{i \vec{k} \cdot \vec{r}_i} \langle p_x | H | p_x' \rangle \tag{9.64}$$

where ε_p is the atomic term value for the O 2p orbital, p_x and p_x' are on next nearest neighbor sites, and the vectors $\vec{r}_i$ are the 12 local position vectors that point from any O atom to an O on a next nearest neighbor site.

The matrix element H_{11} describes only the O $2p_x$ orbitals, which occupy a ccp sublattice in the rock salt structure. In this configuration, the vector pointing from each O atom to its 12 nearest neighbors is not necessarily perpendicular or parallel to the axes of the p_x-orbitals. Therefore, the matrix element $\langle p_x | H | p_x' \rangle$ does not have pure σ or π character. For this general case, the interatomic matrix element is the sum of σ and π components, as illustrated schematically in Fig. 9.14. The most general forms of the interatomic matrix elements ($E_{ll'}$), which depend on angle, have been tabulated by Slater and Koster [19]; a partial list is given in Table 9.7.

For the O next nearest neighbor interactions in MgO, we are interested in matrix elements of the type $E_{x,x}$:

$$E_{x,x} = l^2 V_{pp\sigma} + (1 - l^2) V_{pp\pi}. \tag{9.65}$$

In Eqn. 9.65, the values of l are the direction cosines of the vectors pointing from the central atom in the cluster to the next nearest neighbors. Taking $\vec{r}_i$ to be the vector pointing from atom 0 to atom i, we find that for the first eight vectors (labeled $\vec{r}_1$ through $\vec{r}_8$), the projection of the position vector on the x axis is $\pm 1/\sqrt{2}$, so

$$E_{x,x} = 1/2 V_{pp\sigma} + 1/2 V_{pp\pi}. \tag{9.66}$$

For the last four position vectors, labeled $\vec{r}_9$ through $\vec{r}_{12}$, the position vectors are perpendicular to the x axis so that $l = 0$ and

$$E_{x,x} = V_{pp\pi}. \tag{9.67}$$

Table 9.7. *Slater and Koster interatomic matrix elements* [19].

matrix element	value	matrix element	value
$E_{s,s}$	$V_{ss\sigma}$	$E_{s,3z^2-r^2}$	$[n^2-1/2(l^2+m^2)]V_{pd\sigma}$
$E_{s,x}$	$lV_{sp\sigma}$	$E_{x,xy}$	$\sqrt{3}\,l^2 m V_{pd\sigma} + m(1-2l^2)V_{pd\pi}$
$E_{x,x}$	$l^2 V_{pp\sigma}+(1-l^2)V_{pp\pi}$	$E_{x,yz}$	$\sqrt{3}\,lmn V_{pd\sigma} - 2lmn V_{pd\pi}$
$E_{x,y}$	$lm V_{pp\sigma}-lm V_{pp\pi}$	$E_{x,zx}$	$\sqrt{3}\,l^2 n V_{pd\sigma} + n(1-2l^2)V_{pd\pi}$
$E_{x,z}$	$ln V_{pp\sigma}-ln V_{pp\pi}$	E_{x,x^2-y^2}	$(\sqrt{3}/2)l(l^2-m^2)V_{pd\sigma}+l(1-l^2+m^2)V_{pd\pi}$
$E_{s,xy}$	$\sqrt{3}\,lm\,V_{sd\sigma}$	$E_{x,3z^2-r^2}$	$l[n^2-\frac{1}{2}(l^2+m^2)]V_{pd\sigma}-\sqrt{3}\,ln^2 V_{pd\pi}$
E_{s,x^2-y^2}	$1/2\sqrt{3}\,(l^2-m^2)\,V_{sd\sigma}$		

Note:
*l, m, and n are the direction cosines of the vector that points from the left state to the right state, where α is the left state for the matrix element $\langle\alpha|H|\beta\rangle$.

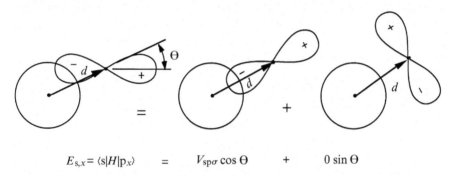

$$E_{s,x} = \langle s|H|p_x\rangle \quad = \quad V_{sp\sigma}\cos\Theta \quad + \quad 0\sin\Theta$$

Figure 9.14. The interatomic matrix element between an arbitrary arrangement of two orbitals is given by the sum of the σ and π components [20]. See Table 9.6 for general forms.

Noting the following relationships:

$$\vec{r}_1 = (x+y)d = -\vec{r}_2, \qquad \vec{r}_4 = (x-y)d = -\vec{r}_3$$
$$\vec{r}_5 = (x+z)d = -\vec{r}_7, \qquad \vec{r}_6 = (x-z)d = -\vec{r}_8$$
$$\vec{r}_{11} = (y+z)d = -\vec{r}_9, \qquad \vec{r}_{12} = (y-z)d = -\vec{r}_{10}, \qquad (9.68)$$

Eqn. 9.64 can be rewritten in the following way:

$$H_{11} = \varepsilon_p + \sum_{i=1}^{12} e^{i\vec{k}\cdot\vec{r}_i} E^{(i)}_{x,x}$$

$$= \varepsilon_p + \sum_{i=1,4,5,6,11,12} 2\cos(\vec{k}\cdot\vec{r}_i) E^{(i)}_{x,x}$$

$$= \varepsilon_p + g_0^{(i)} E^{(i)}_{x,x}. \tag{9.69}$$

In Eqn. 9.69, note that the exponential terms, which contain the information about the crystal structure, are in the prefactor g_0. The matrix element $E_{x,x}$, is sensitive to the interatomic separation and ε_p depends on the atom type. Equation 9.69 can be written out more explicitly in the following way:

$$\begin{aligned}
H_{11} = \varepsilon_p &+ (V_{pp\sigma} + V_{pp\pi})[\cos(k_x + k_y)d + \cos(k_x - k_y)d \\
&+ \cos(k_x + k_z)d + \cos(k_x - k_z)d] \\
&+ 2V_{pp\pi}[\cos(k_y - k_z)d + \cos(k_y + k_z)d]. \tag{9.70}
\end{aligned}$$

The remaining matrix elements can be determined using a similar procedure. Although there are 15 additional matrix elements, the task is not as tedious as it appears. First, there are only 10 unique matrix elements. Furthermore, the results for H_{22} and H_{33} will be very similar in form to H_{11}, H_{13} and H_{23} will be similar to H_{12}, and H_{24} and H_{34} will be similar to H_{14}. Therefore, we only need to determine four matrix elements (H_{11}, H_{12}, H_{14}, and H_{44}) to draw conclusions about the electronic structure of a rock salt structured compound. It is left as an exercise to show that:

$$\begin{aligned}
H_{12} &= (V_{pp\sigma} - V_{pp\pi})[\cos(k_x + k_y)d - \cos(k_x - k_y)d] \\
H_{44} &= \varepsilon_s + 2V_{ss\sigma}[\cos(k_x + k_y)d + \cos(k_x - k_y)d + \cos(k_y + k_z)d \\
&\quad + \cos(k_y - k_z)d + \cos(k_x + k_z)d + \cos(k_x - k_z)d] \\
H_{14} &= -2iV_{sp\sigma}\sin k_x d. \tag{9.71}
\end{aligned}$$

If we limit our analysis to the $k_x - k_y$ plane, where $k_z = 0$, H_{43}, H_{13}, and H_{23} go to zero. Thus, the 4×4 Hamiltonian matrix now has the form:

$$\begin{vmatrix}
H_{11} - \varepsilon & H_{12} & 0 & H_{14} \\
H_{21} & H_{22} - \varepsilon & 0 & H_{24} \\
0 & 0 & H_{33} - \varepsilon & 0 \\
H_{41} & H_{42} & 0 & H_{44} - \varepsilon
\end{vmatrix} = 0. \tag{9.72}$$

Determining the dispersion relations for an arbitrary value of $\vec{k}$ is challenging because of the size of the matrix. However, at special points along the [100] direction, the problem becomes easy. For example, at the center of the Brillouin zone ($\vec{k} = [000]$ or the 'Γ' point), the matrix elements H_{14} and H_{24} vanish. In the $k_x k_y$ plane (where $k_z = 0$), we have:

$$\varepsilon_{1,2} = \frac{1}{2}(H_{11} + H_{22}) \pm \left[\frac{1}{2}(H_{11} - H_{22})^2 + H_{12}^2\right]^{1/2}$$

$$\varepsilon_3 = H_{33}$$

$$\varepsilon_4 = H_{44}. \tag{9.73}$$

At the zone center, we find:

$$\varepsilon_1 = \varepsilon_2 = \varepsilon_3 = \varepsilon_p + 4V_{pp\sigma} + 8V_{pp\pi}$$

$$\varepsilon_4 = \varepsilon_s + 12V_{ss\sigma} \tag{9.74}$$

and at $\vec{k} = [\pi/d, 0, 0]$ we find:

$$\varepsilon_1 = \varepsilon_p - 4V_{pp\sigma}$$

$$\varepsilon_2 = \varepsilon_3 = \varepsilon_p - 4V_{pp\pi}$$

$$\varepsilon_4 = \varepsilon_s - 4V_{pp\sigma}. \tag{9.75}$$

To plot the dispersion along [001], we will make the assumption that the two end points are connected by a smooth curve with a sinusoidal shape. Based on Table 9B.2 in Appendix 9B, $\varepsilon_s = -6.86$ eV and $\varepsilon_p = -14.13$ eV. According to Pantelides [21], we can take $V_{pp\pi} = -(1/8)V_{pp\sigma}$ and $V_{pp\sigma}$ as approximately 1 eV for the rock salt structure. Assuming that the interactions between the magnesium cations are small, then $V_{ss\sigma}$ is also small. For the purposes of demonstration only, we arbitrarily assign $V_{ss\sigma}$ to 0.1 eV. We can then use these values for the parameters to construct the energy level diagram (band structure) in Fig. 9.15.

The three lowest energy bands of states will hold six electrons per formula unit. Four of these states are filled by the valence electrons originally on the O atoms. This means that there are two remaining states to accommodate the s electrons from the Mg. Thus, the valence electrons from the Mg transfer to the valence band states formed by the overlap of the O 2p orbitals. This means that there is an energy gap between the filled valence band states and the empty conduction band states. If we compute the size of this gap, at Γ, we find that it is about 4 eV (the exact value depends on $V_{ss\sigma}$, which can not be less than 0). Based on this value, we would conclude (correctly) that MgO is a transparent insulator. We should also mention that the manner in which the LCAO model was applied

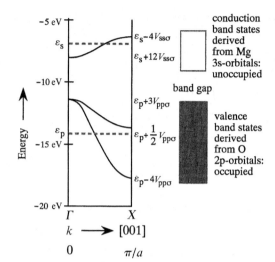

Figure 9.15 Dispersion of highest occupied and lowest unoccupied bands in MgO. The energies on the left-hand side are with respect to the vacuum level, and $V_{ss\sigma} = -0.1$ eV and $V_{pp\sigma} = 1$ eV.

was not intended to be quantitatively accurate, but to produce the correct qualitative trends. In this case, the measured band gap of MgO is 7.8 eV [22] and the width of the valence band is only about 6 eV [23]. Considering the approach that was used to get our matrix elements, these differences are not surprising. An alternative approach (which we will not pursue) would be to use the experimental values for the band gap and width to determine values for $V_{pp\sigma}$ and $V_{ss\sigma}$.

iii. *The density of states*

Before moving on, we should note that one of the most important ways to represent the band structure is by showing the density of states as a function of energy. In other words, at any point along the vertical energy axis of the band structure diagram in Fig. 9.15, how many energy levels are there in a given range of energy? You can think of this as a new kind of energy level diagram that does not distinguish the states based on their wave vector. This is important for several reasons. First, many optical and electrical properties are influenced by the density of states at the Fermi level. Second, many experiments can not discriminate states based on the wave vector. So, the interpretation of spectral data (which are determined by allowed transitions of electrons between energy levels) is often based on the density of states computed from the band structure.

From Chapter 8, you will remember that the energy density of states is $dN/d\varepsilon$. We can rewrite this differential in the following way [24]:

$$\frac{dN}{d\varepsilon} = \frac{dN}{dk}\left|\frac{dk}{d\varepsilon}\right| = \frac{dN}{dk}\left|\frac{d\varepsilon}{dk}\right|^{-1}. \tag{9.76}$$

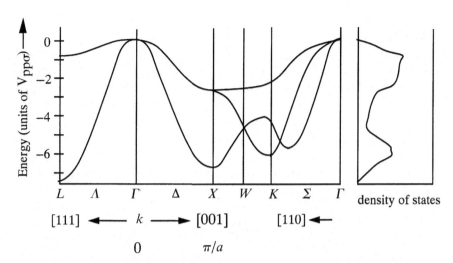

Figure 9.16. Valence band energy levels for a compound with the rock salt structure. On the left, dispersion along high symmetry directions is shown. On the right, the density of states spectrum is shown [26].

This form is instructive since dN/dk (the density of states in k-space) is simply $Na/2\pi$ and the second term in the product is the derivative of the dispersion relation (the absolute value is used because the density of states is positive definite). So, even without doing any calculations, one can see that the density of states will be inversely proportional to the slope of the dispersion. While the expression above is straightforward for one dimension, the extension to three dimensions is not simple. To calculate the total density of states, the dispersion must be known at all points in k-space, not just along a single high symmetry direction, such as is shown in Fig. 9.15. Common solutions to this problem involve partitioning k-space into a fixed number of cells, approximating the dispersion in each one as being linear, and then accurately determining how each cell contributes to the total density of states [25].

As an example, consider the valence band density of states shown in Fig. 9.16 for a rock salt structured compound. On the left-hand side of the diagram, the dispersion relations are shown more completely than in Fig. 9.15. Note that along lower symmetry directions, each p-orbital has a distinct energy. On the right-hand side, the full density of states spectrum is shown [26].

Example 9.3

Consider an fcc metallic crystal with one atom at each lattice site and a nearest neighbor separation of d. Show that for small values of k, the functional form of the LCAO band structure is identical to the free electron band structure and that the density of states is the same [27]. Explain how the free electron dispersion and the LCAO dispersion differ at the Brillouin zone boundary (where k is large).

1. If we assume that each atom has a single s electron, it can be shown that the LCAO dispersion relation is (see Problem 9.6):

$$\varepsilon(\vec{k}) = \varepsilon_s + 4V_{ss\sigma}[\cos(k_x d)\cos(k_y d) + \cos(k_x d)\cos(k_z d) + \cos(k_y d)\cos(k_z d)]. \quad (9.77)$$

2. If, for small values of k, we make the assumption that k-space is isotropic, Eqn. 9.77 reduces to:

$$\varepsilon(\vec{k}) = \varepsilon_s + 4V_{ss\sigma}[3\cos^2(kd)]. \quad (9.78)$$

3. Substituting $1 - \sin^2(kd)$ for $\cos^2(kd)$ and approximating $\sin^2(kd)$ (for small kd) as $(kd)^2$, we have

$$\varepsilon(\vec{k}) = \varepsilon_s + 12V_{ss\sigma} - 12V_{ss\sigma}(kd)^2. \quad (9.79)$$

4. In Eqn. 9.79, we see that for small k, the dispersion is parabolic, just as it was for the free electron model.

5. To compute the density of states, we can differentiate Eqn. 9.80 and apply Eqn. 9.76.

$$\frac{dN}{d\varepsilon} = \frac{V}{\pi^2}k^2 \times \frac{1}{24d^2 V_{ss\sigma}k} = \frac{V(2m)^{1/2}\hbar}{24\pi^2 d^2 V_{ss\sigma}}\varepsilon^{1/2}. \quad (9.80)$$

The result in Eqn. 9.80 shows that the density of states is proportional to $\sqrt{\varepsilon}$, just as it is in the free electron model.

6. The major difference between the LCAO and free electron dispersion at the zone boundary is that the free electron dispersion continuously increases while the slope of the LCAO dispersion (which is a trigonometric function) is zero at the boundary.

---

iv. *Sphalerite*

In the closely related diamond, sphalerite, and wurtzite structures, all of the atoms are in tetrahedral coordination. These materials are not only excellent

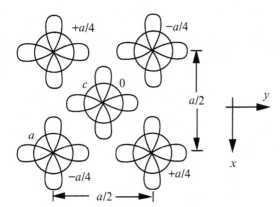

Figure 9.17. Configuration of the orbitals in the sphalerite structure. The central atom is the group III or metallic atom (the cation) and the neighbors, above and below the plane of the cation, are the anions. The $|p_z\rangle$ orbitals are not shown. In Eqn. 9.81, $\vec{r}_1 = [111]a/4$, $\vec{r}_2 = [1\bar{1}\bar{1}]a/4$, $\vec{r}_3 = [\bar{1}1\bar{1}]a/4$, and $\vec{r}_4 = [\bar{1}\bar{1}1]a/4$ [28].

examples of covalent (diamond) and polar covalent bonding (sphalerite and wurtzite), but are also highly relevant because of their uses as semiconductors. We will consider the general case of an AB compound that takes the sphalerite structure, assuming A is a group III element and B is a group V element. This model, however, is also appropriate for the diamond and wurtzite structures. The geometric arrangement of the orbitals in the sphalerite structure is illustrated in Fig. 9.17.

The model we adopt [29, 30] is based on a linear combination of the s- and p-orbitals on both the anion and the cation. Therefore, we will have a linear combination of eight Bloch sums and the energy levels will be given by an 8×8 matrix. Considering only nearest neighbor overlaps, the matrix has the form shown in Fig. 9.18. Since only nearest neighbor interactions are considered, the diagonal elements are atomic term values. The nondiagonal matrix elements are written as the product of an interatomic matrix element from Table 9.6 and a sum of exponential factors (g_i) in a manner analogous to Eqn. 9.69. In this case, the values of g_i are:

$$g_0 = e^{i\vec{k}\cdot\vec{r}_1} + e^{i\vec{k}\cdot\vec{r}_2} + e^{i\vec{k}\cdot\vec{r}_3} + e^{i\vec{k}\cdot\vec{r}_4}$$
$$g_1 = e^{i\vec{k}\cdot\vec{r}_1} + e^{i\vec{k}\cdot\vec{r}_2} - e^{i\vec{k}\cdot\vec{r}_3} - e^{i\vec{k}\cdot\vec{r}_4}$$
$$g_2 = e^{i\vec{k}\cdot\vec{r}_1} - e^{i\vec{k}\cdot\vec{r}_2} + e^{i\vec{k}\cdot\vec{r}_3} - e^{i\vec{k}\cdot\vec{r}_4}$$
$$g_3 = e^{i\vec{k}\cdot\vec{r}_1} - e^{i\vec{k}\cdot\vec{r}_2} - e^{i\vec{k}\cdot\vec{r}_3} + e^{i\vec{k}\cdot\vec{r}_4}. \tag{9.81}$$

At any specific value of $\vec{k}$, there are eight energy levels. For arbitrary values of $\vec{k}$, determining the eight solutions is only practical using a computer. We shall instead proceed as we did for the rock salt structure and note that at high

	s^c	s^a	p_x^c	p_y^c	p_z^c	p_x^a	p_y^a	p_z^a
s^c	ε_s^c	$E_{ss}g_0$	0	0	0	$E_{sp}g_1$	$E_{sp}g_2$	$E_{sp}g_3$
s^a	$E_{ss}g_0^*$	ε_s^a	$-E_{sp}g_1^*$	$-E_{sp}g_2^*$	$-E_{sp}g_3^*$	0	0	0
p_x^c	0	$-E_{sp}g_1$	ε_p^c	0	0	$E_{xx}g_0$	$E_{xy}g_3$	$E_{xy}g_2$
p_y^c	0	$-E_{sp}g_2$	0	ε_p^c	0	$E_{xy}g_3$	$E_{xx}g_0$	$E_{xy}g_1$
p_z^c	0	$-E_{sp}g_3$	0	0	ε_p^c	$E_{xy}g_2$	$E_{xy}g_1$	$E_{xx}g_0$
p_x^a	$E_{sp}g_1^*$	0	$E_{xx}g_0^*$	$E_{xy}g_3^*$	$E_{xy}g_2^*$	ε_p^a	0	0
p_y^a	$E_{sp}g_2^*$	0	$E_{xy}g_3^*$	$E_{xx}g_0^*$	$E_{xy}g_1^*$	0	ε_p^a	0
p_z^a	$E_{sp}g_3^*$	0	$E_{xy}g_2^*$	$E_{xy}g_1^*$	$E_{xx}g_0^*$	0	0	ε_p^a

Figure 9.18. 8×8 matrix for the LCAO description of sphalerite [29, 30]. See text for explanation of the notation.

symmetry points the solution is easier. For example, at $\vec{k} = 0$, g_1, g_2, and g_3 vanish. In this case, we have four solutions:

$$E_1 = \frac{\varepsilon_s^c + \varepsilon_s^a}{2} \pm \sqrt{\left(\frac{\varepsilon_s^c - \varepsilon_s^a}{2}\right)^2 + (4E_{ss})^2}$$

$$E_2 = \frac{\varepsilon_p^c + \varepsilon_p^a}{2} \pm \sqrt{\left(\frac{\varepsilon_p^c - \varepsilon_p^a}{2}\right)^2 + (4E_{ss})^2}. \tag{9.82}$$

Note the similarity of these solutions to the simple form of Eqn. 9.20 for the polar diatomic molecule. The two lowest energy solutions give the energies of the valence or bonding bands and the two higher energy solutions give the energies of the conduction or antibonding bands. Note that the second equation (E_2), which comes from the coupling of the p-orbitals, is a triply degenerate state describing three bands. The eight valence electrons fill only the lowest four bands, the valence bands, and a gap exists between these states and the conduction band states. From Eqn. 9.82, we can see that the size of this gap depends on E_{xx} and E_{ss} (V_2) and the differences between the anion and cation atomic term states (V_3). According to this equation, the gap will grow as the interatomic

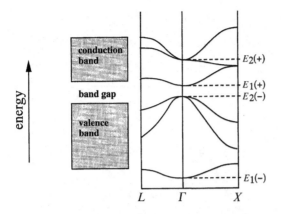

Figure 9.19. The LCAO bands for Ge [31]. The labeled energy levels at the Γ point correspond to Eqn. 9.82.

spacing diminishes or the electronegativity difference increases. These are the trends noted in the first section of the chapter.

Since this model included only nearest neighbor interactions, it does not discriminate between the diamond structure, the sphalerite structure, the wurtzite structure, or the polymorphs of SiC. The LCAO band structure for Ge, determined from this model, is shown in Fig. 9.19.

E) Bands deriving from d-electrons

The description of d-electrons in solids presents a distinct challenge. Electrons in s- and p-orbitals interact strongly with their neighbors. Within the framework of the free electron theory or the LCAO model, this leads to the formation of bands of states in the crystal. Electrons in f-orbitals, on the other hand, are localized in the vicinity of the ion core and are described as atomic states that are weakly perturbed by a crystal field. Electrons in d-orbitals are intermediate between these two cases and their characteristics vary widely from compound to compound.

Within the framework of the LCAO model, the matrix elements between weakly overlapping d-orbitals are small and the bands formed from d–d overlaps are, therefore, relatively narrow. In a compound with a high degree of ionicity, this narrow band falls in the gap between the filled valence band states derived primarily from anionic p-orbitals and the unfilled conduction band states derived primarily from the cation s-orbitals. In a transition metal, on the other hand, this narrow d-band overlaps with a broad, free electron-like s-band. We will first examine the case of the transition metal.

The cohesive energy of the metallic elements varies periodically, as shown in Fig. 9.20. With the exception of the central feature in each curve, the general

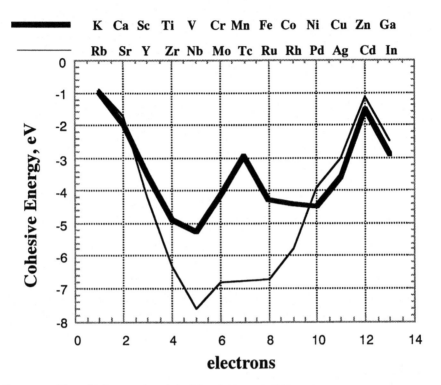

Figure 9.20. The cohesive energies of the 3d and 4d metals [32].

trend is that the crystals initially become more stable as the number of d-electrons increases to the point where the d-levels are half full and then they become less stable as the filling of the d-shell is completed. A simple way to incorporate the d-electron contribution to the cohesion is to apply the Friedel model [33–35]. The schematic energy level diagram in Fig. 9.21 illustrates the principal features of the Friedel model: a narrow band formed from the overlap of d-orbitals and a broad band formed from s-orbitals. The s-band is assumed to have a parabolic density of states and the d-band is assumed to have a rectangular density of states. Note that because the d-band is narrow and can hold ten electrons per atom, the density of states is much higher than in the s-band.

According to the Friedel model, the electrons are partitioned between the s-like free electron band and the localized d-band, with the number of electrons in the s-band equal to Z_s and the number of electrons in the d-band equal to Z_d. The sum of these two quantities must be equal to the total number of valence electrons per atom, Z. We can compute each of these quantities by integrating over the density of states of each band:

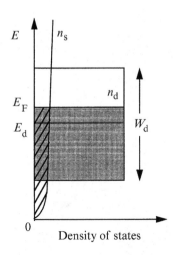

Figure 9.21. The density of states proposed by Friedel to explain the properties of transition metals. A relatively narrow d-band, centered at E_d (width $= W_d$), is superimposed on a much wider free electron (parabolic) band. Filled states in the parabolic band are hatched and filled states in the d-band are shaded. The highest filled state in each band is defined by the Fermi level E_F.

$$Z = \int_0^{E_F} n_s(\varepsilon)d\varepsilon + \int_{E_d - W_d/2}^{E_F} n_d(\varepsilon)d\varepsilon. \tag{9.83}$$

In this expression, n_s is the parabolic free electron density of states and n_d is $10/W_d$, where W_d is the width of the d-band and E_d is the center of the d-band.

We can also use the Friedel model to calculate the energy per atom (ε) when the d-band holds Z_d electrons:

$$\varepsilon = \int_{-W_d/2}^{-W_d/2 + Z_d W_d/10} \left(\frac{10}{W_d}\right)\varepsilon\, d\varepsilon = 5W_d\left[-\frac{Z_d}{10} + \left(\frac{Z_d}{10}\right)^2\right]. \tag{9.84}$$

The quantities needed for this calculation (E_d and W_d) can be determined from the LCAO method. The result in Eqn. 9.84 exhibits a minimum cohesion for $Z_d = 0$ and 10, a maximum cohesion for $Z_d = 5$, and a parabolic shape between the two endpoints. This is very similar to the periodic variation in the cohesive energy that is shown in Fig. 9.20. We interpret this increase and decrease in cohesion in the following way. As the number of d-electrons per atom increases to five, only states with bonding character (in the bottom half of the band) are filled and the crystal becomes more stable. As more electrons are added, the crystal is destabilized by the filling of states with anti-bonding character. Note that for the group III metals (Ga and In), extra stability arises from putting the first electron in the empty p-band. It is worth remembering the general rule that a half filled band is always the most stable electronic configuration.

In its simplest form, however, the Freidel model does not predict the 'hump' that appears in the center of the cohesive energy versus atomic number curve for the 3d metals. This destabilization is due to spin–spin interactions. When

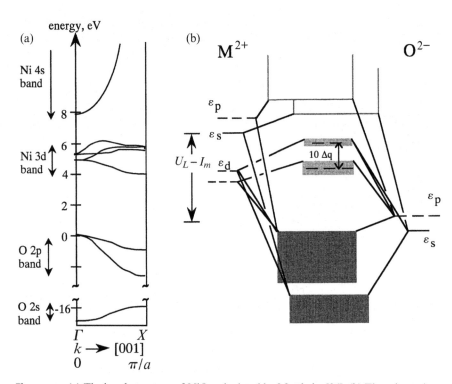

Figure 9.22 (a) The band structure of NiO, calculated by Mattheiss [36]. (b) The schematic band structure of transition metal oxides with the rock salt structure (such as NiO), based on estimates by Goodenough [37].

electrons partially fill a set of degenerate levels, they minimize their energy by maximizing their spin (in other words, electron spins adopt a parallel alignment). However, in a crystal, where the degeneracy of the electronic states is broken, parallel alignment of spins destabilizes the bonding by forcing electrons into higher energy (and sometimes anti-bonding) states. When the spin–spin interactions are large, as they are for the 3d metals, this leads to the diminished cohesion in the center of the row. This effect is not as apparent in the 4d row where the atoms are farther apart and the spin–spin interactions are weaker.

In a compound with a considerable degree of ionicity, such as an oxide, the d-bands do not overlap with a free electron band. Instead, we expect the d-bands to fall in the gap between the anion derived valence band and the cation derived conduction band. For example, two energy level diagrams for NiO are compared in Fig. 9.22. The band structure in Fig. 9.22a was derived by Mattheiss [36] using techniques similar to those described earlier in this chapter. NiO crystallizes in the rock salt structure, as does MgO. Note that the Ni 4s band and the O 2p band in Fig. 9.22a are analogous to the Mg 3s and O 2p bands in Fig. 9.15. The d-bands, which are much narrower, fall in the rather large band gap.

Figure 9.22b show an alternative and more schematic version of the energy levels in NiO proposed by Goodenough [37]. Goodenough has reasoned that as long as detailed band structure calculations are not quantitatively correct, phenomenological energy level diagrams provide a useful substitute. Phenomenological energy level diagrams, such as the one in Fig. 9.22b, are constructed with the constraint that the dispersion curves conform to the symmetry of the crystal. For example, the narrow d-bands are split into two distinct levels by the octahedral crystal field. Furthermore, when specific compounds from an isostructural series are compared, the relative energies of the bands must be consistent with the differences between the energies of the atomic orbitals. The diagrams should also reflect experimental observations. For example, the divalent Ni cation has eight d-electrons and we, therefore, expect NiO to have a partially filled band. The fact that NiO is an insulator indicates that the d-band width is vanishingly small and that the d-electrons are localized in atomic-like states. For comparison, we note that the calculated band structure indicates that the d-bands are about 2 eV wide and that there is no splitting; these features would suggest that NiO is a metal. Because the properties of transition metal compounds are sensitive to features of the band structure that are not accurately reproduced by calculations, the phenomenological energy diagrams constructed by Goodenough [37] are viewed by many as being more valuable.

The nonstoichiometric compound Na_xWO_3 represents a clear example of how shifting the Fermi level within the d-band alters the properties of the solid [38]. Pure WO_3 ($x = 0$) has a distorted ReO_3 structure. In this case, all of the oxygen-derived states are filled and all of the W-derived conduction band states are empty. There is an approximately 2.4 eV band gap between the O states and the W states and, therefore, this material is insulating. The dissolution of an alkali metal, such as Na, in WO_3 changes the position of the highest occupied electron state. Na occupies an interstitial site in the structure. The 3s valence electron on the Na is transferred to the lowest energy levels in the W-derived conduction band. As the Na concentration increases, the occupied and unoccupied states are no longer separated by an energy gap. The electrons added to the t_{2g} band derived from W d-states are free to move and thus able to carry charge and reflect light, just as they do in a metal. The change in properties (from transparent and insulating to opaque and conductive) that occurs as an alkali metal is added to WO_3 (or certain other transition metal oxides) is the basis for a number of interesting devices such as electrochromic windows and ion-exchange battery electrodes [39].

(F) The distinction between metals and nonmetals

Throughout this chapter, the polar covalent bond has been taken to have two distinct components: the covalent energy, represented by V_2, and the polar energy, represented by V_3. For solids with a small polar energy, we must also consider the factors that differentiate metals from nonmetals. From the LCAO point of view, we may say that the simple broadening of atomic-like states leads to the bands of states in a metal, while the covalent solid is distinguished by the splitting of bonding and anti-bonding levels [40]. For example, if we consider the gap (E_g) between the $E_1(+)$ level and the $E_2(-)$ level in Eqn. 9.82 (see Fig. 9.19), we have:

$$E_g = -\frac{1}{2}[(\varepsilon_p^c - \varepsilon_s^c) + (\varepsilon_p^a - \varepsilon_s^a)] + \sqrt{\left(\frac{\varepsilon_s^c - \varepsilon_s^a}{2}\right)^2 + (4E_{ss})^2}$$

$$+ \sqrt{\left(\frac{\varepsilon_p^c - \varepsilon_p^a}{2}\right)^2 + (4E_{xx})^2}. \tag{9.85}$$

We can simplify Eqn. 9.85 by noting that the positive contributions represent the polar and covalent energies for the s and p orbitals and have the form $\sqrt{V_2^2 + V_3^2}$. The magnitude of the negative term is determined by the splitting of the s and p energy levels on the individual atoms. It is the balance of these two quantities that determines the size of E_g. The magnitude of the gap shrinks to zero and the solid becomes metallic when the s–p splitting term is larger than the terms that depend on the polar and covalent energies. Therefore, we define the metallic energy, V_1, in the following way,

$$V_1 = \frac{1}{4}(\varepsilon_p - \varepsilon_s) \tag{9.86}$$

and rewrite Eqn. 9.85 as:

$$E_g = -2(V_{1,c} + V_{1,a}) + \sqrt{\left(\frac{\varepsilon_s^c - \varepsilon_s^a}{2}\right)^2 + (4E_{ss})^2}$$

$$+ \sqrt{\left(\frac{\varepsilon_p^c - \varepsilon_p^a}{2}\right)^2 + (4E_{xx})^2}. \tag{9.87}$$

Based on the above reasoning, the metallicity (α_m) is defined as:

$$\alpha_m = \frac{2(V_{1,c} + V_{1,a})}{\sqrt{\left(\frac{\varepsilon_s^c - \varepsilon_s^a}{2}\right)^2 + (4E_{ss})^2} + \sqrt{\left(\frac{\varepsilon_p^c - \varepsilon_p^a}{2}\right)^2 + (4E_{xx})^2}}. \tag{9.88}$$

Table 9.8. *The metallicity of selected elements [41].*

Element	Metallic energy (V_1)	Covalent energy $V_2 = 4E_{xx}$	Metallicity $\alpha_m = 1.11 V_1/V_2$
C	2.13	6.94	0.34
Si	1.76	2.98	0.66
Ge	2.01	2.76	0.81
Sn	1.64	2.10	0.87
Pb*	1.57	1.57	> 1.00
Al*	1.31	2.02	0.72

Note:
* Computed using the interatomic spacing in the stable ccp phase, but using the matrix elements for the diamond structure.

Considering the data in Tables 9.4 and 9.7, we can say that for the diamond or sphalerite structure, $E_{ss} = 2.59 E_{xx}$ $(E_{xx} = 1/3 V_{pp\sigma} + 2/3 V_{pp\pi})$, and we can assign $4E_{xx}$ to V_2. If we define $\bar{V}_1 = 1/2(V_{1,a} + V_{1,c})$ and note that $(\varepsilon_{s,c} - \varepsilon_{s,a}) \approx 2(\varepsilon_{p,c} - \varepsilon_{p,a})$, Eqn. 9.88 can be written in the following simplified manner:

$$\alpha_m = \frac{1.11 \bar{V}_1}{\sqrt{V_2^2 + V_3^2}}. \tag{9.89}$$

Trends in metallicity, illustrated in Table 9.8, reproduce the division between semiconducting gray tin and metallic Pb. The dividing line between metallic and nonmetallic elements on the periodic chart, proposed in Chapter 1, occurs at a metallicity of approximately 0.7. We now have three quantitative parameters whose relative values determine the bond type: V_1, V_2, and V_3. Note that V_1 and V_3 depend only on the types of atoms in the compounds; V_2 alone is sensitive to the structure.

Ⓖ The distinction between covalent and ionic solids

The distinguishing signature of covalent bonding in crystals is the tetrahedral coordination of the component atoms. In solids which are ionically or metallically bound, the coordination numbers are usually higher and the bonding environment is more isotropic. Before turning to the question of how we can use the LCAO model to decide which atoms will adopt tetrahedral configurations, we turn briefly to the model for hybridized bond orbitals that is used to describe the electronic states of atoms in a tetrahedral environment. In this model, the s- and p-orbitals of a single principal level are hybridized to form four equivalent sp³ bond orbitals. It is then possible to write linear combinations of these bond orbitals and proceed as before.

The sp³ hybrid bond orbitals, $|h_i\rangle$, are a linear combination of the s and p orbitals [42].

$$|h_1\rangle = \frac{1}{2}[|s\rangle + |p_x\rangle + |p_y\rangle + |p_z\rangle], \text{ along } [111]$$

$$|h_2\rangle = \frac{1}{2}[|s\rangle + |p_x\rangle - |p_y\rangle - |p_z\rangle], \text{ along } [1\bar{1}\bar{1}]$$

$$|h_3\rangle = \frac{1}{2}[|s\rangle - |p_x\rangle + |p_y\rangle - |p_z\rangle], \text{ along } [\bar{1}1\bar{1}]$$

$$|h_4\rangle = \frac{1}{2}[|s\rangle - |p_x\rangle - |p_y\rangle + |p_z\rangle], \text{ along } [\bar{1}\bar{1}1]. \tag{9.90}$$

The hybrid orbitals are mutually orthogonal and have their highest electron densities directed along the lines connecting the nearest neighbor atoms. Each of the hybrid bond orbitals has the energy expectation value, ε_h:

$$\varepsilon_h = \frac{1}{4}(\varepsilon_s + 3\varepsilon_p) \tag{9.91}$$

where ε_s and ε_p are the term values for the s- and p-orbitals. By analogy, we can also define a hybrid covalent energy ($V_{2,h}$) and a hybrid polar energy ($V_{3,h}$):

$$V_{3,h} = \frac{1}{2}(\varepsilon_{2,h} - \varepsilon_{1,h}) \tag{9.92}$$

$$-V_{2,h} = \langle h_1 | H | h_2 \rangle = \frac{1}{4}(-V_{ss\sigma} + 2\sqrt{3}\,V_{sp\sigma} + 3V_{pp\sigma}) = \frac{4.37\hbar^2}{md^2}. \tag{9.93}$$

At this point, we could continue as we did earlier in the chapter; the schematic results are summarized in Fig. 9.23. It should be emphasized that bond orbital models and atomic orbital models are equivalent. Depending on the property to be described, one of the two descriptions might have the practical advantage of mathematical convenience. In the present context, the hybrid bond orbital model is important for our understanding of the energetic factors that underpin the tetrahedral coordination state.

From the energy level diagram in Fig. 9.23, we can see that hybridization requires promotion of relatively lower energy s orbitals to the new hybrid orbitals. The promotion energy (E_{pro}) for a group IV atom can be written in the following way:

$$E_{pro} = \varepsilon_s + 3\varepsilon_p - 2\varepsilon_s - 2\varepsilon_p = \varepsilon_p - \varepsilon_s = 4V_1. \tag{9.94}$$

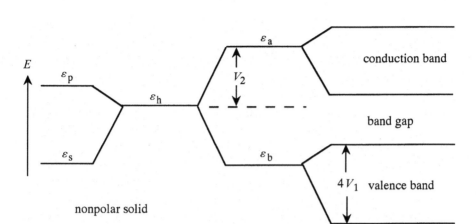

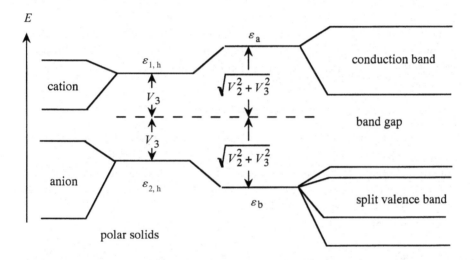

Figure 9.23 Energy level diagram for the formation of solids from hybridized orbitals [43].

So, the energy cost for forming the hybrids is proportional to the metallic energy, or the s–p splitting. This energy cost is offset by the covalent bond energy, $V_{2,h}$, which is inversely proportional to the square of the interatomic separation. For a compound of elements not from group IV, the promotion energy is:

$$E_{pro} = \left(1 + \frac{\Delta Z}{4}\right) 4 V_{1,c} + \left(1 - \frac{\Delta Z}{4}\right) 4 V_{1,a}, \tag{9.95}$$

where the metallic energies on the cation, $V_{1,c}$, and anion, $V_{1,a}$, are considered separately. The bond energy is:

$$E_{bond} = \frac{1}{4}\left[(4+\Delta Z)\left(\sqrt{V_{2,h}^2 + V_{3,h}^2} + V_{3,h}\right) + (4-\Delta Z)\left(\sqrt{V_{2,h}^2 + V_{3,h}^2} - V_{3,h}\right)\right]$$

$$= 2\sqrt{V_{2,h}^2 + V_{3,h}^2} - \frac{1}{2}\Delta Z V_{3,h}. \tag{9.96}$$

Based on these results, we can say the following. If the bond energy is large compared to the promotion energy (which is related to the metallic energy), then the compound will be stable in a tetrahedrally coordinated structure. On the other hand, if the bond energy is low compared to the promotion energy, a higher coordination is likely to be adopted. We can see that the bond energy decreases as the polarity increases $(\Delta Z V_{3,h})$ and as the covalent energy decreases. Therefore, tetrahedral coordination is not favored for highly polar compounds. As we shall see in Chapter 10, Phillips [44] used a quantity similar to the polarity (Eqn. 9.21) to distinguish compounds that form tetrahedral structures from those that form octahedral structures. As we saw in Example 9.2, an increase in polarity leads to a decrease in the band width (V_2). When the band width is small in comparison to the energy difference between the s and p levels, hybridization and tetrahedral bonding geometries are not favorable.

(H) The cohesive energy of a covalently bonded solid

In the sections before this, we concentrated on determining how the electron energy levels of atoms were modified when a solid crystal is formed. For any stable crystalline solid, the average energy of the electrons in the crystal orbitals is lower than the energy of the electrons on the separate atoms. For a solid with a tetrahedral structure, we will take the cohesive energy (E_{coh}) to be the sum of three terms [45]:

$$-E_{coh} = E_{pro} - E_{bond} + V_0(d). \tag{9.97}$$

In Eqn. 9.97, $V_0(d)$ is the repulsive overlap interaction that contains the core–core electrostatic repulsions, the electron kinetic energy, and the electron–electron interactions. This term is similar in spirit to the short range d^{-12} part of the Lennard-Jones potential introduced in Chapter 6 and is difficult to estimate accurately. The promotion and bond energy have already been defined above (Eqns. 9.95 and 9.96).

Considering the fact that we have an equation that describes the cohesive energy as a function of the interatomic spacing, we could perform the same manipulations that we did with our earlier bonding models to determine the cohesive energy and equilibrium spacing. Unfortunately, the uncertainties in the

Table 9.9. *Cohesive energies, per bond, for semiconductors, in eV [46].*

Material	E_{pro}	$V_0(d)$	E_{bond}	E_{coh} (theory)	E_{coh} (exp)
C	4.26	20.14	28.08	3.68	3.68
BN	4.00	20.14	26.17	2.03	3.34
BeO	3.37	20.14	21.21	2.30	3.06
Si	3.52	6.22	12.06	2.32	2.32
AlP	3.31	6.22	11.64	2.11	2.13
Ge	4.02	5.22	11.18	1.94	1.94
GaAs	3.80	5.22	10.78	1.76	1.63
ZnSe	3.24	5.22	9.65	1.19	1.29
CuBr	1.71	5.22	7.12	0.19	1.45
Sn	3.28	3.66	8.50	1.56	1.56
InSb	3.12	3.66	8.21	1.43	1.40
CdTe	2.69	3.66	7.38	1.03	1.03
AgI	1.41	3.66	5.63	0.56	1.18

repulsive term and the neglect of long range interactions make estimates based on this technique questionable. However, it is still possible to get some information from this model. We begin by recognizing that the overlap energy constitutes a considerable fraction of the total energy and is the principal factor that determines the interatomic separation. If we consider the fact that E_{coh} for Si is 2.32 eV, E_{pro} is 3.52, and E_{bond} is 12.06 eV, we find that V_0, the repulsive contribution, is 6.22 eV. Noting that compounds in the same isoelectronic series have nearly the same interatomic spacing, we assume that these same compounds have nearly the same repulsive energy (we assume that this component of the total energy is independent of polarity). For example, since $d_{Si} = 2.35$ Å and $d_{AlP} = 2.36$ Å, we can assume that $V_0(Si) = V_0(AlP)$. When we compute E_{coh} for AlP using $V_0(Si)$, we find that it is 2.11 eV, which compares well with the experimental value of 2.13 eV. Table 9.9 illustrates how this approach can be extended to other compounds.

The simplified version of the LCAO model described here is not ideally suited for determining cohesive energies or predicting interatomic spacings without empirical corrections. In this regard, it is inferior to the ionic bonding model. However, it has the advantage that it allows the energy levels of the electrons in the crystal to be calculated and this information can be used to compute the dynamic properties of electrons. Furthermore, the trends it predicts are often qualitatively correct. For example, at the beginning of the chapter, we observed that for isoelectric tetrahedrally bonded compounds, the cohesive energy decreases with an increase in polarity (see Table 9.3). This is predicted by Eqns. 9.96 and 9.97. For isoelectric compounds, the metallic energy, the repulsive overlap energy, and the covalent energy are approximately constant; it is the increase of ΔZ that decreases the bond energy and the cohesive energy. Interestingly, although the cohesion decreases, the melting point increases with polarity. The origin of this difference is that the cohesive energy measures the stability with reference to the free, neutral atoms. The melting point measures the stability with respect to the liquid, within which the atoms remain polarized; the energy required to transfer charge from the anion to the cation does not influence the melting temperature, but does affect the cohesive energy.

The total energy is only weakly sensitive to the exact configuration of atoms (the crystal structure) and this makes structural predictions difficult, if not impossible. As noted in previous chapters, the crystal structure of a solid influences the total cohesive energy by only a few percent. The implication is that the relative stability of a compound has more to do with the elements from which it is composed (which determine V_1 and V_3) than from their precise arrangement, which influences only V_2. However, the problem that remains for the materials scientist is that the properties are sensitive to the structure.

(I) Overview of the LCAO model and correlation with phenomenological trends

The central idea that underlies the LCAO model is that when a crystal is formed from independent atoms, the total number of orbitals remains the same and the orbitals even retain much of their atomic character. However, there are continuous changes in the energies of the orbitals as the separation between the atoms changes. As an example, we can consider a collection of group IV atoms. When they are initially separated by a large distance, the energies of their valence electrons (in s and p orbitals) will be identical. As the atoms are brought closer together, the orbitals begin to interact with each other and their energy levels must distinguish themselves. The result is a broadening of the individual atomic states into a band of crystalline states. At large separations, where interactions are weak, the crystal will behave as a metal because of the partially filled p-band. However, as the interaction between the orbitals increases (as $1/d^2$) and the covalent energy (V_2) becomes large compared to the s–p splitting, the bonding and anti-bonding levels can split to form an arrangement of energy levels where half of the states are filled and separated from a set of empty states by a band gap. This progression, as a function of d, is illustrated schematically in Fig. 9.24. Note that we can easily position the group IV elements with

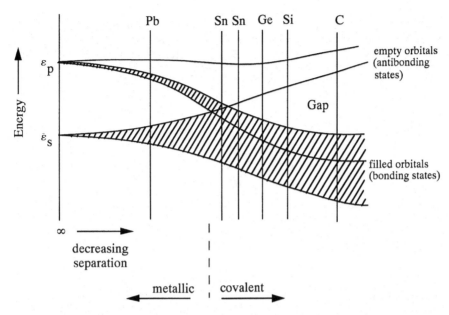

Figure 9.24. Schematic illustration of how the electronic structure is affected by interatomic separation [47].

reference to their interatomic separations to see how different electronic behaviors arise.

The LCAO model contains three quantitative parameters (V_1, V_2, and V_3), the relative values of which determine the class of bonding and the characteristic structure type of the solid. If the polar energy, V_3, which scales with the electronegativity difference, is sufficiently small, then the balance between the metallic energy (V_1) and the covalent energy (V_2) determines whether a close-packed metallically conducting structure is adopted or whether a tetrahedrally coordinated semiconducting structure is adopted. If either V_2 or V_3 is significant compared to V_1, then it is the relative values of the polar and covalent energy that determine whether a tetrahedrally coordinated semiconducting structure is adopted or an ionic structure with a higher coordination number. For a polar-covalent structure, V_2 is proportional to the inverse square of the interatomic separation. The bandgap between the highest occupied and lowest unoccupied states $\left(2\sqrt{V_2^2 + V_3^2}\right)$ increases with both the covalency and the polarity. The ability of this model to describe bonds in terms of independent ionic, metallic, and covalent contributions reinforces the idea that there is actually a continuous range of bonding types between the three limits of Ketlaar's triangle (see Chapter 1).

Table 9.10. *Bandgaps of selected tetralide, III–V, and II–VI materials [48].*

material	E_g, eV	T_m (K)	material	E_g, eV	T_m (K)	material	E_g, eV	T_m (K)
C	5.4	4300	AlAs	2.1	2013	ZnO	3.2	2250
SiC(3C)	2.3	3070	AlN	6.2	—	ZnS	3.6	1920
Si	1.11	1685	GaN	3.37	1920	ZnSe	2.58	1790
Ge	0.67	1231	InN	1.87	—	ZnTe	2.26	1510
			GaP	2.25	1750	CdS	2.42	2020
			GaAs	1.43	1510	CdSe	1.74	1530
			GaSb	0.70	980	CdTe	1.45	1370
			InP	1.35	1330			
			InAs	0.35	1215			
			InSb	0.17	798			

Ⓙ The bandgap

Crystalline materials with the diamond, sphalerite, and wurtzite crystal structures are widely applied in semiconducting, solid state devices including transistors, photodetectors, and light emitters. For devices that detect or emit light, it is the band gap between the occupied valence band states and the unoccupied conduction band states that determines the wavelength (color) of the light detected or emitted. Thus, one aspect of engineering a device to detect or emit specific wavelengths amounts to controlling the bandgap. For example, the silica–germania glass fibers that form the backbone of optical communications networks have their maximum transparency at about 0.8 eV (1550 nm). Therefore, this is the most desirable bandgap for emitters and detectors that transmit and receive signals, respectively. If we examine the data in Table 9.10, we note that no element or simple binary compound has this particular bandgap. However, many of the isostructural compounds form a complete range of solid solutions that have intermediate bandgaps. In Fig. 9.25, the compounds that form solutions are connected by lines. You can see that by forming the appropriate semiconductor alloy, any bandgap can be achieved.

It must be noted that making useful light emitters of any wavelength is substantially more complex than simply choosing the appropriate alloy. A diode laser, for example, is composed of several layers with different compositions and bandgaps. Because the devices are produced as multiple thin film layers, maintenance of epitaxy demands that each layer have a similar lattice parameter. So, one must choose several compositions with similar lattice parameters and different

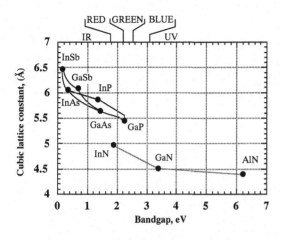

Figure 9.25. The relationship between the bandgap and the lattice constant for III–V compounds. The antinomides, arsenides, and phosphides have the sphalerite structure. The nitrides have the wurtzite structure; in these cases, the effective cubic lattice constant was computed by assuming that $a_s = \sqrt{2}\, a_w$, where a_w is the basal plane lattice constant of the wurtzite structure and a_s is the cubic lattice constant in the sphalerite structure. Compounds that form solid solutions are connected by lines, which are schematic.

bandgaps that can be processed together. The devices that operate in the infrared and red regions of the spectrum comprise a well established technology. Current research is focused on developing devices in the green, blue, and UV regions of the spectrum. Such devices are desired for use in optical data storage systems, flat panel displays, and even lamps. While the ZnSe–CdSe system was intensively studied for these applications [49], it now seems likely that it will be the group III nitride based materials that are used [50].

Throughout this chapter, we have noted that the gap between the occupied and unoccupied states of a crystal should vary as $\sqrt{V_2^2 + V_3^2}$, where V_2 and V_3 are the polar and covalent energy. For compounds with similar polarities, the overall change in the bandgap should be the result of differences in V_2. According to Eqn. 9.46, the covalent energy is inversely proportional to the square of the interatomic separation. Noting that the cubic lattice constant is directly proportional to the interatomic separation, it is satisfying that the bandgap data in Fig. 9.25 reflect the expected trend. Indeed, more complete tabulations of bandgap v. interatomic spacing data for a range of structures verify this trend [14, 15].

(K) Problems

(1) In section C, we determined that along the [100] direction, the energy width of a band of states formed by the overlap of a square planar array of s orbitals was $4V_2$, where V_2 is the covalent energy. Demonstrate that a three-dimensional simple cubic array of s-orbitals has the same band width along [100].

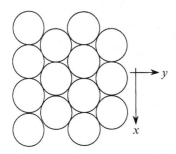

Figure 9.26. The illustration for problem 5.

(2) Calculate the dispersion for p_x orbitals on a square planar array with a lattice constant of a.

    (i) Plot and compare the dispersion in the [100] and [010] directions. What is the band width in each of these directions?

    (ii) Assume that the lattice distorts in such a way that the angle between the a and b lattice vectors is less than 90°. How is the dispersion relation altered? How does the distortion alter the band width?

(3) For a square planar array of s-orbitals, calculate the dispersion and band width along the [110] direction and make sure that your answer agrees with that shown in Fig. 9.7.

(4) In Example 9.1, the dispersion relation for the homopolar square lattice was modified to include distant near neighbors. Show that the terms for the third and fourth neighbors are correct as stated.

(5) Consider a single close-packed plane of atoms (see Fig. 9.26). There is a single identical s orbital associated with each atom. Using the LCAO model and assuming that only the nearest neighbors interact, you can calculate a dispersion relation for the band of states formed by the interaction of s-orbitals.

    (i) Demonstrate that the dispersion relation for this two dimensional structure is:

$$\varepsilon(\vec{k}) = \varepsilon_s + 2V_{ss\sigma}\left[\cos(k_x a) + 2\cos\left(\frac{1}{2}k_x a\right)\cos\left(\frac{\sqrt{3}}{2}k_y a\right)\right].\tag{9.98}$$

In Eqn. 9.98, k_y and k_y are the components of $\vec{k}$ along the x and y axes, respectively. The nearest neighbor spacing between atoms is a. The remaining variables have their usual meanings.

    (ii) What is the free electron dispersion relation for this two dimensional crystal?

    (iii) Show that for small values of k, Eqn. 9.98 has the same functional form (k-dependence) as the free electron dispersion.

(iv) What happens to the width of the LCAO band (along the [100] direction) as the size of the crystal decreases?

(v) What happens to the width of the LCAO band (along the [100] direction) as the spacing between the atoms increases?

(6) Consider an fcc metallic crystal with one atom at each lattice site and a nearest neighbor separation of d. Use the LCAO method and the approximation that only nearest neighbor s orbitals interact, together with our usual definition of variables, to show that the following dispersion relation describes the lowest energy band:

$$\varepsilon(\vec{k}) = \varepsilon_s + 4V_{ss\sigma}[\cos(k_x d)\cos(k_y d) + \cos(k_x d)\cos(k_z d) + \cos(k_y d)\cos(k_z d)].$$

$$(9.99)$$

(7) Use the LCAO method to determine a dispersion relation for a ring containing N hydrogen atoms.

(i) Begin by including only nearest neighbor interactions and assuming that N is a very large number. Plot the dispersion between 0 and $\pm \pi/a$.

(ii) Determine and plot two more dispersion relations, including next-nearest neighbor interactions in the first, and third nearest neighbors in the second.

(iii) Describe the changes that occur as N becomes small, say $N = 6$.

(iv) Determine a new dispersion relation for another large ring, this time composed of two different atoms, A and B (assume that the relevant states derive from s-orbitals). How do the band width and band gap of this ring vary with polarity?

(8) After describing the bonding in GaAs (a III–V compound) within the framework of the LCAO method, Chadi & Cohen [29] arrived at the Hamiltonian matrix shown in Fig. 9.18. Based on this matrix, answer the following questions:

(i) In this description of GaAs, is the overlap between p_x orbitals on neighboring As atoms (next-nearest neighbors in the structure) included? How can you tell?

(ii) Determine the following matrix elements:

$$E_{sp}g_1$$
$$E_{xy}g_3.$$

You can leave your answer in terms of the 'covalent energies', $V_{ll'm}$.

(iii) How many bands of electron states will this model give?

(iv) How many will be filled and how many will be empty?

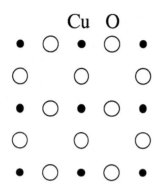

Figure 9.27. Arrangement of the Cu and O atoms in the *a–b* plane of tetragonal cuprate. See Problem 12.

(v) What structural variables affect E_{sp} and E_{xy} and how?

(9) SiC crystallizes in a number of different polymorphic forms. Two of these polymorphs are very common structure types, sphalerite ($a = 4.358$ Å) and wurtzite ($a = 3.076$ Å, $c = 5.048$ Å). Both of these structures were described in Chapter 4. While examining a single crystal of each of these specimens, you find that they have a different band gap. Can you explain which of the two has the larger band gap using the LCAO model?

(10) Fig. 9.15 shows the band structure of MgO. Using quantitative details whenever possible, answer the following questions.

(i) How are the dispersion relations affected by a hydrostatic compression?

(ii) How are the dispersion relations affected by a uniaxial compression, along z?

(11) The Mg ions in MgO have an ideal octahedral coordination. In many compounds, the coordination environment is far from ideal and the metal atoms occupy sites that are distorted from the ideal octahedral symmetry. Recompute the band structure in Fig. 9.15 assuming that the Mg ion is displaced by a small amount (less than 10% of the ideal interatomic spacing) along the z axis of the crystal. Thus, instead of six identical Mg–O distances, there will be three different distances.

(12) In all of the high critical temperature superconducting cuprates (the so-called 'high T_c' materials), the single most important structural feature is thought to be the two dimensional Cu–O planes. The structure of these planes can be described as a square net (the *a–b* plane of a tetragonal structure), with Cu atoms at the vertices and O at the midpoints along the edges (this structure is shown in Fig. 9.27). Assume that the only atomic orbitals participating in bonding are the $2p_x$ and the $2p_y$ on the O and the $3d_{x^2-y^2}$ on the Cu [51].

(i) Draw a sketch showing the orientation of these atomic orbitals.

(ii) Next, use the LCAO method to determine the dependence of the band energy on the wave vector. You may assume that p_x and p_y orbitals do not interact, and that the only matrix elements that are nonzero are the on-site (H_{ii}) matrix elements and those between nearest neighbors atoms.

(iii) Plot the dispersion along $\vec{k} = [100]$ and $[110]$. Make sure that you label relative values on the energy axis.

(iv) Where is the Fermi level? Assume that copper is divalent, has a d^9 configuration, and that the d-orbitals not included in the model are fully occupied. The apparent charge imbalance is because this plane is only one unit of a larger structure.

(v) Finally, compare the LCAO bands to the free electron bands from the same lattice. (You should do this by plotting the bands.)

(13) In Section D(*ii*) of this chapter, we discussed the formulation of the LCAO bands of MgO. Our description of the bonding involved the s orbital on the cation and the p orbitals on the anion. In this problem, we consider the isostructural compound TiO. Because Ti(II) has d-electrons in the valence shell, our bonding description should include the 3d orbitals (which we take to be at a lower energy than the 4s or 4p).

(i) The free Ti atom has five degenerate d orbitals (degenerate orbitals have the same energy). When the compound crystallizes in the rock salt structure, this degeneracy is broken by the crystal field. Which set of orbital have the lowest energy and why?

(ii) We can formulate a simplified LCAO description of TiO using a six orbital basis that consists of the highest energy anion states and the lowest energy cation states. Construct a Hamiltonian matrix using this basis and the assumption that only nearest neighbors interact. The matrix should only contain term values and cation–anion overlap terms.

(iii) Determine the dispersion in the k_x direction. Explain any approximations that you apply.

(iv) Based on your analysis, is TiO an insulator or a conductor?

(14) Draw the two lowest energy 'free electron' bands between Γ and X, as in Fig. 9.7b, but separate the upper band from the lower one by adding a constant energy, U. This creates an energy gap at the Brillouin zone boundary. Note that the energy width of the bands are unchanged by this addition. We could select any value for U to produce any desired bandgap. Next, assume that free electron theory is altered so that the electron mass is now a variable. The only alteration to the band dispersion is that the fixed electron mass, m, is replaced by the variable, m^*. Explain how we could use this new variable, m^*, together with U, to model the bands in semiconductors of varying covalency and polarity.

(L) References and sources for further study

[1] Y.W. Yang and P. Coppens, On the experimental electron distribution in Si, *Solid State Comm.* **15** (1974) 1555–9. Fig. 9.1 was drawn after Fig. 2 on p. 1557 of this paper.

[2] M.T. Yin and M.L. Cohen, Theory of static structural properties, crystal stability, and phase transformations: Application to Si and Ge, *Phys. Rev.* B **26** (1982) 5668–87.

[3] J.R. Chelikowsky and M.L. Cohen, Nonlocal pseudopotential calculations for the electron structure of eleven diamond and zinc blende semiconductors, *Phys. Rev. B* **14** (1976) 556–82.

[4] F. Bloch, Uber die Quantenmechanik der Elektronen in Kristallgittern, *Z. Phyzik* **52** (1928) 555.

[5] L. Pauling, *The Nature of the Chemical Bond* (Cornell University Press, Ithaca, 1960). The data in Table 9.1 comes from Table 7.13 on p. 246.

[6] J.C. Phillips, *Bonds and Bands in Semiconductors* (Academic Press, New York, 1973). The data in Table 9.1 comes from Table 1.5 on p. 22.

[7] W.A. Harrison, *Electronic Structure and the Properties of Solids: The Physics of the Chemical Bond* (Dover Publications, Inc., New York, 1989). Much of the discussion in Section B(i) is based on Chapter 1 of Harrison's book. Furthermore, throughout Chapter 9, Harrison's nomenclature is largely preserved. This book should be considered mandatory reading for anybody with a deeper interest in the LCAO method in particular and the electronic structure of solids in general.

[8] W.A. Harrison, *Electronic Structure and the Properties of Solids: The Physics of the Chemical Bond* (Dover Publications, Inc., New York, 1989). Appendix B, p. 536.

[9] W.A. Harrison, *Electronic Structure and the Properties of Solids: The Physics of the Chemical Bond* (Dover Publications, Inc., New York, 1989). Fig. 9.3 is drawn after Fig. 1.10a on p. 19.

[10] W.A. Harrison, *Electronic Structure and the Properties of Solids: The Physics of the Chemical Bond* (Dover Publications, Inc., New York, 1989). Fig. 9.4 is drawn after Fig. 1.10b on p. 19.

[11] N.W. Ashcroft and N.D. Mermin, *Solid State Physics* (Holt Rinehart and Winston, New York, 1976) pp. 133–41. Several proofs of Bloch's theorem can be found in this source.

[12] C.S. Nichols, *Structure and Bonding in Condensed Matter* (Cambridge University Press, Cambridge, 1995) pp. 127–9.

[13] S. Froyen and W.A. Harrison, Elementary Prediction of Linear Combination of Atomic Orbitals Matrix Elements, *Phys. Rev. B* **20** (1979) 2420–2.

[14] W.A. Harrison and S. Ciraci, Bond Orbital Method. II, *Phys. Rev. B* **10** (1974) 1516–27.

[15] S.T. Pantelides and W.A. Harrison, Structure of Valence Bands of Zinc-Blende-Type Semiconductors, *Phys. Rev. B* **11** (1975) 3006–21.

[16] W.A. Harrison, *Electronic Structure and the Properties of Solids: The Physics of the Chemical Bond* (Dover Publications, Inc., New York, 1989). Fig. 9.8 is drawn after Fig. 2.8 on p. 52.

[17] W.A. Harrison, *Electronic Structure and the Properties of Solids: The Physics of the Chemical Bond* (Dover Publications, Inc., New York, 1989), pp. 72–4.

[18] W.A. Harrison, *Electronic Structure and the Properties of Solids: The Physics of the Chemical Bond* (Dover Publications, Inc., New York, 1989). pp. 50–1, Table 2.2. Harrison cites the following source for the values: F. Herman and S. Skillman, *Atomic Structure Calculations* (Prentice Hall, Englewood Cliffs, NJ, 1963). I was not able to obtain the original source.

[19] J.C. Slater and G.F. Koster, Simplified LCAO Method for the Periodic Potential Problem, *Phys. Rev.* **94** (1954) 1498–1524. Table 9.6 is based on Table I, p. 1503. This clear exposition of the LCAO method is recommended reading.

[20] W.A. Harrison, *Electronic Structure and the Properties of Solids: The Physics of the Chemical Bond* (Dover Publications, Inc., New York, 1989). Fig. 9.14 is drawn after Fig. 2.8 on p. 52.

[21] S. Pantelides, Universal Valence Bands for Rocksalt-Type Compounds and their Connection with those of Tetrahedral Crystals, *Phys. Rev. B* **11** (1975) 5082–93.

[22] Y.-M. Chiang, D. Birnie III, and W.D. Kingery, *Physical Ceramics* (John Wiley & Sons, New York, 1997) p. 120 (Table 2.3).

[23] V.E. Henrich and P.A. Cox, *The Surface Science of Metal Oxides* (Cambridge University Press, Cambridge, 1994) p. 129.

[24] C.S. Nichols, *Structure and Bonding in Condensed Matter* (Cambridge University Press, Cambridge, 1995) pp. 141–2.

[25] W.A. Harrison, *Electronic Structure and the Properties of Solids: The Physics of the Chemical Bond* (Dover Publications, Inc., New York, 1989). pp. 55–6.

[26] S. Pantelides, Universal Valence Bands for Rocksalt-Type Compounds and their Connection with those of Tetrahedral Crystals, *Phys. Rev. B* **11** (1975) pp. 5082–93. Figure 9.16 is drawn after Fig. 4, p. 5085.

[27] C.S. Nichols, *Structure and Bonding in Condensed Matter* (Cambridge University Press, Cambridge, 1995) p. 167. This example is based on problem 10.6.

[28] W.A. Harrison, *Electronic Structure and the Properties of Solids: The Physics of the Chemical Bond* (Dover Publications, Inc., New York, 1989). Fig. 9.17 is drawn after Fig. 3.7 on p. 75.

[29] D.J. Chadi and M.L. Cohen, Tight-Binding Calculations of the Valence Bands of Diamond and Zincblende Crystals, *Phys. Stat. Sol.* (b) **68** (1975) 405.

[30] W.A. Harrison, *Electronic Structure and the Properties of Solids: The Physics of the Chemical Bond* (Dover Publications, Inc., New York, 1989), pp. 75–80.

[31] W.A. Harrison, *Electronic Structure and the Properties of Solids: The Physics of the Chemical Bond* (Dover Publications, Inc., New York, 1989). Fig. 9.19 is drawn after Fig. 3.8 on p. 79.

[32] V.L. Morruzi, A.R. Williams, and J.F. Janak, Local Density Theory of Metallic Cohesion, *Phys. Rev. B* **15** (1977) 2854–7. That data in Fig. 9.20 is from Fig. 1.

[33] J. Friedel, Transition Metals. Electronic Structure of the d-Band. Its role in the Crystalline and Magnetic Structures, in *The Physics of Metals*, ed. J.M. Zinman, (Cambridge University Press, New York, 1969) pp. 340–408.

[34] W.A. Harrison, *Electronic Structure and the Properties of Solids: The Physics of the Chemical Bond* (Dover Publications, Inc., New York, 1989) Chapter 20.

[35] C.S. Nichols, *Structure and Bonding in Condensed Matter* (Cambridge University Press, Cambridge, 1995) p. 167. Chapter 15.

[36] L.F. Mattheiss, Electronic structure of the 3d transition metal monoxides. I. Energy-Band Results, *Phys. Rev. B* **5** (1972) 290–315. Fig. 9.22a is drawn after Fig. 2. on p. 295.

[37] J.B. Goodenough, Metallic Oxides, in: *Progress in Solid State Chemistry*, Vol. 5, ed. H. Reiss (Pergamon Press, New Jersey, 1971). Fig. 9.22b is drawn after Fig. 32. on p. 224.

[38] D.W. Bullett, Bulk and surface electronic states in WO_3 and tungsten bronzes, *J. Phys. C: Solid State Phys.* **16** (1983) 2197–207.

[39] K-.C. Ho, T.G. Rukavina, and C.B. Greenberg, Tungsten Oxide–Prussian Blue Electrochromic System Based on a Proton Conducting Polymer, *J. Electrochem. Soc.* **141** (1994) 2061–67.

[40] W.A. Harrison, *Electronic Structure and the Properties of Solids: The Physics of the Chemical Bond* (Dover Publications, Inc., New York, 1989) p. 88.

[41] W.A. Harrison, *Electronic Structure and the Properties of Solids: The Physics of the Chemical Bond* (Dover Publications, Inc., New York, 1989). Some data in Table 9.8 are from Table 3.2 on p. 90.

[42] W.A. Harrison, *Electronic Structure and the Properties of Solids: The Physics of the Chemical Bond* (Dover Publications, Inc., New York, 1989) pp. 61–74.

[43] W.A. Harrison, *Electronic Structure and the Properties of Solids: The Physics of the Chemical Bond* (Dover Publications, Inc., New York, 1989). Fig. 9.23 is drawn after Fig. 3.3 on p. 66.

[44] J.C. Phillips, *Bonds and Bands in Semiconductors* (Academic Press, New York, 1973) p. 43.

[45] W.A. Harrison, *Electronic Structure and the Properties of Solids: The Physics of the Chemical Bond* (Dover Publications, Inc., New York, 1989) Chapter 7.

[46] W.A. Harrison, *Electronic Structure and the Properties of Solids: The Physics of the Chemical Bond* (Dover Publications, Inc., New York, 1989). Table 9.8 is adapted from Table 7.3 on p. 176.

[47] W.A. Harrison, *Electronic Structure and the Properties of Solids: The Physics of the Chemical Bond* (Dover Publications, Inc., New York, 1989). Fig. 9.24 is drawn after Fig. 2.3 on p. 39.

[48] L. Solymar and D. Walsh, *Electrical Properties of Materials* 6th ed. (Oxford University Press, Oxford, 1998) p. 138. The data on Table 9.8 is largly based on Table 8.2 on p. 138.

[49] R.L. Gunshor and A.V. Nurmikko, II–VI Blue-Green Laser Diodes: a Frontier of Materials Research, *MRS Bulletin*, vol. 20, no. 7 (July 1995) p. 15.

[50] S. Nakamura, InGaN/GaN/AlGaN-Based Laser Diodes with an Estimated Lifetime of Longer than 10000 hours *MRS Bulletin*, vol. 23, no. 5 (May 1998) p. 37.

[51] J.D. Jorgensen, H.-B. Schüttler, D.G. Hinks, D.W. Capone, II, K. Zhang, and M.B. Brodsky, Lattice Instability and High-Tc Superconductivity in $La_{2-x}Ba_xCuO_4$, *Phys. Rev. Lett.* **58** (1987) 1024–7.

Chapter 10
Models for Predicting Phase Stability and Structure

(A) Introduction

We began Chapter 1 by noting that the central, fundamental question that motivates continued materials research is, how can elements be combined to produce a solid with specified properties? Previously, we divided this problem into three separate issues that we restate here to provide appropriate context for the final chapter. First, when any given elements are combined under some controlled conditions, will they react to form a compound, will they dissolve in one another, or will they be immiscible? Second, what structure will the product of this combination have and how will it be influenced by the conditions under which the elements were combined? Third, given the product phase or phases and the structure, what are the properties of this material? In this chapter, we will discuss approaches that have been developed to predict answers to at least parts of the first two questions. For the purposes of this course, we limit our structural discussion to the atomic structure. However, we note that to fully answer these two questions, the defect structure and microstructure must also be addressed.

It was noted earlier that the quantitative physical models for bonding described in Chapters 6–9 were not able to reliably predict the equilibrium crystal structure of a compound. Despite impressive theoretical achievements and the development of more advanced models during the past few decades, phase stability, crystal structures, and the properties of solids must still be determined experimentally. Conventional bonding models allow the total energy of the solid to be computed and compared with the total energy of the free atoms so that the cohesive energy can be determined. The weakness of this approach lies in the fact that the cohesive energy is typically small in comparison to either of the total energies. Furthermore, the energy differences between alternative configurations of the atoms (different polymorphs) are even smaller, so the required accuracy is generally not attainable. In each of the quantitative models, the most important feature determining the relative stability of a compound is the elements in the compound, not the precise configuration of the atoms.

There are, however, a number of phenomenological models which have impressive predictive capabilities. In contrast to the physical models, which are based on some known, fundamental principle, the phenomenological models are based on experimental observations. The success of a phenomenological model is determined by its ability to reproduce known data and accurately predict unknown data. In most cases, such models are formulated without reference to a mechanism or fundamental principle. Nevertheless, these models are an important part of the scientific process because they usually precede an advanced mechanistic understanding of the phenomena. Consider, for example, Mendeleev's law of periods, which states that the properties of atoms vary in a systematic way according to their mass. Later knowledge of the electronic structure of the atom provided a mechanistic basis for this phenomenological law.

As the various phenomenological models are reviewed in this chapter, we will attempt to understand the reasons for their success by making comparisons, when appropriate, to the physical models. We shall see that the most successful of the models share two characteristics. One is that predictions are based on properties associated with the solid state, on a per atom basis, rather than the properties of free atoms. The second is that, instead of using well defined, fixed physical quantities, the parameters used in phenomenological models are variable and fit to the existing, relevant experimental data.

(B) Models for predicting phase stability

i.   Hume-Rothery rules and electron compounds

Based on experimental observations, Hume-Rothery proposed a set of rules to predict the relative solubility of metallic elements and how the composition of an alloy affects its structure [1–3]. The first rule is that the solvent phase is unstable if the solute atoms differ in size from the solvent atoms by more than 15%. In other words, a solid solution is unstable if it contains atoms with vastly different sizes and we should, therefore, expect the elements to co-exist as two separate phases or to form an intermetallic compound. If the size difference, δ, is given by:

$$\delta = \frac{r_B - r_A}{r_A}, \tag{10.1}$$

where r_A is the solvent radius and r_B is the solute radius, then we can estimate the maximum size of an atom that can dissolve in a matrix by using linear elasticity theory. If we equate the energy of an elastic distortion with the available thermal energy, we find that the criterion for stability can be written as:

$$\delta \geq \sqrt{\frac{kT}{3G_A\Omega_A}}. \tag{10.2}$$

In Eqn. 10.2, G_A is the solvent shear modulus and Ω_A is the solvent atomic volume. When typical values are inserted, a limit of about 10% size misfit is obtained.

The second rule is that if two atoms have significantly different electronegativity, then compounds are formed. In this case, electron transfer or polarization allows the compound to be stabilized by the electrostatic forces described in Chapter 7.

The third rule states that specific structures arise preferentially in characteristic ranges of the valence electron concentration, measured as electrons per atom (e/a):

$$\frac{e}{a} = Z_A(1 - \nu_B) + Z_B\nu_B. \tag{10.3}$$

In Eqn. 10.3, Z_i are the metal atom valences and ν_B is the concentration of the solute atoms. While the assignment of valences to metallic elements has always been a contentious issue, the Hume-Rothery scheme works well if the group IB, IIB, IIIA, and IVA metals are assigned valences of 1, 2, 3, and 4, respectively, and transition metal elements such as Fe, Co, and Ni are assigned a valence of zero.

When the size and electronegativity of a solute atom are favorable in a given solvent, its solubility is apparently limited by the electron concentration. The stabilities of a variety of Cu, Au, and Ag alloys are illustrated in Fig. 10.1 as a function of electron concentration. The most notable trend is that the solubility of an alloying element in the fcc phase extends to an electron concentration of approximately 1.4. At higher concentrations, new phases (either the bcc, B2, or the hcp ζ phase) appear. At an electron concentration of 1.62, a complex cubic phase labeled γ is stable and, finally, at an electron concentration of 1.75, the hexagonal ε phase is stable. Additional examples of so-called *electron compounds* are listed in Table 10.1.

The axial ratio (c/a) of hexagonal structures is also related to the electron concentration. The ε phase, with a lower than ideal c/a ratio, is stable for concentrations between 1.7 and 1.9. The ζ phase, which has a nearly ideal c/a ratio, is stable for electron concentrations between 1.2 and 1.75. The η phase, which has a higher than ideal c/a ratio, is stable in the 1.92 to 2.0 range.

Although the observed relationship between the axial ratio of hexagonal structures and the electron concentration has not been explained, the relationship between electron concentration and crystal structure can be understood in terms of band filling by recalling the basic elements of the nearly free electron theory introduced in Chapter 8. The two relevant aspects of the nearly free electron theory are that the electronic energy is proportional to k_F^2 and that the

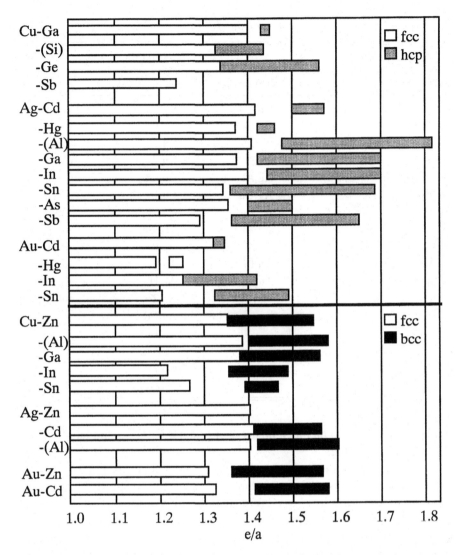

Figure 10.1. Extent of stability for Cu, Ag, and Au solid solutions with B metal atoms [2].

crystal potential opens up a gap at the zone boundary. As the electron concentration of a solid is increased, higher and higher electronic energy levels are filled and the total electronic energy increases. However, because the band flattens near the zone boundary, the rate at which the energy increases with electron concentration diminishes as the first band is nearly filled. The energy is then sharply increased as the electron wave vector crosses the first zone boundary and

Table 10.1. *Examples of electron compounds [1].*

	$e/a = 1.5$		$e/a = 1.62$	$e/a = 1.75$
B2 related	β-Mn	hcp related	cubic γ-brass	hcp related
CuBe	Cu_5Si	Cu_3Ga	Cu_5Zn_8	$CuZn_3$
CuZn	AgHg	Cu_5Ge	Cu_5Cd_8	$CuCd_3$
Cu_3Al	Ag_3Al	AgZn	Cu_5Hg_8	Cu_3Sn
Cu_5Sn	Au_3Al	AgCd	Cu_9Al_4	Cu_3Ge
AgMg	$CoZn_3$	Ag_3Al	Cu_9Ga_4	Cu_3Si
AuMg		Ag_3Ga	Cu_9In_4	$AgZn_3$
AuZn		Ag_3In	$Cu_{31}Si_8$	$AgCd_3$
AuCd		Ag_5Sn	$Cu_{31}Sn_8$	Ag_3Sn
FeAl		Ag_5Sb	Ag_5Zn_8	$AuZn_3$
CoAl		Au_3In	Ag_5Cd_8	$AuCd_3$
NiIn		Au_3Sn	Ag_5Hg_8	Au_3Sn
PdIn			Au_5Zn_8	Au_5Al_3
			Au_5Cd_8	
			Au_9In_4	
			Fe_5Zn_{21}	
			Co_5Zn_{21}	
			Ni_5Zn_{21}	

electrons are forced to fill much higher energy states in the second band. This is illustrated in Fig. 10.2.

It has been hypothesized that when the electron concentration surpasses a critical density that just fills the first band, it might be energetically favorable for the solid to transform to a less dense structure with a larger Brillouin zone that can accommodate more electrons rather than promote electrons to a higher energy band [2]. This hypothesis can be supported by computing the electron concentrations at which the Fermi sphere contacts the first Brillouin zone boundary. In other words, this is the electron concentration above which electrons must be promoted across the gap. These data are summarized in Table 10.2.

While there are clearly some realistic elements to this theory, it has also received some sound criticism for its reliance on the rigid band assumption. The rigid band assumption is the idea that electrons simply fill the empty levels without changing the distribution of the energy levels. However, more recently developed models that are more consistent with the modern theory of the metallic bond do little to increase the predictive capabilities of the model.

Table 10.2. *Observed and experimental critical electron concentrations.*

structure	calc. contact concentration	observed transformation
fcc	1.36	1.4
bcc	1.48	1.5
γ	1.54	1.62
ζ	1.72	1.75

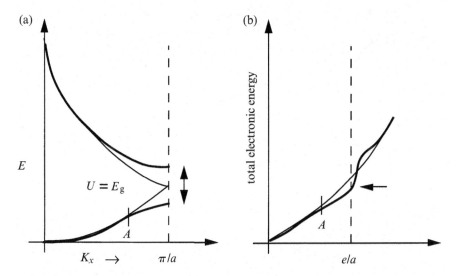

Figure 10.2. (a) Dispersion for the free electron model (normal line) and the nearly free electron model (bold line). (b) Total electronic energy as a function of electronic concentration for filling free electron bands (normal line) and nearly free electron bands (bold line). The rate of energy increase is high after filling the first band.

Finally, we comment that electron counting rules are applied to transition metal alloys with caution and skepticism.

ii. *Miedema's rules for alloy formation*
Miedema's model for alloy formation is an example of a phenomenological model that uses per atom quantities of solid state properties and is fitted to a large body of existing phase equilibria data [4–13]. At the heart of Miedema's model is the assumption that an elemental metal can be disassembled into single Wigner–Seitz (WS) units (recall that the WS unit cell is the volume of space about each lattice site closer to that site than any other site) and that these

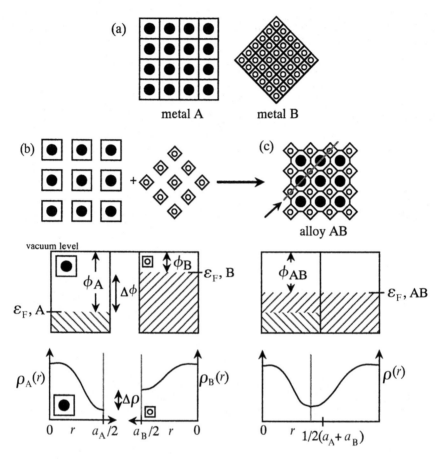

Figure 10.3. Schematic representation of Miedema's model for alloy formation. (a) Individual metals A and B. (b) Each metal is subdivided into Wigner–Seitz units with a characteristic Fermi level and electron density. (c) When the cells are brought together to form the alloy, charge flows from B to A and the cells change shape so that the charge density at the interfaces between the cells is continuous. The plot of the electron density of the alloy is along the direction indicated by the arrow in (c).

microscopic pieces of metal have the same properties as the bulk metals. An alloy or compound is then formed by assembling the atomic scale volumes from different elements into a new, bulk solid. This process is illustrated schematically in Fig. 10.3.

The enthalpy change associated with assembling an alloy from microscopic pieces of the elemental metals is then assumed to have two contributions. The first is a negative or attractive component that arises from charge redistribution;

to equilibrate the chemical potential of electrons in the crystal, charge flows from the microscopic pieces of metal with the higher Fermi energy to those with the lower Fermi energy. The second contribution is positive or repulsive and arises from changes in the electron density that must occur at the boundary between the WS cells from different metals. Since the electron density is required to be a continuous function of position, the electron density at the boundaries of all of the WS cells must exactly match. Assuming that the original (unmatched) electron densities are the minimum energy values, the changes necessary to achieve continuity will have a positive contribution to the enthalpy.

To construct a quantitative model, Miedema chose two physical properties of the metals as measures of the positive and negative contributions to the enthalpy. In a macroscopic sense, the magnitude of the charge transfer across the interface between two adjacent WS cells from the filled states of the metal with the higher Fermi level to the empty states of the metal with the lower Fermi level should be proportional to the contact potential or the difference between the chemical potentials of the electrons in each metal. Since the chemical potential (the Fermi level) must be uniform in the single phase material, charge will flow from the higher potential to the lower until they are equal. As illustrated in Fig. 10.3, the difference in the work function (the energy required to remove an electron from the highest filled state of the metal to the vacuum level) of the two metals ($\Delta\phi$) is equal to the difference in the chemical potentials. The work function might also be thought of as the ionization energy of the crystal and it scales with the electronegativity of the constituent atoms. The charging of the two WS cells lowers the energy by a factor of $(\Delta\phi)^2$. In Fig. 10.4, the work functions used in the Miedema scheme are labeled ϕ^*; the asterisk is used to indicate that the Miedema work functions are actually adjusted values that have the same trends as the accepted values, but are quantitatively different.

The second parameter is the difference in the charge densities (ρ) at the edge of the WS cells. This discontinuity can be removed through the expansion of the cell with higher density and/or the contraction of the cell with lower charge density, as illustrated in Fig. 10.3. The increase in energy is proportional to $(\Delta\rho^{1/3})^2$. The values of the charge densities for each element can be calculated using methods similar to those described in Chapters 8 and 9; the values used by Miedema are given in Fig. 10.4. The charge density at the WS cell boundary is measured in *density units* (d.u.), with the charge density at the WS cell boundary of Li defined to be equal to 1.

Based on these two parameters, Miedema wrote that the heat of formation of a metallic alloy is:

$$\Delta H = K[Q(\Delta\rho^{1/3})^2 - P(\Delta\phi^*)^2]. \tag{10.4}$$

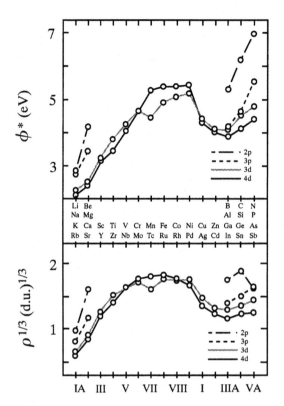

Figure 10.4. The numerical parameters used in the Miedema scheme [4].

In Eqn. 10.4, K is a positive definite constant that includes geometric factors which describe the area of contact between the two adjacent WS cells. Within the framework of this model, the heat of formation is proportional to the surface area shared by the dissimilar cells. Since K is positive, the sign of the heat of formation is determined by the ratio of the difference in the work functions to the difference in the electron densities at the WS cell boundary. Specifically, the heat of formation is negative when $(\Delta\phi^*/\Delta\rho^{1/3}) < (Q/P)^{1/2}$.

The validity of this approach is demonstrated in Fig. 10.5. Each point on the plot in Fig. 10.5 represents a binary transition metal alloy or an alloy between a transition metal and an alkali, alkaline earth, or noble metal. Combinations of elements that form compounds are represented by minus signs, indicating that they have a negative heat of formation. If the components do not react and their solid solubility is limited to less than 10 a/o, then a plus sign is used to indicate a positive heat of formation. Note that the separation between the plus and minus signs is nearly perfect. The straight line at the boundary is defined by a constant ratio of $\Delta\phi^*$ to $\Delta\rho^{1/3}$, as predicted by Eqn. 10.4.

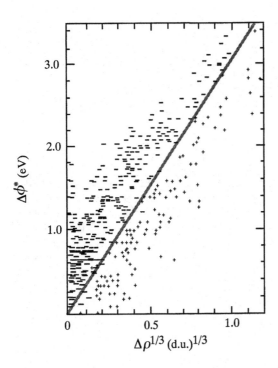

Figure 10.5. As predicted by Miedema, alloys with a negative heat of formation are separated from those with a positive heat of formation by a constant ratio. The line on the figure corresponds to $Q/P = 9.4$ $(eV)^2/(d.u.)^{2/3}$. See text for further explanation [4].

Figure 10.6 illustrates the periodic nature of alloy formation. The numbers at the edges of the matrix are the 'Mendeleev numbers', as originally defined by Pettifor [4]. While the atomic numbers increase with mass, the Mendeleev numbers illustrate periodicity in an alternative way. The numbers increase as you go from the bottom to the top of each group and sequentially through all the groups on the periodic chart, beginning with the inert gases, the s-fillers, the d-fillers, and then the p-fillers (a key is given in the lower right portion of the chart). A large domain of negative heats of formation occurs in the d–d region and is centered at a position where the d-band is half filled. Recall that in Chapter 9, we used the Friedel model to explain the increased stability of transition metals with half filled d-bands.

In addition to predicting the sign of the heat of formation, Miedema's method can be used to quantitatively calculate heats of formation, the enthalpy of mixing as a function of composition, and even heats of adsorption. However, for such calculations, the geometric parameters in the constant K are required.

$$K = 2c_A c_B[(V_A^{2/3} V_B^{2/3})/(c_A V_A^{2/3} + c_B V_B^{2/3})][(\rho_A)^{-1/3} + (\rho_B)^{-1/3}]^{-1}. \quad (10.5)$$

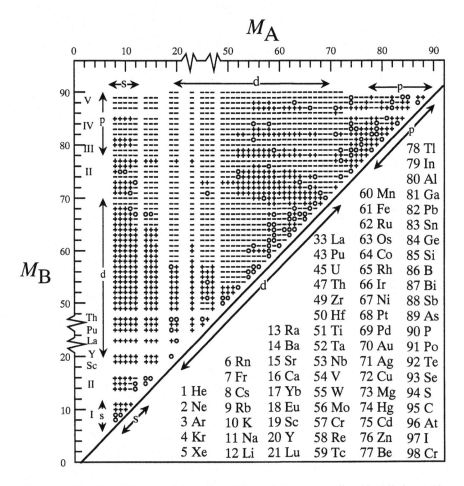

Figure 10.6 The sign of the heat of formation of binary alloys, as predicted by Miedema. The circles indicate alloys for which $|\Delta H| \leq 0.01$ eV [4].

In Eqn. 10.5, the values of c are the concentrations of the constituents in the alloy and the values of V are the volumes of the WS cells. The first term in square brackets describes the deviation from regular solution behavior caused by volumetric differences, and the second term in brackets is included to account for electronic screening at the interface between different WS cells. Equation 10.4 defines the heat of formation for a disordered alloy. For an ordered alloy, an additional multiplicative pre-factor must be included.

$$K_{\text{ordered}} = K[1 + 8(c_{s,A}c_{s,B})^2].$$
(10.6)

Here, the terms $c_{s,i}$ are the surface area concentrations computed from the volumes and bulk concentrations.

$$c_{s,i} = c_i V_i^{2/3} / (c_A V_A^{2/3} + c_B V_B^{2/3}). \tag{10.7}$$

The behavior of alloys between transition metals and post-transition metals does not fit the model described above. For these systems, Miedema added an additional term to account for p–d bonding.

Although very successful, Miedema's model should be regarded as phenomenological. From a mechanistic point of view, it is difficult to justify its foundation. We know, for example, that atomic-sized pieces of a metal do not behave as if they were bulk metal nor can we expect electrons to simply flow from one set of rigid bands to another. However, one might compare the Miedema model to a thermochemical process such as the Born–Haber cycle, where we use the notion of path independence to break a complex process up into a set of fundamental steps. In this case, the fact that the model works so well should not lead us to conclude that the mechanistic description is accurate. However, it does provoke us to ask why it works so well and how the phenomenological parameters can be related to established physical parameters.

iii. Villars' method for predicting compound formation

To predict whether or not two elements will react to form a compound, Villars established a method that is similar in spirit to that of Miedema, but is based on graphic representations of relative atomic properties rather than specific equations. Villars classified 182 different sets of atomic properties into five classes: those which quantify an atom's size (A), those which quantify the directional character of its homopolar bonds (B), those which quantify the strength of its homopolar bonds (C), those which quantify an atom's affinity for other atoms (D), and those which represent the atom's electronic structure (E). Examples of each of these properties are listed in Table 10.3. From each of these classes, a single property, judged to be the most accurate and complete, was selected as the representative property (see Table 10.3).

The moduli of differences, products, sums, and ratios of these properties were then calculated for experimentally known systems to see which three quantities best distinguished those systems that form compounds from those that do not. The three parameters which worked best were the magnitude of the difference in the number of valence electrons, $|\Delta VE_{AB}|$, the absolute difference between the Zunger pseudopotential radii sums, $|\Delta(r_s + r_p)^z_{AB}|$, and the ratio of the melting points, T_A/T_B, such that $T_A/T_B > 1$. The values of the relevant parameters are listed in Fig. 10.7. The criterion for distinguishing the

Table 10.3 *Selected atomic properties, according to Villars' classification system.*

Class	Properties	Representative property
A	covalent radius atomic volume density	Zunger pseudopotential radii sum
B	atomic number atomic electron scattering factor specific heat	periodic group number
C	heat of sublimation Young's modulus cohesive energy	melting point
D	electron affinity first ionization potential Herman–Skillman term values	Martynov–Batsanov electronegativity
E	holes in the d-band above Fermi level	number of valence electrons

two types of systems is graphical. Based on these three parameters, it is possible to make three-dimensional plots where each of the three parameters is a spatial coordinate and every possible binary system is represented by a point in this space. The points corresponding to systems that form compounds fall into a distinct region of this space. The points corresponding to the systems which do not form compounds occupy the remainder of the volume. The result is a map of the parameter space where boundaries between compound forming regions and non-compound forming regions are clearly defined (see Fig. 10.8).

Villars considered the 3486 binary systems that result from the combination of all elements with atomic number less than 97, excluding the rare gases and the halogens. The possible binary systems were divided into two classes: the 1107 that form between isostructural elements (if the two elements have a common polymorphic structure, they are taken to be isostructural) and the 2379 that form between elements that do not share a common structure. The maps corresponding to the first group are shown in Fig. 10.6. Note that Villars distinguished several different types of non-compound forming binaries and that these are also well separated on the maps (see the dashed lines). The solubility type exhibits a continuous solid solution in some region of the phase diagram and the insoluble type exhibits complete immiscibility. Two additional types of insolubility are

Periodic table of atomic parameters (each entry lists: element symbol, number of valence electrons; Zunger pseudopotential radii sum; melting temperature; crystal structure):

Element	Valence	Radii sum	Melting T	Structure
H	1	1.25	14	b
Li	1	1.61	454	h,c,b
Be	2	1.08	1556	u,h
Na	1	2.65	372	h,b
Mg	2	2.03	923	h
K	1	3.69	337	b
Ca	2	3.00	1123	c,b
Sc	3	2.75	1811	h,b
Ti	4	2.58	1938	h,c,b
V	5	2.43	2190	b
Cr	6	2.44	2176	b
Mn	7	2.22	1517	u,u,c,b
Fe	8	2.11	1812	b,f,b
Co	9	2.02	1768	h,c
Ni	10	2.18	1728	c
Cu	11	2.04	1356	c
Zn	12	1.88	693	h
B	3	0.795	2300	u
C	4	0.64	3500	u,d
N	5	0.54	63	u,b
O	6	0.465	54	u,u
Al	3	1.675	832	c
Si	4	1.42	1683	d
P	5	1.24	870	u
S	6	1.1	392	u
Ga	3	1.695	303	u
Ge	4	1.56	1210	d
As	5	1.415	1090	As
Se	6	1.285	490	Se
Rb	1	4.10	312	b
Sr	2	3.21	1043	c,h,b
Y	3	2.94	1773	h,b
Zr	4	2.825	2128	h,b
Nb	5	2.76	2770	b
Mo	6	2.72	2890	b
Tc	7	2.65	2473	h
Ru	8	2.605	2700	h
Rh	9	2.52	2239	c
Pd	10	2.45	1823	c
Ag	11	2.375	1234	c
Cd	12	2.215	594	h
In	3	2.05	1773	b
Sn	4	1.88	505	b
Sb	5	1.765	903	As
Te	6	1.67	723	Se
Cs	1	4.31	302	b
Ba	2	3.402	983	b
La	3	3.08	1193	L,c,b
Hf	4	2.31	2250	h,b
Ta	5	2.79	3270	b,u
W	6	2.735	3650	b
Re	7	2.68	3308	h
Os	8	2.65	3500	h
Ir	9	2.628	2727	c
Pt	10	2.70	2043	c
Au	11	2.66	1336	c
Hg	12	2.41	234	u
Tl	3	2.235	2300	h,c
Pb	4	2.09	601	c
Bi	5	1.997	545	As
Po	6	1.90	527	u,u
Fr	1	4.37	293	b
Ra	2	3.53	973	u
Ac	3	3.13	1470	u

Lanthanides:

Element	Valence	Radii sum	Melting T	Structure
Ce	3	4.50	1077	c,L,c,b
Pr	3	4.48	1208	L,b
Nd	3	3.99	1297	L,b
Pm	3	3.99	1300	L,b
Sm	3	4.14	1345	u,h,b
Eu	3	3.94	1100	b
Gd	3	3.91	1585	h,b
Tb	3	3.89	1029	h,b
Dy	3	3.67	1680	h,b
Ho	3	3.65	1734	h,b
Er	3	3.63	1770	h,b
Tm	3	3.60	1818	h,b
Yb	3	3.59	1087	f,b
Lu	3	3.37	1928	h,b

Actinides:

Element	Valence	Radii sum	Melting T	Structure
Th	3	4.98	1968	c,b
Pa	3	4.96	1500	u
U	3	4.72	1406	u,u,u,b
Np	3	4.93	913	u,u,b
Pu	3	4.91	913	u,u,c,b
Am	3	4.89	1103	h

Figure 10.7. Atomic parameters used by Villars. The number to the right of the symbol is the number of valence electrons, the number below the symbol is the Zunger pseudopotential radii sum, the number below that is the melting temperature, and the lowest line gives the crystal structures: b = bcc, c = ccp, h = hcp, d = diamond, L = La, As = As, Se = Se, u = a structure unique to that element [14].

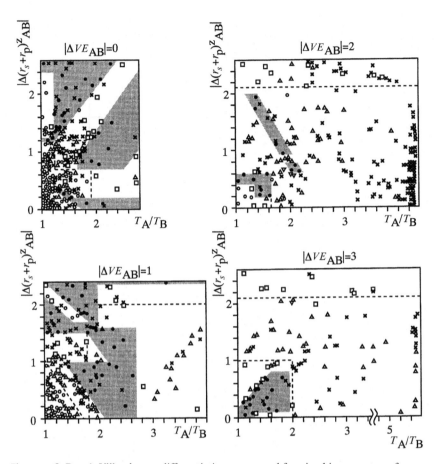

Figure 10.8. Part 1. Villars' maps differentiating compound forming binary systems from non-compound forming binary systems for 1107 isostructural systems. Each plot represents a 'slice' from the $|\Delta VE_{AB}|$ space and is labeled. The compound forming regions of the space are shaded. Boundaries between the two regions were drawn to maximize the number of systems in the correct domain. Each possible binary system is represented by a point: ● is for a system that forms a compound, ○ is for solid solutions, △ is for insoluble systems, □ is for the eutectic systems, ▽ is for the peritectic systems. The ✗ represents a system that has not been experimentally investigated. The dashed lines mark boundaries between regions where non-compound forming systems have similar types of phase diagrams. The 28 systems known to violate the boundaries are excluded [14].

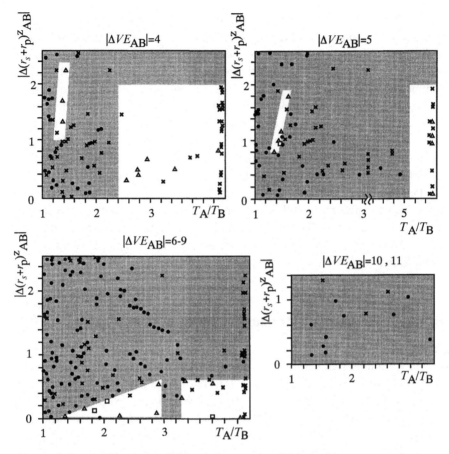

Figure 10.8. Part 2. Villars' maps differentiating compound forming binary systems from non-compound forming binary systems for 1107 isostructural systems. For a complete description, see the caption under part 1 on the previous page [14].

distinguished: eutectic systems melt at a temperature lower than either of the two pure elements and the peritectic systems melt at temperatures intermediate between the two elements.

Of the 1107 systems on the maps in Fig. 10.6, 707 have been experimentally investigated and 400 remain unknown. Of the 707 known systems, only 28 (4%) lie in the incorrect domain. When the remaining 2379 systems are subjected to a similar analysis, there are only 36 incorrect predictions in the group of 1260 experimentally known systems, an error rate of only 3%. The isostructural and nonisostructural groups have very different characteristics; 66% of the isostructural combinations do not form compounds, while more than 90% of the nonisostructural systems do form compounds. Furthermore, 80% of the isostructural combinations that do not form compounds have phase diagrams with regions of complete solubility or complete immiscibility. So, one rule of thumb that can be taken from these results is that if two elements do not share a common structure, they are likely to form a compound. If they do share a common structure, then compound formation is less likely. Based on this observation, Villars postulated that elements tend to retain the structure of the pure form whenever possible. In comparison to models such as Miedema's, it is interesting to note that Villars found the best separation between the two domains when the electronegativity was not considered as a parameter.

Perhaps the most interesting facet of Villars' work is its completeness and from this we can state the following trends. In general, compound formation is most likely when the difference in the number of valence electrons is large, the size difference is large, and the melting points are similar. The trends differ little from those found by Hume Rothery and provide us with few useful quantitative guidelines.

Ⓒ Factors that determine structure in polar-covalent crystals

i. Pauling's rules

By 1929, Pauling [15] had already realized that electrostatic bonding models, such as the one described in Chapter 7, were of little use for the prediction of crystal structures. Although theoretical tools have improved over the decades and it is now possible to compute the energies of many different structural configurations in a relatively short period of time, the accuracy needed to differentiate between very similar structures has yet to be achieved. Recognizing these limitations, Pauling proposed a set of rules (that have come to be known as 'Pauling's rules') which are 'neither rigorously derived nor universally applicable'. Based on experimental observations, these empiricisms allowed Pauling and many researchers since to deduce possible stable structures for solids. The rules are to be applied to materials with a significant degree of ionic bonding.

The first rule is that structures comprise coordinated polyhedra of anions that surround each cations, where the anion–cation bond length is determined by the sum of the ionic radii and the cation coordination number is determined by the ratio of the univalent ionic radii. This statement is simply an affirmation of the hard sphere packing model and radius ratio rules described in earlier chapters. Thus, from this rule, one can predict the cation coordination number and bond length for any arbitrary combination of two atomic species.

The second rule is used to determine the anion coordination number. Pauling's electrostatic valence principle states that the electrostatic bond valence strength, s, is:

$$s = \frac{Z_c}{NN_c},$$

(10.8)

where Z_c is the cation charge and NN_c is the cation coordination number. In a stable structure, the electrostatic charge of each anion compensates the strength of the electrostatic valence bonds with neighboring cations. So:

$$Z_a = \sum_{i=1}^{NN_a} s_i,$$

(10.9)

where Z_a is the anion charge and NN_a is the anion coordination number. If we consider an ideal structure where all s_i are equal (the cation is symmetrically coordinated), then:

$$NN_a = \frac{Z_a}{s} = \frac{Z_a NN_c}{Z_c}.$$

(10.10)

A more general statement of this rule is that the sum of the bond valences (s_i) around an atom is equal to the atom's formal charge. To see how the first and second rule can be used to predict coordination numbers, consider ReO_3. Based on stoichiometry, we know that Re is oxidized to the $+6$ state when O is a divalent anion. Examination of the radius ratios leads to the conclusion that Re is octahedrally coordinated by O, so $NN_c = 6$. According to Eqn. 10.8, the electrostatic bond strength of the Re–O bond (s) is 1. Therefore, according to Eqn. 10.10, the anion coordination number is two. These predictions conform to the known structure of ReO_3, shown in Fig. 10.9.

Pauling's third rule states that the presence of shared polyhedral edges and especially shared polyhedral faces decreases the stability of a crystal structure. This is due to the fact that in these configurations, the cations are closer together and there is a significant increase in the repulsive energy. For example, the metal–metal separation in structures built from octahedra that share corners (such as the one shown in Fig. 10.9) is twice the metal–oxygen distance ($2d_{MO}$). If the octahedra share edges, this decreases to $\sqrt{2}d_{MO}$. If they share faces, the distance decreases to $1.16d_{MO}$. The destabilizing effect is greater for cations with high charges and lower coordination numbers. For example, in silicate structures, SiO_4 tetrahedra are found only in corner-sharing arrangements. When cations are found in close proximity to each other, one should suspect that metal–metal bonds have formed.

Pertaining to ternary compounds, the fourth rule states that in crystals

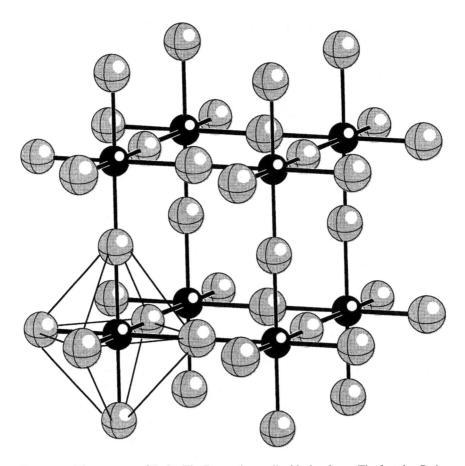

Figure 10.9. The structure of ReO_3. The Re are the smaller black spheres. The fact that Re is octahedrally coordinated and O is in two-fold coordination was deduced from Pauling's first two rules.

containing different cations, those with large valence and small coordination number tend not to share polyhedral elements with each other. In effect, this rule says that highly charged cations stay as far apart as possible. As an example, consider the arrangement of cations in the wurtzite related structures that are shown in Fig. 4.28.

The fifth rule, or the rule of parsimony, states that the number of different constituents in a crystal tends to be small. For example, bonds shared by chemically similar anions should be as nearly equal as possible and polyhedra circumscribed about chemically similar cations should, when possible, be chemically similar and similar in their contiguous environments (the manner in which they

Table 10.4. *Metal–metal distances for different polyhedral arrangements.*

M–M distances for connected octahedra		M–M distances for connected tetrahedra	
corner-sharing	$2\,d_{MO}$	corner-sharing	$2\,d_{MO}$
edge-sharing	$1.41\,d_{MO}$	edge-sharing	$1.16\,d_{MO}$
face-sharing	$1.16\,d_{MO}$	face-sharing	$0.67\,d_{MO}$

share corners, edges, and faces). An example of this parsimony can be found in the structures of TiO_2. By applying the first two rules, you can find that Ti should be octahedrally coordinated by O and that O should be coordinated by three Ti. This is achieved in several ways in different polymorphs. In rutile, each octahedron shares two edges. In brookite, each shares three edges and in anatase, each shares four edges. In none of these structures are there some octahedra that share two edges and some that share three or four, even though the first two rules could be satisfied in this way. The fact that these linkage types are not mixed is an example of the rule of parsimony.

Finally, acknowledging the fact that coordination polyhedra are rarely ideal and are often distorted, the sixth rule states that cation–cation repulsions shrink shared edges and the edges of shared faces. In the case of strong repulsions, which occur near the lower limit of radius ratio stability, cations are displaced from the centers of the polyhedra. These effects are most common in compounds of high oxidation state cations such as Nb^{5+} and W^{6+}.

During the decades since Pauling proposed these rules, they have served as useful guidelines for selecting probable structures of ionic compounds of known composition. There have, however, been some changes and some elaborations, described below.

ii.   *The bond valence method*
While it can still be assumed that bond lengths are the sum of ionic radii, the recognition that ionic radii are not fixed allows bond lengths to be predicted with increased accuracy. Specifically, use of Shannon's radii (given in Table 7B.1) allows the accurate prediction of most bond lengths. However, the most general method for the prediction of bond lengths is the so-called *bond valence theory*.

Bond valence theory is a quantitative application of Pauling's rules that can be used to predict bond lengths. Specifically, it was suggested by Zachariasen [16] that the electrostatic bond strength introduced by Pauling (Eqn. 10.8) is given by the following empirical relationship:

$$s_{ij} = e^{B(R_{ij}-d_{ij})} \tag{10.11}$$

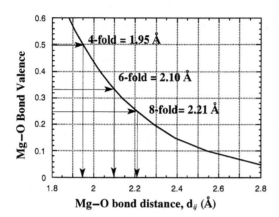

Figure 10.10. Predicted Mg–O bond lengths for different Mg coordinations.

where s_{ij} is the bond valence between atoms i and j, R_{ij} is a constant specific to that pair of elements, B is a constant $(1/0.37)$, and d_{ij} is the bond length. Thus, the bond valence, which can be taken as a reflection of the bond strength, increases as the bond distance decreases, as shown in Fig. 10.10. This rule, together with the rule that the sum of the valences at a site is constant, predicts that bond lengths decrease with decreasing coordination number, as we have observed before. However, Eqn. 10.11 is of little practical value without a reliable set of empirical constants, R_{ij}.

In 1985, Brown and Altermatt [17] determined a list of R_{ij} by fitting Eqn. 10.11 to all recorded crystal structures in the international data base. A more general set of parameters has recently been proposed by O'Keeffe and Brese [18] which allow R_{ij} for any two atoms to be computed based on atomic size (r) and electronegativity (c) parameters according to Eqn. 10.12.

$$R_{ij} = r_i + r_j - \frac{r_i r_j \left(\sqrt{c_i} - \sqrt{c_j}\right)^2}{c_i r_i + c_j r_j}. \tag{10.12}$$

These parameters also were fitted to reproduce as nearly as possible the existing crystal structure data. Note that this is simply the sum of the size parameters with an adjustment for the electronegativity difference. The empirically derived size and electronegativity parameters are given in Table 10.5.

While the value of Pauling's rules lies in the ability to predict the qualitative arrangement of ions, the value of the bond valence theory lies in the ability to quantitatively predict the lengths of bonds and dimensions of unit cells. Consider, for example, $SrTiO_3$, which crystallizes in the cubic perovskite structure. The Ti^{4+} are in octahedral coordination and the Sr^{2+} is in 12-fold

Table 10.5. *Atomic parameters for the calculation of bond valence parameters [18].*

Z		c	r	Z		c	r	Z		c	r	Z		c	r
1	H	0.89	0.38	23	V	1.45	1.21	44	Ru	1.42	1.21	65	Tb	1.10	1.56
3	Li	0.97	1.00	24	Cr	1.56	1.16	45	Rh	1.54	1.18	66	Dy	1.10	1.54
4	Be	1.47	0.81	25	Mn	1.60	1.17	46	Pd	1.35	1.11	67	Ho	1.10	1.53
5	B	1.60	0.79	26	Fe	1.64	1.16	47	Ag	1.42	1.12	68	Er	1.11	1.51
6	C	2.00	0.78	27	Co	1.70	1.09	48	Cd	1.46	1.28	69	Tm	1.11	1.50
7	N	2.61	0.72	28	Ni	1.75	1.04	49	In	1.49	1.34	70	Yb	1.06	1.49
8	O	3.15	0.63	29	Cu	1.75	0.87	50	Sn	1.72	1.37	71	Lu	1.14	1.47
9	F	3.98	0.58	30	Zn	1.66	1.07	51	Sb	1.72	1.41	72	Hf	1.23	1.42
11	Na	1.01	1.36	31	Ga	1.82	1.14	52	Te	2.72	1.40	73	Ta	1.33	1.39
12	Mg	1.23	1.21	32	Ge	1.51	1.21	53	I	2.38	1.33	74	W	1.40	1.38
13	Al	1.47	1.13	33	As	2.23	1.21	55	Cs	0.86	2.05	75	Re	1.46	1.37
14	Si	1.58	1.12	34	Se	2.51	1.18	56	Ba	0.97	1.88	77	Ir	1.55	1.37
15	P	1.96	1.09	35	Br	2.58	1.13	57	La	1.08	1.71	80	Hg	1.44	1.32
16	S	2.35	1.03	37	Rb	0.89	1.84	58	Ce	1.08	1.68	81	Tl	1.44	1.62
17	Cl	2.74	0.99	38	Sr	0.99	1.66	59	Pr	1.07	1.66	82	Pb	1.55	1.53
19	K	0.91	1.73	39	Y	1.11	1.52	60	Nd	1.07	1.64	83	Bi	1.67	1.54
20	Ca	1.04	1.50	40	Zr	1.22	1.43	62	Sm	1.07	1.61	90	Th	1.11	1.70
21	Sc	1.20	1.34	41	Nb	1.23	1.40	63	Eu	1.01	1.62	92	U	1.22	1.59
22	Ti	1.32	1.27	42	Mo	1.30	1.37	64	Gd	1.11	1.58				

Table 10.6. *Al–O bond lengths in Na-β″-alumina.*

atom	coordination	bond length	valence
Al(1)	O(2) × 6	1.889 Å	0.52
			sum = 3.12
Al(3)	O(1) × 2	1.844 Å	0.59
	O(2) × 2	1.986 Å	0.40
	O(3) × 1	1.848 Å	0.59
	O(4) × 1	1.947 Å	0.45
			sum = 3.02
Al(4)	O(1) × 3	1.763 Å	0.74
	O(5) × 1	1.681 Å	0.92
			sum = 3.14
Al(2)	O(2) × 3	1.842 Å	0.60
	O(4) × 1	1.856 Å	0.57
			sum = 2.37

coordination, so $s_{Ti-O} = 2/3$ and $s_{Sr-O} = 1/6$. By referring to the values in Table 10.5 and Eqn. 10.12, we find that $R_{Ti-O} = 1.814$ and $R_{Sr-O} = 2.068$. Substituting into Eqn. 10.11, we find that $d_{Ti-O} = 1.96$ and $d_{Sr-O} = 2.73$. Each of the values of d can be used to compute the lattice constant, a_0 (3.92 Å and 3.86 Å). The average of these two (3.89 Å) differs from the actual value (3.90 Å) by less than 0.3 %.

This method can be used to test the validity of proposed crystal structures, locate likely sites for atoms in substitutionally disordered materials, and locate weakly scattering atoms (H, Li) in crystals. As an example, consider Na-β″-alumina, which has the formula $Na_{1.67}Mg_{0.67}Al_{10.33}O_{17}$. In this crystal structure, a small amount of Mg(II) (atomic number 12) is substituted on the Al(III) (atomic number 13) sites. Although these two atoms are nearly identical in the way that they scatter X-rays, an analysis of the bond lengths (see Table 10.6) reveals the location of the Mg through the reduced bond valence sum. The partial substitution of divalent Mg on the trivalent Al(2) site lowers the total bond valence as implied by the average lengthening of the remaining bonds.

The previous example illustrates how the valence sum rule can be applied to learn about the character of specific sites in a static structure with known atomic positions and bond lengths. It is also possible to extend this method to determine the relative atomic positions and bond lengths in a pre-established network by combining additional constraints in the form of the loop sum rule. Brown quantified Pauling's rule of parsimony (bonds between similar pairs of atoms will be

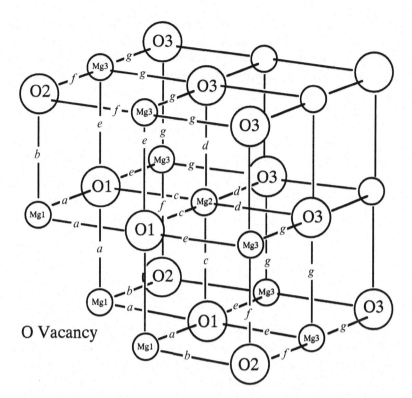

Figure 10.11. Model for the structure of MgO near an O vacancy. The other seven octants of the structure around the defect are identical. The unlabeled bonds are the same as *g*.

as nearly equal as possible) in the loop sum rule: the sum of bond valences around any loop in the crystal structure (taken with alternating signs) is equal to zero. The proper application of the valence sum rule and loop sum rule allows the formulation of an independent set of linear equations that can be solved simultaneously to give the valence of each bond in a crystal and, therefore, its bond length. This method is particularly useful for modeling the local structure near defects, as is illustrated in the example below.

Consider the introduction of an O vacancy in an otherwise perfect crystal of MgO, as is illustrated in Fig. 10.11. Although every Mg–O bond in the perfect crystal is identical, we expect the O vacancy to alter the structure locally. For example, the coordination of the neighboring Mg cations is lowered from six to five and their positive charges are no longer shielded from one another by the intervening O anion. We therefore expect the bonds between the neighboring Mg and the next nearest neighbor O to be somewhat different from bonds far from the defect. There also must be adjustments in the positions of other atoms to

447

make the transition from the distorted nearest neighbor bonds to the ideal bulk bonds. To account for this in our model, we label all of the distinct atom species with a number and all of the distinct bonds with a letter. To terminate the model, we assume that bonds between fifth and greater near neighbors are ideal or bulk-like.

Using the labeling scheme shown in Fig. 10.11, we can then write a matrix of the valences associated with each distinct bond. The rows are labeled by the cation types and the columns are labeled by the anion types. The variable in each position of the matrix stands for the valence of the bond between the corresponding atom pair. The subscript of the valence indicates the multiplicity of that bond type around the cation species.

$$
\begin{array}{cccc}
 & O_1 & O_2 & O_3 \\
Mg_1 & a_4 & b_1 & - \\
Mg_2 & c_3 & - & d_3 \\
Mg_3 & e_1 & f_1 & g_4
\end{array}
\tag{10.13}
$$

There are seven unknown bond valences which we can determine by applying the appropriate number of constraints. First, we know that the sum of the bond valences around each atom is equal to the formal valence. For any N independent atoms in a bonded network, it is possible to write $N-1$ independent sums. In this case, it allows us to write five independent equations for the sums around Mg_1, Mg_2, Mg_3, O_2, and O_3, respectively. The last equation was written to fix the bonds labeled g as ideal.

$$
\begin{aligned}
4a + b &= 2 \\
3c + 3d &= 2 \\
e + f + 4g &= 2 \\
b + 5f &= 2 \\
6g &= 2.
\end{aligned}
\tag{10.14}
$$

We can get two additional constraints by applying the loop rule. The matrix in Eqn. 10.13 is useful in this regard since any rectangular path in the matrix corresponds to a loop in the crystal. Identification of two independent loops allows us to write the following constraints:

$$
\begin{aligned}
(a - b) + (f - e) &= 0 \\
(c - d) + (g - e) &= 0.
\end{aligned}
\tag{10.15}
$$

Table 10.7. *Bond valences and lengths around an O vacancy in MgO.*

	contact	bond valence	bond length*
a	Mg_1-O_1	0.40404	2.028 Å
b	Mg_1-O_2	0.38384	2.094 Å
c	Mg_2-O_1	0.33838	2.094 Å
d	Mg_2-O_3	0.32828	2.105 Å
e	Mg_3-O_1	0.34343	2.088 Å
f	Mg_3-O_2	0.32323	2.111 Å
g	Mg_3-O_3	0.33333	2.099 Å

Note:
* Determined from 1.693 Å $- 0.37\ln(s)$.

These seven equations can then be solved for the seven unknown bond valences and used to determine bond lengths via Eqn. 10.11. The results are shown in Table 10.7.

Note that the predicted length of the bulk Mg_3-O_3 bond differs from the experimentally determined length (2.106) by less than one hundredth of an angstrom. The bonds to the Mg nearest the vacancy are shortened with respect to the ideal length, as we would expect from the reduction in the coordination number. To compensate for the increased bond valences of the shorter Mg_1-O_1 bonds, some of the bonds to the Mg_2 and Mg_3 are lengthened with respect to the ideal, so that the bond valence sums are maintained at 2. One fascinating aspect of this model is that although only pairwise nearest neighbor interactions are considered, it still predicts that a point defect will perturb the structure beyond the nearest neighbor distance.

iii.   *The maximum volume principle*
One change in the conventional wisdom regarding crystal structures is the so-called maximum volume principle. Because ionic solids take eutactic structures, it is generally assumed that ions pack into crystals in a manner that fills space most efficiently. However, one must remember that 'eutactic' means well-arranged; the atoms occupy the sites of a close packed lattice, but are not actually in closest packed positions. In fact, in structures where the bonding is dominated by interionic electrostatic forces, the repulsions of next nearest neighbor ions (of the same charge) favor a volume expansion. This fact, combined with the observation that the nearest neighbor distances between cations and

anions are fixed as a function of coordination number, suggests that ions should arrange themselves so that the unit cell volume is maximized subject to the constraint of a fixed anion–cation separation. This maximum volume principle, originally stated by O'Keeffe [19], can be illustrated by the following example.

Both ZnO and BeO crystallize in the hexagonal wurtzite structure, with the space group P6$_3$mc. If we say that the Be occupy the (2a) sites given as $(0,0,u)$ and $(1/3,2/3,1/2+u)$, and the O occupy (2a) sites given by $(0,0,0)$ and $(1/3,2/3,1/2)$, the structure can be characterized by two free parameters: $\gamma = c/a$ and u. To apply the maximum volume principle, we need to maximize the cell volume:

$$V = \frac{\sqrt{3}}{2} a^2 c \qquad (10.16)$$

with the constraint that the bond length (L) is fixed. To get an expression for γ as a function of u, we equate two bond lengths.

$$L_{Be-O(1)}(0,0,u \rightarrow 0,0,0) = L_{Be-O(2)}\left(0,0,u \rightarrow \frac{1}{3},\frac{2}{3},\frac{1}{2}\right)$$

$$u^2 c^2 = \frac{1}{3}a^2 + \left(\frac{1}{2} - u\right)^2 c^2 = L^2. \qquad (10.17)$$

From this we can find that:

$$\gamma^2 = \frac{4}{(12u - 3)}. \qquad (10.18)$$

Using this equation, and the fact that $L = cu$, we can rewrite the volume (Eqn. 10.16) as:

$$V = \sqrt{27}\, L^3 \frac{\left(u - \frac{1}{4}\right)}{2u^3}. \qquad (10.19)$$

After maximizing the volume with repect to u, we find that $u = 3/8$ and $\gamma = 1.633$. The favorable comparison of the derived parameters with those typical for compounds with the wurtzite structure (see Table 10.8) suggests the validity of the maximum volume principle.

iv. *Cation eutaxy*

In Chapter 4, we identified structures as being eutactic arrays of close packed anions with cations occupying interstitial positions. While the use of this conventional description of crystal structure is wide-spread, it has a number of weaknesses. The principal flaw of attempting to describe all structures according to anion eutaxy is that there are many structures where the anion packing is

Table 10.8. *Structural parameters from compounds with the wurtzite structure.*

Compound	u	γ
ideal wurtzite	0.375	1.633
BeO	0.378	1.623
ZnO	0.345	1.602
AlN	0.385	1.600
NH_4F	0.365	1.600

Table 10.9. *Examples of structures and anti-structures [20].*

Structure type	Compound with structure	Compound with anti-structure
C1	CaF_2	OLi_2
C4	TiO_2	NTi_2
C23	$PbCl_2$	$SiBa_2$
C6	CdI_2	FAg_2
C19	$CdCl_2$	OCs_2
DO_3	BiF_3	CF_3
DO_9	ReO_3	NCu_3
$D5_2$	La_2O_3	Sb_2Mg_3
$D5_3$	Y_2O_3	N_2Mg_3
$E2_1$	$CaTiO_3$	$GaNCr_3$
—	K_2NiF_4	Sb_2OCa_4

not regular in any easily recognizable way. In others, the distortions of the anion positions are large enough to make recognition difficult.

However, we are not constrained to describe structures only according to anion eutaxy. Recall that we described the fluorite structure as ccp cations with the anions filling the tetrahedral voids. There is a less common anti-fluorite structure (for example, the structure of Li_2O and Na_2O) which could be described according to anion eutaxy. In fact, this situation should not be viewed as uncommon. Many of the prototypes described in Chapter 4 in terms of anion eutaxy have anti-structures that could be described in terms of cation eutaxy; a partial list is given in Table 10.9. In addition to the structure/anti-structure relationships, there are other structures where anions and cations mix on the same eutactic sublattice. For example, in the $CaTiO_3$ structure, the Ca and O together make up the ccp framework.

Table 10.10. *Examples of ternary structures where the cation sublattice has the same arrangement as an intermetallic structure [20].*

intermetallic structure	intermetallic structure with filled interstitial sites	intermetallic structure	intermetallic structure with filled interstitial sites
Cu(A1)	MgO, CaF_2, Y_2O_3	CuZn(B2)	$SrTiO_3$, $Ca_2Fe_2O_5$
Cu_3Au (L1$_2$)	K_3SiF_7, Ba_3SiO_5	$MoSi_2(C11)_b$	Sr_2TiO_4, Bi_2MoO_6
CuPt (L1$_1$)	α-NaFeO$_2$	$Fe_3Al(L2_1)$	BaW_3O_6, Na_3PO_4
MoPt$_2$	Li_2CuO_2	UCo	$K_2Pb_2O_3$
WAl$_5$	Li_5ReO_6	$Cu_2MnAl(L2_1)$	K_2LaNaF_6
$TiAl_3(DO_{22})$	Na_3PaF_6	$MgCu_2(C15)$	$MgAl_2O_4$
Mg(A3)	La_2O_3, ZnO	CrB(B33)	$CsBeF_3$, β-Ga$_2$O$_3$
Sm[(hc)h]	$YbFe_2O_4$	FeB(B27)	$BaSO_4$
[(hc)^{2}h]	$Yb_2Fe_3O_7$	$Mn_5Si_3(D8_8)$	$Ca_5P_3O_{12}OH$ (apatite)
[(hc)^{3}h]	$Yb_3Fe_4O_{10}$	$NiAs(B8_a)$	$CaCO_3$ (aragonite)
[(hc)^{4}h]	$Yb_4Fe_5O_{13}$	$Ni_2Si(C37)$	Ba_2TiO_4, β-Ca$_2$SiO$_4$
BaPb$_3$	Ca_3SiO_5	$Ni_2In(B8_b)$	Mg_2SiO_4
$Ni_3Sn(DO_{19})$	Li_3PO_4	$FeS_2(C2_a)$	ZrP_2O_7
W(A2)	χ-La$_2$O$_3$	$CuAl_2(C16)$	Pb_3O_4 $(Pb^{2+}_2Pb^{4+}O_4)$
β-Hg	TiO_2, $ZrSiO_4$ (zircon)	$Cr_3Si(A15)$	$Ca_3Al_2Si_3O_{12}$ (garnet)
NaTl(B32)	$KFeO_2$	PtPb$_4$	$BaMn_4O_8$ (hollandite)

O'Keeffe and Hyde [20] noted that in many ternary oxide structures, the arrangement of the cations is identical to the arrangements of atoms in a known alloy structure. Consider, for example, spinel, $MgAl_2O_4$. Here, the O have approximate ccp eutaxy and the metal cations are distributed over tetrahedral and octahedral interstices. However, one might also say that the $MgAl_2$ sublattice has a structure identical to the C15 Laves phase, $MgCu_2$, and that the O fill tetrahedral interstices in this structure. This structure is, therefore, built from corner and edge-sharing $OMgAl_3$ groups. O'Keeffe and Hyde recently argued that the cation eutaxy description was superior to the anion eutaxy description and cited numerous examples where known ternary metal oxide structures have metal arrangements that are identical to the structures of known intermetallic phases (see Table 10.10).

The cation eutaxy point of view is most valuable when it can be used to describe complex structures in which the anions have no regular arrangement. For example, the Cu_2O, CuO, and Cu_3O_4 structures can all be visualized as ccp

Cu with different concentrations of O filling a portion of the tetrahedral sites such that the Cu^+ are in linear two-fold coordination and the Cu^{2+} are in four-fold planar coordination. The bixbyite structure (M_2O_3) can be visualized as ccp metal with O in three-quarters of the tetrahedral sites. The pyrochlore structure, with the general formula $A_2B_2O(1)_6O(2)$ has the cations in a ccp arrangement with the O in seven-eighths of the tetrahedral sites; the $O(1)$ fill A_2B_2 tetrahedra, the $O(2)$ fill a single A_4 tetrahedron, and a single B_4 tetrahedron remains vacant. Due to the ordering of the A and B atoms, the ccp unit cell is doubled along each edge.

Cation eutaxy offers the only reasonable description of the structure of the mineral baryte, $BaSO_4$. Earlier descriptions concentrated on the SO_4 tetrahedral groups. However, O'Keeffe and Hyde realized that the structure could be described as a BaS array (in the FeB, B15 structural arrangement) with the O occupying sites in the centers of SBa_3 tetrahedra. At higher temperatures, baryte and many of the sulfates and perchlorates that take this structure transform to the cubic HO_5 structure type where the Ba and S are in a rock salt arrangement, as in pure BaS, but with interstices filled by O.

O'Keeffe and Hyde used the cation eutaxy approach to explain a number of phase transformations and some structural parameters (cell dimensions and atomic positions) in systems where the anions had no simple arrangement. There are two valuable aspects of this approach. First, for some structures, such as baryte, this is simply the most reasonable description of the structure. Second, the coincidence of intermetallic alloy structures and cation arrangements in oxides suggests the importance of nonbonded interactions in polar compounds. Repulsions from species of like charge lead to arrangements that have easily recognizable eutaxy and this is just as likely to occur on the electropositive sublattice as on the electronegative sublattice. Thus, while conventional crystal structure descriptions concentrate on anion eutaxy, descriptions based on cation arrangements should also be considered as equally appropriate alternatives.

v.   Sanderson's description of the polar covalent bond
In Chapter 7, we noted a number of obvious flaws in the model for ionic bonding in crystals. Specifically, the model assumes full charges on the ions and ignores polarization effects. However, because these assumptions lead to compensating errors, calculated interatomic distances and lattice energies are still reasonably accurate. Sanderson has strongly criticized this model and asserted that full ionization is not only unreasonable, but can also lead to mistaken impressions. For example, assuming full ionization of the Mg cations and oxide anions in MgO, we are led to believe a very strong electron acceptor (Mg^{2+}) could be situated 2 Å away from a very strong electron donor (O^{2-}) without charge equalizing.

Sanderson also pointed out that observed atomic radii in crystals are not consistent with full ionization. As was mentioned in Chapter 5, it is possible to construct an electron density map of a crystal from careful measurements of the diffracted X-ray peak intensities. The electron density maximizes at the atomic positions and goes through a (nonzero) minimum along the interatomic axis. This minimum can be taken as the boundary between two adjacent 'ions'. In NaCl, the interatomic separation is 2.81 Å. Based on the electron density, the apparent size of the Na atom is 1.18 Å and the apparent size of the Cl atom is 1.64 Å. In each case, these sizes are intermediate between the ionic radii (r_{Na^+} = 0.99 Å, r_{Cl^-} = 1.81 Å) and the atomic radii, suggesting that the ions do not have a full unit charge.

Sanderson concludes that all bonds in nonmolecular crystals should be considered polar-covalent and suggests that the term *coordinated polymeric* be used instead of ionic. He has quantified his model by proposing new scales for electronegativity and atomic size that are based on the concepts of effective nuclear charge and partial atomic charges.

Although atoms are globally neutral, at distances on the order of the atomic radius, the electrons do not completely shield the nucleus and a portion of the positive nuclear charge is felt. It is this attraction between the nucleus of one atom and the electrons on another that leads to bond formation. This is similar in spirit to the definition of electronegativity used by Allred and Rochow (Chapter 7, Section E). Sanderson defines the effective nuclear charge as the positive charge that would be felt by a foreign electron arriving at the periphery of the atom. The effective charge has the same periodicity as electronegativity; it increases from left to right across the chart. This is because the outermost electrons provide the lowest screening efficiency. Calculations of screening constants indicate that the effective nuclear charge increases in steps of two-thirds as a period is traversed from Na to Cl. Thus, alkali metals have the lowest effective nuclear charge and halogens have the greatest.

The concept of electronegativity, already discussed in Chapters 1 and 7, obviously has a close relationship to the effective nuclear charge. Because the electronegativities discussed earlier do not have absolute scales, it is difficult to use them as a basis for calculations. Sanderson reasoned that electronegativity should be a measure of the attractive force between the effective nuclear charge and the outermost electrons, which he quantified as the 'compactness' of the atom. If we assume that inert gas atoms have a minimum effective nuclear charge, then the sizes of hypothetical 'minimum effective charge' atoms (those with a minimum electronegativity) can be defined for any atomic number by using a linear interpolation between the points designating the inert gas atom values. The effective nuclear charge usually leads to a contraction relative to the inert

Table 10.11. *Atomic size and electronegativity parameters for Sanderson's model [21].*

element	r_c	charge	S	B_{gas}	B_{solid}	element	r_c	charge	S	B_{gas}	B_{solid}
H	0.32	+	3.55	0.974	–	Ge	1.22	+	3.59	0.577	–
Li	1.34	+	0.74	1.201	0.812	As	1.19	+	3.90	0.417	–
Be	0.91	+	1.99	0.597	0.330	Se	1.16	+	4.21	0.395	–
B	0.82	+	2.93	0.591	–	Se	1.16	–	4.21	0.364	0.665
C	0.77	+	3.79	0.486	–	Br	1.14	–	4.53	0.695	1.242
N	0.74	+	4.49	0.311	–	Rb	2.16	+	0.36	1.192	1.039
N	0.74	–	4.49	0.063	–	Sr	1.91	+	1.06	0.649	0.429
O	0.70	–	5.21	0.240	0.401	Ag	1.50	+	2.57	–	0.208
F	0.68	–	5.75	0.536	0.925	Cd	1.46	+	2.59	–	0.132
Na	1.54	+	0.70	0.972	0.763	Sn^{+2}	1.40	+	2.31	0.331	–
Mg	1.38	+	1.56	0.638	0.349	Sn^{+4}	1.40	+	3.09	0.542	–
Al	1.26	+	2.22	0.580	–	Sb	1.38	+	3.34	0.350	–
Si	1.17	+	2.84	0.587	–	Te	1.35	–	3.59	–	0.692
P	1.10	+	3.43	0.404	–	I	1.33	–	3.84	0.705	1.384
S	1.04	+	4.12	0.681	–	Cs	1.35	+	0.28	1.273	0.963
S	1.04	–	4.12	0.222	0.675	Ba	1.98	+	0.78	0.517	0.348
Cl	0.99	–	4.93	0.727	1.191	Tl	1.48	+	1.89	0.701	–
K	1.96	+	0.42	1.097	0.956	Pb^{+2}	1.47	+	2.38	0.371	–
Ca	1.74	+	1.22	0.691	0.550	Pb^{+4}	1.47	+	3.08	0.301	–
Ti	1.32	+	1.40	0.330	–	Bi	1.46	+	3.36	0.132	–

gas atoms, so real atoms are smaller than the hypothetical minimum effective charge atoms. Sanderson used the amount of this contraction to define the electronegativities as the ratio of the atomic electron density (the atomic volume divided by the atomic number) to the atomic electron density of the hypothetical zero effective charge atom. This electronegativity scale (S) is reproduced in Table 10.11. Thus, the Sanderson electronegativity is meant to be a quantitative measure of the effective nuclear charge.

To compute these electronegativities, we obviously need a reliable size scale. We have already seen that the atomic radius apparently varies with the bond type and coordination number. While many different sets of radii have been tabulated, most have not gained universal acceptance. The best point of reference, and the one adopted by Sanderson, is the set of non-polar covalent radii (r_c) listed in Table 10.11. Our confidence in this set of radii stems from the fact that they appear to be constant for any one atom in a wide variety of structures from molecular to crystalline.

When a bond forms, charge is transferred from the atom of lower electronegativity to the one with higher electronegativity and the atoms acquire partial charges. This implies that the effective nuclear charge, the electronegativities, and the sizes of the atoms must change. To account for these changes, Sanderson proposed the principle of electronegativity equalization: '*When two or more atoms initially different in electronegativity combine chemically, they become adjusted to the same intermediate electronegativity within the compound*'. The compound electronegativity was taken to be equal to the geometric mean of the atomic electronegativities.

$$S_{compound} = \sqrt{S_A S_B}.$$ (10.20)

To account for the partial charge, Sanderson postulated that the '*partial charge is defined as the ratio of the change in electronegativity undergone by an atom on bond formation to the change it would have undergone on becoming completely ionic with a charge of +1 or −1*'. To calibrate this system, it was assumed that bonds in NaF are 75% ionic and that electronegativities change linearly with charge. So, any atom, on undergoing a unit positive or negative change in charge, has a change in electronegativity (ΔS) of:

$$\Delta S = 2.08 \sqrt{S},$$ (10.21)

and the partial charge, δ, is:

$$\delta = \frac{S - S_{compound}}{2.08\sqrt{S}}.$$ (10.22)

It is instructive to consider the partial charges predicted by this model. Table 10.12 shows the partial charges on O for a variety of oxides. The first thing to note is that the partial charge never reaches −1, even though we conventionally regard the oxide anion as having a charge of −2. Secondly, note that the partial charge or ionicity increases for larger metal atoms and for structures with higher coordination numbers. For example, wurtzite structured monoxides (CN = 4) have a lower O partial charge than rock salt structured monoxides (CN = 6), and rutile structured dioxides (CN = 6) have a lower O partial charge than fluorite structured dioxides (CN = 8). The partial charges have been found which correlate to the electron donor/acceptor properties of these solids. Below a partial charge of about 0.3, the oxides are acidic, and above this point, they are basic.

Table 10.12. *Partial charges on selected solid oxides [22].*

compound	$-\delta_O$	compound	$-\delta_O$	compound	$-\delta_O$	compound	$-\delta_O$
Cu_2O, cu	0.41	HgO, ci	0.27	Ga_2O_3, c	0.19	CO_2, m	0.11
Ag_2O, cu	0.41	ZnO, w	0.29	Tl_2O_3, b	0.21	GeO_2, q	0.13
Li_2O, af	0.80	CdO, rs	0.32	In_2O_3, b	0.23	SnO_2, r	0.17
Na_2O, af	0.81	CuO, co	0.32	B_2O_3, bo	0.24	PbO_2, r, f	0.18
K_2O, af	0.89	BeO, w	0.36	Al_2O_3, c	0.31	SiO_2, q	0.23
Rb_2O, af	0.92	PbO, rpo	0.36	Fe_2O_3, c	0.33	MnO_2, r	0.29
Cs_2O, acc	0.94	SnO, rpo	0.37	Cr_2O_3, c	0.37	TiO_2, r	0.39
		FeO, rs	0.40	Sc_2O_3, b	0.47	ZrO_2, f	0.44
		CoO, rs	0.40	Y_2O_3, b	0.52	HfO_2, f	0.45
		NiO, rs	0.40	La_2O_3, l	0.56		
		MnO, rs	0.41				
		MgO, rs	0.50				
		CaO, rs	0.56				
		SrO, rs	0.60				
		BaO, rs	0.68				

Notes:

cu = cuprite, af = anti-fluorite, acc = anti-$CdCl_2$, ci = cinnabar, w = wurtzite, rs = rock salt, co = cupric oxide, rpo = red plumbus oxide, c = corundum, bo = boron sesquioxide, b = bixbyite, l = lanthanum sesquioxide, m = molecular, q = quartz, r = rutile, f = fluorite.

Acquisition of a positive partial charge is expected to lead to an atomic contraction and acquisition of a negative charge is expected to lead to an expansion. To account for this, Sanderson used the expression:

$$r = r_c - B\delta. \tag{10.23}$$

In Eqn. 10.23, the values of B are empirical constants listed in Table 10.11. This equation for the size of an atom is analogous to the one used later by O'Keeffe and Brese to predict characteristic bond lengths (Eqn. 10.12). Table 10.13 illustrates the accuracy of this scheme. The average error between the known and computed bond lengths for 203 trial compounds is 0.023 Å.

Sanderson extended his model of polar covalent bonding to compute bond energies. Specifically, he suggested that all bonds between dissimilar atoms have a covalent component related to the strength of the homopolar bonds

Table 10.13. *Accuracy of bond length prediction for binary compounds [21].*

compound type	number of compounds		Aver. $|$calc. $-$ exp.$|$ length, Å	
	Gas	Solid	Gas	Solid
Fluorides	39–7[a]	12–1	0.03	0.06
Chlorides	36–2	13	0.02	0.02
Bromides	35–4	13	0.01	0.01
Iodides	31–3	13	0.02	0.04
Oxides	—	5	—	0.03
Sulfides	—	5	—	0.03
Selenides	—	5–1	—	0.04
Tellurides	—	5–2	—	0.03
Hydrides	11	—	0.01	—

Note:
[a] The second number indicates the number of predicted compounds for which an experimental comparison is not possible.

and an ionic component related to the partial charges on the atoms. The ionicity fraction, t_i, is determined from the partial charges on the two atoms, δ_A and δ_B.

$$t_i = \frac{\delta_A - \delta_B}{2}.$$ (10.24)

The fractional covalent part, t_c, is $1 - t_i$, such that $t_i + t_c = 1$. Note that as the ionicity grows, it replaces rather then augments the covalent portion. Since the ionic component is always larger than the covalent, increased ionicity will lead to greater cohesion.

The ionic part of the bond energy (E_i) can be computed using the method described in Chapter 7. The covalent component of the energy (E_c) is taken to be the weighted geometric mean of homopolar bond energies, E_{AA} and E_{BB}.

$$E_c = \frac{R_c}{R_{obs}} \sqrt{E_{AA}E_{BB}}.$$ (10.25)

In Eqn. 10.25, the weighting factor is the sum of the covalent radii (R_c) over the observed bond length (R_{obs}). The homopolar bond energies are computed by the empirical relationship that $E_{AA} = C_A r_c S_A$, where r_c and S_A are the homopolar covalent radius and the electronegativity, respectively, and C is an empirical constant determined from the observed linearity of bond energy versus the product of r_c and S. The value of C is constant for each row of the periodic chart.

For any crystal, the cohesive energy, E, is:

$$E = t_c E_c n + \frac{t_i U}{f}. \qquad (10.26)$$

In Eqn. 10.26, n is a correction factor to account for the difference in the number of pairs of electrons involved in the bonding. For a simple diatomic molecule, it is one. In a crystal with four or more nearest neighbors, n is set equal to four. In some cases, especially when smaller atoms are involved, better agreement is achieved with values of three or six. In the second part of Eqn. 10.26, U is the lattice energy. Since this differs from the cohesive energy (recall that the lattice energy is the energy difference between the crystal and the free ions, while the cohesive energy is the energy difference between the crystal and the free atoms), a correction factor (f) is introduced. For halides, f is taken to be equal to 1 and for oxides it is 0.63. For a discrete molecule, the bond energy is:

$$E = m \left(t_c E_c + t_i \frac{332}{R_{obs}} \right) \qquad (10.27)$$

where m is a factor to account for the differences between single bonds ($m = 1$), double bonds ($m = 1.5$), and triple bonds ($m = 1.75$).

While the empirical nature of Sanderson's model and the apparent arbitrary nature of the correction factors might be unsatisfying, it is remarkably successful at quantitatively reproducing bond energies (see Table 10.14) and at distinguishing the compounds that form molecular structures from those that form extended or coordinated polymeric structures. As an example, consider CO_2 and SiO_2, the dioxides of two adjacent, isoelectronic, group IV atoms. Carbon dioxide forms a discrete molecule with two carbon–oxygen bonds; solid CO_2 is a molecular crystal that sublimes below room temperature. Silica, on the other hand, is a coordinated polymeric structure with four single Si–O bonds per Si atom; crystalline SiO_2 is stable to more than 1600 °C. Sanderson's model can be used to compute the bond energies of each compound in the coordinated polymeric form and in the molecular form and the results are illustrated in Table 10.15. There is a significant energy difference between the molecular and extended forms and the lowest energy state for each compound is reproduced correctly.

This analysis led Sanderson to propose an explanation for why some compounds take molecular forms, while others crystallize in extended or coordinated polymeric structures. In many cases, the choice between extended structures and molecular structures amounts to the choice between double bonds and single

Table 10.14. *Computed and experimental cohesive energies [22].*

Compound	n	$-\delta_{anion}$	E_{calc} (eV)	E_{exp} (eV)
Li_2O	6	0.80	12.14	12.10
Na_2O	4	0.80	8.82	9.13
K_2O	4	0.84	8.05	8.20
Rb_2O	4	0.86	7.90	7.70
Cs_2O	4	0.90	7.47	7.49
BeO	4	0.42	12.16	12.18
MgO	4	0.50	10.57	10.78
CaO	4	0.57	10.76	11.00
SrO	4	0.60	10.44	10.38
BaO	4	0.67	9.93	10.18
LiF	3	0.74	9.00	8.83
NaF	3	0.75	7.94	7.88
KF	4	0.84	7.63	7.62
RbF	4	0.86	7.37	7.36
CsF	4	0.90	7.13	7.13
BeF_2	3	0.34	15.15	15.56
MgF_2	4	0.41	15.53	15.24
CaF_2	4	0.47	16.02	16.05
SrF_2	4	0.50	15.77	15.90
BaF_2	4	0.57	15.88	15.89

Table 10.15. *Bond energies of SiO_2 and CO_2 in extended and molecular forms.*

	Molecular, M = O	Extended crystal, M–O
CO_2	16.7 eV	14.6 eV
SiO_2	13.1 eV	19.3 eV

bonds. For example, CO_2 has two C–O double bonds per C while SiO_2 has four Si–O single bonds per Si. Normally, two single bonds are stronger than one double bond. So, why does C form two double bonds to O rather than four single bonds? One difference between the two situations is the disposition of the nonbonded valence electron pairs on the O (lone pairs). In single bonds, the nonbonded pairs provide extra shielding along the internuclear axis, lowering the effective nuclear charge and the electronegativity. This reduces the single bond energy with respect to the double bond energy and, therefore, the homopolar component of the bonding. Thus, it is the 'lone pair weakening' effect on single bonds that leads to the formation of double bonds and discrete molecular structures. Since this affects only the covalent portion of the bonding, the weakening is less significant in more polar compounds. For example, the more polar Si–O single bonds are favored over double bonds, while the less polar C–O bonds are destabilized with respect to the C=O bonds. The same line of reasoning can be used to explain why O and N form gaseous diatomic molecules, while S and P (the adjacent atoms of the same groups) form extended solids. In this case, the lone pairs of the larger atoms have reduced shielding efficiencies and do not weaken the single bond as much. With seven valence electrons, the halogens have no opportunity to form double bonds and the lone pair weakening effect can be seen in the measured single bond energies. As the atoms get smaller and the internuclear distance diminishes, one expects bond energies to get progressively larger. While this is true for I_2, Br_2, and Cl_2, the lone pair weakening effect dominates in the smallest atom and the measured strength of the F–F bond is weaker than the Cl–Cl bond.

(D) Structure stability diagrams

i. Mooser–Pearson Plots

The Mooser–Pearson [23] plot is a successful graphical scheme that can be used to understand what factors determine which structure a compound will take. The primary assumption in this method is that the bond polarity (the electronegativity difference) and the principal quantum number of the valence electrons (size) are the primary factors that influence the crystal structure. The reasoning can be explained in the following way.

First, as the principal quantum number (n) increases, the atomic orbitals involved in bond formation, and thus the bonds themselves, gradually lose their directional character because they are spatially less localized and less influenced by the nucleus. Because bonding combinations typically involve two atoms with valence electrons in two different principal shells, we define the variable $\bar{n}$ as the average principal quantum number.

$$\bar{n} = \frac{\sum\limits_i c_i n_i}{\sum\limits_i c_i}. \tag{10.28}$$

In Eqn. 10.28, n_i is the principal quantum number for the atom of the i-th kind and c_i is the number, per formula unit, of atoms of this kind.

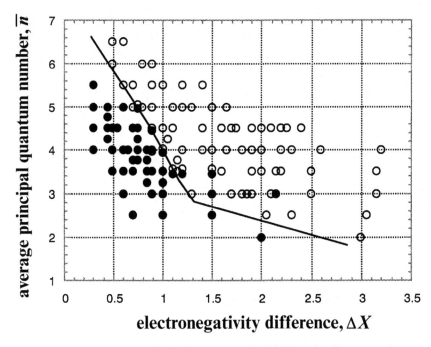

Figure 10.12. Mooser–Pearson plot for AX compounds with the rock salt structure (open circles) and compounds with the sphalerite, wurtzite, and chalcopyrite structures (filled circles). In the rock salt structure, atoms have an octahedral coordination while in the other three structures, atoms are tetrahedrally coordinated [23].

The second factor that affects structure selection is the electronegativity difference. As the electronegativity difference increases, the bonds become more ionic, less covalent, and less directional. This is quantified by the factor ΔX,

$$\Delta X = |\bar{X}_{anion} - \bar{X}_{cation}|,\qquad(10.29)$$

where the values for $\bar{X}$ are the arithmetic means (in ternary and quaternary structures) of the cation and anion electronegativities.

Among the most common structures for AX compounds are rock salt, sphalerite, and wurtzite. The Mooser–Pearson plot in Fig. 10.12 shows that AX compounds with different coordination numbers clearly map onto separate regions of the diagram. The black line on the figure separates those with octahedral coordination and those with tetrahedral coordination. The chalcopyrite phase, ABX_2, that is included in this analysis can be regarded as a ZnS superlattice structure (see Fig. 4.27).

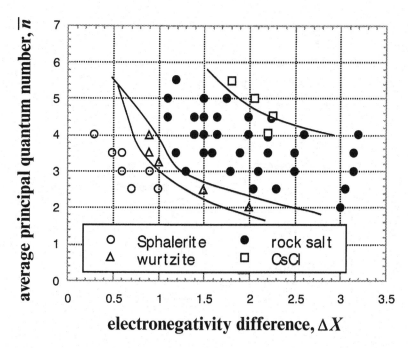

Figure 10.13. A Mooser–Pearson plot limited to the AX compounds where A is an 'A group' cation (cations with no partially filled d-orbitals) [23].

Figure 10.12 does not include several exceptions, which fall into two categories and are, perhaps, more interesting than those that follow the trend. First, there are 12 compounds that crystallize in the rock salt structure (or a distorted version of it) despite being in the tetrahedral structure field. These are, however, all heavy tetralide compounds that retain a nonbonding pair of s-electrons. The other exceptions are the interstitial hydrides which also take the rock salt structure in spite of falling in the tetrahedral structure field.

Figure 10.13 shows a particularly clear trend when the chemistry is limited to A group cations (those without partially filled d-orbitals). On the plot, the eight-coordinate, six-coordinate, and four-coordinate regions are clearly defined. One interesting thing is that several alkali halides which, on the basis of radius ratio rules, should crystallize as CsCl fall well within the rock salt field, consistent with the experimental determination. The AX compounds containing transition metal elements are shown on the Mooser–Pearson plot in Fig. 10.14. There are clearly two distinct fields: one where the rock salt structure dominates and one where the NiAs and MnP structures dominate. All three of these structures place the anions and cations in six-fold coordination, but the metal atoms are

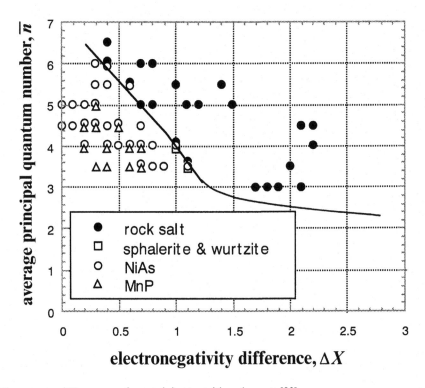

Figure 10.14. AX compounds containing transition elements [23].

closer together in the NiAs and MnP structures. NiAs and MnP compounds are usually metallic. Finally, note that those compounds that take structures with tetrahedral coordination are found at the boundary between the two fields.

The Mooser–Pearson plot for the AX_2 compounds is also divided into distinct fields (see Fig. 10.15). The upper right-hand field contains structures that have atoms in high coordination number environments. The $PbCl_2$ structure contains four-, five-, and nine-coordinate atoms and the fluorite structure has four- and eight-coordinate atoms. The next field contains structures that have six-coordinate cations and three-coordinate anions (these include rutile, CdI_2, MoS_2, and $CdCl_2$). It is interesting to note that the rutile compounds, which dominate the right-hand side of this field, have three-dimensional bonding, while the three structures in the left-hand side of the field are all layered compounds. The compounds in the lower left field have four-coordinate cations and two-coordinate anions. These compounds crystallize in the quartz and GeS_2 structures. It is clear that covalency increases toward the lower left-hand side of the diagram.

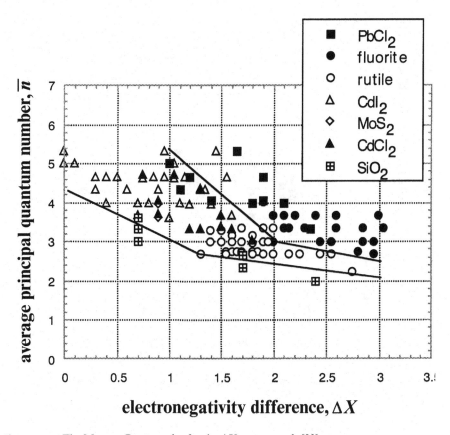

Figure 10.15. The Mooser–Pearson plot for the AX$_2$ compounds [23].

To illustrate the fact that these empirical plots can be used to predict properties as well as structure, an example is shown in Fig. 10.16. We find that the metallic and nonmetallic ternary fluorites (with the composition ABX) are clearly separated into two different fields.

ii.   Villars' structure stability diagrams
Villars [24–26] developed a set of three-dimensional structure stability diagrams that are similar to the Mooser–Pearson plots in that the coordinates are based on atomic properties. The specific parameters for this more comprehensive mapping scheme were determined using the method already outlined in Section B(*iii*) of this chapter. By sequentially trying all possible combinations of the atomic parameters, Villars found three that best separate the structures into distinct fields in a three-dimensional plot. In this case, the three parameters are the

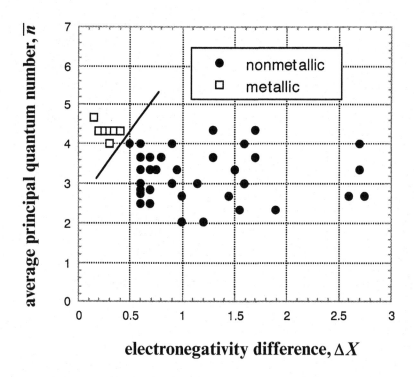

Figure 10.16. Mooser–Pearson plot for ternary ABX phases that take the fluorite structure [23].

sum of the number of valence electrons, ΣVE_{AB}, the absolute difference between the Zunger pseudopotential radii sums, $|\Delta(r_s + r_p)^z_{AB}|$, and the magnitude of the difference in the Martynov–Batsanov electronegativities, $|X^{MB}_A - X^{MB}_B|$. The values needed to calculate the first two parameters are found in Fig. 10.7 and the values of X^{MB} are listed in Table 10.16. Considering that size is a function of the principal level of the valence electrons, then two of the coordinates used by Villars are related to the coordinates used by Mooser and Pearson.

Villars first created structure stability diagrams for AB binaries. He excluded metastable phases, impurity stabilized phases, high pressure phases, and structures that have fewer than six representative phases; this leaves just over 1000 phases that crystallize in 21 structure types (see Table 3.5). Interestingly, of the known binary phases, 55% crystallize in the rock salt or CsCl structure. Phases with the NiAs structure were eventually eliminated from the analysis, since they did not occur in a confined domain on the structure stability diagram. This leaves 998 phases existing in 20 structures; the structure stability maps corresponding

Table 10.16. *Martynov–Batsanov electronegativities [24].*

Element	X^{MB}	Element	X^{MB}	Element	X^{MB}	Element	X^{MB}
H	2.10	Fe	1.67	In	1.63	Hf	1.73
Li	0.90	Co	1.72	Sn	1.88	Ta	1.94
Be	1.45	Ni	1.76	Sb	2.14	W	1.79
B	1.90	Cu	1.08	Te	2.38	Re	2.06
C	2.37	Zn	1.44	I	2.79	Os	1.85
N	2.85	Ga	1.70	Cs	0.77	Ir	1.87
O	3.32	Ge	199	Ba	1.08	Pt	1.91
F	3.78	As	2.27	La	1.35	Au	1.19
Na	0.89	Se	1.54	Ce	1.1	Hg	1.49
Mg	1.31	Br	1.83	Pr	1.1	Tl	1.69
Al	1.64	Rb	0.80	Nd	1.2	Pb	1.92
Si	1.98	Sr	1.13	Pm	1.15	Bi	2.14
P	2.32	Y	1.41	Sm	1.2	Po	2.40
S	2.65	Zr	1.70	Eu	1.15	At	2.64
Cl	2.98	Nb	2.03	Gd	1.1	Fr	0.70
K	0.80	Mo	1.94	Tb	1.2	Ra	0.90
Ca	1.17	Tc	2.18	Dy	1.15	Ac	1.10
Sc	1.50	Ru	1.97	Ho	1.2	Th	1.3
Ti	1.86	Rh	1.99	Er	1.2	Pa	1.5
V	2.22	Pd	2.08	Tm	1.2	U	1.7
Cr	2.00	Ag	1.07	Yb	1.1	Np	1.3
Mn	2.04	Cd	1.40	Lu	1.2	Pu	1.3
						Am	1.3

to these phases are shown in Fig. 10.17. Not plotted in the figure are 22 phases which occur in the incorrect domain. However, this represents an error of only 2.2%. Domains associated with specific structures are permitted to overlap when compounds in the overlapping regions are polymorphic. For example, note in Fig. 10.17 that the CsCl and CrB volumes are allowed to overlap. In most cases, the overlap is only partial. However, GeS, HgS, and wurtzite exist only within the NaCl (GeS) and sphalerite (HgS and wurtzite) domains. The structure stability diagrams also show regions where no compounds form. Villars extended this method to 1011 AB_2 compounds, 648 AB_3 compounds, and 389 A_3B_5 compounds with similar results.

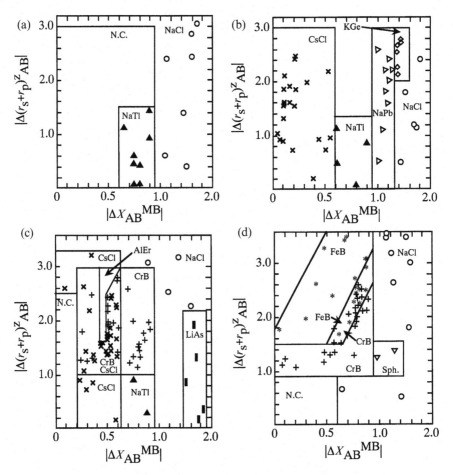

Figure 10.17. (a–d) Villars' structure stability diagrams for 998 AB compounds. Each point on the diagrams represents a known compound. The different symbols correspond to different crystal structures. The domains of a single structure type are labeled, and the meaning of the symbols can be inferred from these labels. The structure stability diagram is three-dimensional and is presented as 16 planar slices with increasing values of ΣVE_{AB}. (a) ΣVE_{AB} = 2 through 4, (b) ΣVE_{AB} = 5, (c) ΣVE_{AB} = 6, (d) ΣVE_{AB} = 7. The remainder of the slices are presented on the following pages. Twenty-two compounds that fall in the wrong structure domain are excluded from the plot. Each structure domain is labeled by the chemical symbols for the formula unit of the prototype, except for Sph. and Wur., which stand for sphalerite and wurtzite, respectively (cubic and hexagonal ZnS). No compounds form in the regions labeled N.C. [24].

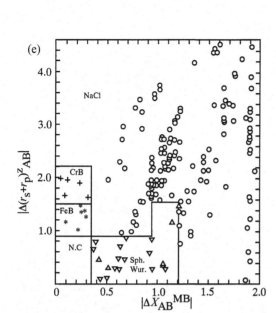

Figure 10.17. (e & f) (e) $\Sigma VE_{AB} = 8$. (f) $\Sigma VE_{AB} = 9$. See caption on Fig. 10.19 (a–d) for details [24].

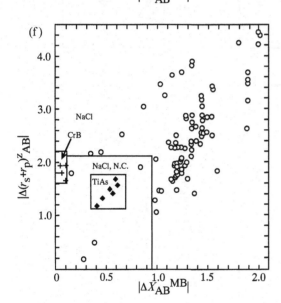

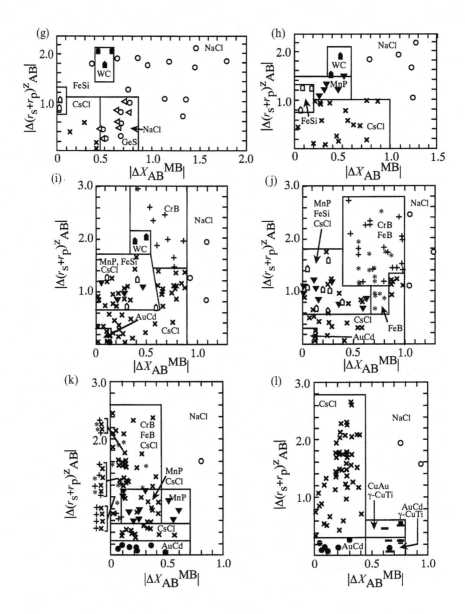

Figure 10.17. (g–l) (g) $\Sigma VE_{AB} = 10$. (h) $\Sigma VE_{AB} = 11$. (i) $\Sigma VE_{AB} = 12$. (j) $\Sigma VE_{AB} = 13$. (k) $\Sigma VE_{AB} = 14$. (l) $\Sigma VE_{AB} = 15$. See caption on Fig. 10.19 (a–d) for details [24].

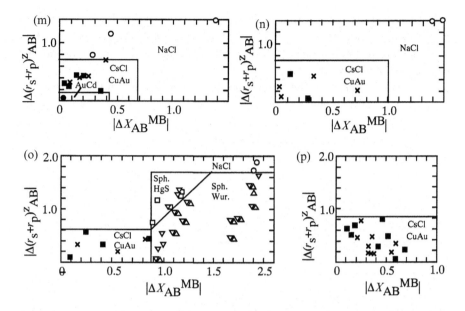

Figure 10.17. (m–p) (m) $\Sigma VE_{AB} = 16$. (n) $\Sigma VE_{AB} = 17$. (o) $\Sigma VE_{AB} = 18$. (p) $\Sigma VE_{AB} = 19$–23. See caption on Fig. 10.19 (a–d) for details [24].

iii.   Phillips ionicity

Phillips [27] developed a method, similar in spirit to the Mooser–Pearson plots, to distinguish compounds with tetrahedral coordination from those with octahedral coordination. However, he assumed that crystal properties, rather than free atom properties, govern structure choice. We have noted earlier that predictive methods based on the properties of solids, rather than atoms, have greater success.

In earlier parts of the book, we noted that tetrahedral bonding arrangements occur when the s and p orbitals hybridize to form four equivalent sp^3 orbitals. In Chapter 9, we saw that the energy required for the hybridization process increases with the energy difference between the s and p orbitals. If the widths of the bands derived separately from the s and p orbitals become large (in terms of the quantities defined in Chapter 9, this occurs at relatively small values of V_3 and larger values of V_2) with respect to the s–p splitting on the anion, then the orbitals hybridize. On the other hand, if the bands remain narrow (high V_3), then separate s- and p-bands form and the octahedral coordination characteristic of the p-orbital configuration is adopted. Phillips proposed a method for differentiating between these two cases that is based on quantum mechanical methods. However, he asserted that the algebraic relations derived from quantum mechanical calculations were more important than the numerical details of variational solutions to the wave equation.

Phillips used spectral measurements to determine the band gap, which he interprets as a measurement of the 'homopolar energy', E_h, and a 'charge transfer energy', C. The charge transfer energy can be thought of as a polar or ionic component to the bonding and it is related to the energy difference in the term levels. According to Phillips,

$$E_g^2 = E_h^2 + C^2. \tag{10.30}$$

The quantities C and E_h differ from the polar energy (V_3) and the covalent energy (V_2) by a factor of two. Recall from Chapter 9 that:

$$E_g = 2\sqrt{V_2^2 + V_3^2}. \tag{10.31}$$

Phillips went on to define the ionicity as:

$$f_i = \frac{C^2}{E_g^2}. \tag{10.32}$$

This is equivalent to the results described in Chapter 9 for the polarity (see Eqn. 9.21) which is related to the probability of finding a valence electron on the anion:

$$f_i = \frac{V_3^2}{V_2^2 + V_3^2}. \tag{10.33}$$

Phillips asserted that it is the ratio of the polar energy to the covalent energy that controls the band width and, therefore, determines whether or not hybrid bond orbitals occur and a crystal adopts tetrahedral coordination. When spectral data from 68 binary compounds are plotted as a function of E_h and C, a perfect separation between octahedrally coordinated and tetrahedrally coordinated compounds is found (see Fig. 10.18). All compounds with $f_i > 0.785$ have octahedral coordination and all those with $f_i < 0.785$ have tetrahedral coordination. The line separating the coordination fields passes through $E_h = C = 0$. Again, we find the result that higher coordination numbers are associated with greater ionicity. Phillips' study is particularly interesting because it provides a link between the parameters used in the quantitative physical model described in Chapter 9 and the factors that determine the structure in which a compound will crystallize.

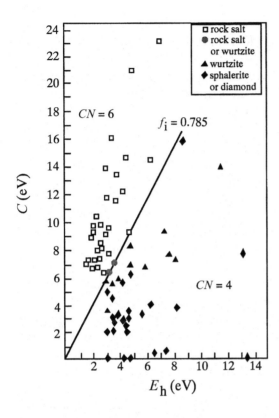

Figure 10.18. Each point on the diagram represents one of 68 binary compounds. There is a critical ratio of the polar to covalent energy that separates structures with octahedral coordination from those with tetrahedral coordination [27].

(E) Problems

(1) Villars excluded NiAs compounds from his structure stability maps. Using the data in Table 4.18, show where these compounds fall on the structure stability maps. Discuss plausible reasons why these compounds do not cluster in a single field.

(2) Villars carefully examined the possibility of using atomic or elemental properties to predict the structure and properties of compound binary solids. He identified five separate classes of atomic or elemental properties that characterize:

    A. atomic sizes

    B. the directionality of homoatomic bonds (for example, A–A)

    C. the strength of homoatomic bonds

    D. the affinity between different atoms (the preference to form A–B rather than A–A bonds)

    E. the electronic structure of the atom

For each of these classes of properties, choose a specific property that belongs to the class and explain how variations in this property influence the crystal structures of binary compounds.

(3) Why do you think that models based on per atom properties of crystals lead to better predictions than models based on free atom properties?

(4) As a general rule for structure and bonding in crystals, we observe that as the number of nearest neighbors increases, so do the nearest neighbor bond lengths.

> (i) Using our model for the energy of an ionically bound crystal as a function of nearest neighbor separation, show that the electrostatic model for the ionic bond reproduces this observation.
>
> (ii) Can Pauling's rule for the electrostatic bond valence be used to reproduce this observation? If so, explain how.
>
> (iii) Provide a geometric explanation for why bond lengths increase as the number of nearest neighbors increases.

(5) In references [14] and [24], Villars lists the compounds that violate the structure fields. Are there any similarities among these compounds?

(6) The γ-LaOF structure is described in Table 3B.31. Because of the similarity of their atomic numbers, the positions of O and F would be very difficult to distinguish simply by measuring X-ray intensities. Explain how you might use Pauling's bond valence concept to differentiate the O and F sites in this compound.

(7) In 1988, Pettifor proposed a set of two dimensional structure stability maps [28]. Briefly describe Pettifor's scheme and compare the maps to those produced by Mooser and Pearson and Villars.

(F) References and sources for further study

[1] W. Hume-Rothery and G. Raynor, *The Structure of Metals and Alloys*, (Institute of Metals, London, 1962) Chapter V. The data in Table 10.1 come from Table XXII on p. 198.

[2] P. Haasen, *Physical Metallurgy* (Cambridge University Press, Cambridge, 1978) Sections 6.3 and 6.4. Figure 10.1 is drawn after Fig. 6.19 on p. 123. Haasen in turn cites Barrett and Massalski ([4] in Chapter 4) as the source.

[3] R.J. Borg and G.J. Dienes, *The Physical Chemistry of Solids* (Academic Press, San Diego, 1992), Chapter 8, Section 5c, pp. 308–10.

[4] D.G. Pettifor, A Quantum Mechanical Critique of the Miedema Rules for Alloy Formation, *Solid State Physics* **40** (1987) 43–92. In this paper, the author cites

Miedema's work, refs. 6–13, as primary sources. Fig. 10.4 is drawn after Fig. 3, p. 48; Fig. 10.4 is drawn after Fig. 2, p. 47; Fig. 10.6 is drawn after Fig. 23, p. 88.

[5] R.J. Borg and G.J. Dienes, *The Physical Chemistry of Solids* (Academic Press, San Diego, 1992), Chapter 8, Section 5b, p. 294–9. An overview of Miedema's work.

[6] A.R. Miedema, F.R. de Boer, and P.F. de Chatel, *J. Phys. F* **3** (1973) 1558.

[7] A.R. Miedema, *J. Less-Common Met.* **32** (1973) 117.

[8] A.R. Miedema, R. Boom, and F.R. de Boer, *J. Less-Common Met.* **41** (1975) 283.

[9] A.R. Miedema, *J. Less-Common Met.* **46** (1976) 67.

[10] A.R. Miedema, and F.R. de Boer, *Metall. Soc. AIME Proc.* (1980).

[11] A.R. Miedema, P.F. de Chatel, and F.R. de Boer, *Physica* **100** (1980) 1.

[12] A.R. Miedema and A.K. Niessen, CALPHAD: *Comput. Coupling Phase Diagrams Thermochem.* **7** (1983) 27.

[13] A.K. Niessen, F.R. de Boer, R. Boom, P.F. de Chatel, W.C.M. Mattens, and A.R. Miedema, CALPHAD: *Comput. Coupling Phase Diagrams Thermochem.* **7** (1983) 51.

[14] P. Villars, *J. Less-Common Met.* **109** (1985) 93–115. Figs. 10.7 and 10.8 are drawn after Table 1 and Fig. 4, respectively.

[15] L. Pauling, The principles determining the structure of complex ionic crystals, *J. Am. Chem. Soc.* **51** (1929) 1010.

[16] W.H. Zachariasen, *Acta. Cryst.* **7** (1954) 795.

[17] I. D. Brown and D. Altermatt, Bond valence parameters obtained from a systematic analysis of the inorganic crystal structure data base, *Acta Cryst.* **B41** (1985) 244–7.

[18] M. O'Keeffe and N. E. Brese, Atom sizes and bond lengths in molecules and crystals, *J. Am. Chem. Soc.* **113** (1991) 3226.

[19] M. O'Keeffe, On the Arrangements of Ions in Crystals, *Acta Cryst.* **A33** (1977) 924.

[20] M. O'Keeffe and B.G. Hyde, *Structure and Bonding* **61** (1985) 77.

[21] R.T. Sanderson, *Chemical Bonds and Bond Energy* (Academic Press, New York, 1976).

[22] A.R. West, *Solid State Chemistry and its Applications* (J. Wiley & Sons, Chichester, 1984) pp. 290–301. A summary of Sanderson's model.

[23] E. Mooser and W. B. Pearson, On the Crystal Chemistry of Normal Valence Compounds, *Acta Cryst.* **12** (1959) 1015. Figs. 10.12 to 10.16 are drawn after Figs. 1, 3, 5, 7, and 11.

[24] P. Villars, *J. Less-Common Met.*, **92** (1983) 215–38. Fig. 10.19 is drawn after Fig. 3 starting on p. 330.

[25] P. Villars, *J. Less-Common Met.*, **99** (1984) 33–43.

[26] P. Villars, *J. Less-Common Met.*, **102** (1984) 199–211.

[27] J. C. Phillips, Ionicity of the Chemical Bond, *Rev. Mod. Phys.* **42** (1970) 317. Fig. 10.18 is drawn after Fig. 10 on p. 331.

[28] D.G. Pettifor, Structure Stability Maps for Pseudobinary and Ternary Phases, *Mat. Sci. Tech.* **4** (1988) 675.

Appendix 1A
Crystal and univalent radii

Table 1A.1 and 1A.2 list Pauling's [1] empirical crystal radii and the univalent radii, respectively. The empirical crystal radii were determined under the assumption that the radius of the oxide (O^{--}) ion is 1.4 Å. These radii follow earlier tabulations by Wasastjerna [2], and Goldschmidt [3], whose tables were based on the assumption that the radius of F^- was 1.33 Å and the radius of O^{--} was 1.32 Å. Bragg [4] also developed a table of radii based on the assumption that both F^- and O^{--} had radii of 1.35 Å. While Pauling's [1] radii are still used, Shannon's [5] radii (discussed in Chapter 7 and compiled in Table 7A.1) are now widely accepted.

Table 1A.1. *Pauling's empirical crystal radii $(\mathring{A})$.*

NH_4^+	1.48	Fe^{2+}	0.76	Ce^{3+}	1.01	H^-	2.08
Li^+	0.60	Fe^{3+}	0.64	Ce^{4+}	1.11	F^-	1.36
Na^+	0.95	Co^{2+}	0.74	Pr^{3+}	1.09	Cl^-	1.81
K^+	1.33	Co^{3+}	0.63	Pr^{4+}	0.92	Br^-	1.95
Rb^+	1.48	Ni^{2+}	0.62	Nd^{3+}	1.08	I^-	2.16
Cs^+	1.69	Ni^{3+}	0.72	Pm^{3+}	1.06	O^{2-}	1.40
Fr^+	1.76	Pd^{2+}	0.86	Sm^{3+}	1.04	S^{2-}	1.84
Be^{2+}	0.31	Cu^+	0.96	Eu^{2+}	1.12	Se^{2-}	1.98
Mg^{2+}	0.65	Ag^+	1.26	Eu^{3+}	1.03	Te^{2-}	2.21
Ca^{2+}	0.99	Au^+	1.37	Gd^{3+}	1.02	N^{3-}	1.71
Sr^{2+}	1.13	Zn^{2+}	0.74	Tb^{3+}	1.00	P^{3-}	2.12
Ba^{2+}	1.35	Cd^{2+}	0.97	Dy^{3+}	0.99	As^{3-}	2.22
Ra^{2+}	1.40	Hg^{2+}	1.10	Ho^{3+}	0.97	Sb^{3-}	2.45
Sc^{3+}	0.81	Ga^+	1.13	Er^{3+}	0.96	C^{4-}	2.60
Y^{3+}	0.93	Ga^{3+}	0.62	Tm^{3+}	0.95	Si^{4-}	2.71
La^{3+}	1.15	In^+	1.32	Yb^{2+}	1.13	Ge^{4-}	2.72
Ti^{2+}	0.90	In^{3+}	0.81	Yb^{3+}	0.94	Sn^{4-}	2.94
Ti^{3+}	0.76	Tl^+	1.40	Lu^{3+}	0.93		
Ti^{4+}	0.68	Tl^{3+}	0.95	Ac^{3+}	1.18		
Zr^{4+}	0.80	C^{4+}	0.15	Th^{3+}	1.14		
Hf^{4+}	0.81	Si^{4+}	0.41	Pa^{3+}	1.12		
V^{2+}	0.88	Ge^{2+}	0.93	Pa^{4+}	0.98		
V^{3+}	0.74	Ge^{4+}	0.53	U^{3+}	1.11		
V^{4+}	0.60	Sn^{2+}	1.12	U^{4+}	0.97		
Nb^{5+}	0.70	Sn^{4+}	0.71	Np^{3+}	1.09		
Cr^{2+}	0.84	Pb^{2+}	1.20	Np^{4+}	0.95		
Cr^{3+}	0.69	Pb^{4+}	0.84	Pu^{3+}	1.07		
Cr^{4+}	0.56	N^{5+}	0.11	Pu^{4+}	0.93		
Mo^{6+}	0.62	P^{5+}	0.34	Am^{3+}	1.06		
Mn^{2+}	0.80	As^{5+}	0.47	Am^{4+}	0.92		
Mn^{3+}	0.66	Sb^{5+}	0.62				
Mn^{4+}	0.54	Bi^{5+}	0.74				
Mn^{7+}	0.46						

Table 1A.2. *Univalent radii ($\mathring{A}$).*

			H^-	He	Li^+	Be^+	B^+	C^+	N^+	O^+	F^+
			2.08	0.93	0.60	0.44	0.35	0.29	0.25	0.22	0.19
C^-	N^-	O^-	F^-	Ne	Na^+	Mg^+	Al^+	Si^+	P^+	S^+	Cl^+
4.14	2.47	1.76	1.36	1.12	0.95	0.82	0.72	0.65	0.59	0.53	0.49
Si^-	P^-	S^-	Cl^-	Ar	K^+	Ca^+	Sc^+	Ti^+	V^+	Cr^+	Mn^+
3.84	2.79	2.19	1.81	1.54	1.33	1.18	1.06	0.96	0.88	0.81	0.75
					Cu^+	Zn^+	Ga^+	Ge^+	As^+	Se^+	Br^+
					0.96	0.88	0.81	0.76	0.71	0.66	0.62
Ge^-	As^-	Se^-	Br^-	Kr	Rb^+	Sr^+	Y^+	Zr^+	Nb^+	Mo^+	
3.71	2.85	2.32	1.95	1.69	1.48	1.32	1.20	1.09	1.00	0.93	
					Ag^+	Cd^+	In^+	Sn^+	Sb^+	Te^+	I^+
					1.26	1.14	1.04	0.96	0.89	0.82	0.77
Sn^-	Sb^-	Te^-	I^-	Xe	Cs^+	Ba^+	La^+	Ce^+			
3.70	2.95	2.50	2.16	1.90	1.69	1.53	1.39	1.27			
					Au^+	Hg^+	Tl^+	Pb^+	Bi^+		
					1.37	1.25	1.15	1.06	0.98		

References

[1] L. Pauling, *The Nature of the Chemical Bond* (Cornell University Press, Ithaca, 1960), p. 518. The data in Table 1A.1 are reproduced from Tables 13.3 and 13.5. The data in Table 1A.2 are reproduced from Table 13.2.

[2] J.A. Wasastjerna, *Soc. Sci. Fenn. Comm. Phys. Math.* **38** (1923) 1.

[3] V.M. Goldschmidt, Geochemische Verteilungsgesetze der Elemente, *Skrifter Norske Videnskaps-Akad. Oslo, I, Mat. Naturv. Kl.*, 1926.

[4] W.L. Bragg, *The Atomic Structure of Minerals* (Cornell University Press, 1937).

[5] R.D. Shannon, Revised Effective Ionic Radii and Systematic Studies of Interatomic Distances in Halides and Chalcogenides, *Acta. Cryst.* **A32** (1976) 751.

Appendix 2A
Computing distances using the metric tensor

The length (L) of any vector ($\vec{R}$) with components $x\vec{a}$, $y\vec{b}$, and $z\vec{c}$ is:

$$L = (\vec{R} \cdot \vec{R})^{1/2}. \tag{2A.1}$$

With reference to the unit cell parameters ($\alpha, \beta, \gamma, a, b, c$):

$$L^2 = [x\,y\,z] \begin{bmatrix} a^2 & ab\cos\gamma & ac\cos\beta \\ ba\cos\gamma & b^2 & bc\cos\alpha \\ ca\cos\beta & cb\cos\alpha & c^2 \end{bmatrix} \begin{bmatrix} x \\ y \\ z \end{bmatrix} = [x\,y\,z][\Gamma]\begin{bmatrix} x \\ y \\ z \end{bmatrix}. \tag{2A.2}$$

The 3×3 matrix in Eqn. 2A.2 is known as the metric tensor, Γ. Note Eqn. 2A.2 expresses the metric tensor for the most general (triclinic) case; for other crystal systems, Γ reduces to simpler forms, given below:

$$\Gamma_{\text{monoclinic}} = \begin{bmatrix} a^2 & 0 & ac\cos\beta \\ 0 & b^2 & 0 \\ ca\cos\beta & 0 & c^2 \end{bmatrix} \tag{2A.3}$$

$$\Gamma_{\text{rhombohedral}} = \begin{bmatrix} a^2 & a^2\cos\alpha & a^2\cos\alpha \\ a^2\cos\alpha & a^2 & a^2\cos\alpha \\ a^2\cos\alpha & a^2\cos\alpha & a^2 \end{bmatrix} \tag{2A.4}$$

$$\Gamma_{\text{hexagonal}} = \begin{bmatrix} a^2 & -\dfrac{a^2}{2} & 0 \\ -\dfrac{a^2}{2} & a^2 & 0 \\ 0 & 0 & c^2 \end{bmatrix} \tag{2A.5}$$

$$\Gamma_{orthorhombic} = \begin{bmatrix} a^2 & 0 & 0 \\ 0 & b^2 & 0 \\ 0 & 0 & c^2 \end{bmatrix} \qquad (2A.6)$$

$$\Gamma_{tetragonal} = \begin{bmatrix} a^2 & 0 & 0 \\ 0 & a^2 & 0 \\ 0 & 0 & c^2 \end{bmatrix} \qquad (2A.7)$$

$$\Gamma_{cubic} = \begin{bmatrix} a^2 & 0 & 0 \\ 0 & a^2 & 0 \\ 0 & 0 & a^2 \end{bmatrix}. \qquad (2A.8)$$

The distance between two points, (x_1, y_1, z_1) and (x_2, y_2, z_2) can also be determined with Eqn. 2A.2. In this case, the components of the vector $\vec{R}$ are $\Delta x \vec{a} = (x_2 - x_1)\vec{a}$, $\Delta y \vec{b} = (y_2 - y_1)\vec{b}$, and $\Delta z \vec{c} = (z_2 - z_1)\vec{c}$. The expanded form for the triclinic case gives:

$$L^2 = \left[(\Delta xa)^2 + (\Delta yb)^2 + (\Delta zc)^2 + 2ab\Delta x \Delta y \cos\gamma \right.$$
$$\left. + 2ac\Delta x \Delta z \cos\beta + 2bc\Delta y \Delta z \cos\alpha\right]. \qquad (2A.9)$$

For the crystal systems that have orthogonal axes, the distance formula in Eqn. 2A.9 reduces to the simpler and more familiar form:

$$L^2 = (\Delta xa)^2 + (\Delta yb)^2 + (\Delta zc)^2. \qquad (2A.10)$$

Appendix 2B
Computing unit cell volumes

The volume (V) of a unit cell is $\mathbf{a} \cdot (\vec{b} \times \vec{c})$. For the different crystal classes, this reduces to the following scalar forms that depend on the lattice parameters.

cubic	$V = a^3$	(2B.1)
tetragonal	$V = a^2 c$	(2B.2)
orthorhombic	$V = abc$	(2B.3)
monoclinic	$V = abc\sin\beta$	(2B.4)
hexagonal	$V = \dfrac{\sqrt{3}\, a^2 c}{2}$	(2B.5)
rhombohedral	$V = a^3 \sqrt{1 - 3\cos^2\alpha + 2\cos^3\alpha}$	(2B.6)

triclinic $V = abc\sqrt{1 - \cos^2\alpha - \cos^2\beta - \cos^2\gamma + 2\cos\alpha\cos\beta\cos\gamma}$ (2B.7)

Appendix 2C
Computing interplanar spacings

The magnitudes of the reciprocal lattice vectors, $\vec{G}_{hkl}$, are proportional to the inverse of the separation of the planes (hkl), or d_{hkl}. Analogous to the way that we computed the length of real space vectors using the metric tensor, Γ, (see Eqn. 2.17) we compute the length of reciprocal lattice vectors $(2\pi/d_{hkl})$ using the reciprocal metric tensor, Γ^*.

$$\frac{1}{d_{hkl}^2} = \frac{\vec{G}_{hkl} \cdot \vec{G}_{hkl}}{(2\pi)^2} = \frac{1}{(2\pi)^2}[h\ k\ l]\begin{bmatrix} \vec{a}^* \cdot \vec{a}^* & \vec{a}^* \cdot \vec{b}^* & \vec{a}^* \cdot \vec{c}^* \\ \vec{b}^* \cdot \vec{a}^* & \vec{b}^* \cdot \vec{b}^* & \vec{b}^* \cdot \vec{c}^* \\ \vec{c}^* \cdot \vec{a}^* & \vec{c}^* \cdot \vec{b}^* & \vec{c}^* \cdot \vec{c}^* \end{bmatrix}\begin{bmatrix} h \\ k \\ l \end{bmatrix}$$

$$\frac{1}{d_{hkl}^2} = \frac{1}{(2\pi)^2}[h\ k\ l]\begin{bmatrix} a^{*2} & a^*b^*\cos\gamma^* & a^*c^*\cos\beta^* \\ b^*a^*\cos\gamma^* & b^{*2} & b^*c^*\cos\alpha^* \\ c^*a^*\cos\beta^* & c^*b^*\cos\alpha^* & c^{*2} \end{bmatrix}\begin{bmatrix} h \\ k \\ l \end{bmatrix}. \quad (2C.1)$$

The matrix expression in Eqn. 2C.1 is general and works in all crystal systems. The expanded forms of this equation, specific to the different crystal systems, are given below:

cubic
$$\frac{1}{d_{hkl}^2} = \frac{h^2 + k^2 + l^2}{a^2} \quad (2C.2)$$

tetragonal
$$\frac{1}{d_{hkl}^2} = \frac{h^2 + k^2}{a^2} + \frac{l^2}{c^2} \quad (2C.3)$$

orthorhombic
$$\frac{1}{d_{hkl}^2} = \frac{h^2}{a^2} + \frac{k^2}{b^2} + \frac{l^2}{c^2} \quad (2C.4)$$

monoclinic
$$\frac{1}{d_{hkl}^2} = \frac{1}{\sin^2\beta}\left(\frac{h^2}{a^2} + \frac{k^2\sin^2\beta}{b^2} + \frac{l^2}{c^2} - \frac{2hl\cos\beta}{ac}\right) \quad (2C.5)$$

hexagonal
$$\frac{1}{d_{hkl}^2} = \frac{4}{3}\left(\frac{h^2 + hk + k^2}{a^2}\right) + \frac{l^2}{c^2} \quad (2C.6)$$

rhombohedral $\dfrac{1}{d_{hkl}^2} = \dfrac{(h^2 + k^2 + l^2)\sin^2\alpha + 2(hk + kl + hl)(\cos^2\alpha - \cos\alpha)}{a^2(1 - 3\cos^2\alpha + 2\cos^3\alpha)}$ (2C.7)

triclinic $\dfrac{1}{d_{hkl}^2} = \dfrac{1}{V^2}\left((b^2c^2\sin^2\alpha)h^2 + (a^2c^2\sin^2\beta)k^2 + (a^2b^2\sin^2\gamma)l^2\right)$

$+ \dfrac{1}{V^2}\left(2abc^2(\cos\alpha\cos\beta - \cos\gamma)hk\right)$

$+ \dfrac{1}{V^2}\left(2a^2bc(\cos\beta\cos\gamma - \cos\alpha)kl\right)$

$+ \dfrac{1}{V^2}\left(2ab^2c(\cos\gamma\cos\alpha - \cos\beta)hl\right)$ (2C.8)

Appendix 3A
The 230 space groups

Table 3A.1. *The triclinic and monoclinic space groups.*

Triclinic

P 1 (1)	P$\bar{1}$ (2)	

Monoclinic (with *b* as unique axis)

P 2 (3)	P m (6)	C c (9)	C 2/m (12)	C 2/c (15)
P 2$_1$ (4)	P c (7)	P 2/m (10)	P 2/c (13)	
C 2 (5)	C m (8)	P 2$_1$/m (11)	P 2$_1$/c (14)	

Table 3A.2. *The orthorhombic space groups.*

P 2 2 2 (16)	P m a 2 (28)	A m a 2 (40)	P n n a (52)	C m c a (64)
P 2 2 2$_1$ (17)	P c a 2$_1$ (29)	A b a 2 (41)	P m n a (53)	C m m m (65)
P 2$_1$ 2$_1$ 2 (18)	P n c 2 (30)	F m m 2 (42)	P c c a (54)	C c c m (66)
P 2$_1$ 2$_1$ 2$_1$ (19)	P m n 2$_1$ (31)	F d d 2 (43)	P b a m (55)	C m m a (67)
C 2 2 2$_1$ (20)	P b a 2 (32)	I m m 2 (44)	P c c n (56)	C c c a (68)
C 2 2 2 (21)	P n a 2$_1$ (33)	I b a 2 (45)	P b c m (57)	F m m m (69)
F 2 2 2 (22)	P n n 2 (34)	I m a 2 (46)	P n n m (58)	F d d d (70)
I 2 2 2 (23)	C m m 2 (35)	P m m m (47)	P m m n (59)	I m m m (71)
I 2$_1$ 2$_1$ 2$_1$ (24)	C m c 2$_1$ (36)	P n n n (48)	P b c n (60)	I b a m (72)
P m m 2 (25)	C c c 2 (37)	P c c m (49)	P b c a (61)	I b c a (73)
P m c 2$_1$ (26)	A m m 2 (38)	P b a n (50)	P n m a (62)	I m m a (74)
P c c 2 (27)	A b m 2 (39)	P m m a (51)	C m c m (63)	

Table 3A.3. *The tetragonal space groups.*

P 4 (75)	P 4 2 2 (89)	P 4 c c (103)	P $\bar{4}$ b 2 (117)	P 4$_2$/m m c (131)
P 4$_1$ (76)	P 4 2$_1$ 2 (90)	P 4 n c (104)	P $\bar{4}$ n 2 (118)	P 4$_2$/m c m (132)
P 4$_2$ (77)	P 4$_1$ 2 2 (91)	P 4$_2$ m c (105)	I $\bar{4}$ m 2 (119)	P 4$_2$/n b c (133)
P 4$_3$ (78)	P 4$_1$ 2$_1$ 2 (92)	P 4$_2$ b c (106)	I $\bar{4}$ c 2 (120)	P 4$_2$/n n m (134)
I 4 (79)	P 4$_2$ 2 2 (93)	I 4 m m (107)	I $\bar{4}$ 2 m (121)	P 4$_2$/m b c (135)
I 4$_1$ (80)	P 4$_2$ 2$_1$ 2 (94)	I 4 c m (108)	I $\bar{4}$ 2 d (122)	P 4$_2$/m n m (136)
P $\bar{4}$ (81)	P 4$_3$ 2 2 (95)	I 4$_1$ m d (109)	P 4/m m m (123)	P 4$_2$/n m c (137)
I $\bar{4}$ (82)	P 4$_3$ 2$_1$ 2 (96)	I 4$_1$ c d (110)	P 4/m c c (124)	P 4$_2$/n c m (138)
P 4/m (83)	I 4 2 2 (97)	P $\bar{4}$ 2 m (111)	P 4/n b m (125)	I 4/m m m (139)
P 4$_2$/m (84)	I 4$_1$ 2 2 (98)	P $\bar{4}$2 c (112)	P 4/n n c (126)	I 4/m c m (140)
P 4/n (85)	P 4 m m (99)	P $\bar{4}$ 2$_1$ m (113)	P 4/m b m (127)	I 4$_1$/a m d (141)
P 4$_2$/n (86)	P 4 b m (100)	P $\bar{4}$ 2$_1$ c (114)	P 4/m n c (128)	I 4$_1$/a c d (142)
I 4/m (87)	P 4$_2$ c m (101)	P $\bar{4}$ m 2 (115)	P 4/n m m (129)	
I 4$_1$/a (88)	P 4$_2$ n m (102)	P $\bar{4}$ c 2 (116)	P 4/n c c (130)	

Table 3A.4. *The trigonal space groups.*

P 3 (143)	R $\bar{3}$ (148)	P 3$_2$ 1 2 (153)	P 3 c 1 (158)	P $\bar{3}$ 1 c (163)
P 3$_1$ (144)	P 3 1 2 (149)	P 3$_2$ 2 1 (154)	P 3 1 c (159)	P $\bar{3}$ m 1 (164)
P 3$_2$ (145)	P 3 2 1 (150)	R 3 2 (155)	R 3 m (160)	P $\bar{3}$ c 1 (165)
R 3 (146)	P 3$_1$ 1 2 (151)	P 3 m 1 (156)	R 3 c (161)	R $\bar{3}$ m (166)
P $\bar{3}$ (147)	P 3$_1$ 2 1 (152)	P 3 1 m (157)	P $\bar{3}$ 1 m (162)	R $\bar{3}$ c (167)

Table 3A.5. *The hexagonal space groups.*

P 6 (168)	P $\bar{6}$ (174)	P 6$_2$ 2 2 (180)	P 6$_3$ m c (186)	P 6/m c c (192)
P 6$_1$ (169)	P 6/m (175)	P 6$_4$ 2 2 (181)	P $\bar{6}$ m 2 (187)	P 6$_3$/m c m (193)
P 6$_5$ (170)	P 6$_3$/m (176)	P 6$_3$ 2 2 (182)	P $\bar{6}$ c 2 (188)	P 6$_3$/m m c (194)
P 6$_2$ (171)	P 6 2 2 (177)	P 6 m m (183)	P $\bar{6}$ 2 m (189)	
P 6$_4$ (172)	P 6$_1$ 2 2 (178)	P 6 c c (184)	P $\bar{6}$ 2 c (190)	
P 6$_3$ (173)	P 6$_5$ 2 2 (179)	P 6$_3$ c m (185)	P 6/m m m (191)	

Table 3A.6. *The cubic space groups.*

P 2 3 (195)	F d $\bar{3}$ (203)	I 4 3 2 (211)	F $\bar{4}$ 3 c (219)	F d $\bar{3}$ m (227)
F 2 3 (196)	I m $\bar{3}$ (204)	P 4_3 3 2 (212)	I $\bar{4}$ 3 d (220)	F d $\bar{3}$ c (228)
I 2 3 (197)	P a $\bar{3}$ (205)	P 4_1 3 2 (213)	P m $\bar{3}$ m (221)	I m $\bar{3}$ m (229)
P 2_1 3 (198)	I a $\bar{3}$ (206)	I 4_1 3 2 (214)	P n $\bar{3}$ n (222)	I a $\bar{3}$ d (230)
I 2_1 3 (199)	P 4 3 2 (207)	P $\bar{4}$ 3 m (215)	P m $\bar{3}$ n (223)	
P m $\bar{3}$ (200)	P 4_2 3 2 (208)	F $\bar{4}$ 3 m (216)	P n $\bar{3}$ m (224)	
P n $\bar{3}$ (201)	F 4 3 2 (209)	I $\bar{4}$ 3 m (217)	F m $\bar{3}$ m (225)	
F m $\bar{3}$ (202)	F 4_1 3 2 (210)	P $\bar{4}$ 3 n (218)	F m $\bar{3}$ c (226)	

Appendix 3B
Selected crystal structure data

Index

Table	Structure
3B.31	The structure of gamma lanthanum oxyfluoride, γ-LaOF
3B.32	The structure of chalcopyrite, $CuFeS_2$
3B.33	The structure of Cr_2Al
3B.34	The structure of Ga_2Zr
3B.35	The structure of graphite
3B.36	The structure of β-copper mercury tetraiodide, Cu_2HgI_4
3B.37	The C14 structure
3B.38	The structure of potassium aluminum fluoride, $KAlF_4$
3B.39	The structure of $TiAl_2$
3B.40	The structure of potassium magnesium tetrafluoride, K_2MgF_4
3B.41	The structure of zircon, $ZrSiO_4$
3B.42	The structure of β-gold cadmium, β-AuCd
3B.43	The structure of anatase

Table 3B.1. *The structure of SiO_2 (α-quartz) [1].*

Formula unit:	SiO_2
Space group:	$P3_121$ (no. 152)
Cell dimensions:	$a = 4.913$ Å, $c = 5.405$ Å
Cell contents:	3 formula units
Atomic positions:	Si in (3a) $(x, 0, 0)$; $(0, x, 1/3)$; $(\bar{x}, \bar{x}, 2/3)$
	$\qquad x = 0.465$
	O in (6c)
	$\qquad (x, y, z)$; $(y, x, 1/3 - z)$; $(\bar{y}, x - y, 1/3 + z)$;
	$\qquad (\bar{x}, y - x, 2/3 - z)$; $(y - x, \bar{x}, 2/3 + z)$;
	$\qquad (x - y, \bar{y}, \bar{z})$; $x = 0.27$, $y = 0.42$, and $z = 0.45$

Table 3B.2. *The structure of SiO_2 (tridymite).*

Formula unit:	SiO_2		
Space group:	$P6_3/mmc$ (no. 194)		
Cell dimensions:	$a = 5.03$ Å, $c = 8.22$ Å		
Cell contents:	4 formula units		
Atomic positions:	Si in (4f)	$\pm(1/3, 2/3, z; 2/3, 1/3, z + 1/2)$	
		$z = 0.44$	
	O(1) in (2c)	$\pm(1/3, 2/3, 1/4)$	
	O(2) in (6g)	$(1/2, 0, 0)$; $\quad (0, 1/2, 0)$; $\quad (1/2, 1/2, 0)$;	
		$(1/2, 0, 1/2)$; $(0, 1/2, 1/2)$; $(1/2, 1/2, 1/2)$	

Table 3B.3. *The structure of scheelite, $CaWO_4$.*

Formula unit:	$CaWO_4$
Space group:	$I4_1/a$ (no. 88)
Cell dimensions:	a = 5.24 Å, c = 11.38 Å
Cell contents:	4 formula units
Atomic positions:	

Ca (4b) (0,0,1/2); (1/2,0,1/4); + I

W (4a) (0,0,0); (0,1/2,1/4); + I

O (16f) (x, y, z); $(\bar{x}, \bar{y}, z)$; $(x, y + 1/2, 1/4 - z)$; $(\bar{x}, 1/2 - y, 1/4 - z)$; $(\bar{y}, x, \bar{z})$; $(y, \bar{x}, \bar{z})$; $(\bar{y}, x + 1/2, z + 1/4)$; $(y, 1/2 - x, z + 1/4)$; + I

$x = 0.25$, $y = 0.15$, $z = 0.075$

Examples of isostructural compounds

compound	a(Å)	c(Å)	compound	a(Å)	c(Å)
$AgReO_4$	5.378	11.805	$NaReO_4$	5.362	11.718
$AgTcO_4$	5.319	11.875	$NaTcO_4$	5.339	11.869
$BaMoO_4$	5.56	12.76	$PbMoO_4$	5.47	12.18
$BaWO_4$	5.64	12.70	$PbWO_4$	5.44	12.01
$BiAsO_4$	5.08	11.70	$RbReO_4$	5.803	13.167
$CaMoO_4$	5.23	11.44	$SrMoO_4$	5.36	11.94
$CdMoO_4$	5.14	11.17	$SrWO_4$	5.40	11.90
$CeGeO_4$	5.045	11.167	$ThGeO_4$	5.14	11.54
$HfGeO_4$	4.849	10.50	$TlReO_4$	5.761	13.33
$KReO_4$	5.675	12.70	$UGeO_4$	5.084	11.226
$KRuO_4$	5.609	12.991	$YNbO_4$	5.16	10.91
$KTcO_4$	5.654	13.03	$ZrGeO_4$	4.871	10.570

Table 3B.4. *The structure of PdS (palladous sulfide).*

Formula unit:	PdS
Space group:	$P4_2/m$ (no. 84)
Cell dimensions:	$a = 6.429$ Å, $c = 6.608$Å
Cell contents:	8 formula units
Atomic positions:	Pd(1) in (2e) (0, 0, 1/4); (0, 0, 3/4)
	Pd(2) in (2c) (0, 1/2, 0); (1/2, 0, 1/2)
	Pd(3) in (4j) $(x, y, 0)$; $(\bar{y}, x, 1/2)$; $(\bar{x}, \bar{y}, 0)$; $(y, \bar{x}, 1/2)$
	$x = 0.48$ and $y = 0.25$
	S in (8k)
	(x, y, z); $(\bar{y}, x, 1/2 + z)$; $(\bar{x}, \bar{y}, z)$; $(y, \bar{x}, 1/2 + z)$
	$(x, y, \bar{z})$; $(\bar{y}, x, 1/2 - z)$; $(\bar{x}, \bar{y}, \bar{z})$; $(y, \bar{x}, 1/2 - z)$
	$x = 0.19$, $y = 0.32$, and $z = 0.23$

Table 3B.5. *The structure of α-MoO$_3$ (molybdenum trioxide) [2].*

Formula unit:	MoO_3 (molybdite)
Space group:	Pnma (no. 62)
Cell dimensions:	$a = 13.94$ Å, $b = 3.66$ Å, $c = 3.92$ Å
Cell contents:	4 formula units
Atomic positions:	All atoms are in 4c sites at $\pm(x, 1/4, z)$; $(x + 1/2, 1/4, 1/2 - z)$

	x	y	z
Mo	0.0998	1/4	0.084
O(1)	0.23	1/4	0.015
O(2)	0.10	1/4	0.56
O(3)	0.435	1/4	0.526

Table 3B.6. *The high pressure structure of Rh_2O_3 (rhodium sesquioxide) [3].*

Formula unit:	Rh_2O_3 II
Space group:	*Pbcn* (no. 60)
Cell dimensions:	$a = 7.2426$ Å, $b = 5.1686$ Å, $c = 5.3814$ Å
Cell contents:	4 formula units
Atomic positions:	

(8d) (x, y, z); $(\bar{x} + 1/2, \bar{y} + 1/2, z + 1/2)$; $(\bar{x}, y, \bar{z} + 1/2)$; $(x + 1/2, \bar{y} + 1/2, \bar{z})$
$(\bar{x}, \bar{y}, \bar{z})$; $(x + 1/2, y + 1/2, \bar{z} + 1/2)$; $(x, \bar{y}, z + 1/2)$; $(\bar{x} + 1/2, y + 1/2, z)$

(4c) $(0, y, 1/4)$; $(1/2, \bar{y} + 1/2, 3/4)$; $(0, \bar{y}, 3/4)$; $(1/2, y + 1/2, 1/4)$

		x	y	z
Rh	(8d)	0.1058	0.7498	0.0312
O(1)	(8d)	0.8494	0.6037	0.1161
O(2)	(4c)		0.0505	

Table 3B.7. *The Millerite structure of NiS (nickel sulfide) [4].*

Formula unit:	NiS (millerite)
Space group:	R3*m* (no. 160)
Cell dimensions:	$a = 9.612$ Å, $c = 3.259$ Å
Cell contents:	9 formula units
Atomic positions:	Both atoms are in 9(b) sites:

(9b) $(x, \bar{x}, z)$; $(x, 2x, z)$; $(2\bar{x}, \bar{x}, z) + R$

	x	z
Ni	−0.088	0.088
S	0.114	0.596

Table 3B.8. *The structure of V_2O_5 (vanadium pentoxide) [5].*

Formula unit:	V_2O_5
Space group:	P*mmn* (no. 59)
Cell dimensions:	$a = 11.519$Å, $b = 3.564$Å, $c = 4.373$Å
Cell contents:	2 formula units
Atomic positions:	Both (2a) and (4f) sites are occupied
	For the (2a): $(0, 0, z)$; $(1/2, 1/2, \bar{z})$
	For the (4f): $(x, 0, z)$; $(\bar{x}, 0, z)$
	$(x + 1/2, 1/2, \bar{z})$; $(1/2 - x, 1/2, \bar{z})$
	V in (4f) with $x = 0.1487$; $z = 0.1086$
	O(1) in (4f) with $x = 0.1460$; $z = 0.4713$
	O(2) in (4f) with $x = 0.3191$; $z = -0.0026$
	O(3) in (2a) with $z = -0.0031$

Table 3B.9. *The crystal structure of $YBaCuFeO_5$ [6].*

Formula unit:	$YBaCuFeO_5$		
Space group:	P4*mm* (no. 99)		
Cell dimensions:	$a = 3.86$ Å, $c = 7.64$ Å		
Cell contents:	1 formula unit		
Atomic positions:	Ba	(1a) $(0, 0, z)$	$z = 0$
	Y	(1a) $(0, 0, z)$	$z = 0.49$
	Cu	(1b) $(1/2, 1/2, z)$	$z = 0.72$
	Fe	(1b) $(1/2, 1/2, z)$	$z = 0.25$
	O1	(1b) $(1/2, 1/2, z)$	$z = 0.99$
	O2	(2c) $(1/2, 0, z)$; $(0, 1/2, z)$	$z = 0.3$
	O3	(2c) $(1/2, 0, z)$; $(0, 1/2, z)$	$z = 0.67$

Table 3B.10. *The crystal structure of* $YBa_2Cu_3O_7$ *[7]*.

Formula unit:	$YBa_2Cu_3O_7$
Space group:	P*mmm* (no. 47)
Cell dimensions:	$a = 3.8227$ Å, $b = 3.8872$ Å, $c = 11.6802$ Å
Cell contents:	1 formula unit
Atomic positions:	

Y	(1h) (1/2, 1/2, 1/2)	
Ba	(2t) (1/2, 1/2, z); (1/2, 1/2, $\bar{z}$)	$z = 0.183$
Cu1	(1a) (0, 0, 0)	
Cu2	(2q) (0, 0, z); (0, 0, $\bar{z}$)	$z = 0.3556$
O1	(1e) (0, 1/2, 0)	
O2	(2s) (1/2, 0, z); (1/2, 0, $\bar{z}$)	$z = 0.3779$
O3	(2r) (0, 1/2, z); (0, 1/2, $\bar{z}$)	$z = 0.3790$
O4	(2q) (0, 0, z); (0, 0, $\bar{z}$)	$z = 0.1590$

Table 3B.11. *The manganese phosphide crystal structure, B31 [8]*.

Formula unit:	MnP
Space group:	P*bnm* (no. 62)
Cell dimensions:	$a = 5.916$ Å, $b = 5.260$ Å, $c = 3.173$ Å
Cell contents:	4 formula units
Atomic positions:	

Mn	(4c)	$\pm(x, y, 1/4)$; $(1/2 - x, y + 1/2, 1/4)$ $x = 0.20$; $y = 0.005$
P	(4c)	$\pm(x, y, 1/4)$; $(1/2 - x, y + 1/2, 1/4)$ $x = 0.57$; $y = 0.19$

MX	a	b	c	MX	a	b	c
AuGa	6.397	6.267	3.421	PdGe	6.259	5.782	3.481
CoAs	5.869	5.292	3.458	PdSi	6.133	5.599	3.381
CoP	5.599	5.076	3.281	PdSn	6.32	6.13	3.87
CrAs	6.222	5.741	3.486	PtGe	6.088	5.732	3.701
CrP	5.94	5.366	3.13	PtSi	5.932	5.595	3.603
FeAs	6.028	5.439	3.373	RhGe	6.48	5.70	3.25
FeP	5.793	5.187	3.093	RhSb	6.333	5.952	3.876
IrGe	6.281	5.611	3.490	RuP	6.120	5.520	3.168
MnAs	6.39	5.64	3.63	VAs	6.317	5.879	3.334
NiGe	5.811	5.381	3.428	WP	6.219	5.717	3.238

Table 3B.12. *The ferric boride crystal structure, B27 [9].*

Formula unit:	FeB		
Space group:	P*bnm* (no. 62)		
Cell dimensions:	$a = 4.053$ Å, $b = 5.495$ Å, $c = 2.946$ Å		
Cell contents:	4 formula units		
Atomic positions:	Fe	(4c)	$\pm(x, y, 1/4)$; $(1/2 - x, y + 1/2, 1/4)$
			$x = 0.125$; $y = 0.180$
	B	(4c)	$\pm(x, y, 1/4)$; $(1/2 - x, y + 1/2, 1/4)$
			$x = 0.61$; $y = 0.04$

Table 3B.13. *The structure of diamond [10].*

Formula unit:	C, diamond	
Space group:	$F d\bar{3}m$ (no. 227)	
Cell dimensions:	$a = 3.56679$ Å	
Cell contents:	8 formula units per cell	
Atom positions:	C in (8a)	(0,0,0); (1/4, 1/4, 1/4)

Table 3B.14. *The structure of lithium metagallate, $LiGaO_2$ [11].*

Formula unit:	$LiGaO_2$		
Space group:	P*na*2$_1$ (no. 33)		
Cell dimensions:	$a = 5.402$ Å, $b = 6.372$ Å, $c = 5.007$ Å		
Cell contents:	4 formula units per cell		
Atomic positions:	All atoms are in (4a) positions:		
	(x, y, z); $\quad(\bar{x}, \bar{y}, z + 1/2)$;		
	$(x + 1/2, \bar{y} + 1/2, z)$; $(\bar{x} + 1/2, y + 1/2, z + 1/2)$		

atom	x	y	z
Ga	0.082	0.126	0.000
Li	0.421	0.127	0.494
O(1)	0.407	0.139	0.893
O(2)	0.070	0.112	0.371

Table 3B.15. *The ReO_3 structure, $D0_9$.*

Formula unit:	ReO_3		
Space group:	$Pm\bar{3}m$ (no. 221)		
Cell dimensions:	$a = 3.8$ Å		
Cell contents:	1 formula units per cell		
Atomic positions:	Re in 1(a)	$m\bar{3}m$	(0, 0, 0)
	O in 3(d)	$4/mmm$	(0, 0, 1/2); (0, 1/2, 0); (1/2, 0, 0)

Table 3B.16. *The hexagonal tungsten bronze structure, HTB.*

Formula unit:	$Rb_{0.33}WO_3$	
Space group:	$P6_3/mcm$ (no. 193)	
Cell dimensions:	$a = 7.3875$ Å, $c = 7.5589$ Å	
Cell contents:	6 formula units per cell	
Atomic positions:	W in (6g)	$\pm(x, 0, 1/4)$; $(0, x, 1/4)$; $(\bar{x}, \bar{x}, 1/4)$
		$x = 0.48$
	O(1) in (6f)	(1/2, 0, 0); (0, 1/2, 0); (1/2, 1/2, 0)
		(1/2, 0, 1/2); (0, 1/2, 1/2); (1/2, 1/2, 1/2)
	O(2) in (12j)	$\pm(x, y, 1/4)$; $(\bar{y}, x-y, 1/4)$; $(y-x, \bar{x}, 1/4)$
		$(y, x, 1/4)$; $(\bar{x}, y-x, 1/4)$; $(x-y, \bar{y}, 1/4)$
		$x = 0.42$; $y = 0.22$
	Rb in (2b)	(0, 0, 0); (0, 0, 1/2)

Table 3B.17. *The tetragonal tungsten bronze structure, TTB.*

Formula unit:	$Na_{0.1}WO_3$	
Space group:	$P4/nmm$ (no. 193)	
Cell dimensions:	$a = 5.248$ Å, $c = 3.895$ Å	
Cell contents:	2 formula units per cell	
Atomic positions:	W in (2c)	$(0, 1/2, z)$; $(1/2, 0, \bar{z})$
		$z = 0.435$
	0.2 Na in (2a)	(0, 0, 0); (1/2, 1/2, 0)
	O(1) in (2c)	$(0, 1/2, z)$; $(1/2, 0, \bar{z})$
		$z = 0.935$
	O(2) in (4e)	(1/4, 1/4, 1/2); (3/4, 3/4, 1/2);
		(1/4, 3/4, 1/2); (3/4, 1/4, 1/2)

Table 3B.18. *The tetragonal close packed arrangement [12].*

Formula unit:	O (ideal spheres)
Space group:	$P4_2/mnm$ (no. 136)
Cell dimensions:	$a = 1.7071$ c (dimensionless)
Cell contents:	4 spheres per cell
Atomic positions:	O in 4(f) mm $(x, x, 0)$; $(\bar{x}, \bar{x}, 0)$; $(1/2 + x, 1/2 - x, 1/2)$; $(1/2 - x, 1/2 + x, 1/2)$ $x = 0.2929$

Table 3B.19. *The corundum (sapphire) structure, aluminum sesquioxide, $D5_1$.*

Formula unit	Al_2O_3, aluminum sesquioxide
Space group:	$R\bar{3}c$ (no. 167)
Cell dimensions:	$a = 4.7626$ Å, $c = 13.0032$ Å
Cell contents:	6 formula units per hexagonal cell
Atomic positions:	Al in (12c) 3 $(0, 0, z)$; $(0, 0, \bar{z} + 1/2)$; $(0, 0, \bar{z})$; $(0, 0, z + 1/2)$; +rh $z = 0.352$
	O in (18e) 2 $(x, 0, 1/4)$; $(0, x, 1/4)$; $(\bar{x}, \bar{x}, 1/4)$; $(\bar{x}, 0, 3/4)$; $(0, \bar{x}, 3/4)$; $(x, x, 3/4)$; +rh $x = 0.306$

Examples

compound	a (Å)	c (Å)	compound	a (Å)	c (Å)
V_2O_3	15.105	14.449	$FeAlO_3$		
Cr_2O_3	4.954	13.584	$AlGaO_3$		
Fe_2O_3	5.035	13.72	$CrVO_3$	4.982	13.752
Rh_2O_3	5.11	13.82	$GaFeO_3$		
Al_2O_3	4.7626	13.0032	$InTiO_3$		
Ga_2O_3	4.9793	13.429	$CrAlO_3$		
Ti_2O_3	5.184	13.636	$FeCrO_3$		

Table 3B.20. *The ilmenite structure, iron titanate.*

Formula unit:	$FeTiO_3$, iron titanate			
Space group:	$R\bar{3}$ (no. 148)			
Cell dimensions:	$a = 5.082$ Å, $c = 14.026$ Å			
Cell contents:	6 formula units per hexagonal cell			
Atomic positions:	Ti in (6c)	3	$\pm(0, 0, z)$ + rh	$z = 0.142$
	Fe in (6c)	3	$\pm(0, 0, z)$ + rh	$z = 0.358$
	O in (18f)	1	$\pm(x, y, z)$; $(\bar{y}, x-y, z)$; $(y-x, \bar{x}, z)$ + rh	
			$x = 0.305$, $y = 0.015$, $z = 0.250$	

Examples

compound	a (Å)	c (Å)	compound	a (Å)	c (Å)
$CdTiO_3$	5.428	14.907	$NiTiO_3$	5.031	13.785
$CdSnO_3$	5.454	14.968	$MgTiO_3$	5.054	13.898
$NiMnO_3$	4.905	13.59	$FeTiO_3$	5.082	14.026
$MgGeO_3$	4.936	13.76	$MnGeO_3$	5.015	14.331
$CoMnO_3$	4.933	13.71	$MnTiO_3$	5.1374	14.284
$MgMnO_3$	4.945	13.73			

Table 3B.21. *The C16 structure.*

Formula unit:	khatyrkite, $CuAl_2$		
Space group:	$I4/mcm$ (no. 140)		
Cell dimensions:	$a = 6.07$ Å, $c = 4.87$ Å		
Cell contents:	4 formula units		
Atomic positions:	Cu in (4a)	422	$(0,0,1/4)$; $(0,0,3/4)$; + I
	Al in (8h)	$m2m$	$(x, 1/2 + x, 0)$; $(\bar{x}, \bar{x} + 1/2, 0)$;
			$(\bar{x} + 1/2, x, 0)$; $(1/2 + x, \bar{x}, 0)$; + I
			$x = 0.158$

Examples

compound	a (Å)	c (Å)	compound	a (Å)	c (Å)
$CuAl_2$	6.07	4.87	BFe_2	5.1317	8.5321
$FeSn_2$	6.539	5.325	BMo_2	5.547	4.739
BCr_2	5.185	4.316	BNi_2	4.991	4.247
BCo_2	5.015	4.220	BW_2	5.568	4.744

Table 3B.22. *The structure of titanium carbosulfide.*

Formula unit:	Ti_2CS		
Space group:	$P6_3/mmc$ (no. 194)		
Cell dimensions:	$a = 3.210$ Å, $c = 11.20$ Å		
Cell contents:	2 formula units per cell		
Atomic positions:	Ti in (4f)	$3m$	$(1/3, 2/3, z)$; $(2/3, 1/3, z+1/2)$
			$(2/3, 1/3, \bar{z})$; $(1/3, 2/3, \bar{z}+1/2)$
			$z = 0.1$
	C in (2a)	$\bar{3}m$	$(0, 0, 0)$; $(0,0,1/2)$
	S in (2d)	$\bar{6}m2$	$(1/3, 2/3, 3/4)$; $(2/3, 1/3, 1/4)$

Table 3B.23. *The garnet structure,* $Al_2Ca_3(SiO_4)_3$.

Formula unit:	$Al_2Ca_3(SiO_4)_3$, grossularite
Space group:	$Ia\bar{3}d$
Cell dimensions:	$a = 11.855$ Å
Cell contents:	8 formula units per cell
Atomic positions:	Ca in (24c) $\pm(1/8, 0, 1/4) + \text{tr} + \text{I}$; $\pm(5/8, 0, 1/4) + \text{tr} + \text{I}$
	Al in (16a) $(0, 0, 0)$; $(1/4, 1/4, 1/4) + \text{F} + \text{I}$
	Si in (24d) $\pm(3/8, 0, 1/4) + \text{tr} + \text{I}$; $\pm(7/8, 0, 1/4) + \text{tr} + \text{I}$
	O in (96h) $\pm(x, y, z) + \text{tr} + \text{I}$; $\pm(x+1/2, 1/2-y, \bar{z}) + \text{tr} + \text{I}$;
	$\pm(\bar{x}, y+1/2, 1/2-z) + \text{tr} + \text{I}$; $\pm(1/2-x, \bar{y}, z+1/2) + \text{tr} + \text{I}$
	$\pm(y+1/4, x+1/4, z) + \text{tr} + \text{I}$; $\pm(y+3/4, 1/4-x, 3/4-z) + \text{tr} + \text{I}$
	$\pm(3/4-y, x+3/4, 1/4-z) + \text{tr} + \text{I}$; $\pm(1/4-y, 3/4-x, z+3/4) + \text{tr} + \text{I}$
	$x = -0.0382(1)$, $y = 0.0457(1)$, $z = 0.1512(1)$

Table 3B.23. (*cont.*)

Examples					
compound	a (Å)	compound	a (Å)	compound	a (Å)
$Al_2Mg_3(SiO_4)_3$	11.459	$Eu_3Fe_5O_{12}$	12.498	$Bi_4(GeO_4)_3$	10.52
$Ca_3Fe_2(SiO_4)_3$	12.048	$Gd_3Fe_5O_{12}$	12.44	$CaGd_2Mn_2(GeO_4)_3$	12.555
$Fe_3Al_2(SiO_4)_3$	11.526	$Ho_3Fe_5O_{12}$	12.380	$CaNa_2Sn_2(GeO_4)_3$	12.430
$Mn_3Al_2(SiO_4)_3$	11.621	$Lu_3Fe_5O_{12}$	12.277	$CaNa_2Ti_2(GeO_4)_3$	12.359
$Ca_3V_2(SiO_4)_3$	12.070	$Nd_3Fe_5O_{12}$	12.60	$CaY_2Mn_2(GeO_4)_3$	12.475
$Cd_3Al_2(SiO_4)_3$	11.82	$Sm_3Fe_5O_{12}$	12.53	$Ca_3Al_2(GeO_4)_3$	12.117
$Cd_3Cr_2(SiO_4)_3$	11.999	$Tb_3Fe_5O_{12}$	12.447	$Ca_3Cr_2(GeO_4)_3$	12.262
$Y_3Fe_2(AlO_4)_3$	12.161	$Tm_3Fe_5O_{12}$	12.352	$Ca_3Fe_2(GeO_4)_3$	12.325
$Y_3Al_5O_{12}$	12.01	$Y_3Fe_5O_{12}$	12.376	$Ca_3Ga_2(GeO_4)_3$	12.251
$Y_4Al_4O_{12}$	11.989	$YNd_2Fe_5O_{12}$	12.530	$Ca_3TiCo(GeO_4)_3$	12.356
$NaCa_2Co_2(VO_4)_3$	12.431	$Y_2NdFe_5O_{12}$	12.454	$Ca_3TiMg(GeO_4)_3$	12.35
$NaCa_2Cu_2(VO_4)_3$	12.423	$Y_2PrFe_5O_{12}$	12.478	$Ca_3TiNi(GeO_4)_3$	12.341
$NaCa_2Mg_2(VO_4)_3$	12.446	$Yb_3Fe_5O_{12}$	12.291	$Ca_3SnCo(GeO_4)_3$	12.47
$NaCa_2Ni_2(VO_4)_3$	12.373	$Y_3Ga_2(GaO_4)_3$	12.277	$Ca_3V_2(GeO_4)_3$	12.324
$NaCa_2Zn_2(VO_4)_3$	12.439	$Yb_3Ga_2(GaO_4)_3$	12.200	$Cd_3Al_2(GeO_4)_3$	12.08
$Na_3Al_2Li_3F_{12}$	12.16	$Nd_4Ga_4O_{12}$	12.54	$Cd_3Cr_2(GeO_4)_3$	12.20
$Dy_3Fe_5O_{12}$	12.414	$Sm_4Ga_4O_{12}$	12.465	$Cd_3Fe_2(GeO_4)_3$	12.26
$Er_3Fe_5O_{12}$	12.349	$Y_4Ga_4O_{12}$	12.30	$Cd_3Ga_2(GeO_4)_3$	12.19

Table 3B.24. *The structure of diaspore, AlO(OH), aluminum oxyhydroxide.*

Formula unit:	AlO(OH), diaspore
Space group:	$Pbnm$ (no. 62)
Cell dimensions:	$a = 4.396$ Å, $b = 9.426$ Å, $c = 2.844$ Å
Cell contents:	4 formula units per cell
Atomic positions:	all atoms in (4c) $\pm(x, y, 1/4)$; $\pm(1/2 - x, y + 1/2, 1/4)$
	Al coords $x = -0.0451$, $y = 0.1446$
	O(1) coords $x = 0.2880$, $y = -0.1989$
	O(2) coords $x = -0.1970$, $y = -0.0532$

Table 3B.25. *The structure of boehmite, AlO(OH), aluminum oxyhydroxide.*

Formula unit:	AlO(OH), *boehmite*
Space group:	B*bmm*
Cell dimensions:	$a = 12.227$ Å, $b = 3.700$Å, $c = 2.866$ Å
Cell contents:	4 formula units per cell
Atomic positions:	all atoms in (4c) $\pm(x, 1/4, 0)$; $\pm(x + 1/2, 1/4, 1/2)$
	Al coord $x = -0.166$
	O(1) coord $x = 0.213$
	O(2) coord $x = 0.433$

Table 3B.26. *The chromium boride crystal structure [13].*

Formula unit	CrB
Space group:	C*mcm* (no. 63)
Cell dimensions:	$a = 2.969$ Å, $b = 7.858$ Å, $c = 2.932$ Å
Cell contents:	4 formula units
Atomic positions:	Cr (4c) $\pm(0, y, 1/4)$; $\pm(1/2, y + 1/2, 1/4)$
	$y = 0.146$
	B (4c) $\pm(0, y, 1/4)$; $\pm(1/2, y + 1/2, 1/4)$
	$y = 0.440$

Table 3B.27. *The structure of lithium ferrite, LiFeO$_2$.*

Formula unit:	LiFeO$_2$, lithium ferrite
Space group:	I4$_1$/*amd* (no. 141)
Cell dimensions:	$a = 4.057$ Å, $c = 8.579$ Å
Cell contents:	4 formula units per cell
Atomic positions:	Fe: (4a) (0, 0, 0); (0, 1/2, 1/4); +I
	Li: (4b) (0, 0, 1/2); (0, 1/2, 3/4); +I
	O: (8e) (0, 0, z); (0, 1/2, z + 1/4);
	(1/2, 0, $\bar{z}$ + 3/4); (1/2, 1/2, $\bar{z}$ + 1/2); +I
	$z = 0.25$

Table 3B.27. (*cont.*)

isostructural compounds					
compound	*a* (Å)	*c* (Å)	compound	*a* (Å)	*c* (Å)
LiFeO$_2$	4.057	8.579	LiLuO$_2$	4.37	9.95
LiInO$_2$	4.316	9.347	LiTmO$_2$	4.405	10.15
ß-LiTlO$_2$	4.547	9.255	LiYO$_2$	4.44	10.35
LiScO$_2$	4.182	9.318	LiYbO$_2$	4.39	10.06
LiErO$_2$	4.42	10.20	LiPN$_2$	4.566	7.145

Table 3B.28. *The structure of pyrochlore, $A_2B_2O_7$*.*

Formula unit:	$A_2B_2O_7$, pyrochlore		
Space group:	Fd3m (no. 227)		
Cell dimensions:	$a = 10.397$ Å		
Cell contents:	8 formula units per cell		
Atomic positions:	A:	(16c)	(1/8, 1/8, 1/8); (1/8, 3/8, 3/8);
			(3/8, 1/8, 3/8); (3/8, 3/8, 1/8); + F
	B:	(16d)	(5/8, 5/8, 5/8); (5/8, 7/8, 7/8)
			(7/8, 5/8, 7/8); (7/8, 7/8, 5/8); + F
	O1:	(8b)	(1/2, 1/2, 1/2); (3/4, 3/4, 3/4); + F
	O2:	(48f)	(x, 0, 0); ($x + 1/4$, 1/4, 1/4);
			($\bar{x}$, 0, 0); (1/4 − z, 1/4, 1/4); + rh; + F

*The mineral pyrochlore is (Na,Ca)$_2$(Nb,Ti)$_2$(O,F)$_7$. This structure is alternatively known as the atopite structure, from the mineral (Ca,Mn,Na)$_2$Sb$_2$(O,OH,F)$_7$.

Table 3B.28. (*cont.*)

isostructural compounds

compound	a (Å)	compound	a (Å)	compound	a (Å)
$Ca_2Sb_2O_7$	10.32	$Ho_2Ru_2O_7$	10.150	$Sm_2Sn_2O_7$	10.507
$Cd_2Nb_2O_7$	10.372	$Ho_2Sn_2O_7$	10.374	$Ta_2Sn_2O_7$	10.48
$Cd_2Sb_2O_7$	10.18	$La_2Hf_2O_7$	10.770	$Tb_2Ru_2O_7$	10.200
$Cd_2Ta_2O_7$	10.376	$La_2Sn_2O_7$	10.702	$Tb_2Sn_2O_7$	10.428
$Dy_2Ru_2O_7$	10.175	$Lu_2Ru_2O_7$	10.103	$Tm_2Ru_2O_7$	10.096
$Dy_2Sn_2O_7$	10.389	$Lu_2Sn_2O_7$	10.294	$Tm_2Sn_2O_7$	10.330
$Er_2Ru_2O_7$	10.120	$Nd_2Hf_2O_7$	10.648	$Y_2Ru_2O_7$	10.144
$Er_2Sn_2O_7$	10.350	$Nd_2Ru_2O_7$	10.331	$Y_2Sn_2O_7$	10.371
$Eu_2Ru_2O_7$	10.252	$Nd_2Sn_2O_7$	10.573	$Y_2Ti_2O_7$	10.095
$Eu_2Sn_2O_7$	10.474	$Pr_2Ru_2O_7$	10.355	$Yb_2Ru_2O_7$	10.087
$Gd_2Ru_2O_7$	10.230	$Pr_2Sn_2O_7$	10.604	$Yb_2Sn_2O_7$	10.304
$Gd_2Sn_2O_7$	10.460	$Sm_2Ru_2O_7$	10.280	$Zr_2Ce_2O_7$	10.699

Table 3B.29. *The* $(D5_3)$ *structure of bixbyite,* $(Fe,Mn)_2O_3$.

Formula unit:	M_2O_3, bixbyite	
Space group:	$Ia3$ (no. 206)	
Cell dimensions:	$a = 9.356$ Å	
Cell contents:	16 formula units per cell	
Atomic positions:	M1:	(8a)	(1/4, 1/4, 1/4); (1/4, 3/4, 3/4); (3/4, 1/4, 3/4); (3/4, 3/4, 1/4); +I

Atomic positions:

M1: (8a) (1/4, 1/4, 1/4); (1/4, 3/4, 3/4);
        (3/4, 1/4, 3/4); (3/4, 3/4, 1/4); +I

M2: (24d) $\pm(x, 0, 1/4)$; $\pm(1/4, x, 0)$; $\pm(0, 1/4, x)$;
        $\pm(\bar{x}, 1/2, 1/4)$; $\pm(1/4, \bar{x}, 1/2)$; $\pm(1/2, 1/4, \bar{x})$; +I
        $x = -0.034$

O: (48e) $\pm(x, y, z)$; $\pm(x, \bar{y}, 1/2-z)$; $\pm(1/2-x, y, \bar{z})$; $\pm(\bar{x}, 1/2-y, z)$;
        $\pm(z, x, y)$; $\pm(1/2-z, x, \bar{y})$; $\pm(\bar{z}, 1/2-x, y)$; $\pm(x, \bar{x}, 1/2-y)$;
        $\pm(y, z, x)$; $\pm(\bar{y}, 1/2-z, x)$; $\pm(y, \bar{z}, 1/2-x)$; $\pm(1/2-y, z, \bar{x})$; +I
        $x = 0.375$; $y = 0.162$; $z = 0.400$

Table 3B.29. (*cont.*)

isostructural compounds					
compound	a (Å)	compound	a (Å)	compound	a (Å)
Cm_2O_3	11.00	β-Mn_2O_3	9.408	Yb_2O_3	10.439
Dy_2O_3	10.66	Nd_2O_3	11.048	Be_2N_3	8.13
Er_2O_3	10.547	Pr_2O_3	11.136	Ca_2N_3	10.40
Eu_2O_3	10.866	Pu_2O_3	11.04	Cd_2N_3	10.79
β-Fe_2O_3	9.40	Sc_2O_3	9.845	Mg_2N_3	9.95
Gd_2O_3	10.813	Sm_2O_3	10.932	U_2N_3	10.678
Ho_2O_3	10.607	Tb_2O_3	10.728	Zn_2N_3	9.743
In_2O_3	10.118	Tl_2O_3	10.543	Be_2P_3	10.15
La_2O_3	11.38	Tm_2O_3	10.488	Mg_2P_3	12.01
Lu_2O_3	10.391	Y_2O_3	10.604	Mg_2As_3	12.33

Table 3B.30. *The TiO_2 II structure.*

Formula unit:	TiO_2	
Space group:	$Pbcn$ (no. 60)	
Cell dimensions:	$a = 4.531$ Å, $b = 5.498$ Å, $c = 4.900$ Å	
Cell contents:	4 formula units	
Atomic positions:	Ti in (4c)	$\pm(0, y, 1/4); \pm(1/2, y + 1/2, 1/4)$
		$y = 0.171$
	O in (8d)	$\pm(x, y, z); \pm(1/2 - x, 1/2 - y, z + 1/2)$
		$\pm(x + 1/2, 1/2 - y, \bar{z}); \pm(\bar{x}, y, 1/2 - z)$
		$x = 0.286, y = 0.376, z = 0.412$

Table 3B.31. *The structure of gamma lanthanum oxyfluoride, γ-LaOF.*

Formula unit	LaOF		
Space group:	P4/*nmm* (no. 129)		
Cell dimensions:	$a = 4.091$ Å, $c = 5.852$ Å		
Cell contents:	2 formula units per cell		
Atomic positions:	F in (2a)	(0, 0, 0);	(1/2, 1/2, 0)
	O in (2b)	(0, 0, 1/2)	(1/2, 1/2, 1/2)
	La in (2c)	(0, 1/2, z)	(1/2, 0, $\bar{z}$)
		$z = 0.778$	

Table 3B.32. *The structure of chalcopyrite, $CuFeS_2$.*

Formula unit:	$CuFeS_2$			
Space group:	I$\bar{4}$2d (no. 122)			
Cell dimensions:	$a = 5.24$ Å, $c = 10.30$ Å			
Cell contents:	4 formula units per cell			
Atomic positions:	Cu in (4a)	(0, 0, 0);	(0, 1/2, 1/4)	+I
	Fe in (4b)	(0, 0, 1/2);	(0, 1/2, 3/4)	+I
	S in (8d)	(x, 1/4, 1/8);	($\bar{x}$, 3/4, 1/8);	
		(3/4, x, 7/8);	(1/4, $\bar{x}$, 7/8)	+I
		$x = 0.25$		

Table 3B.33. *The structure of Cr_2Al.*

Formula unit:	Cr_2Al (C11$_b$)	
Space group:	I4/*mmm* (no. 139)	
Cell dimensions:	$a = 3.0045$ Å, $c = 8.6477$ Å	
Cell contents:	2 formula units per cell	
Atomic positions:	Cr in (4e)	(0, 0, z); (0, 0, $\bar{z}$)
		$z = 0.32$
	Al in (2a)	(0, 0, 0)

Table 3B.34. *The structure of Ga$_2$Zr.*

Formula unit:	Ga$_2$Zr			
Space group:	*Cmmm* (no. 65)			
Cell dimensions:	$a = 12.0944$ Å, $b = 3.9591$, $c = 4.0315$ Å			
Cell contents:	4 formula units per cell			
Atomic positions:				
	A/Zr in (4g)	$(x, 0, 0)$;	$(\bar{x}, 0, 0)$	$+C$
		$x = 0.351$		
	B/Ga(1) in (2a)	$(0, 0, 0)$		$+C$
	B/Ga(2) in (2c)	$(1/2, 0, 1/2)$		$+C$
	B/Ga(3) in (4h)	$(x, 0, 1/2)$;	$(\bar{x}, 0, 1/2)$	$+C$
		$x = 0.176$		

Table 3B.35. *The structure of graphite.*

Formula unit:	C, graphite
Space group:	P6$_3$*mc* (no. 186)
Cell dimensions:	$a = 2.456$ Å; $c = 6.696$ Å
Cell contents:	4 formula units
Atomic positions:	C(1) in (2a) $\quad$ $(0, 0, z)$; $(0, 0, z + 1/2)$; $z \approx 0$
	C(2) in (2b) $\quad$ $(1/3, 2/3, z)$; $(2/3, 1/3, z + 1/2)$; $z \approx 0$

Table 3B.36. *The structure of β-copper mercury tetraiodide, Cu$_2$HgI$_4$.*

Formula unit:	β-Cu$_2$HgI$_4$
Space group:	I$\bar{4}$2*m* (no. 121)
Cell dimensions:	$a = 6.078$ Å, $c = 12.254$ Å
Cell contents:	2 formula units per cell
Atomic positions:	Hg in (2a) $\quad$ $(0, 0, 0) + I$
	Cu in (4d) $\quad$ $(0,1/2,1/4)$; $(0,1/2,3/4)$; $+I$
	I in (8i) $\quad$ (x, x, z); $(\bar{x}, \bar{x}, z)$; $(x, \bar{x}, \bar{z})$; $(\bar{x}, x, \bar{z})$; $+I$
	$\quad\quad\quad\quad\quad$ $x = 0.27$; $z = 0.13$

Table 3B.37. *The C14 structure.*

Formula unit	$MgZn_2$		
Space group:	$P6_3 / mmc$ (no. 194)		
Cell dimensions:	$a = 5.18$ Å, $c = 8.52$ Å		
Cell contents:	4 formula units		
Atomic positions:	Mg in (4f)	$3m$	$(1/3, 2/3, z)$; $(2/3, 1/3, \bar{z})$;
			$(2/3, 1/3, 1/2 + z)$;
			$(1/3, 2/3, 1/2 - z)$
			$z = 0.062$
	Zn in (2a)	$\bar{3}$	$(0, 0, 0)$; $(0, 0, 1/2)$;
			$(x, 2x, 1/4)$; $(2\bar{x}, \bar{x}, 1/4)$; $(x, \bar{x}, 1/4)$
	Zn in (6h)	$mm2$	$(\bar{x}, 2\bar{x}, 3/4)$; $(2x, x, 3/4)$; $(\bar{x}, x, 3/4)$
			$x = 0.833$

Examples

compound	a (Å)	c (Å)	compound	a (Å)	c (Å)
$MgZn_2$	5.18	8.52	$CaCd_2$	5.993	9.654
$TiZn_2$	5.064	8.210	$CaMg_2$	6.2386	10.146
$TiFe_2$	4.785	7.799	$CdCu_2$	4.96	7.98
$ZrAl_2$	5.275	8.736	$TaFe_2$	4.816	7.868
$MoBe_2$	4.434	7.275	WFe_2	4.727	7.704
$MoFe_2$	4.73	7.72	$SmOs_2$	5.336	8.879

Table 3B.38. *The structure of potassium aluminum fluoride, $KAlF_4$.*

Formula unit:	$KAlF_4$	
Space group:	$P4/mmm$ (no. 123)	
Cell dimensions:	$a = 3.350$ Å, $c = 6.139$ Å	
Cell contents:	1 formula unit per cell	
Atomic positions:	K in (1a)	$(0, 0, 0)$
	Al in (1d)	$(1/2, 1/2, 1/2)$
	F(1) in (2e)	$(0, 1/2, 1/2)$; $(1/2, 0, 1/2)$
	F(2) in (2h)	$(1/2, 1/2, z)$; $(1/2, 1/2, \bar{z})$
		$z = 0.21$

Table 3B.38. (*cont.*)

Examples					
compound	a	c	compound	a	c
$KAlF_4$	3.350	6.139	$TlAlF_4$	3.61	6.37
NH_4AlF_4	3.587	6.346	NH_4ScF_4	4.06	6.67
$NaAlF_4$	3.48	6.29	NH_4GaF_4	3.71	6.39
$RbAlF_4$	3.615	6.37			

Table 3B.39. *The structure of* $TiAl_2$.

Formula unit:	$TiAl_2$
Space group:	*Cmmm* (no. 65)
Cell dimensions:	$a = 12.0944$ Å, $b = 3.9591$ Å, $c = 4.0315$ Å
Cell contents:	4 formula units per cell
Atomic positions:	Al(1) in (2a) $(0, 0, 0) + C$
	Al(2) in (2d) $(1/2, 0, 1/2) + C$
	Al(3) in (4h) $(x, 0, 1/2); (\bar{x}, 0, 1/2); + C$
	$x = 0.176$
	Ti in (4g) $(x, 0, 0); (\bar{x}, 0, 0); + C$
	$x = 0.351$

Table 3B.40. *The structure of potassium magnesium tetrafluoride,* K_2MgF_4.

Formula unit:	K_2MgF_4
Space group:	I4/*mmm* (no. 139)
Cell dimensions:	$a = 3.955$ Å, $c = 13.706$ Å
Cell contents:	2 formula units per cell
Atomic positions:	Mg in (2a) $(0, 0, 0) + I$
	K in (4e) $(0, 0, z); (0, 0, \bar{z}); + I$
	$z = 0.35$
	F(1) in (4c) $(0, 1/2, 0); (1/2, 0, 0); + I$
	F(2) in (4e) $(0, 0, z); (0, 0, \bar{z}); + I$
	$z = 0.15$

Table 3B.40. (*cont.*)

Examples of compounds with the R_2MX_4 structure

compound	a	c	$z(R)$	$z(X)$	compound	a	c	$z(R)$	$z(X)$
K_2CoF_4	4.074	13.08			K_2UO_4	4.335	13.10	0.36	0.145
K_2CuF_4	4.155	12.74	0.356	0.153	La_2NiO_4	3.855	12.652	0.360	0.170
K_2NiF_4	4.006	13.076	0.352	0.151	Nd_2CuO_4	3.94	12.15		
$(NH_3)_2NiF_4$	4.084	13.79			Nd_2NiO_4	3.81	12.31		
Rb_2CoF_4	4.135	13.67			Rb_2UO_4	4.345	13.83		
Rb_2NiF_4	1.087	13.71			Sm_2CuO_4	3.91	11.93		
Tl_2CoF_4	4.10	14.1			Sr_2IrO_4	3.89	12.92		
Tl_2NiF_4	4.054	14.22			$SrLaAlO_4$	3.75	12.5		
Cs_2CrCl_4	5.215	16.460			Sr_2MnO_4	3.79	12.43		
Ba_2PbO_4	4.305	13.273	0.355	0.155	Sr_2MoO_4	3.92	12.84		
Ba_2SnO_4	4.140	13.295	0.355	0.155	Sr_2RhO_4	3.85	12.90		
Ca_2MnO_4	3.67	12.08			Sr_2RuO_4	3.87	12.74		
Cs_2UO_4	4.38	14.79			Sr_2SnO_4	4.037	12.53	0.353	0.153
Gd_2CuO_4	3.89	11.85			Sr_2TiO_4	3.884	12.60	0.355	0.152

Table 3B.41. *The structure of zircon, $ZrSiO_4$.*

Formula unit:	$ZrSiO_4$	
Space group:	$I4_1/amd$ (no. 141)	
Cell dimensions:	$a = 6.6164$ Å, $c = 6.0150$ Å	
Cell contents:	4 formula units per cell	
Atomic positions:	Zr in (4a)	$(0,0,0); (0,1/2,1/4); +I$
	Si in (4b)	$(0,0,1/2); (0,1/2,3/4); +I$
	O in (16h)	$(0, y, z); (1/2, \bar{y}+1/2, z+1/2);$
		$(\bar{y}, 1/2, z+1/4); (y+1/2, 0, z+3/4);$
		$(1/2, y, \bar{z}+3/4); (0, \bar{y}+1/2, \bar{z}+1/4);$
		$(y+1/2, 1/2, \bar{z}+1/2); (\bar{y}, 0, \bar{z}); +I$
		$y = 0.20; z = 0.34$

Table 3B.41. (*cont.*)

Examples					
compound	*a*	*c*	compound	*a*	*c*
DyAsO$_4$	7.0733	6.3133	YbPO$_4$	6.824	5.98
ErAsO$_4$	7.0203	6.2761	BiVO$_4$	7.2999	6.4573
EuAsO$_4$	7.1541	6.3953	CeVO$_4$	7.399	6.496
GdAsO$_4$	7.1326	6.3578	DyVO$_4$	7.1434	6.313
HoAsO$_4$	7.0548	6.3159	ErVO$_4$	7.100	6.279
LuAsO$_4$	6.952	6.230	EuVO$_4$	7.2365	6.3675
ScAsO$_4$	6.7101	6.1126	GdVO$_4$	7.211	6.350
SmAsO$_4$	7.1865	6.3999	LuVO$_4$	7.026	6.231
TbAsO$_4$	7.1025	6.3536	NdVO$_4$	7.3290	6.4356
TlAsO$_4$	6.9939	6.2595	PrVO$_4$	7.367	6.468
TmAsO$_4$	7.000	6.256	SmVO$_4$	7.266	6.394
YbAsO$_4$	6.9716	6.2437	TbVO$_4$	7.179	6.324
DyPO$_4$	6.917	6.053	TmVO$_4$	7.071	6.263
ErPO$_4$	6.863	6.007	YVO$_4$	7.123	6.291
HoPO$_4$	6.891	6.031	YbVO$_4$	7.043	6.248
LuPO$_4$	6.798	5.961	ThGeO$_4$	7.238	6.520
TbPO$_4$	6.941	6.070	ThSiO$_4$	7.104	6.296
TmPO$_4$	6.847	5.994			

Table 3B.42. *The structure of β-gold cadmium, β-AuCd.*

Formula unit:	β-AuCd (low temperature phase)
Space group:	P*mma* (no. 51)
Cell dimensions:	$a = 4.7654$ Å, $b = 3.1540$ Å, $c = 4.8644$ Å
Cell contents:	2 formula units per cell
Atomic positions:	Cd in (2e) (1/4, 0, z); (3/4, 0, $\bar{z}$)
	$z = 0.312$
	Au in (2f) (1/4, 1/2, z); (3/4, 1/2, $\bar{z}$)
	$z = 0.812$

Table 3B.43. *The structure of anatase, TiO_2.*

Formula unit:	anatase, TiO_2	
Space group:	$I4_1/amd$ (no. 141)	
Cell dimensions:	$a = 3.785$ Å, $c = 9.514$ Å	
Cell contents:	4 formula units	
Atomic positions:	Ti in (4a)	$(0, 0, 0)$; $(0, 1/2, 1/4)$; $+I$
	O in (8e)	$(0, 0, z)$; $(0, 0, \bar{z})$;
		$(0, 1/2, z + 1/4)$; $(0, 1/2, 1/4 - z)$; $+I$
		$z = 0.2$

References

[1] R.W.G. Wyckoff, *Crystal Structures* Volume 1 (John Wiley & Sons, New York, 1964). Tridymite, p. 315, α-quartz, p. 312. The structure of SiO_2.

[2] R.W.G. Wyckoff, *Crystal Structures* Volume 1 (John Wiley & Sons, New York, 1964) p. 81. The structure of MoO_3.

[3] R.D. Shannon and C.T. Prewitt, Synthesis and Structure of a New High-Pressure Form of Rh_2O_3, *J. Solid State Chemistry* **2** (1970) 134–6. The structure of Rh_2O_3 II.

[4] R.W.G. Wyckoff, *Crystal Structures* Volume 1 (John Wiley & Sons, New York, 1964) p. 122. The structure of Millerite (NiS).

[5] R.W.G. Wyckoff, *Crystal Structures* Volume 2 (John Wiley & Sons, New York, 1964) p. 185. The structure of V_2O_5 (vanadium pentoxide).

[6] J.T. Vaughey and K.R. Poeppelmeier, Structural Diversity in Oxygen Deficient Perovskites, *NIST Special Publication 804, Chemistry of Electronic Ceramic Materials,* Proceedings of the International Conference, Jackson, WY, Aug. 17–22, 1990 (1991). The crystal structure of $YBaCuFeO_5$.

[7] J.D. Jorgensen, B.W. Veal, A.P. Paulikas, L.J. Nowicki, G.W. Crabtree, H. Claus, and W.K. Kwok, *Phys. Rev. B* **41** (1990) 1863. The crystal structure of $YBa_2Cu_3O_7$.

[8] R.W.G. Wyckoff, *Crystal Structures* Volume 1 (John Wiley & Sons, New York, 1964) pp. 127–8. The structure of MnP.

[9] R.W.G. Wyckoff, *Crystal Structures* Volume 1 (John Wiley & Sons, New York, 1964) pp. 127–8. The structure of FeB.

[10] R.W.G. Wyckoff, *Crystal Structures* Volume 1 (John Wiley & Sons, New York, 1964) pp. 25–6. The structure of diamond.

[11] M. Marezio and J.P. Remeika, *J. Chem. Phys.* **44** (1966) 3348. The structure of $LiGaO_2$.

[12] W.H. Baur, *Mat. Res. Bull.* **16** (1981) 339. The tetragonal close packed arrangement.

[13] R.W.G. Wyckoff, *Crystal Structures* Volume 1 (John Wiley & Sons, New York, 1964) p. 130. The structure of CrB.

Appendix 5A
Introduction to Fourier series

It is convenient to write the electron density as a Fourier series. For more details about Fourier series [1] and its relevance to diffraction [2–5], several sources are cited at the conclusion of this appendix.

Most periodic functions can be represented as a sum of trigonometric functions of increasing frequency. For example:

$$f(x) = \frac{a_0}{2} + \sum_{n=1}^{\infty} (a_n \cos nx + b_n \sin nx). \tag{5A.1}$$

Using this form, a function can be completely specified by knowing the coefficients a_n and b_n. If $f(x)$ is known, the coefficients are computed in the following way:

$$a_0 = \frac{1}{\pi} \int_{-\pi}^{\pi} f(x) dx$$

$$a_n = \frac{1}{\pi} \int_{-\pi}^{\pi} f(x) \cos nx \, dx \quad (n \geq 0)$$

$$b_n = \frac{1}{\pi} \int_{-\pi}^{\pi} f(x) \sin nx \, dx \quad (n > 0). \tag{5A.2}$$

As an example, we examine the Fourier representation of a square wave:

$$f(x) = \begin{cases} -1 & (x < 0) \\ +1 & (x \geq 0) \end{cases}. \tag{5A.3}$$

Substitution into Eqn. 5A.2 and integration yields the result that all $a_n = 0$ and

$$b_n = \begin{cases} 4/n\pi & \text{if } n \text{ odd} \\ 0 & \text{if } n \text{ even} \end{cases}. \tag{5A.4}$$

So, the Fourier series is:

$$g(x) = \frac{4}{\pi} \sum_{n=\text{odd}}^{\infty} \frac{1}{n} \sin x$$

$$g(x) = \frac{4}{\pi} \sin x + \frac{4}{3\pi} \sin 3x + \frac{4}{5\pi} \sin 5x + \frac{4}{7\pi} \sin 7x + \dots \tag{5A.5}$$

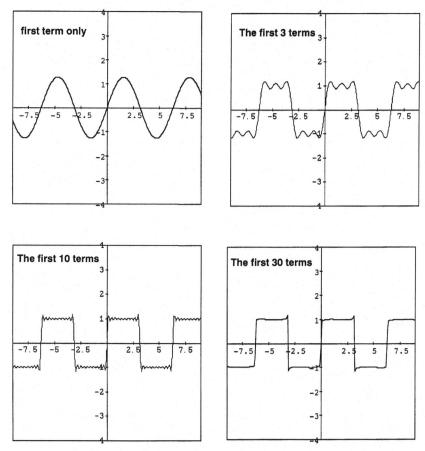

Figure 5A.1. Adding additional terms to the Fourier representation of the square wave improves the approximation.

In practice, it is impossible to deal with an infinite number of terms. However, as more terms are used, $g(x)$ more nearly approximates $f(x)$. This is illustrated in Fig. 5A.1.

It turns out that a more compact exponential series is usually more useful.

$$f(x) = \sum_{h=-\infty}^{\infty} S_n e^{i(2\pi n\, x/a)} \qquad (5A.6)$$

where $0 \le x \le a$ and the coefficients are:

$$S_n = \frac{1}{a} \int_0^a f(x) e^{i(2\pi n\, x/a)} dx. \qquad (5A.7)$$

In this case, a is the periodicity of the function.

References

[1] E. Butkov, *Mathematical Physics* (Addison-Wesley, Reading, Mass, 1968) Chapter 4.

[2] C. Kittel, *Introduction to Solid State Physics*, 5th edition (J. Wiley & Sons, New York, 1976) pp. 45–6.

[3] B.E. Warren, *X-ray Diffraction* (Addison-Wesley, Reading, Mass, 1969) Chapter 9.

[4] D.E. Sands, *Introduction to Crystallography* (W.A. Benjamin, New York, 1969) pp. 100–3.

[5] M.J. Burger, *Contemporary Crystallography* (McGraw-Hill, New York, 1970) pp. 229–45.

Appendix 5B
Coefficients for atomic scattering factors

Table 5B.1. *The coefficients for Eqn. 5.55.*

atom	Z	a_1	a_2	a_3	a_4	b_1	b_2	b_3	b_4
Ac	89	006.278	028.323	005.195	004.949	002.321	000.557	000.000	000.000
Ag	47	002.036	061.497	003.272	011.824	002.511	002.846	000.837	000.327
Al	13	002.276	072.322	002.428	019.773	000.858	003.080	000.317	000.408
Am	95	006.378	029.156	005.495	005.102	002.495	000.565	000.000	000.000
Ar	18	001.274	026.682	002.190	008.813	000.793	002.219	000.326	000.307
As	33	002.399	045.718	002.790	012.817	001.529	002.280	000.594	000.328
At	85	006.133	028.047	005.031	004.957	002.239	000.558	000.000	000.000
Au	79	002.388	042.866	004.226	009.743	002.689	002.264	001.255	000.307
B	5	000.945	046.444	001.312	014.178	000.419	003.223	000.116	000.377
Ba	56	007.821	117.657	006.004	018.778	003.280	003.263	001.103	000.376
Be	4	001.250	060.804	001.334	018.591	000.360	003.653	000.106	000.416
Bi	83	003.841	050.261	004.679	011.999	003.192	002.560	001.363	000.318
Bk	97	006.502	028.375	005.478	004.975	002.510	000.561	000.000	000.000
Br	35	002.166	033.899	002.904	010.497	001.395	002.041	000.589	000.307
C	6	000.731	036.995	001.195	011.297	000.456	002.814	000.125	000.346
Ca	20	004.470	099.523	002.971	022.696	001.970	004.195	000.482	000.417
Cd	48	002.574	055.675	003.259	011.838	002.547	002.784	000.838	000.322
Ce	58	005.007	028.283	003.980	005.183	001.678	000.589	000.000	000.000
Cf	98	006.548	028.461	005.526	004.965	002.520	000.557	000.000	000.000
Cl	17	001.452	030.935	002.292	009.980	000.787	002.234	000.322	000.323
Cm	96	006.460	028.396	005.469	004.970	002.471	000.554	000.000	000.000
Co	27	002.367	061.431	002.236	014.180	001.724	002.725	000.515	000.344
Cr	24	002.307	078.405	002.334	015.785	001.823	003.157	000.490	000.364
Cs	55	006.062	155.837	005.986	019.695	003.303	003.335	001.096	000.379
Cu	29	001.579	062.940	001.820	012.453	001.658	002.504	000.532	000.333
Dy	66	005.332	028.888	004.370	005.198	001.863	000.581	000.000	000.000
Er	68	005.436	028.655	004.437	005.117	001.891	000.577	000.000	000.000
Eu	63	006.267	100.298	004.844	016.066	003.202	002.980	001.200	000.367
F	9	000.387	020.239	000.811	006.609	000.475	001.931	000.146	000.279
Fe	26	002.544	064.424	002.343	014.880	001.759	002.854	000.506	000.350

Table 5B.1 (*cont.*)

atom	Z	a_1	a_2	a_3	a_4	b_1	b_2	b_3	b_4
Fr	87	006.201	028.200	005.121	004.954	002.275	000.556	000.000	000.000
Ga	31	002.321	065.602	002.486	015.458	001.688	002.581	000.599	000.351
Gd	64	005.225	029.158	004.314	005.259	001.827	000.586	000.000	000.000
Ge	32	002.447	055.893	002.702	014.393	001.616	002.446	000.601	000.342
H	1	000.202	030.868	000.244	008.544	000.082	001.273	000.000	000.000
He	2	000.091	018.183	000.181	006.212	000.110	001.803	000.036	000.284
Hf	72	005.588	029.001	004.619	005.164	001.997	000.579	000.000	000.000
Hg	80	002.682	042.822	004.241	009.856	002.755	002.295	001.270	000.307
Ho	67	005.376	028.773	004.403	005.174	001.884	000.582	000.000	000.000
I	53	003.473	039.441	004.060	011.816	002.522	002.415	000.840	000.298
In	49	003.153	066.649	003.557	014.449	002.818	002.976	000.884	000.335
Ir	77	005.754	029.159	004.851	005.152	002.096	000.570	000.000	000.000
K	19	003.951	137.075	002.545	022.402	001.980	004.532	000.482	000.434
Kr	36	002.034	029.999	002.927	009.598	001.342	001.952	000.589	000.299
La	57	004.940	028.716	003.968	005.245	001.663	000.594	000.000	000.000
Li	3	001.611	107.638	001.246	030.480	000.326	004.533	000.099	000.495
Lu	71	005.553	028.907	004.580	005.160	001.969	000.577	000.000	000.000
Mg	12	002.268	073.670	001.803	020.175	000.839	003.013	000.289	000.405
Mn	25	002.747	067.786	002.456	015.674	001.792	003.000	000.498	000.357
Mo	42	003.120	072.464	003.906	014.642	002.361	003.237	000.850	000.366
N	7	000.572	028.847	001.043	009.054	000.465	002.421	000.131	000.317
Na	11	002.241	108.004	001.333	024.505	000.907	003.391	000.286	000.435
Nb	41	004.237	027.415	003.105	005.074	001.234	000.593	000.000	000.000
Nd	60	005.151	028.304	004.075	005.073	001.683	000.571	000.000	000.000
Ne	10	000.303	017.640	000.720	005.860	000.475	001.762	000.153	000.266
Ni	28	002.210	058.727	002.134	013.553	001.689	002.609	000.524	000.339
Np	93	006.323	029.142	005.414	005.096	002.453	000.568	000.000	000.000
O	8	000.455	023.780	000.917	007.622	000.472	002.144	000.138	000.296
Os	76	005.750	028.933	004.773	005.139	002.079	000.573	000.000	000.000
P	15	001.888	044.876	002.469	013.538	000.805	002.642	000.320	000.361
Pa	91	006.306	028.688	005.303	005.026	002.386	000.561	000.000	000.000
Pb	82	003.510	052.914	004.552	011.884	003.154	002.571	001.359	000.321
Pd	46	004.436	028.670	003.454	005.269	001.383	000.595	000.000	000.000
Pm	61	005.201	028.079	004.094	005.081	001.719	000.576	000.000	000.000
Po	84	006.070	028.075	004.997	004.999	002.232	000.563	000.000	000.000
Pr	59	005.085	028.588	004.043	005.143	001.684	000.581	000.000	000.000
Pt	78	005.803	029.016	004.870	005.150	002.127	000.572	000.000	000.000

Table 5B.1 (*cont.*)

atom	Z	a_1	a_2	a_3	a_4	b_1	b_2	b_3	b_4
Pu	94	006.415	028.836	005.419	005.022	002.449	000.561	000.000	000.000
Ra	88	006.215	028.382	005.170	005.002	002.316	000.562	000.000	000.000
Rb	37	004.776	140.782	003.859	018.991	002.234	003.701	000.868	000.419
Re	75	005.695	028.968	004.740	005.156	002.064	000.575	000.000	000.000
Rh	45	004.431	027.911	003.343	005.153	001.345	000.592	000.000	000.000
Rn	86	004.078	038.406	004.978	011.020	003.096	002.355	001.326	000.299
Ru	44	004.358	027.881	003.298	005.179	001.323	000.594	000.000	000.000
S	16	001.659	036.650	002.386	011.488	000.790	002.469	000.321	000.340
Sb	51	003.564	050.487	003.844	013.316	002.687	002.691	000.864	000.316
Sc	21	003.966	088.960	002.917	020.606	001.925	003.856	000.480	000.399
Se	34	002.298	038.830	002.854	011.536	001.456	002.146	000.590	000.316
Si	14	002.129	057.775	002.533	016.476	000.835	002.880	000.322	000.386
Sm	62	005.255	028.016	004.113	005.037	001.743	000.577	000.000	000.000
Sn	50	003.450	059.104	003.735	014.179	002.118	002.855	000.877	000.327
Sr	38	005.848	104.972	004.003	019.367	002.342	003.737	000.880	000.414
Ta	73	005.659	028.807	004.630	005.114	002.014	000.578	000.000	000.000
Tb	65	005.272	029.046	004.347	005.226	001.844	000.585	000.000	000.000
Tc	43	004.318	028.246	003.270	005.148	001.287	000.590	000.000	000.000
Te	52	004.785	027.999	003.688	005.083	001.500	000.581	000.000	000.000
Th	90	006.264	028.651	005.263	005.030	002.367	000.563	000.000	000.000
Ti	22	003.565	081.982	002.818	019.049	001.893	003.590	000.483	000.386
Tl	81	005.932	029.086	004.972	005.126	002.195	000.572	000.000	000.000
Tm	69	005.441	029.149	004.510	005.264	001.956	000.590	000.000	000.000
U	92	006.767	085.951	006.729	015.642	004.014	002.936	001.561	000.335
V	23	003.245	076.379	002.698	017.726	001.860	003.363	000.486	000.374
W	74	005.709	028.782	004.677	005.084	002.019	000.572	000.000	000.000
Xe	54	003.366	035.509	004.147	011.117	002.443	002.294	000.829	000.289
Y	39	004.129	027.548	003.012	005.088	001.179	000.591	000.000	000.000
Yb	70	005.529	028.927	004.533	005.144	001.945	000.578	000.000	000.000
Zn	30	001.942	054.162	001.950	012.518	001.619	002.416	000.543	000.330
Zr	40	004.105	028.492	003.144	005.277	001.229	000.601	000.000	000.000

Reference

[1] M. DeGraef and M. McHenry, *Crystallography, Symmetry, and Diffraction*, to be published by Cambridge University Press, 2002.

Appendix 7A
Evaluation of the Madelung constant

Evaluation of the Madelung constant amounts to computing the sum:

$$A = \frac{1}{2} \sum_{i=1}^{n} \left[\sum_{\substack{j=1 \\ j \neq i}}^{n} \frac{Z_i Z_j}{\rho(\vec{r}_{ij})} + \sum_{k=1}^{N_L-1} \left[\sum_{j=1}^{n} \frac{Z_i Z_j}{\rho(\vec{R}_k + \vec{r}_{ij})} \right] \right]. \tag{7A.1}$$

Attempting this computation on even a simple structure, such as rock salt, is instructive. The first thing one discovers is that because of the alternating charges, the sum does not converge rapidly. One trick that can be used to speed convergence is to sum over concentric shells bounded by charge neutral planes. However, this method is only easily applied to rather simple, high symmetry structures. Since the Madelung constants for these structures are already known, the method has only historical importance. More useful for realistic computations is the Ewald method, where the summation is partially carried out over the reciprocal lattice.

The Ewald method involves envisioning the lattice of discrete point charges as a superposition of two spatially continuous charge distributions. Briefly summarizing Tosi's full explanation of the method [1], the first distribution is a periodic lattice of positive Gaussian distributions neutralized by a constant homogeneous negative charge. The second is a periodic lattice of positive point charges neutralized by negative Gaussian distributions centered on the lattice points. Splitting the charge in this way allows one to write the slowly and conditionally converging Coulombic energy as the total of two quickly converging sums, one carried out over the first charge distribution in reciprocal space via Fourier transformation and the other carried out over the localized second distribution in real space.

For the mathematical representation of this method, we summarize an equivalent derivation by Jackson and Catlow [2]. The function $1/r$ in the Coulomb energy is rewritten as:

$$\frac{1}{r} = \frac{2}{\sqrt{\pi}} \int_0^{\infty} e^{-r^2 t^2} \, dt = \frac{2}{\sqrt{\pi}} \left[\int_0^{\alpha} e^{-r^2 t^2} \, dt + \int_{\alpha}^{\infty} e^{-r^2 t^2} \, dt \right]. \tag{7A.2}$$

Following [2], the second part of the total integral from α to infinity can be evaluated by utilizing the variable change $s = rt$:

$$\frac{2}{\sqrt{\pi}} \int_{\alpha}^{\infty} e^{-r^2 t^2} \, dt = \frac{\text{erfc}(\alpha r)}{r}. \tag{7A.3}$$

Unlike the simple original $1/r$ function, this function decreases rapidly with r for large values of α and can be incorporated directly into the original total energy sum with a short-range spatial cut-off determined by the desired accuracy for the calculation.

To complete the evaluation of $1/r$, the first part of the total integral above from 0 to α is carried out by a Fourier transformation and the application of the variable change $s = -K^2/(4t^2)$:

$$\frac{2}{\sqrt{\pi}} \int_{0}^{\alpha} e^{-r^2 t^2} \, dt = \frac{1}{2\pi^2} \int_{-\infty}^{\infty} d^3 K \frac{\exp(-K^2/4\alpha^2)}{K^2} \exp(-i\vec{K}\bullet\vec{r}). \tag{7A.4}$$

The vectors $\vec{K}$ in the above equation are general reciprocal lattice wave vectors comprised of sums of the primitive reciprocal lattice vectors defined in Chapter 2 scaled by $1/N^{1/3}$. N is equal to the number of unit cells in the crystal.

$$\vec{K} = n_1 \vec{k}_1 + n_2 \vec{k}_2 + n_3 \vec{k}_3$$
$$\vec{k}_1 = \vec{a}*/N^{1/3}, \ \vec{k}_2 = \vec{b}*/N^{1/3}, \ \vec{k}_3 = \vec{c}*/N^{1/3}. \tag{7A.5}$$

(Note that the vectors $\vec{k}_i$ in [2] are our vectors $\vec{k}_i/2\pi$.)

Converting the reciprocal space integral into a discrete sum, the electrostatic energy can now be written as:

$$N_L \kappa c^2 \frac{1}{2} \left[\sum_{i}^{n} \sum_{l=0}^{N_L-1} \sum_{\substack{j \\ j\neq i \text{ for } l=0}}^{n} Z_i Z_j \frac{\text{erfc}(\alpha|\vec{R}_l + \vec{r}_i - \vec{r}_j|)}{|\vec{R}_l + \vec{r}_i - \vec{r}_j|} + \right.$$
$$\left. \frac{1}{2\pi^2 N v_c} \sum_{K} \frac{\exp(-K^2/4\alpha^2)}{K^2} \sum_{i}^{n} \sum_{l=0}^{N_L-1} \sum_{\substack{j \\ j\neq i \text{ for } l=0}}^{n} Z_i Z_j \exp(-i\vec{K}\bullet\vec{R}_l) \exp(-i\vec{K}\bullet(\vec{r}_i - \vec{r}_j)) \right]$$

$$\tag{7A.6}$$

where v_c is the volume per unit cell. The $l=0$ lattice site is the chosen origin; thus, the distance $R_0 = 0$. Continuing to follow [2], summing over all $\vec{K}$ vectors and lattice sites will result in a cancellation of all $\exp(-i\vec{K}\bullet\vec{R}_l)$ terms where $\vec{K}$ is not equal to $\vec{G}$. The remaining terms will sum to N_L due to the relationship between $\vec{G}$ and R_l. Now Eqn. 7A.6 can be written in final form as:

$$
N_{\mathrm{L}}\kappa e^2 \frac{1}{2}
\begin{bmatrix}
\sum_i^n \sum_{l=0}^{N_{\mathrm{L,max}}} \sum_{\substack{j \\ j \neq i \text{ for } l=0}}^n Z_i Z_j \dfrac{\mathrm{erfc}(\alpha|\vec{R}_l + \vec{r}_i - \vec{r}_j|)}{|\vec{R}_l + \vec{r}_i - \vec{r}_j|} + \\[2ex]
\dfrac{N_{\mathrm{L,u}}}{2\pi^2 v_{\mathrm{c}}} \sum_G^{G_{\max}} \dfrac{\exp(-G^2/4\alpha^2)}{G^2} \sum_i^n \left[\left[\sum_j^n Z_i Z_j \exp\left(-i\vec{G}\cdot(\vec{r}_i - \vec{r}_j)\right) \right] - \dfrac{2\alpha Z_i Z_i}{\sqrt{\pi}} \right]
\end{bmatrix}
$$

$$(7\text{A}.7)$$

where $N_{\mathrm{L,u}}$ is the number of lattice sites per unit cell and the subtracted term explicitly removes all self interactions now included in the reciprocal space sum due to the lattice site sum simplification.

The values of α and the cut-off values in real and reciprocal space, $N_{\mathrm{L,max}}$ and $G_{\max}$, are chosen such that terms in the respective sums beyond $N_{\mathrm{L,max}}$ and $G_{\max}$ are less than some desired accuracy, a, (10^{-4} is more than sufficient) and the number of terms in both the real and reciprocal space sums is minimized. The resulting values are:

$$\alpha = \left(\frac{s}{8v_{\mathrm{c}}^2}\right)^{1/6} \tag{7A.8}$$

$$r_{\max} = \frac{(-\ln a)^{1/2}}{\alpha} \tag{7A.9}$$

$$N_{\mathrm{L,max}} = \frac{4\pi\, r_{\max}^3}{3} \frac{s}{v_{\mathrm{c}}} = \left(\frac{8\pi\sqrt{2}s}{3}\right)(-\ln(a))^{3/2} \tag{7A.10}$$

$$G_{\max} = 2\alpha(-\ln a)^{1/2} \tag{7A.11}$$

where s is the number of ions per unit cell. A thorough explanation can be found in [2].

References

[1] M.P. Tosi, *Solid State Physics*, **16** (1964) 1. The Madelung constant.
[2] R.A. Jackson and C.R.A. Catlow, *Molecular Simulation*, **1** (1988) 207. The Madelung constant.

Appendix 7B
Ionic radii for halides and chalcogenides

Table 7B.1. *Ionic radii for halides and chalcogenides*

ion	e.c.	CN	r	ion	e.c.	CN	r	ion	e.c.	CN	r
Ac+3	6p 6	6	1.12	B+3	1s 2	3	0.01			8	1.12
Ag+1	4d 10	2	0.67			4	0.11			9	1.18
		4	1.00			6	0.27			10	1.23
		4 sq	1.02	Ba+2	5p 6	6	1.35			12	1.34
		5	1.09			7	1.38	Cd+2	4d 10	4	0.78
		6	1.15			8	1.42			5	0.87
		7	1.22			9	1.47			6	0.95
		8	1.28			10	1.52			7	1.03
Ag+2	4d 9	4 sq	0.79			11	1.57			8	1.10
		6	0.94			12	1.61			12	1.31
Ag+3	4d 8	4 sq	0.67	Be+2	1s 2	3	0.16	Ce+3	6s 1	6	1.01
		6	0.75			4	0.27			7	1.07
Al+3	2p 6	4	0.39			6	0.45			8	1.143
		5	0.48	Bi+3	6s 2	5	0.96			9	1.196
		6	0.535			6	1.03			10	1.25
Am+2	5f 7	7	1.21			8	1.17			12	1.34
		8	1.26	Bi+5	5d 10	6	0.76	Ce+4	5p 6	6	0.87
		9	1.31	Bk+3	5f 8	6	0.96			8	0.97
Am+3	5f 6	6	0.975	Bk+4	5f 7	6	0.83			10	1.07
		8	1.09			8	0.93			12	1.14
Am+4	5f 5	6	0.85	Br−1	4p 6	6	1.96	Cf+3	6d 1	6	0.95
		8	0.95	Br+3	4p 2	4 sq	0.59	Cf+4	5f 8	6	0.821
As+3	4s 2	6	0.58	Br+5	4s 2	3 py	0.31			8	0.92
As+5	3d 10	4	0.335	Br+7	3d 10	4	0.25	Cl−1	3p 6	6	1.81
		6	0.46			6	0.39	Cl+5	3s 2	3 py	0.12
At+7	5d 10	6	0.62	C+4	1s 2	3	−0.08	Cl+7	2p 6	4	0.08
Au+1	5d 10	6	1.37			4	0.15			6	0.27
Au+3	5d 8	4 sq	0.68			6	0.16	Cm+3	5f 7	6	0.97
		6	0.85	Ca+2	3p 6	6	1.00	Cm+4	5f 6	6	0.85
Au+5	5d 6	6	0.57			7	1.06			8	0.95

Table 7B.1. (*cont.*)

ion	e.c.	CN	r	ion	e.c.	CN	r	ion	e.c.	CN	r
Co+2	3d 7	4 HS	0.58	Dy+3	4f 9	6	0.912			6	0.620
		5	0.67			7	0.97	Gd+3	4f 7	6	0.938
		6 LS	0.65			8	1.027			7	1.00
		6 HS	0.745			9	1.083			8	1.053
		8	0.90	Er+3	4f 11	6	0.89			9	1.107
Co+3	3d 6	6 LS	0.545			7	0.945	Ge+2	4s 2	6	0.73
		6 HS	0.61			8	1.004	Ge+4	3d 10	4	0.390
Co+4	3d 5	4	0.40			9	1.062			6	0.530
		6 HS	0.53	Eu+2	4f 7	6	1.17	H+1	1s 0	1	−0.38
Cr+2	3d 4	6 LS	0.73			7	1.20			2	−0.18
		6 HS	0.80			8	1.25	Hf+4	4f 14	4	0.58
Cr+3	3d 3	6	0.615			9	1.30			6	0.71
Cr+4	3d 2	4	0.41			10	1.35			7	0.76
		6	0.55	Eu+3	4f 6	6	0.947			8	0.83
Cr+5	3d 1	4	0.345			7	1.01	Hg+1	6s 1	3	0.97
		6	0.49			8	1.066			6	1.19
		8	0.57			9	1.120	Hg+2	5d 10	2	0.69
Cr+6	3p 6	4	0.26	F−1	2p 6	2	1.285			4	0.96
		6	0.44			3	1.30			6	1.02
Cs+1	5p 6	6	1.67			4	1.31			8	1.14
		8	1.74			6	1.33	Ho+3	4f 10	6	0.901
		9	1.78	F+7	1s 2	6	0.08			8	1.015
		10	1.81	Fe+2	3d 6	4 HS	0.63			9	1.072
		11	1.85			4sqHS	0.64			10	1.12
		12	1.88			6 LS	0.61	I−1	5p 6	6	2.20
Cu+1	3d 10	2	0.46			6 HS	0.78	I+5	5s 2	3 py	0.44
		4	0.60			8 HS	0.92			6	0.95
		6	0.77	Fe+3	3d 5	4 HS	0.49	I+7	4d 10	4	0.42
Cu+2	3d 9	4	0.57			5	0.58			6	0.53
		4 sq	0.57			6 LS	0.55	In+3	4d 10	4	0.62
		5	0.65			6 HS	0.645			6	0.800
		6	0.73			8 HS	0.78			8	0.92
Cu+3	3d 8	6 LS	0.54	Fe+4	3d 4	6	0.585	Ir+3	5d 6	6	0.68
D+1	1s 0	2	−0.10	Fe+6	3d 2	4	0.25	Ir+4	5d 5	6	0.625
Dy+2	4f 10	6	1.07	Fr+1	6p 6	6	1.80	Ir+5	5d 4	6	0.57
		7	1.13	Ga+3	3d 10	4	0.47	K+1	3p 6	4	1.37
		8	1.19			5	0.55			6	1.38

Table 7B.1. (*cont.*)

ion	e.c.	CN	r	ion	e.c.	CN	r	ion	e.c.	CN	r
		7	1.46	Mo+4	4d 2	6	0.650	Ni+4	3d 6	6 LS	0.48
		8	1.51	Mo+5	4d 1	4	0.46	No+2	5f 14	6	1.10
		9	1.55			6	0.61	Np+2	5f 5	6	1.10
		10	1.59	Mo+6	4p 6	4	0.41	Np+3	5f 4	6	1.01
		12	1.64			5	0.50	Np+4	5f 3	6	0.87
La+3	4d 10	6	1.032			6	0.59			8	0.98
		7	1.10			7	0.73	Np+5	5f 2	6	0.75
		8	1.160	N−3	2p 6	4	1.46	Np+6	5f 1	6	0.72
		9	1.216	N+3	2s 2	6	0.16	Np+7	6p 6	6	0.71
		10	1.27	N+5	1s 2	3	−0.104	O−2	2p 6	2	1.35
		12	1.36			6	0.13			3	1.36
Li+1	1s 2	4	0.590	Na+1	2p 6	4	0.99			4	1.38
		6	0.760			5	1.00			6	1.40
		8	0.92			6	1.02			8	1.42
Lu+3	4f 14	6	0.861			7	1.12	OH−		2	1.32
		8	0.977			8	1.18			3	1.34
		9	1.032			9	1.24			4	1.35
Mg+2	2p 6	4	0.57			12	1.39			6	1.37
		5	0.66	Nb+3	4d 2	6	0.72	Os+4	5d 4	6	0.630
		6	0.720	Nb+4	4d 1	6	0.68	Os+5	5d 3	6	0.575
		8	0.89			8	0.79	Os+6	5d 2	5	0.49
Mn+2	3d 5	4 HS	0.66	Nb+5	4p 6	4	0.48			6	0.545
		5 HS	0.75			6	0.64	Os+7	5d 1	6	0.525
		6 LS	0.67			7	0.69	Os+8	5p 6	4	0.39
		6 HS	0.830			8	0.74	P+3	3s 2	6	0.44
		7 HS	0.90	Nd+2	4f 4	8	1.29	P+5	2p 6	4	0.17
		8	0.96			9	1.35			5	0.29
Mn+3	3d 4	5	0.58	Nd+3	4f 3	6	0.983			6	0.38
		6 LS	0.58			8	1.109	Pa+3	5f 2	6	1.04
		6 HS	0.645			9	1.163	Pa+4	6d 1	6	0.90
Mn+4	3d 3	4	0.39			12	1.27			8	1.01
		6	0.530	Ni+2	3d 8	4	0.55	Pa+5	6p 6	6	0.78
Mn+5	3d 2	4	0.33			4 sq	0.49			8	0.91
Mn+6	3d 1	4	0.255			5	0.63			9	0.95
Mn+7	3p 6	4	0.25			6	0.690	Pb+2	6s 2	4 py	0.98
		6	0.46	Ni+3	3d 7	6 LS	0.56			6	1.19
Mo+3	4d 3	6	0.69			6 HS	0.60			7	1.23

Table 7B.1. (*cont.*)

ion	e.c.	CN	r	ion	e.c.	CN	r	ion	e.c.	CN	r
		8	1.29			7	1.56			8	1.27
		9	1.35			8	1.61			9	1.32
		10	1.40			9	1.63	Sm+3	4f 5	6	0.958
		11	1.45			10	1.66			7	1.02
		12	1.49			11	1.69			8	1.079
Pb+4	5d 10	4	0.65			12	1.72			9	1.132
		5	0.73			14	1.83			12	1.24
		6	0.775	Re+4	5d 3	6	0.63	Sn+4	4d 10	4	0.55
		8	0.94	Re+5	5d 2	6	0.58			5	0.62
Pd+1	4d 9	2	0.59	Re+6	5d 1	6	0.55			6	0.690
Pd+2	4d 8	4 sq	0.64	Re+7	5p 6	4	0.38			7	0.75
		6	0.86			6	0.53			8	0.81
Pd+3	4d 7	6	0.76	Rh+3	4d 6	6	0.665	Sr+2	4p 6	6	1.18
Pd+4	4d 6	6	0.651	Rh+4	4d 5	6	0.60			7	1.21
Pm+3	4f 4	6	0.97	Rh+5	4d 4	6	0.55			8	1.26
		8	1.093	Ru+3	4d 5	6	0.68			9	1.31
		9	1.143	Ru+4	4d 4	6	0.620			10	1.36
Po+4	6s 2	6	0.94	Ru+5	4d 3	6	0.565			12	1.44
		8	1.08	Ru+7	4d 1	4	0.38	Ta+3	5d 2	6	0.72
Po+6	5d 10	6	0.67	Ru+8	4p 6	4	0.36	Ta+4	5d 1	6	0.68
Pr+3	4f 2	6	0.99	S−2	3p 6	6	1.84	Ta+5	5p 6	6	0.64
		8	1.126	S+4	3s 2	6	0.37			7	0.69
		9	1.179	S+6	2p 6	4	0.12			8	0.74
Pr+4	4f 1	6	0.85			6	0.29	Tb+3	4f 8	6	0.923
		8	0.96	Sb+3	5s 2	4 py	0.76			7	0.98
Pt+2	5d 8	4 sq	0.60			5	0.80			8	1.040
		6	0.80			6	0.76			9	1.095
Pt+4	5d 6	6	0.625	Sb+5	4d 10	6	0.60	Tb+4	4f 7	6	0.76
Pt+5	5d 5	6	0.57	Sc+3	3p 6	6	0.745			8	0.88
Pu+3	5f 5	6	1.00			8	0.870	Tc+4	4d 3	6	0.645
Pu+4	5f 4	6	0.86	Se−2	4p 6	6	1.98	Tc+5	4d 2	6	0.60
		8	0.96	Se+4	4s 2	6	0.50	Tc+7	4p 6	4	0.37
Pu+5	5f 3	6	0.74	Se+6	3d 10	4	0.28			6	0.56
Pu+6	5f 2	6	0.71			6	0.42	Te−2	5p 6	6	2.21
Ra+2	6p 6	8	1.48	Si+4	2p 6	4	0.26	Te+4	5s 2	3	0.52
		12	1.70			6	0.400			4	0.66
Rb+1	4p 6	6	1.52	Sm+2	4f 6	7	1.22			6	0.97

Table 7B.1. (*cont.*)

ion	e.c.	CN	r	ion	e.c.	CN	r	ion	e.c.	CN	r
Te + 6	4d 10	4	0.43	U + 3	5f 3	6	1.025			6	0.60
		6	0.56	U + 4	5f 2	6	0.89	Xe + 8	4d 10	4	0.40
Th + 4	6p 6	6	0.94			7	0.95			6	0.48
		8	1.05			8	1.00	Y + 3	4p 6	6	0.900
		9	1.09			9	1.05			7	0.96
		10	1.13			12	1.17			8	1.019
		11	1.18	U + 5	5f 1	6	0.76			9	1.075
		12	1.21			7	0.84	Yb + 2	4f 14	6	1.02
Ti + 2	3d 2	6	0.86	U + 6	6p 6	2	0.45			7	1.08
Ti + 3	3d 1	6	0.670			4	0.52			8	1.14
Ti + 4	2p 6	4	0.42			6	0.73	Yb + 3	4f 13	6	0.868
		5	0.51			7	0.81			7	0.925
		6	0.605			8	0.86			8	0.985
		8	0.74	V + 2	3d 3	6	0.79			9	1.042
Tl + 1	6s 2	6	1.50	V + 3	3d 2	6	0.640	Zn + 2	3d10	4	0.60
		8	1.59	V + 4	3d 1	5	0.53			5	0.68
		12	1.70			6	0.58			6	0.740
Tl + 3	5d 10	4	0.75			8	0.72			8	0.90
		6	0.885	V + 5	3p 6	4	0.335	Zr + 4	4p 6	4	0.59
		8	0.98			5	0.46			5	0.66
Tm + 2	4f 13	6	1.03			6	0.54			6	0.72
		7	1.09	W + 4	5d 2	6	0.66			7	0.78
Tm + 3	4f 12	6	0.880	W + 5	5d 1	6	0.62			8	0.84
		8	0.994	W + 6	5p 6	4	0.42			9	0.89
		9	1.052			5	0.51				

Appendix 7C
Pauling electronegativities

IA																nonmetals	0
1 H 2.20	IIA																2 He
3 Li 0.98	4 Be 1.57		Metals									5 B 2.04	6 C 2.55	7 N 3.04	8 O 3.44	9 F 3.98	10 Ne
11 Na 0.93	12 Mg 1.31	IIIB	IVB	VB	VIB	VIIB	VIII			IB	IIB	13 Al 1.61	14 Si 1.90	15 P 2.19	16 S 2.58	17 Cl 3.16	18 Ar
19 K 0.82	20 Ca 1.00	21 Sc 1.36	22 Ti 1.54	23 V 1.63	24 Cr 1.66	25 Mn 1.55	26 Fe 1.83	27 Co 1.88	28 Ni 1.91	29 Cu 1.90	30 Zn 1.65	31 Ga 1.81	32 Ge 2.01	33 As 2.18	34 Se 2.55	35 Br 2.96	36 Kr
37 Rb 0.82	38 Sr 0.95	39 Y 1.22	40 Zr 1.33	41 Nb 1.6	42 Mo 2.16	43 Tc 1.9	44 Ru 2.2	45 Rh 2.28	46 Pd 2.20	47 Ag 1.93	48 Cd 1.69	49 In 1.78	50 Sn 1.96	51 Sb 2.05	52 Te 2.1	53 I 2.66	54 Xe
55 Cs 0.79	56 Ba 0.89	57 La 1.10	72 Hf 1.3	73 Ta 1.5	74 W 2.36	75 Re 1.9	76 Os 2.2	77 Ir 2.20	78 Pt 2.28	79 Au 2.54	80 Hg 2.00	81 Tl 2.04	82 Pb 2.33	83 Bi 2.02	84 Po 2.0	85 At 2.2	86 Rn
87 Fr 0.7	88 Ra 0.9	89 Ac 1.1															

Lanthanides

58 Ce 1.12	59 Pr 1.13	60 Nd 1.14	61 Pm 1.2	62 Sm 1.17	63 Eu 1.2	64 Gd 1.20	65 Tb 1.2	66 Dy 1.22	67 Ho 1.23	68 Er 1.24	69 Tm 1.25	70 Yb 1.2	71 Lu 1.27
90 Th 1.3	91 Pa 1.5	92 U 1.38	92 Np 1.36	94 Pu 1.28	95 Am 1.3	96 Cm 1.3	97 Bk 1.3	98 Cf 1.3	99 Es 1.3	100 Fm 1.3	101 Md 1.3	102 No 1.3	103 Lr 1.3

Actinides

Figure 7C.1 Updated Pauling electronegativities, based on A.L. Allred, Electronegativity values from thermochemical data, *J. Inorg. Nucl. Chem.* **17** (1961) 215–21.

Appendix 9A
Cohesive energies and band gap data

Experimentally determined cohesive energies for crystals with the diamond, sphalerite, and wurtzite crystal structures are listed in Table 9A.1. Band gap data for selected polar compounds are listed in Table 9A.2.

Table 9A.1. *Cohesive energies, per bond, for semiconductors, in eV.*

Material	E_{coh}	Material	E_{coh}	Material	E_{coh}
C	3.68	InSb	1.40	GaP	1.78
BN	3.34	CdTe	1.03	ZnS	1.59
BeO	3.06	AgI	1.18	CuCl	1.58
Si	2.32	SiC	3.17	InP	1.74
AlP	2.13	BP	2.52	MgTe	1.43
Ge	1.94	AlN	2.88	CdS	1.42
GaAs	1.63	GaN	2.24	GaSb	1.48
ZnSe	1.29	ZnO	1.89	InAs	1.55
CuBr	1.45	InN	1.93	ZnTe	1.14
Sn	1.56	AlAs	1.89	CuI	1.33

W.A. Harrison, *Electronic Structure and the Properties of Solids: The Physics of the Chemical Bond* (Dover Publications, Inc., New York, 1989). The data in this table come from Table 7.3 on p. 176.

Table 9A.2. *Bandgaps for polar compounds, in eV.*

Material	E_g	Material	E_g	Material	E_g
PbSe	0.27	BaF_2	8.9	VO	0.3
PbTe	0.29	SrF_2	9.5	Fe_2O_3	3.1
PbS	0.34–0.37	CaF_2	10.0	Ga_2O_3	4.6
AgI	2.8	MgF_2	11.8	Al_2O_3	8.8
AgCl	3.2	SrO	5.7	$BaTiO_3$	2.8
BN	4.8	MgO	7.8	TiO_2	3.0
NaF	6.7	NiO	4.2	UO_2	5.2
KCl	7	CoO	4.0	SiO_2	8.5
NaCl	7.3	MnO	3.7	$MgAl_2O_4$	7.8
LiF	12.0	FeO	2		

Y.-M. Chiang, D. Birnie III, and W.D. Kingery, *Physical Ceramics* (John Wiley & Sons, 1997, New York). These data are taken from Table 2.3 on p. 120.

Appendix 9B
Atomic orbitals and the electronic structure of the atom

Atomic orbitals are specified by a principal quantum number, n, and an orbital angular momentum, l. The number of different orbital angular momentum states in each principal level increases with n. Each state with the same angular momentum is degenerate (in other words, each state has the same energy) and the degeneracy increases with the orbital angular momentum. This information is summarized in Table 9B.1. As l increases, the electrons are increasingly localized near the atomic core. The number of states in each l level goes as $2l+1$ and each state can hold two electrons, one with spin up and the other with spin down.

Table 9B.1 *Angular momentum of different atomic orbitals.*

l	notation	degenerate states	principal levels
0	s	1	all n
1	p	3	all $n > 1$
2	d	5	all $n > 2$
3	f	7	all $n > 3$

Figure 9B.1 shows the electronic structure of the H atom. The electronic states are filled from lowest to highest energy according to Hund's rules. Note that for multi-electron atoms, states with the same principal quantum number are, in general, non-degenerate. As shown in Fig. 9B.2, the increase in the nuclear charge that occurs as the atomic number increases leads to an increased electrostatic attraction and a significant lowering of the energy levels. In some cases, there are unexpected variations, such as the filling of the 4s states before the 3d states. The atomic energy levels are called term values.

In Chapter 9, we construct the crystal orbitals from linear combinations of atomic orbitals. While knowledge of the functional form of these orbitals (specified by Eqns. 9B.1–9B.3) are not critical to our analysis, knowledge of the geometry of these atomic orbitals is important. The s orbitals are spherically symmetric about the nucleus. Each of the three p orbitals is directed along one of the three Cartesian axes, as shown in Fig. 9B.3. The geometries of the five d-orbitals were illustrated earlier in Fig. 7.6. The analytical descriptions of the s, p, and d orbitals (the wave functions) are given in Eqns. 9B.1, 9B.2, and 9B.3, respectively.

$$\Psi_{n0} = A_{n0} e^{-r/a_0} \tag{9B.1}$$

$$\Psi_{n1m} = \left(\frac{3}{4\pi}\right)^{1/2} R_{n1}(r) \begin{cases} x/r \\ y/r \\ z/r \end{cases} \tag{9B.2}$$

$$\Psi_{n2m} = \left(\frac{15}{4\pi}\right)^{1/2} R_{n2}(r) \begin{cases} yz/r^2 \\ xz/r^2 \\ xy/r^2 \\ (x^2 - y^2)/2r^2 \\ (3z^2 - r^2)/2\sqrt{3}\, r^2 \end{cases} \tag{9B.3}$$

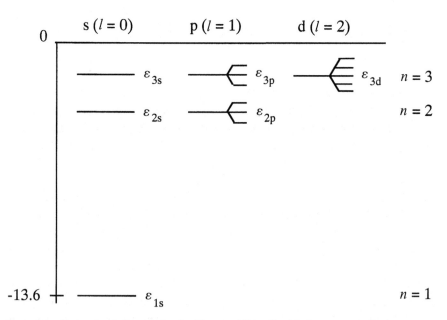

Figure 9B.1. Energy level diagram for the H atom. All levels with the same principal quantum number are degenerate (have the same energy) [1].

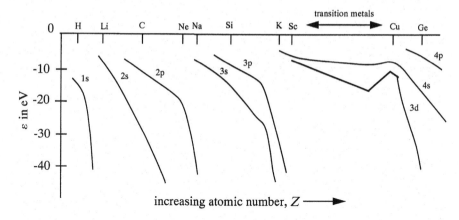

Figure 9B.2. Schematic illustration of how the energy levels of different atomic orbitals (term values) change with the atomic number [1].

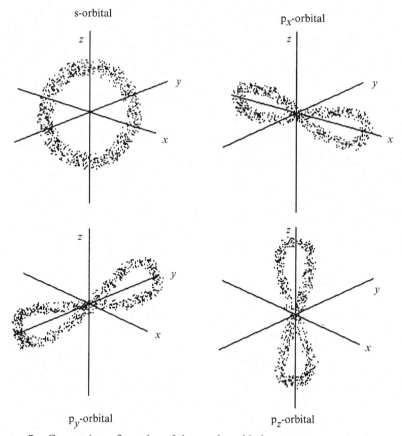

Figure 9B.3. Geometric configuration of the s and p orbitals.

Table 9B.2. *Atomic term values for selected elements [2].*

Atom	$-\varepsilon_s$, eV	$-\varepsilon_p$, eV	Atom	$-\varepsilon_s$, eV	$-\varepsilon_p$, eV
Li	5.48		Rb	3.94	
Be	8.17	4.14	Sr	5.00	
B	12.54	6.64	Ag	6.41	2.05
C	17.52	8.97	Cd	7.70	3.38
N	23.04	11.47	In	10.12	4.69
O	29.14	14.13	Sn	12.50	5.94
F	35.80	16.99	Sb	14.80	7.24
Na	5.13		Te	17.11	8.59
Mg	6.86	2.99	I	19.42	9.97
Al	10.11	4.86	Cs	3.56	
Si	13.55	6.52	Ba	4.45	
P	17.10	8.33	Au	6.48	2.38
S	20.80	10.27	Hg	7.68	3.48
Cl	24.63	12.31	Tl	9.92	4.61
K	4.19		Pb	12.07	5.77
Ca	5.41		Bi	14.15	6.97
Cu	6.92	1.83	Po	16.21	8.19
Zn	8.40	3.38	At	18.24	9.44
Ga	11.37	4.90	Fr	3.40	
Ge	14.38	6.36	Ra	4.24	
As	17.33	7.91			
Se	20.32	9.53			
Br	23.35	11.20			

References

[1] W.A. Harrison, *Electronic Structure and the Properties of Solids: The Physics of the Chemical Bond* (Dover Publications, Inc., New York, 1989) pp. 8–16. Fig. 9B.1 is drawn after Fig. 1.7 on p. 14. Fig. 9A.2 is drawn after Fig. 18. on p. 15.

[2] W.A. Harrison, *Electronic Structure and the Properties of Solids: The Physics of the Chemical Bond* (Dover Publications, Inc., New York, 1989). The data in Table 9B.2 were taken from Table 2.2 on pp. 50–1. The original source is: F. Herman and S. Skillman, *Atomic Structure Calculations* (Prentice Hall, Englewood Cliffs, NJ, 1963).

Index